Biology and Ecology of Aquatic Animals

Biology and Ecology of Aquatic Animals

Edited by John Frost

SYRAWOOD
PUBLISHING HOUSE

New York

Published by Syrawood Publishing House,
750 Third Avenue, 9th Floor,
New York, NY 10017, USA
www.syrawoodpublishinghouse.com

Biology and Ecology of Aquatic Animals
Edited by John Frost

International Standard Book Number: 978-1-68286-726-6 (Hardback)

Cataloging-in-Publication Data

Biology and ecology of aquatic animals / edited by John Frost.
    p. cm.
Includes bibliographical references and index.
ISBN 978-1-68286-726-6
1. Aquatic animals. 2. Aquatic organisms. 3. Aquatic biology. 4. Aquatic ecology.
I. Frost, John.
QL120 .B56 2019
591.76--dc23

TABLE OF CONTENTS

PREFACE

This book was inspired by the evolution of our times; to answer the curiosity of inquisitive minds. Many developments have occurred across the globe in the recent past which has transformed the progress in the field.

An aquatic animal is a vertebrate or invertebrate animal, which lives in water. They can be classified as fresh water or marine aquatic animals. The animals inhabiting salt-water ecosystems include corals, echinoderms, cephalopods and sharks. Fresh water ecosystems comprise of lake, river and wetland ecosystems. The salinity, nutrient levels, water depth, flow, etc. influence the aquatic ecosystem. They are characterized by low oxygen levels and slow diffusion of oxygen and carbon dioxide. Aquatic organisms exhibit a range of physiological and morphological adaptations that allows them to thrive in such environments. This book elucidates the concepts and innovative models around prospective developments with respect to the biology and ecology of aquatic animals. It consists of contributions made by international experts. It will serve as a valuable source of reference for marine biologists, ecologists, environmentalists and other professionals involved in this area of study.

This book was developed from a mere concept to drafts to chapters and finally compiled together as a complete text to benefit the readers across all nations. To ensure the quality of the content we instilled two significant steps in our procedure. The first was to appoint an editorial team that would verify the data and statistics provided in the book and also select the most appropriate and valuable contributions from the plentiful contributions we received from authors worldwide. The next step was to appoint an expert of the topic as the Editor-in-Chief, who would head the project and finally make the necessary amendments and modifications to make the text reader-friendly. I was then commissioned to examine all the material to present the topics in the most comprehensible and productive format.

I would like to take this opportunity to thank all the contributing authors who were supportive enough to contribute their time and knowledge to this project. I also wish to convey my regards to my family who have been extremely supportive during the entire project.

Editor

An integrative systematic framework helps to reconstruct skeletal evolution of glass sponges (Porifera, Hexactinellida)

Martin Dohrmann[1][*], Christopher Kelley[2], Michelle Kelly[3], Andrzej Pisera[4], John N. A. Hooper[5,6] and Henry M. Reiswig[7,8]

Abstract

Background: Glass sponges (Class Hexactinellida) are important components of deep-sea ecosystems and are of interest from geological and materials science perspectives. The reconstruction of their phylogeny with molecular data has only recently begun and shows a better agreement with morphology-based systematics than is typical for other sponge groups, likely because of a greater number of informative morphological characters. However, inconsistencies remain that have far-reaching implications for hypotheses about the evolution of their major skeletal construction types (body plans). Furthermore, less than half of all described extant genera have been sampled for molecular systematics, and several taxa important for understanding skeletal evolution are still missing. Increased taxon sampling for molecular phylogenetics of this group is therefore urgently needed. However, due to their remote habitat and often poorly preserved museum material, sequencing all 126 currently recognized extant genera will be difficult to achieve. Utilizing morphological data to incorporate unsequenced taxa into an integrative systematics framework therefore holds great promise, but it is unclear which methodological approach best suits this task.

Results: Here, we increase the taxon sampling of four previously established molecular markers (18S, 28S, and 16S ribosomal DNA, as well as cytochrome oxidase subunit I) by 12 genera, for the first time including representatives of the order Aulocalycoida and the type genus of Dactylocalycidae, taxa that are key to understanding hexactinellid body plan evolution. Phylogenetic analyses suggest that Aulocalycoida is diphyletic and provide further support for the paraphyly of order Hexactinosida; hence these orders are abolished from the Linnean classification. We further assembled morphological character matrices to integrate so far unsequenced genera into phylogenetic analyses in maximum parsimony (MP), maximum likelihood (ML), Bayesian, and morphology-based binning frameworks. We find that of these four approaches, total-evidence analysis using MP gave the most plausible results concerning congruence with existing phylogenetic and taxonomic hypotheses, whereas the other methods, especially ML and binning, performed more poorly. We use our total-evidence phylogeny of all extant glass sponge genera for ancestral state reconstruction of morphological characters in MP and ML frameworks, gaining new insights into the evolution of major hexactinellid body plans and other characters such as different spicule types.

(Continued on next page)

* Correspondence: m.dohrmann@lrz.uni-muenchen.de
[1]Department of Earth & Environmental Sciences, Palaeontology & Geobiology, Molecular Geo- & Palaeobiology Lab, Ludwig-Maximilians-University Munich, Richard-Wagner-Str. 10, 80333 Munich, Germany
Full list of author information is available at the end of the article

(Continued from previous page)

Conclusions: Our study demonstrates how a comprehensive, albeit in some parts provisional, phylogeny of a larger taxon can be achieved with an integrative approach utilizing molecular and morphological data, and how this can be used as a basis for understanding phenotypic evolution. The datasets and associated trees presented here are intended as a resource and starting point for future work on glass sponge evolution.

Keywords: Ancestral state reconstruction, Character evolution, Classification, Hexactinellida, Integrative systematics, Phylogeny, Porifera, Total evidence

Background

Glass sponges (Hexactinellida; Fig. 1) constitute one of the four classes of Porifera, being distinguished from the other three classes (Demospongiae, Homoscleromorpha, and Calcarea) by having siliceous skeletal elements (spicules) with triaxonic symmetry (i.e., six-rayed spicules [hexactins] and their derivatives with reduced rays; Fig. 2a) and a largely syncytial soft tissue organization [1, 2]. Within Porifera, they are most closely related to Demospongiae [3, 4] and their monophyly is strongly supported by both morphological and molecular data [3]. Although in terms of known extant diversity they represent a relatively minor group (625 valid species as of May 2016 [5]), glass sponges are of great importance for the ecology of the deep-sea benthos (the habitat they are mostly restricted to) and are geologically relevant as they contributed to the formation of massive reefs, especially in the Mesozoic, which are still preserved as rock formations throughout Europe (e.g., [2, 6–9]). Furthermore, their spicules have remarkable physical properties, which make them highly interesting study objects for materials scientists (e.g., [10–12]). Glass sponges can be aesthetically appealing in terms of their unusual morphology and astonishing variety of spicule forms [2] (Figs. 1, 2 and 3). The high diversity and complexity of morphological features of hexactinellids provide ample characters for morphology-based systematics, and as a result there is relatively good agreement between molecular phylogenies and taxonomy in comparison to other sponge groups [3, 13, 14]. For example, monophyly is supported by molecular data for all except one of the families sampled so far for more than one genus, as well as for almost all genera sampled so far for more than one species [3, 15, 16] (the only exceptions were Euretidae, a clear "waste-bin" taxon [17], *Rossella*, which has subsequently been split into two separate genera [18], and *Aphrocallistes* and *Heterochone*, whose reciprocal monophyly might be difficult to reconstruct due to gene-tree species-tree conflicts [19]).

Fig. 1 Some examples of glass sponges (Porifera: Hexactinellida). **a-c**, **e-h** from off Hawaii (images **a** and **e-h** captured by the *Deep Discoverer* ROV onboard the NOAA ship *Okeanos Explorer*, courtesy of NOAA OER; images **b-c** captured by the *Pisces 4 and 5* submersibles onboard the R/V *Kaimikai-o-Kanaloa*, courtesy of HURL); **d** from off New Zealand, Chatham Rise (image captured by DTIS [Deep Towed Imaging System] onboard RV Tangaroa, courtesy of NIWA). **a-d** examples of dictyonal sponges, **e-h** examples of lyssacine sponges (see text). **a-g** subclass Hexasterophora, **h** subclass Amphidiscophora. **a** *Farrea occa* (Farreidae), specimen 10–30 cm high, 2026 m depth. **b** *Heterorete* sp. (Euretidae), specimen 30–50 cm (?) diameter, 1559 m depth. **c** *Tretopleura* sp. (Uncinateridae), specimen 51 cm high, 888 m depth. **d** *Aulocalyx australis* (Aulocalycidae), specimen ~6 cm in diameter, 770–919 m depth. **e** *Lophocalyx* sp. (Rossellidae Lanuginellinae), specimens 10–50 cm high, depth 2247 m. **f** *Regadrella* sp. (Euplectellidae Corbitellinae, showing the iconical "venus-flower basket" body shape), specimen 5–30 cm high, 2132 m depth. **g** *Saccocalyx* sp. (Euplectellidae Bolosominae; note fleshy stalk below main body), specimen 30–50 cm (?) high, 1557 m depth. **h** *Hyalonema* sp. (Hyalonematidae; note stalk of naked anchor spicules below main body), specimen 5–10 cm high, 4824 m depth

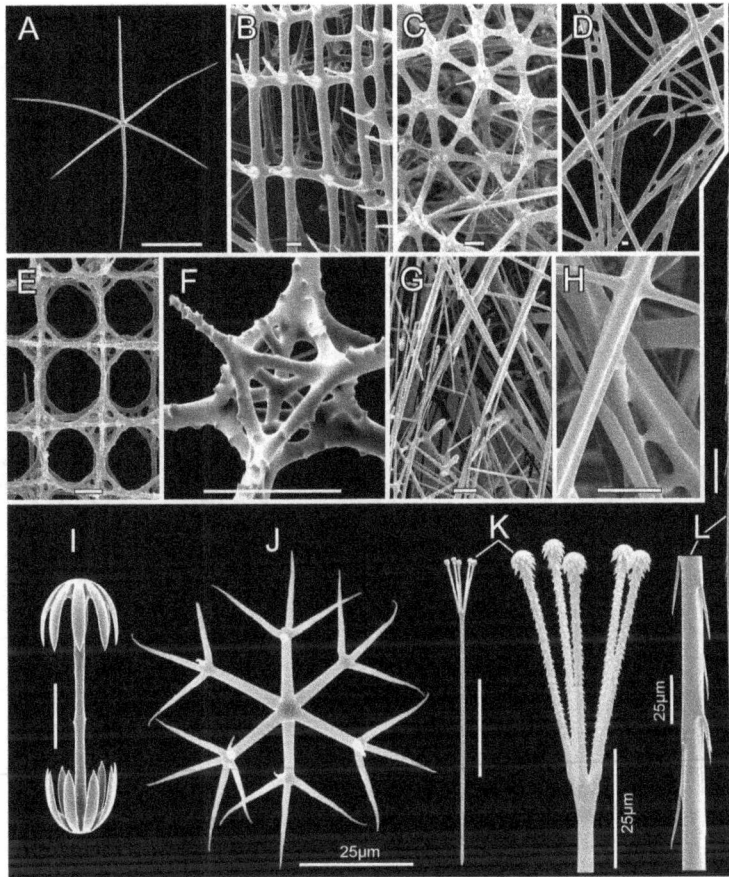

Fig. 2 Hexactinellid framework and basic spicule types. Scanning Electron Micrographs (SEM). **a** hexactine megasclere (*Dictyocalyx* sp., Euplectellidae Corbitellinae), the eponymous character for the class. Other megasclere (structural spicule) types are derived from hexactins by reduction of rays. **b** dictyonal framework of the "regular" (euretoid) type (*Conorete gordoni*, Euretidae), as found in most sceptrulophorans. **c** dictyonal framework with "haphazard" connection of hexactins (*Dactylocalyx pumiceus*), as found in Dactylocalycidae. **d** dictyonal framework of the aulocalycoid type (*Aulocalyx australis*), as found in Aulocalycidae. Note that in addition to proper ray fusion (as in b, c, and e; see [2] and Additional file 4 for details), synapticular fusion – cementation of spicules by siliceous bridges – is also common in this type of framework construction. **e** dictyonal framework of the Lychniscosida (*Neoaulocystis zitteli*, Aulocystidae; facial view). Instead of regular hexactins, lantern-like spicules (lychniscs) are the building-blocks of the frameworks in this paleontologically important relict-group. **f** oblique view of surface lychnisc of *N. zitteli*. **g** lyssacine construction type of parenchymal skeleton (*Atlantisella* sp., Euplectellidae Corbitellinae), as found in Lyssacinosida and Amphidiscophora. In this type of body plan, spicules either do not fuse at all, or (often older) parts of the skeleton fuse by synapticular bridging (only in Lyssacinosida); proper ray fusion as in the dictyonal body plan never occurs. **h** detail of g, showing synapticular bridging (*lower right*). **i** amphidisc (*Hyalonema populiferum*, Hyalonematidae), the diagnostic microsclere and defining autapomorphy of subclass Amphidiscophora. **j** hexaster (*Farrea omniclavata*, Farreidae), the diagnostic microsclere and defining autapomorphy of subclass Hexasterophora. The pictured spicule is an oxyhexaster, meaning that the secondary ray tips are pointed (without ornamentation); for further examples of hexasters see Fig. 3. **k** septrule (*C. gordoni*), the diagnostic spicule type and defining autapomorphy of Sceptrulophora. The pictured spicule is a scopule; for further examples of sceptrules and discussion of their evolution, see [19]. **l** unicate (*F. omniclavata*), a spicule type found in most species of Sceptrulophora and Amphidiscophora. All *scale bars* without lettering = 100 μm

The division of Hexactinellida into the two subclasses Amphidiscophora and Hexasterophora [20, 21] is well supported by the mutually exclusive occurrence of amphidiscs and hexasters in these groups (Fig. 2i, j), and is also highly corroborated by molecular data [3]. Amphidiscophora contains a single extant order, Amphidiscosida (Hyalonematidae, Pheronematidae, Monorhaphididae). Hexasterophora currently comprises four orders: Hexactinosida (Aphrocallistidae, Auloplacidae, Craticulariidae, Cribrospongiidae, Dactylocalycidae, Euretidae, Farreidae, Fieldingiidae, Tretodictyidae), Aulocalycoida (Aulocalycidae, Uncinateridae), Lychniscosida (Aulocystidae, Diapleuridae), and Lyssacinosida (Euplectellidae, Leucopsacidae, Rossellidae). Members of the first three orders have so-called dictyonal frameworks, which are rigid internal skeletons composed of fused hexactins, whereas members of Lyssacinosida have internal skeletons composed of mostly unfused spicules, a condition called lyssacine that is also characteristic of Amphidiscophora (Fig. 2b-h). Molecular phylogenetic analyses based on ribosomal

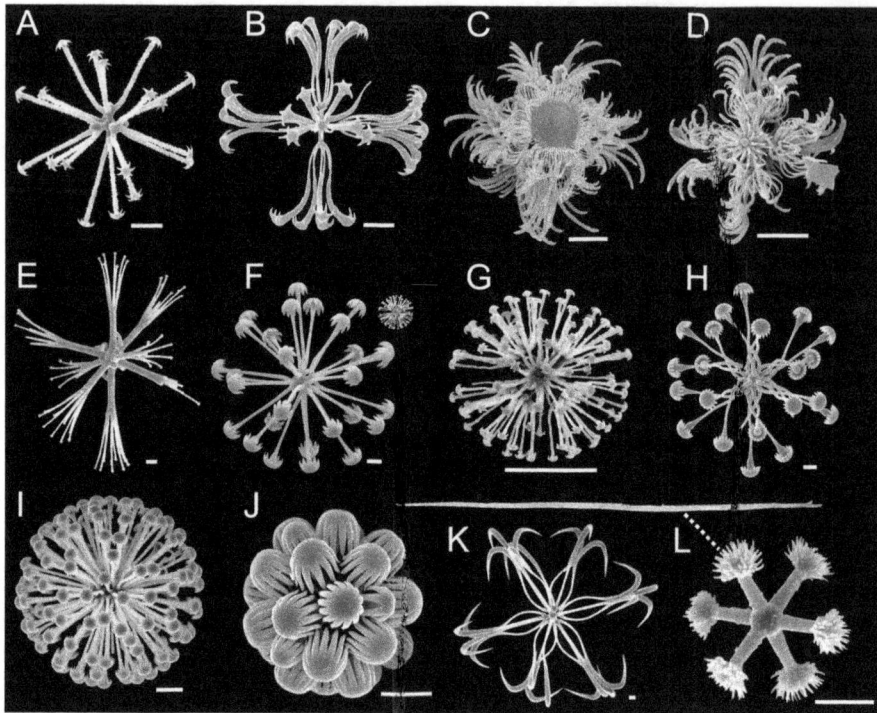

Fig. 3 Some examples of hexasters. SEM. **a** spherical discohexaster with few terminal rays (*Hyalascus* sp., Rossellidae Rossellinae). **b** floricome (*Regadrella* sp., Euplectellidae Corbitellinae). **c** discoplumicome (*Saccocalyx pedunculatus*, Euplectellidae Bolosominae). **d** strobiloplumicome (*Doconesthes dustinchiversi*, Rossellidae Lanuginellinae). **e** discoctaster (*Acanthascus malacus*, Rossellidae Acanthascinae). **f** comparison of macrodiscohexaster (*Amphidiscella lecus*, Euplectellidae Bolosominae) and microdiscohexaster (*Schaudinnia* sp., Rossellidae Rossellinae) *upper right* at same scale. **g** microdiscohexaster, enlarged (*Schaudinnia* sp.). **h** spirodiscohexaster (*Saccocalyx pedunculatus*, Euplectellidae Bolosominae). **i** discaster (*Walteria flemmingi*, Euplectellidae Corbitellinae). **j** spherical discohexaster with large anchorate discs (*Rhabdopectella tintinnus*, Euplectellidae Bolosominae). **k** drepanocome (*Amphidiscella* sp.). **l** graphiocome, center and terminal ray (*Regadrella* sp., Euplectellidae Corbitellinae). All *scale bars* = 10 μm

DNA (rDNA) and cytochrome oxidase subunit I (COI) sequences have supported monophyly of Lyssacinosida [15, 16] but have found that Dactylocalycidae is more closely related to that order than to the remaining hex-actinosidans (= Sceptrulophora; a clade well-supported by possession of sceptrules and uncinates [Fig. 2k, l]), rendering Hexactinosida paraphyletic [3, 15, 16]. This suggests that dictyonal skeletons could have either evolved independently in Sceptrulophora and Dactylo-calycidae, or alternatively that the lyssacine body plan "re-evolved" in Lyssacinosida [15]. However, a more comprehensive phylogeny, especially including Aulocalycoida and Lychniscosida, is necessary to understand the evolution of the dictyonal and lyssacine body plans. Understanding the evolution of other aspects of the glass sponge skeleton, such as the myriad different types of hexasters (see Fig. 3 for a few examples), would also greatly benefit from a phylogeny including as many genera as possible.

Prior to the present study, sequence coverage of genera was only ~36%. Although we here increase this number to ~45%, this is still low, and given the difficulties of

obtaining suitable material from targeted taxa of this deep-sea group, is not likely to increase substantially in the near future. Therefore, utilization of morphological character data to integrate unsequenced genera into a phylogenetic analysis framework holds some promise to obtain a, albeit somewhat provisional, comprehensive phylogeny of Hexactinellida. In this study, we have in-creased molecular taxon sampling of the four markers established by Dohrmann et al. [3, 16] by 12 additional genera, which include representatives of one additional order (Aulocalycoida), and three additional families (Aulocalycidae, Craticulariidae, Uncinateridae). Further-more, we compiled morphological character matrices from all described extant genera of the two subclasses. Molecular and morphological datasets were first analyzed separately and then combined to incorporate the unse-quenced genera into the molecular phylogeny by using "total-evidence" approaches in maximum-parsimony (MP), maximum-likelihood (ML), and Bayesian analysis frame-works, as well as a "morphology-based phylogenetic binning" approach recently developed by Berger and Stamatakis [22]. Comparisons of the results of these four methods revealed that MP yielded trees in better congruence with previous taxonomic and phylogenetic hypotheses than the other methods, at least for Hexasterophora. We then used this comprehensive, total-evidence phylogeny to investigate hexactinellid skeletal evolution by means of ancestral state reconstruction in MP and ML frameworks. Based on our phylogenetic results, we also propose several changes to the current higher-level Linnean classification.

Methods

Morphological data assembly and phylogenetic analysis

Analyses of morphological data were performed on genus-level, i.e., monophyly of all genera was assumed a priori. Although a simplifying assumption, it is reasonable because most hexactinellid genera are morphologically well delineated. Furthermore, almost half of them (59/126 = 46.8%; counted May 2016) are monospecific (see Appendix 1). In any case, a comprehensive analysis on species-level would have been too time-consuming and is left for future projects.

A genus-level matrix comprising 105 taxa and 154 morphological characters was recently compiled by Henkel et al. [23] to reconstruct hexactinellid phylogeny, with a focus on Hexasterophora. This matrix was kindly provided to us by Daniela Henkel and used as a starting point for building our own morphological datasets. Upon inspection of this matrix, we noticed several errors and also that taxonomic literature published after the 2002 book *Systema Porifera* [1] was not taken into consideration by these authors. Thus, in order to incorporate post-2002 taxonomic work and correct errors, we re-checked every entry against the relevant literature [1, 17–19, 24–55], as well as some personal observations. We also excluded some uninformative or in our opinion not useful characters, while including others that we deemed informative. Our coding philosophy followed Dohrmann et al. [3], i.e., we used hierarchical, presence-absence (0/1) coding of inferred ground states, thereby avoiding an excess of character states, missing data, and polymorphisms. Matrix editing was performed in Mesquite 2.75 [56].

We also added eight genera that were described or resurrected after 2002 (*Acanthascus, Asceptrulum, Dictyoplax, Homoieurete, Indiella, Nodastrella, Pinulasma, Staurocalyptus*), eight genera of the subclass Amphidiscophora (*Chalaronema, Compsocalyx, Lophophysema, Platylistrum, Poliopogon, Schulzeviella, Sericolophus, Tabachnickia*), three genera that were not included by Henkel et al. [23] for unknown reasons (*Clathrochone, Hyaloplacoida, Ijimaiella*), and two so far undescribed genera for which DNA sequence data are available: Rossellinae n. gen. from New Zealand (Reiswig & Kelly, in prep.) and Bolosominae n. gen. from Hawaii (MD, unpubl. obs.). For rooting purposes, we also included an "artificial" outgroup taxon with all characters coded as "0" – using an actual non-hexactinellid sponge genus as an outgroup would have been useless since most characters used here are not comparable to characters of other sponge classes. The only taxa we did not include are three very poorly known genera (*Deanea, Diaretula, Hyalocaulus*), which would have mostly introduced missing data without contributing much information. A taxonomic overview of the included genera is given in Appendix 1.

Initial phylogenetic analysis of this matrix yielded quite unsatisfactory results (not shown), as many groups whose monophyly seems highly plausible were not recovered, and conversely some clades emerged that cannot reasonably be accepted as real. We therefore excluded some characters suspected to be overly homoplastic, included additional characters that we hoped might be informative, and/or recoded certain characters. Finally, we decided to construct separate matrices for the two subclasses. This was necessary because, although monophyly of Amphidiscophora was always recovered, this clade consistently nested within Hexasterophora – apparently because it shares some important characters with the hexasterophoran family Rossellidae (Lyssacinosida). Although these similarities are striking, it is highly unlikely that they represent synapomorphies of the two taxa – they are better interpreted as symplesiomorphies or convergences, because the two subclasses are highly supported as being reciprocally monophyletic by other characters and molecular data (see Introduction). The final Amphidiscophora matrix has 13 taxa and 29 characters, and the final Hexasterophora matrix has 114 taxa and 108 characters; these datasets can be found in Nexus format in Additional files 1 and 2 and as tables with annotated character lists in Additional files 3 and 4 (all Additional files are available at figshare [https://doi.org/10.6084/m9.figshare.3120130.v3]).

Phylogenetic analysis of the morphological data matrices was performed under maximum parsimony (MP) as implemented in TNT v1.1 [57], using "new technology" searches (sectorial search, ratchet, drift, and tree fusing; *init. addseqs* = 100, *find min. length* = 10) with implied weighting (default function, concavity constant $K = 3.0$), which is a technique to down-weight overly homoplastic characters [58]. Preliminary analyses under equal weighting were also performed but generally produced longer trees that were less congruent with morphology-based taxonomy and/or molecular evidence (results not shown). Assessment of clade support using resampling techniques was performed in provisional analyses, but these support values were generally very low, even for well-established taxa (results not shown). We here take the position that quantitative, especially resampling, metrics are of limited value in studies of relatively small morphological matrices with an expected high amount of homoplasy. We therefore took a qualitative approach to evaluating clade support, by looking for potential synapomorphies of sets of genera, using the MP-based tracing function in MacClade 4.08a [59].

It has been argued that model-based approaches for analyzing discrete morphological data are superior to MP [60, 61]. Therefore, we also analyzed the morphological data matrices in the Bayesian framework of MrBayes 3.2.3 [62] using the Markov-k model with a four-rate category gamma correction for among-site rate variation ($Mk + G_4$) and *coding = informative* to account for the fact that we only included parsimony-informative characters [63, 64]. We ran 2×4 Markov Chain Monte Carlo (MCMC) chains in parallel using Metropolis coupling [65] for 10^6 generations (sampling every 100), checked for convergence using Tracer 1.6 [66] and discarded the first 10% of samples as burn-in before calculating 50% majority-rule consensus trees (MRCs). For comparison, we also used the maximum likelihood (ML) implementation of the same model as provided by RAxML 8.2.4 [67], using the *-f a* option to perform rapid bootstrapping [68] followed by search for the ML tree. Bootstrapping was automatically stopped using the *auto-MRE* option [69].

Molecular data assembly and phylogenetic analysis

The datasets of Dohrmann et al. [16] – consisting of 18S ribosomal DNA (rDNA), 28S rDNA, mitochondrial (mt)

16S rDNA and mt cytochrome oxidase subunit I (COI) fragments (see [3, 16]) – were supplemented with subsequently published hexactinellid sequences [17, 18, 70]; sequences for 16 additional specimens were newly generated for this study (Table 1).

DNA was extracted by boiling small pieces of tissue for 20 min in 20% Chelex (Sigma-Aldrich) (detailed protocol available upon request from MD). Polymerase chain reaction (PCR) was performed with GoTaq (Promega) or MyTaq (Bioline) according to manufacturers' instructions. Thermal regimes and primers are described in [3, 16]. PCR products were purified either with ExoSAP-IT (Affymetrix) in case of clear single bands, or otherwise cut out from 1% agarose gels and cleaned with a "freeze-squeeze" method adopted from [71] or a QIAquick Gel Extraction kit (QIAGEN). Sanger sequencing was performed with BigDye Terminator chemistry (Applied Biosystems) at the sequencing facility of the University of Alabama, Birmingham, AL, USA and the Sequencing Service of the Department of Biology at LMU Munich. Chromatograms were edited using Codon Code Aligner (Codon Code Corporation) or Geneious 6.1.6 (Biomatters) and consensus sequences manually aligned to the datasets described above.

In addition, 16S rDNA and COI sequences from *Lophophysema eversa*, *Tabachnickia* sp. (Amphidiscophora: Hyalonematidae), and *Vazella pourtalesii* (Lyssacinosida: Rossellidae) were extracted from mt genome sequencing

Table 1 Specimen information and sequence accession numbers for newly sampled species

Species	Family	Origin	Voucher	Source	18S	28S	16S	COI
Schulzeviella n. sp.	Pheronematidae	Hawaii	P4-224 sp5	HURL	–	LT627545[b]	LT627531	–
Doconesthes dustinchiversi	Rossellidae	B.C.	014-00412-001	RBCM	–	–	LT627517	LT627550[a]
Asconema fristedti	Rossellidae	Florida	17-XI-05-2-2	HBOI	–	LT627532[b]	LT627516	–
Bolosominae n. gen. n. sp.	Euplectellidae	Hawaii	P4-224 sp7	HURL	–	LT627534[b]	LT627520	LT627552[a]
Atlantisella sp.	Euplectellidae	Galapagos	22-X-95-1-7	HBOI	LT627547[b]	LT627533	LT627519	–
Iphiteon sp.	Dactylocalycidae	Bahamas	24-V-93-1-7	HBOI	–	LT627537[b]	LT627522	LT627553
Dactylocalyx sp.	Dactylocalycidae	Bahamas	22-XI-02-3-13	HBOI	–	LT627538[b]	LT627525	–
Dactylocalyx pumiceus	Dactylocalycidae	Bahamas	12-IV-05-1-10	HBOI	LT627548	LT627539	LT627523	–
Dactylocalyx pumiceus	Dactylocalycidae	Bonaire	RMNH POR 9215	RMNH	–	LT627540	LT627524	LT627554[a]
Euryplegma auriculare	Aulocalycidae	NZ	NIWA 43457	NIWA	LT627546[a]	LT627535[b]	LT627518	LT627551[a]
Tretopleura n. sp. 1	Uncinateridae	Hawaii	P5-701 sp4	HURL	–	LT627543[b]	LT627530	LT627555[b]
Tretopleura n. sp. 2	Uncinateridae	Hawaii	P4-229 sp10	HURL	–	LT627542[b]	LT627529	LT627556[b]
Heterorete sp.	Euretidae	Hawaii	P4-224 sp1	HURL	–	LT627536[b]	LT627521	–
Homoieurete macquariense	Euretidae	MR	QM G331848	QM	–	–	LT627528	LT627559[a]
Cyrtaulon sigsbeei	Tretodictyidae	Bonaire	RMNH POR 9219	RMNH	LT627549[b]	LT627544[a]	LT627526	LT627557[b]
Laocoetis perion	Craticulariidae	Madagascar	DW 3213	MIRIKY	–	LT627541	LT627527	LT627558[b]

Notes: *B.C.* British Columbia, Canada, *MR* Macquarie Ridge, *NZ* New Zealand (specimen supplied by NIWA Invertebrate Collection, NIWA, Wellington), *HURL* Hawaiian Undersea Research Laboratory (samples collected with submersible PISCES), *HBOI* Harbor Branch Oceanographic Institution (samples collected with submersible Johnson-Sea-Link II; most subsamples taken during August 2011 PorToL Integrative Taxonomy Workshop at Ft. Pierce, FL, USA), *QM* Queensland Museum, Brisbane, Australia, *NIWA* National Institute of Water and Atmospheric Research, New Zealand, *RBCM* Royal British Columbia Museum, *RMNH* Naturalis, Leiden, The Netherlands (samples provided by R.W.M. van Soest), *MIRIKY* French Expedition MIRIKY 2009. Underwater photographs of *Heterorete* sp. and *Tretopleura* n. sp. 1 are given in Fig. 1b and c, respectively. [a]only 5' half; [b]only 3' half data [72, 73] by aligning the whole-genome sequences to the 16S and COI alignments, respectively, with the profile alignment option in ClustalX 2.1 [74], followed by manual trimming and correction. Full single-gene alignments, including RNA structure information for 18S and 28S rDNA, are available in Additional files 5, 6, 7 and 8.

To check for conflicting phylogenetic signal between markers, single-gene alignments were first analyzed separately in RAxML 8.0.26 [67], after removal of unalignable regions and sites with excessive numbers of gaps. For COI and 16S rDNA, general-time-reversible (GTR) + G_4 models [64, 75] were employed, and in the 18S and 28S rDNA analyses the S16 + G_4 paired-sites model (see [76]) was assigned to stem-encoding regions in addition to GTR + G_4 for loop-encoding regions. We used the *-f a* option to perform rapid bootstrapping [68] with the *autoMRE* option to automatically determine the sufficient number of pseudoreplicates [69]. For the final analyses, all four markers were concatenated in SeaView [77]. For *Dactylocalyx pumiceus*, a hybrid sequence was constructed from specimens HBOI 12-IV-05-1-10 (rDNA) and RMNH POR 9215 (COI) to maximize marker coverage (see Table 1). The concatenated matrix (supermatrix hereafter) consists of 73 taxa and 4806 base pairs (bp), and features 1926 distinct alignment patterns and ~30% missing data (Additional file 9; for RNA structure and partitioning information, see Additional files 10 and 11). Phylogenies were inferred from the supermatrix using ML and Bayesian methods as follows.

For ML analyses, we used RAxML under the models and options described above, assuming a single topology and set of branch lengths across partitions but independent model parameters for each partition. For Bayesian analyses, we used MrBayes 3.2.4 [62] under a model-partitioning scheme analogous to the ML analyses. For 18S and 28S rDNA stem-encoding regions we employed the Doublet model (based on [78]). Structure information was converted from dot-bracket format to MrBayes format using a perl script written by Oliver Voigt (see [79]). We ran 2 × 4 MCMC chains in parallel for 5 × 10^6 generations, sampling every 100th. Convergence was checked in Tracer 1.6 [66] and 50% of samples were discarded as burn-in before calculating the MRC.

Combined analysis of molecular and morphological data ("total evidence")

To investigate whether addition of morphological characters to the molecular dataset would influence tree topologies, we first conducted analyses restricted to those genera with molecular data. For this purpose we split the supermatrix into submatrices representing Hexasterophora and Amphidiscophora, respectively (see above), excluded all but one species each of genera represented by multiple species, and renamed the remaining terminal

taxa to match the taxon names in the morphological dataset. Species retained from multi-species genera were chosen as to minimize missing data; those were *C. weddelli* for *Caulophacus*, *R. nuda* for *Rossella*, *N. asconemaoida* for *Nodastrella*, *B. spinosus* for *Bathydorus*, *E.* sp. 1 for *Euplectella*, *I. panicea* for *Iphiteon*, *D. pumiceus* for *Dactylocalyx*, *H. calyx* for *Heterochone*, *A. vastus* for *Aphrocallistes*, *A. australia* for *Aspidoscopulia*, *T.* n. sp. 2 for *Tretopleura*, and *H.* sp. 3 for *Hyalonema*. *Iphiteon* and *Semperella* were included as outgroups for Amphidiscophora and Hexasterophora, respectively. Morphological characters for the outgroups were all coded as absent (0) in order to mimic the conditions of the morphological analyses (see above). Molecular and morphological partitions were then concatenated in SeaView and analyzed in RAxML and MrBayes as described above (using 10% burnin in the MrBayes analyses, and not accounting for ascertainment bias of the morphological partition in the RAxML analyses due to software limitations). To facilitate comparison, the taxon-reduced molecular supermatrices were also analyzed separately in RAxML.

Next, we attempted to reconstruct complete genus-level phylogenies of the two subclasses. For this purpose we first concatenated the taxon-reduced molecular supermatrices with the complete morphological matrices (resulting in datasets with ~62 and ~65% missing data for Amphidiscophora and Hexasterophora, respectively), and analyzed those datasets in RAxML and MrBayes as described above. Second, we analyzed the data with MP in TNT under the same settings as used for the morphology-only analyses (see above; gaps were treated as missing data). Because TNT does not support mixed data types, we first recoded the sequence data such that A was replaced with 0, C with 1, G with 2, and T with 3; all ambiguous characters (N, R, Y etc.) were recoded as gaps (–). For comparability with the ML analysis, we performed bootstrapping [80] with 550 (Hexasterophora) and 1000 (Amphidiscophora) pseudoreplicates (*init. addseqs* = 10, *find min. length* = 5). Third, we used the weighted version of the "fossil placement" or "morphology-based phylogenetic binning" approach developed by Berger and Stamatakis [22] as implemented in RAxML to place those genera without sequence data onto the molecular backbone (reference) phylogeny (see [22, 81] for details of the method). In these analyses, we also used the Mk + G_4 model with the Lewis correction to account for ascertainment bias (see above).

Ancestral state reconstruction

For the purpose of more in-depth investigations of character evolution across Hexactinellida, we first merged the morphological data sets of Amphidiscophora and Hexasterophora, resulting in a matrix of 124 characters for 125 taxa (plus outgroup coded as 0 for all characters; see above) (Additional file 12). We then also included six additional characters that were not used for the purpose of phylogeny reconstruction because they are too prone to homoplasy or overly simplistic representations of complex morphological features (extended matrix in Additional file 13): 1) pinular hexactins (hexactins with a bushy distal ray), 2) basiphytous attachment to the substrate (attachment by a siliceous plate), 3) lophophytous attachment to the substrate (attachment by anchoring spicules), 4) general presence of synapticular fusion (fusion of spicules through siliceous bridges [see Fig. 2d, h]; a merger of characters 23 and 39 in Additional file 2), 3) presence of a lyssacine body plan, and 4) presence of a dictyonal body plan. Finally, we manually combined the total-evidence trees of the two subclasses obtained with TNT (see Results and discussion) and used the resulting tree (included in Additional file 13) for ancestral state reconstruction.

In order to obtain quantitative estimates of the evolution of selected characters and reconstruct ground pattern features of major clades, we used ML ancestral state reconstruction methods as implemented in Mesquite 2.75 [56]. To assess the sensitivity of results to model choice, we employed two different models to calculate proportional likelihoods of ancestral character states: the 1-parameter Mk model (Mk1 [63]) and the asymmetrical Mk 2-parameter model (aMk2 [56]). The Mk1 model assumes equal rates of gains (0 to 1) and losses (1 to 0), whereas under the aMk2 model, the two rates are allowed to differ. For aMk2 analyses, we assumed equilibrium root state frequencies (default in Mesquite 2.75) (we also experimented with equal root state frequencies, although the assumption that the presence and absence of characters in the ancestral glass sponge are equally likely is clearly unrealistic; accordingly, these analyses yielded some contradictory and biologically nonsensical results [not shown]). Because phylogenetic uncertainty can bias ancestral state reconstruction [82] we also evaluated the influence of alternative topological arrangements in crucial parts of the tree on the reconstruction of important characters.

Results and discussion
Phylogenies inferred from morphological data
Amphidiscophora
Maximum-parsimony analysis of the Amphidiscophora matrix found one single most parsimonious tree (MPT) (Fig. 4). Congruent with previous results [3], Pheronematidae is resolved as monophyletic. The position of *Monorhaphis* (the sole representative of Monorhaphididae) is here resolved as being inside Hyalonematidae, rendering this family paraphyletic. However, this result has to be viewed with caution (see supplementary

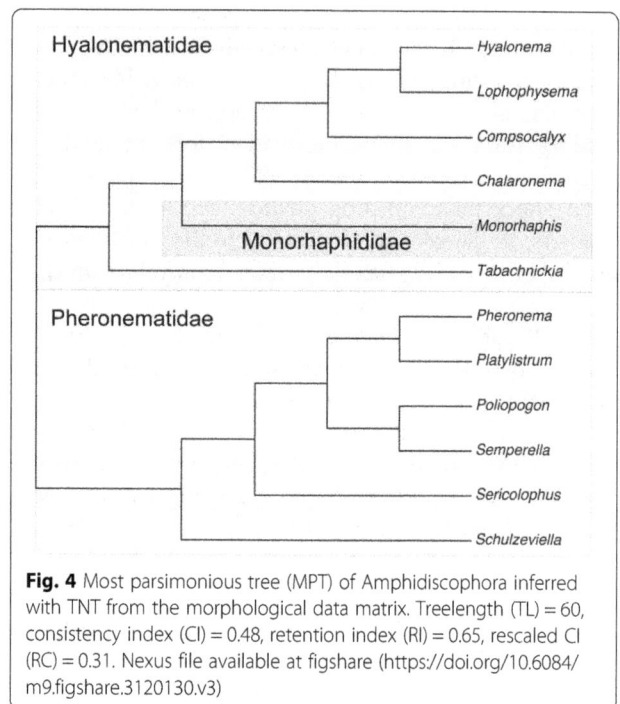

Fig. 4 Most parsimonious tree (MPT) of Amphidiscophora inferred with TNT from the morphological data matrix. Treelength (TL) = 60, consistency index (CI) = 0.48, retention index (RI) = 0.65, rescaled CI (RC) = 0.31. Nexus file available at figshare (https://doi.org/10.6084/m9.figshare.3120130.v3)

discussion in Additional file 14 available at figshare [https://doi.org/10.6084/m9.figshare.3120130.v3]). In the Bayesian tree (Additional file 15: Figure S1), Pheronematidae is also recovered as monophyletic (posterior probability [PP] = 0.96); four of the five genera of Hyalonematidae form a highly supported clade (PP = 0.98), but the positions of *Tabachnickia* and *Monorhaphis* within Amphidiscophora remain unresolved. The ML phylogeny (Additional file 16: Figure S2) is similar to the MP tree, but with somewhat different branching order within families; with two exceptions, bootstrap support (BS) is very low. For the interested reader, a more detailed account of the relationships within Pheronematidae and Hyalonematidae, and potential character support is provided in the supplementary discussion in Additional file 14.

Hexasterophora
Maximum-parsimony analysis of the Hexasterophora matrix resulted in 46 MPTs, the strict consensus of which is shown in Figs. 5 and 6. Of the four orders of Hexasterophora, only Aulocalycoida and the small relict group Lychniscosida are recovered as monophyletic. The genus *Heterorete* (Fig. 1b; currently in Sceptrulophora: Euretidae, although lacking sceptrules and uncinates) is reconstructed as the sister group of Lychniscosida, and Aulocalycoida is deeply nested within Sceptrulophora as the sister group of Auloplacidae. Lychniscosida + *Heterorete* and the Sceptrulophora *sensu lato* (s. l.) clade together form a clade with the second sceptrule- and uncinate-lacking euretid genus, *Myliusia*, the exact placement

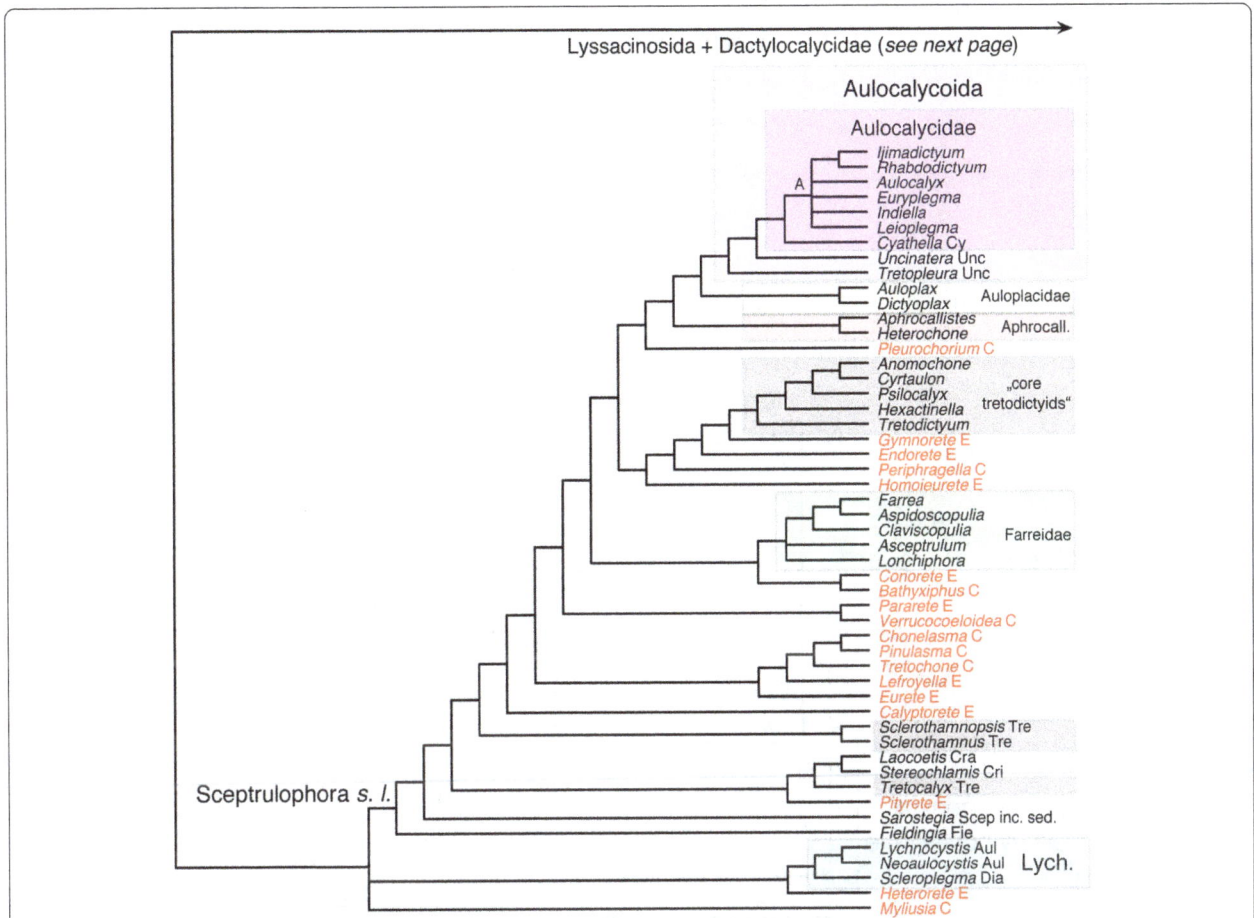

Fig. 5 Strict consensus tree of the 46 MPTs of Hexasterophora inferred with TNT from the morphological data matrix. TL = 650, CI = 0.17, RI = 0.64, RC = 0.11. Part 1: A, Aulocalycinae; Aphrocall., Aphrocallistidae; Aul, Aulocystidae; C, Chonelasmatinae; Cra, Craticulariidae; Cri, Cribrospongiidae; Cy, Cyathellinae; D, Diapluridae; E, Euretinae; Fie, Fieldingiidae; Lych., Lychniscosida; Scep inc. sed., Sceptrulophora *incertae sedis*; s. l., *sensu lato*; Tre, Tretodictyidae; Unc, Uncinateridae. Genera currently classified in Euretidae highlighted in light red

of which is not resolved. In agreement with molecular results [3, 15, 16], Dactylocalycidae comes out closer to Lyssacinosida than to Sceptrulophora. However, it is here reconstructed as the sister group to a Euplectellidae + Rossellidae clade to the exclusion of Leucopsacidae and the two Lyssacinosida *incertae sedis* (inc. *sed.*) genera, rendering Lyssacinosida paraphyletic.

Of the 12 hexasterophoran families with more than one genus, nine are recovered as monophyletic groups (Aphrocallistidae, Aulocalycidae, Aulocystidae, Auloplacidae, Dactylocalycidae, Euplectellidae, Farreidae, Leucopsacidae, Rossellidae) and three are inferred to be para- or polyphyletic (Uncinateridae, Euretidae, Tretodictyidae; for the latter family, five of the eight genera form a clade ["core tretodictyids"]). Within Rossellidae, only subfamilies Lanuginellinae (*sensu* [83]) and Acanthascinae (cf. [51]) are monophyletic; Rossellinae is not recovered as a natural group, congruent with previous results [3]. Boury-Esnault et al. [83] moved *Caulophacus* and *Caulophacella* from Rossellinae to Lanuginellinae, mostly based on molecular

evidence [16, 83], which we confirm here with our cladistic analysis of morphological data (since the revised diagnosis of Lanuginellinae provided by [83] is rather vague, we provide a more concise and comprehensive summary based on our character analysis in Appendix 2). Within Euplectellidae, a clade of genera with the iconical "venus-flower basket" body shape (Fig. 1f) ("VFB clade") and a clade comprising most genera of the stalked subfamily Bolosominae (Fig. 1g) ("core bolosomins") is recovered; subfamilies Euplectellinae and Corbitellinae are clearly not recovered as natural groups (see also [16]). Overall, the Bayesian tree (Additional file 17: Figure S3) is similar but less well resolved than the MP tree. In contrast, the ML tree (Additional file 18: Figure S4) displays a vastly different topology (and branch lengths), and appears in large parts highly incongruent with current taxonomy and/or molecular evidence (e.g., paraphyletic Rossellidae basal to the remaining taxa). For a more detailed account, the interested reader is referred to the supplementary discussion in Additional file 14.

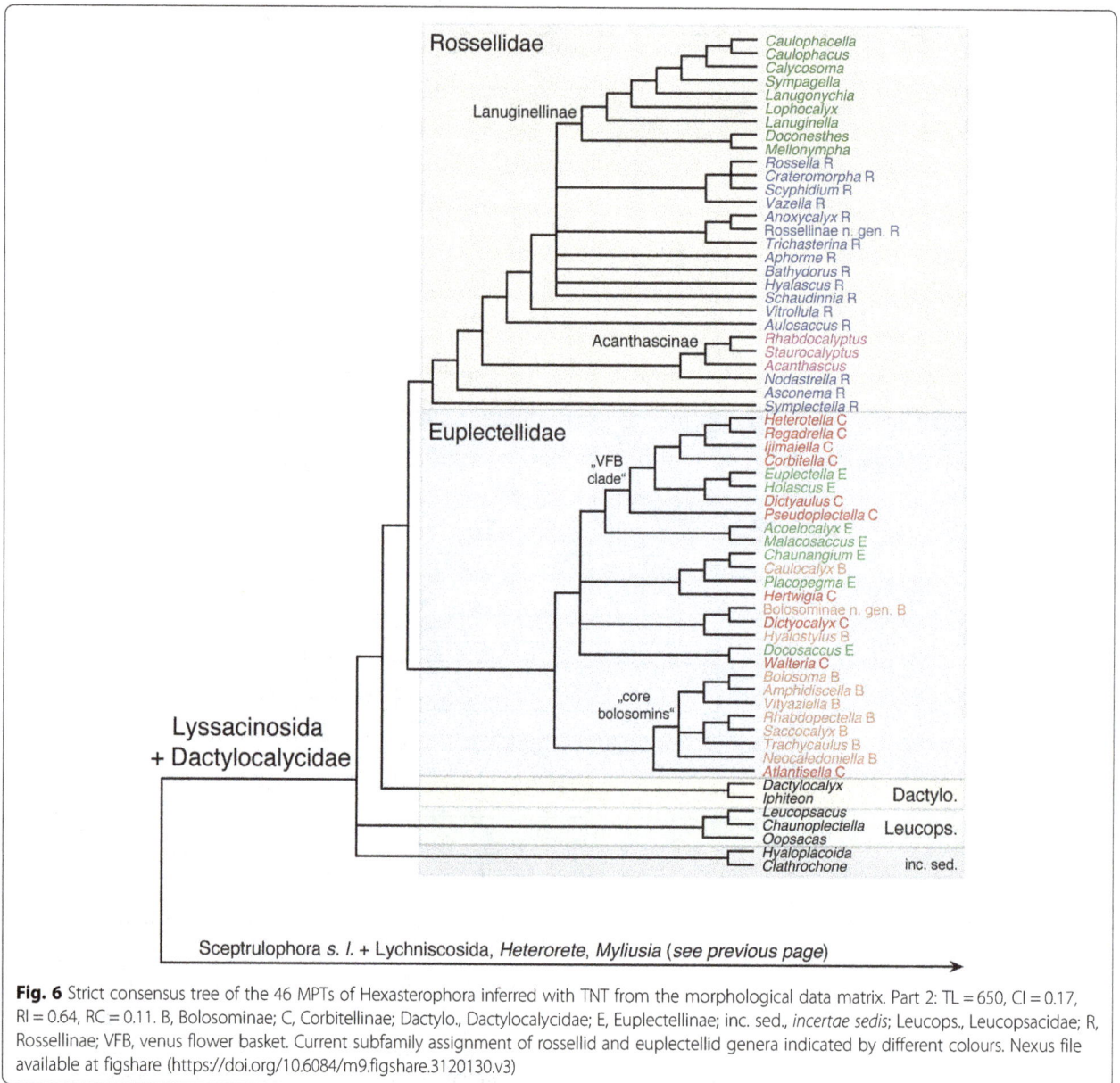

Fig. 6 Strict consensus tree of the 46 MPTs of Hexasterophora inferred with TNT from the morphological data matrix. Part 2: TL = 650, CI = 0.17, RI = 0.64, RC = 0.11. B, Bolosominae; C, Corbitellinae; Dactylo., Dactylocalycidae; E, Euplectellinae; inc. sed., *incertae sedis*; Leucops., Leucopsacidae; R, Rossellinae; VFB, venus flower basket. Current subfamily assignment of rossellid and euplectellid genera indicated by different colours. Nexus file available at figshare (https://doi.org/10.6084/m9.figshare.3120130.v3)

Phylogenies inferred from molecular data

Congruence between markers

Topologies of the four single-gene trees were largely congruent, only differing in poorly supported regions (Additional files 19, 20, 21 and 22: Figures S5–S8). A notable exception to this was monophyly of *Aphrocallistes* in the 28S phylogeny (Additional file 21: Figure S7), as discussed previously [19]. Another conflict involved the position of *Dactylocalyx* sp., which was placed inside *D. pumiceus* (BS = 85%) in the 16S tree (Additional file 19: Figure S5), whereas *D. pumiceus* was monophyletic (BS = 70%) in the 28S tree (Additional file 21: Figure S7). Currently, there are only two accepted species of *Dactylocalyx*, although more nominal species exist [5, 84]. Our *Dactylocalyx* sp. might represent a so far undescribed

species (HMR, pers. obs.), but our results further demonstrate that the genus is in urgent need of revision, preferably using a combined morphological/molecular approach.

Phylogenetic analyses of concatenated markers

Figure 7 shows the ML phylogram obtained from the supermatrix. The Bayesian tree is largely congruent with this phylogeny and is given in Additional file 23: Figure S9. Corroborating previous analyses (reviewed in [14]), Hexactinellida is divided into three major, well-supported clades: Amphidiscophora, Sceptrulophora, and a clade containing Lyssacinosida and Dactylocalycidae ("LD clade" hereafter). Below we discuss relationships within these clades, but for the sake of brevity, we will refrain from

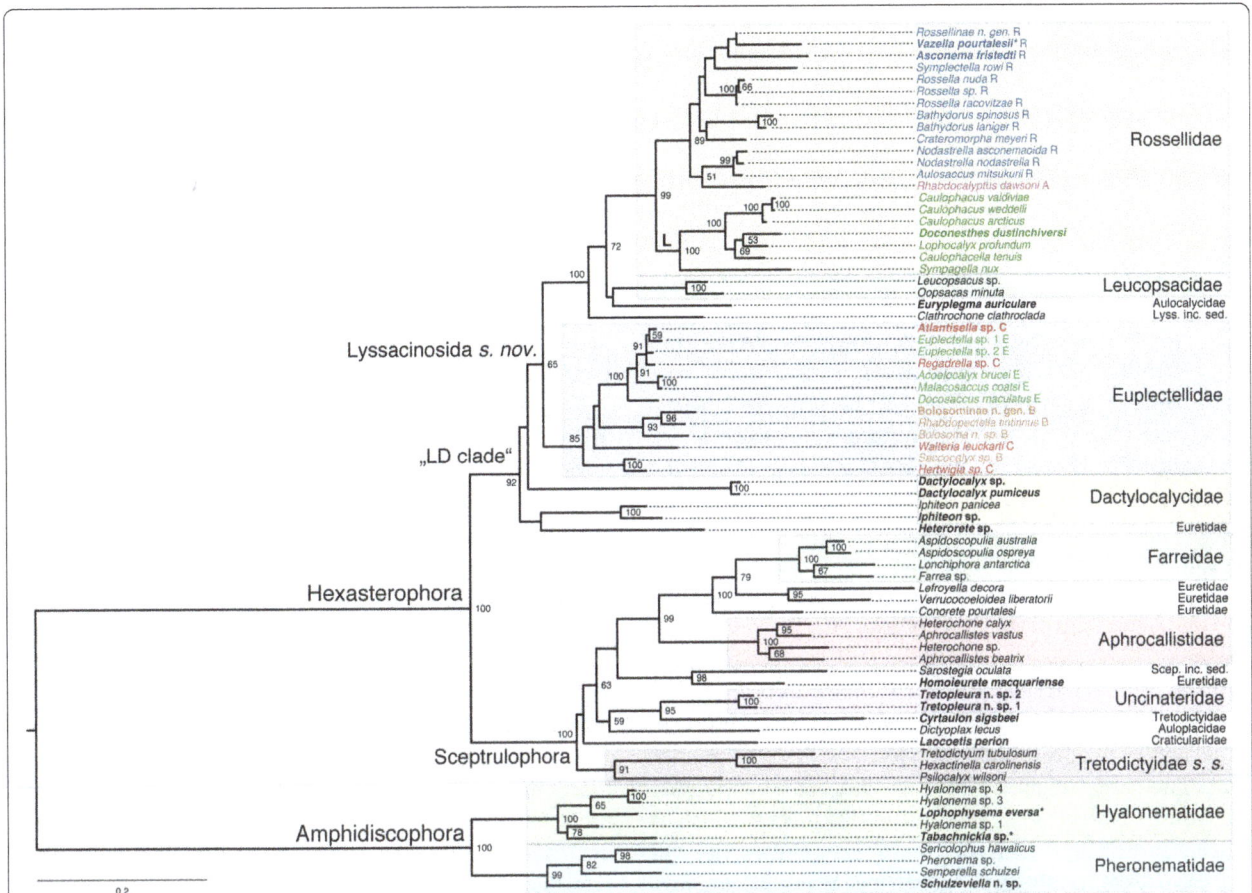

Fig. 7 Phylogeny of Hexactinellida inferred from concatenated molecular markers with RAxML. Bootstrap values >50% shown on branches (based on 600 pseudoreplicates). Newly sampled species highlighted in **bold**. *, 16S and COI sequence data from mitochondrial genome sequencing projects [72, 73]. A, Acanthascinae; B, Bolosominae; C, Corbitellinae; L, Lanuginellinae; Lyss inc. sed., Lyssacinosida *incertae sedis*; R, Rossellinae; Scep inc. sed., Sceptrulophora *incertae sedis*; s. nov., *sensu novo*; s. s., *sensu stricto*. Current subfamily assignment of rossellid and euplectellid genera indicated by different colours. Scale bar, expected number of substitutions per site. Nexus file available at figshare (https://doi.org/10.6084/m9.figshare.3120130.v3)

describing results that are unchanged compared to previous studies [3, 15–18, 70].

Amphidiscophora. Addition of *Schulzeviella, Tabachnickia*, and *Lophophysema* confirms monophyly of families Pheronematidae and Hyalonematidae. Among Pheronematidae, *Schulzeviella* n. sp. is sister to the remaining sampled genera, consistent with the morphology-based tree (Fig. 4). Among Hyalonematidae, we find the megadiverse genus *Hyalonema* (120 spp. in 12 subgenera) to be polyphyletic: *Hyalonema* sp. 3 and *Hyalonema* sp. 4 are sister to *Lophophysema eversa*, whereas *Hyalonema* sp. 1 is sister to *Tabachnickia* sp. Because *H.* sp. 3 and 4 are likely members of *H.* (*Cyliconema*) whereas *H.* sp. 1 likely belongs to *H.* (*Corynonema*) (MD, pers. obs.), this result is in line with the view that at least some of the subgenera of *Hyalonema* should actually be classified as separate genera [85]. However, statistical support for the positions of *Lophophysema* and *Tabachnickia* is only moderate to low. Clearly, increased taxon sampling of *Hyalonema* spp., preferably including representatives of all 12 subgenera, will

be necessary to support a revised classification of this large complex of morphologically poorly differentiated species.

Sceptrulophora. In this study, we have added five additional sceptrule-bearing species to the molecular dataset: *Tretopleura* n. sp. 1 and 2 (Uncinateridae; Fig. 1c), *Laocoetis perion* (the sole extant survivor of the paleontologically important Craticulariidae), the recently described euretid *Homoieurete macquariense* [44], and the tretodictyid *Cyrtaulon sigsbeei*. In the ML tree, *Laocoetis* is the sister taxon to a clade containing all sampled sceptrulophorans except *Tretodictyum, Hexactinella*, and *Psilocalyx* (Tretodictyidae *sensu stricto* [s. str.] hereafter). However, BS for this arrangement is very low, and in the Bayesian tree the position of this longest-ranging hexactinellid genus (since the Late Jurassic [86]) with respect to Tretodictyidae *s. str.* and the remaining sceptrulophorans remains unresolved. A clade including all sceptrulophorans except *Laocoetis* and Tretodictyidae *s. str.* is found in both the ML and Bayesian trees, but only receives significant support in the latter (PP = 0.99).

Within this clade, *Cyrtaulon* is reconstructed as the sister group (within the current taxon sampling) of *Tretopleura* with high support, thus rejecting inclusion of *Cyrtaulon* in Tretodictyidae. *Homoieurete* forms a highly supported clade with *Sarostegia* (Sceptrulophora *inc. sed.*) that is sister to the well-established Aphrocallistidae + Euretidae *part.* + Farreidae clade (cf. [17]) in the ML tree, and sister to a *Dictyoplax* (Auloplacidae) + *Cyrtaulon/Tretopleura* clade in the Bayesian tree (both with low statistical support).

The position of *Tretopleura* within Sceptrulophora, combined with the fact that the two new species definitely bear sceptrules and uncinates (MD and HMR, pers. obs.), necessitates that Uncinateridae be moved from Aulocalycoida to Sceptrulophora within the Linnean classification (Appendix 2). The position of *Cyrtaulon* outside Tretodictyidae is not too unexpected given that this taxon lacks some typical morphological features of the family (see Additional file 14). On the other hand, potential synapomorphies with *Tretopleura* remain elusive to us. However, the long branches separating these two genera indicate that they are probably part of a larger clade including many of the so far unsampled genera of Sceptrulophora, making conclusions about morphological similarities (or lack thereof) premature. The same might be true for the highly supported position of *Homoieurete* as sister to *Sarostegia*, which is equally surprising from a morphological point of view. In any case, these results demonstrate that the scope and definition of Tretodictyidae and especially Euretidae are far from being stable – clearly, more genera of both of these families need to be sampled for molecular phylogenetics. As an interim solution, we remove *Homoieurete* and *Cyrtaulon* from their respective families and treat them as Sceptrulophora *inc. sed.* within the Linnean classification (Appendix 2), following Reiswig and Dohrmann [17].

LD clade. Dactylocalycidae (currently in Hexactinosida) was so far only represented by *Iphiteon panicea* in molecular phylogenies (reviewed in [14]). We here included another, possibly new, species of that genus, as well as two species (one possibly new) of the second and type genus of the family, *Dactylocalyx*. Whereas monophyly of *Iphiteon* and *Dactylocalyx*, respectively, is confirmed here, we did not recover monophyly of the family: *Iphiteon* is weakly reconstructed as sister to *Heterorete* (Euretidae, discussed below) and the position of *Dactylocalyx* with respect to *Iphiteon/Heterorete* and the remainder of the LD clade is basically unresolved (polytomy in the Bayesian tree and BS < 50% for a position closer to Lyssacinosida in the ML tree). However, given this poor resolution, our results do not provide strong evidence for non-monophyly of Dactylocalycidae (see also Additional file 14 and next section), so this family should be retained in the Linnean system for the time being. One reason for this lack of resolution could be that so far unsampled or undiscovered taxa might be

related to *Iphiteon*, *Dactylocalyx*, and *Heterorete*, and would have to be included to resolve this part of the phylogeny. In any case, however, our molecular (and morphological; see above) analyses confirm that both *Iphiteon* and *Dactylocalyx* are more closely related to Lyssacinosida than to the remaining Hexactinosida (Sceptrulophora). We therefore here abolish the order Hexactinosida from the Linnean classification and elevate Sceptrulophora from subordinal [19] to ordinal status (Appendix 2). Pending further evidence, and given that recognizing the LD clade as a Linnean taxon is problematic (see below), we here treat Dactylocalycidae as Hexasterophora *inc. sed.* (Appendix 2).

The enigmatic dictyonal genus *Heterorete* (Fig. 1b; currently in Euretidae) is here included for the first time in a molecular phylogenetic study. As this taxon lacks sceptrules and uncinates it is not too surprising that it comes out closer to Lyssacinosida than to Sceptrulophora. One reason (but see also above) for our inability to confidently infer the exact position of this genus is likely the low gene coverage (Table 1); additional, better preserved specimens are needed to obtain more sequence data from this important taxon. Regardless of its exact position, however, our results clearly reject an affinity of *Heterorete* to Sceptrulophora, and it is best treated as Hexasterophora *inc. sed.* for the time being (Appendix 2).

The three lyssacinosidan families (Euplectellidae, Rossellidae, Leucopsacidae) and *Clathrochone* (Lyssacinosida *inc. sed.*) group together in a clade (BS = 65%, PP = 0.97). However, this clade also includes *Euryplegma auriculare*, which is the first-ever sampled representative of Aulocalycidae (Aulocalycoida), a family of dictyonal taxa without confirmed sceptrules or uncinates. *Euryplegma* nests within a maximally supported clade that is sister to Euplectellidae; within this clade *Clathrochone* is the earliest-branching genus. In the Bayesian tree, *Euryplegma* is weakly resolved as sister to Leucopsacidae + Rossellidae (PP = 0.51), whereas in the ML tree it weakly groups with Leucopsacidae (BS < 50%). A relationship of *Euryplegma* (and other aulocalycids) to Lyssacinosida, and especially Leucopsacidae, is consistent with some early taxonomic ideas (see historical overview in [87]). The firm placement of *Euryplegma* among lyssacinosidans implies convergent evolution of a dictyonal framework in this taxon (further discussed below) and renders Lyssacinosida (*sensu* [88]) paraphyletic. However, instead of abolishing this order, we here emend its diagnosis and broaden its scope to include Aulocalycidae; consequently, the order Aulocalycoida is abolished from the Linnean classification (Appendix 2).

Monophyly of Euplectellidae is supported in both the ML and Bayesian trees (BS = 85%, PP = 0.99). In line with previous molecular results [16] and the morphological analysis (Figs. 5 and 6), our phylogenies are inconsistent with the current subfamilial division of Euplectellidae (see below and Additional file 14). The topology is somewhat different to that obtained in [16], but this concerns only weakly supported nodes. The new genus of Bolosominae from Hawaii firmly groups with

Rhabdopectella in both trees. The newly sampled *Atlantisella* sp. appears closely related to *Euplectella* and *Regadrella*. In fact, it is sister to *Euplectella* sp. 1 to the exclusion of *Euplectella* sp. 2, rendering *Euplectella* paraphyletic. However, this exact position receives insignificant support in the ML tree (BS = 59%). Moreover, branch lengths in this part of the tree are very short, suggesting that these three genera might be the product of a recent radiation that is likely difficult to resolve with the current set of markers.

Regarding the newly included taxa within Rossellidae, *Vazella pourtalesii* groups with *Symplectella rowi* (see [70]) in the Bayesian tree and with Rossellinae n. gen. (Reiswig & Kelly, in prep.) in the ML tree. However, neither of these positions is significantly supported. *Asconema fristedti* appears to be related to the three aforementioned genera, but its exact position is likewise poorly supported. *Doconesthes dustinchiversi* [55] is firmly nested in Lanuginellinae, further confirming monophyly of this subfamily (*sensu* [83]), although its exact position as sister to *Lophocalyx* remains uncertain due to low support values.

Maximum-likelihood analyses of the reduced supermatrices
The tree inferred from the Amphidiscophora supermatrix reduced to only one species per genus was fully consistent with the tree inferred from the complete matrix (not shown). Likewise, in the tree reconstructed from the reduced Hexasterophora matrix (Additional file 24: Figure S10), only a few minor differences concerning nodes with low BS are observed. Thus, reducing the taxon sampling of the molecular supermatrix to match the taxonomic level of the morphological matrix had no adverse effects on the inferred relationships.

Phylogenies inferred from combined molecular and morphological data
Combined analyses restricted to genera with sequence data
Not surprisingly – given the small number of informative characters available for this subclass – addition of morphological data to the Amphidiscophora matrix had no effect on the tree topology; only some BS values slightly decreased (not shown). In contrast, addition of morphological characters to the Hexasterophora matrix had some noticeable effects: In the ML tree (Additional file 25: Figure S11), the exact positions of *Asconema*, *Rossella*, and *Atlantisella* within Rossellidae and Euplectellidae, respectively, changed (albeit with poor BS), Dactylocalycidae came out monophyletic (also with weak support), and support for monophyly of Lyssacinosida (*sensu novo* [s. nov.]), Euplectellidae, Tretodictyidae *s. str.*, and *Lophocalyx* + *Doconesthes* substantially increased. In the MrBayes tree (Additional file 26: Figure S12), the position of *Homoieurete* + *Sarostegia* changed (PP = 0.88), Dactylocalycidae came out monophyletic with high support (PP = 0.97), *Heterorete* was reconstructed as sister to

the remaining LD clade (PP = 0.97, but note that support for the LD clade as a whole decreased to 0.79), and the topology within Rossellidae changed (similar to the ML analysis, albeit with overall less resolution). These results suggest that the morphological characters indeed harbor additional signal in support of some clades and can have an impact on phylogenetic inference, despite being much smaller in number than the molecular characters.

Combined analyses including all genera
The Bayesian, ML, and MP analyses we used to obtain complete genus-level phylogenies of the two subclasses of Hexactinellida all produced poorly supported trees, i.e., with low (<0.95, < 70%) to very low (<0.5, < 50%) PP and BS values for most branches (for the morphological binning analyses, quantitative support was not assessed). We suspect that these low values are caused by the high amount of missing data (in the molecular partition), which is known to pose challenges for phylogenetic tree space exploration [89, 90]. Since quantitative support measures were not very useful in this situation, we again took a qualitative approach and looked for characters that might provide potential synapomorphies of groups of genera, as well as overall congruence of trees with well-supported taxonomic/phylogenetic hypotheses. Of the four approaches, the MP analyses produced the most plausible results (Figs. 8 and 9). In contrast, except for the Amphidiscophora Bayesian analysis (Additional file 27:

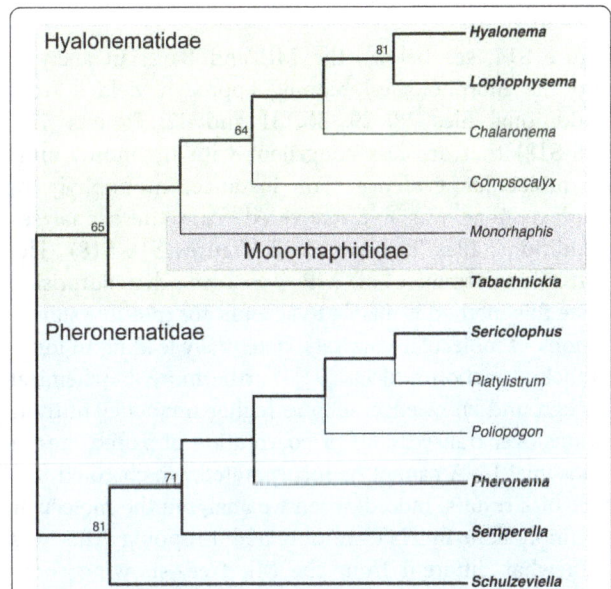

Fig. 8 Phylogeny of Amphidiscophora inferred with TNT from concatenated molecular and morphological data, including all genera. Genera with sequence data highlighted in bold and connected with thick branches. Bootstrap values >50% shown on branches (based on 1000 pseudoreplicates). TL = 2704, CI = 0.78, RI = 0.54, RC = 0.42. Nexus file available at figshare (https://doi.org/10.6084/m9.figshare.3120130.v3)

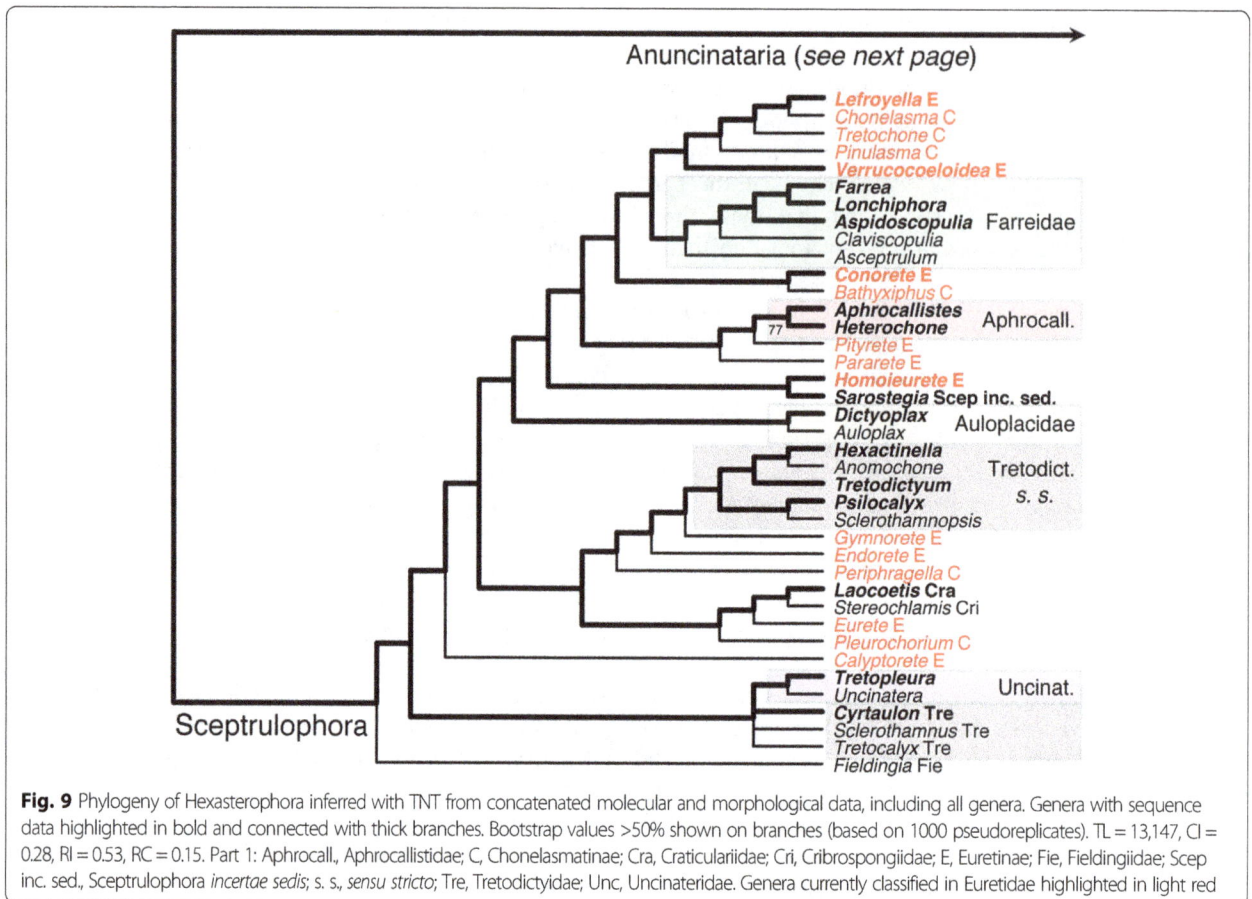

Fig. 9 Phylogeny of Hexasterophora inferred with TNT from concatenated molecular and morphological data, including all genera. Genera with sequence data highlighted in bold and connected with thick branches. Bootstrap values >50% shown on branches (based on 1000 pseudoreplicates). TL = 13,147, CI = 0.28, RI = 0.53, RC = 0.15. Part 1: Aphrocall., Aphrocallistidae; C, Chonelasmatinae; Cra, Craticulariidae; Cri, Cribrospongiidae; E, Euretinae; Fie, Fieldingiidae; Scep inc. sed., Sceptrulophora *incertae sedis*; s. s., *sensu stricto*; Tre, Tretodictyidae; Unc, Uncinateridae. Genera currently classified in Euretidae highlighted in light red

Figure S14; see below), the ML and Bayesian analyses and the morphological binning approach yielded trees (Additional files 28, 29, 30, 31 and 32: Figures S13, S15-S18) that are less congruent with taxonomy and/ or molecular evidence. For instance, monophyly of Euplectellidae was not recovered from these analyses (Additional files 30, 31 and 32: Figures S16-S18). The better performance of MP was somewhat surprising since this method is unable to account for multiple substitutions in molecular data sets, potentially leading to long-branch attraction artifacts [91]. Furthermore, biochemical background knowledge, such as higher frequency of transitions over transversions or coevolution of paired sites in ribosomal RNA cannot be incorporated, which could further bias results. Indeed, when we analyzed the molecular partition alone in TNT, we obtained a topology that was somewhat different from the ML tree, showing some "irregularities" such as *Heterorete* + *Iphiteon* sister to Euplectellidae (results not shown). However, the conflicting nodes had very low BS and the rest of the topology was largely congruent with the ML and Bayesian trees, indicating that phylogenetic signal in the molecular partition is fairly clear and robust to method choice. Apparently, MP was then better at handling the

morphological data added in the total-evidence matrix, and this additional information helped to improve the overall result. In contrast, ML and Bayesian methods might not be able to correctly "model" morphological evolution, which might be causing their poorer performance. In-depth investigations of these issues are beyond the scope of the present paper, but our results (see also above for performance of ML on the morphological matrix) indicate that currently available model-based approaches to phylogenetic analysis of morphological data might not always be the best choice (*contra* [60, 61]). Below, we will only discuss the MP trees and regard these – with some caveats – as the currently best-supported working hypotheses for the phylogenetic relationships between all glass sponge genera. For Amphidiscophora, one single MPT was found (Fig. 8), whereas for Hexasterophora nine MPTs were found, the strict consensus of which is presented here (Figs. 9 and 10).

Amphidiscophora. The total-evidence tree of Amphidiscophora (Fig. 8) is fully congruent with the molecular phylogenies (Fig. 7, Additional file 23: Figure S9). As in the morphology-based tree (Fig. 4), *Monorhaphis* (Monorhaphididae) is nested within Hyalonematidae, rendering the family paraphyletic. However, this result

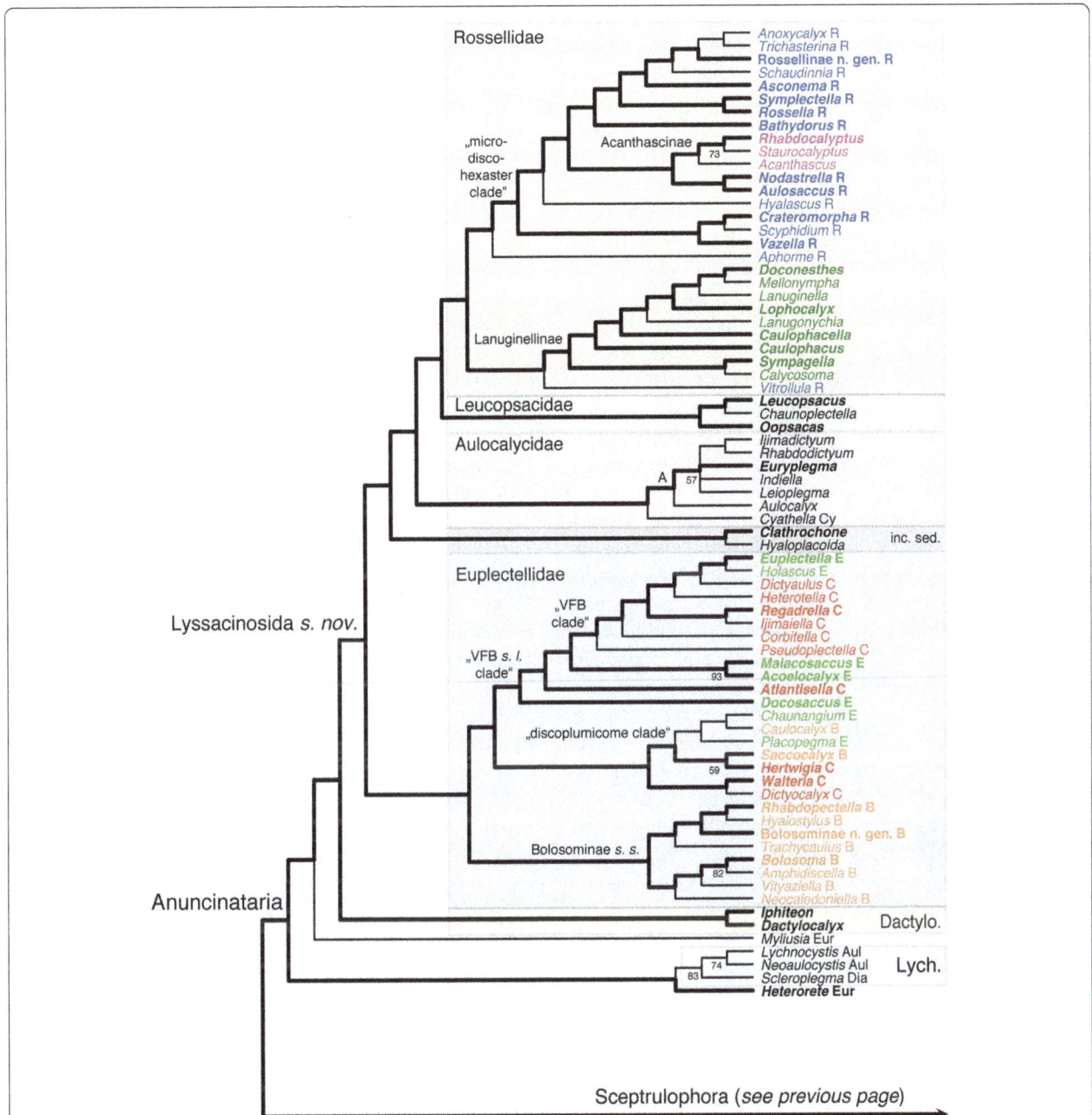

Fig. 10 Phylogeny of Hexasterophora inferred with TNT from concatenated molecular and morphological data, including all genera. Part 2: Genera with sequence data highlighted in bold and connected with thick branches. Bootstrap values >50% shown on branches (based on 1000 pseudoreplicates). TL = 13,147, CI = 0.28, RI = 0.53, RC = 0.15. A, Aulocalycinae; Aul, Aulocystidae; B, Bolosominae; C, Corbitellinae; Cy, Cyathellinae; Dia, Diapluridae; Dactylo., Dactylocalycidae; E, Euplectellinae; Eur, Euretidae; inc. sed., *incertae sedis*; Lych., Lychniscosida; R, Rossellinae; s. l., *sensu lato*; s. nov., *sensu novo*; s. s., *sensu stricto*; VFB, venus flower basket. Current subfamily assignment of rossellid and euplectellid genera indicated by different colours. Nexus file available at figshare (https://doi.org/10.6084/m9.figshare.3120130.v3)

has to be viewed with caution (see Additional file 14). Indeed, the Bayesian analysis weakly resolved Hyalonematidae as monophyletic, with *Monorhaphis* being sister to all remaining amphidiscophorans (Additional file 27: Figure S14). A possible synapomorphy of Hyalonematidae and Pheronematidae to the exclusion of *Monorhaphis* is the presence of anchorate basalia (attachment spicules with anchor-like distal ends). However, molecular data of *Monorhaphis* are required to further test its phylogenetic position within Amphidiscophora. The remaining part of the Hyalonematidae topology is congruent with the morphology-based tree (Fig. 4) in showing a sister-group relationship of *Hyalonema* and *Lophophysema*. However, the positions of *Compsocalyx* and *Chalaronema* are reversed, character support for which remains unclear. In any case, for a comprehensive understanding of character evolution within Hyalonematidae, all subgenera of *Hyalonema* need to be included in future studies, given the molecular evidence for non-monophyly of this genus (see above).

As in the morphology-based tree (Fig. 4), *Schulzeviella* forms the sister group to the remaining pheronematids. In contrast, addition of the molecular data substantially changed the rest of the topology: *Semperella* branches off after *Schulzeviella*, and *Pheronema* and *Poliopogon* are successive sister groups to a *Sericolophus* + *Platylistrum* clade. Regarding its implications for character evolution, this arrangement appears somewhat more plausible than the morphology-only topology. Choanosomal stauractins (four-rayed spicules in the skeleton of the middle tissue layer) are reconstructed as a synapomorphy of all genera except *Schulzeviella* (secondarily lost in *Sericolophus*); choanosomal tauactins (three-rayed spicules) are reconstructed as a synapomorphy (convergent to *Monorhaphis*, where they are the dominant spicules, however) for *Poliopogon*, *Sericolophus*, and *Platylistrum*; and *Sericolophus* and *Platylistrum* share the secondary absence of macramphidiscs (amphidiscs of the largest size class). Only the placement of *Pheronema* remains elusive in terms of character support. Interestingly, *Poliopogon*, *Sericolophus*, and *Platylistrum* share an asymmetric body shape, with the atrial (exhalant) surface exposed and directed to one side. This character was not included in the matrix, but could be another synapomorphy uniting these three genera (although see [92], who suggest that the body shape of *Sericolophus* resulted from an evolutionary pathway independent of the one relating *Poliopogon* and *Platylistrum*).

Deep phylogeny of Hexasterophora. The total-evidence tree of Hexasterophora (Figs. 9 and 10) is also largely congruent with the molecular phylogenies (Fig. 7, Additional file 23: Figure S9) and the trees inferred from the total-evidence matrix restricted to sequenced genera (Additional files 25 and 26: Figures S11-S12). In contrast to the morphology-only tree (Fig. 5), the Lychniscosida + *Heterorete* clade is now resolved as the sister group of the LD clade plus *Myliusia* (currently in Euretidae, but lacking sceptrules and uncinates), and Dactylocalycidae and Lyssacinosida *s. nov.* are reciprocally monophyletic. The latter not only includes *Euryplegma*, but the entire family Aulocalycidae as the sister group of Leucopsacidae + Rossellidae, which is in strong contrast to the morphology-only tree, where this taxon is deeply nested within Sceptrulophora (Fig. 5). Thus, according to these results, dictyonal frameworks of the "aulocalycoid" construction type (Fig. 2d; see [2] and Additional file 4 for details of framework construction) evolved entirely independently from dictyonal skeletons found in other taxa. Furthermore, the major division in Hexasterophora appears to be between Sceptrulophora and a clade containing all taxa that lack sceptrules and

uncinates (and not between dictyonal and lyssacine taxa). For the latter we here propose the name "Anuncinataria". This name is to be preferred over "Asceptrulophora" because uncinates also occur in most species of Amphidiscophora and hence the lack of uncinates might be a derived feature of this group. Even if uncinates evolved convergently in Amphidiscophora and Sceptrulophora (see section Maximum-likelihood ancestral state reconstruction and [93]), the absence of uncinates has at least some diagnostic value. However, we refrain from erecting a Linnean taxon for Anuncinataria for three reasons: 1) no meaningful *positive* morphological diagnosis can be provided for this clade; 2) super- or suborders would have to be introduced, but the number of ranks should be kept at a minimum; and 3) it is very important that the monophyly of this proposed group is first further tested with molecular data from Lychniscosida and *Myliusia*. However, we consider the morphological evidence for the placement of *Myliusia* outside Sceptrulophora sufficient to remove it from Euretidae and therefore re-classify it as Hexasterophora *inc. sed.* within the Linnean system (Appendix 2). On a historical note, it should be pointed out that our Anuncinataria concept was principally long foreshadowed by Schulze [94]. In his "genealogical tree" (his Figs. 9 and 10, p. 495) this author already divided Hexasterophora (although not in his Linnean classification) into a group with uncinates (Uncinataria, which later became Sceptrulophora [93]) and an unnamed group containing lyssacinosidans and his "Maeandrospongidae", which included *Aulocystis* (= *Neoaulocystis*) and *Scleroplegma* (Lychniscosida), *Dactylocalyx*, *Margaritella* (= *Iphiteon*), and *Myliusia* [20].

Parallel fusion of the rays of hexactine choanosomal megascleres is present in the dictyonal frameworks of most sceptrulophorans (except Auloplacidae and Uncinateridae), Lychniscosida, *Heterorete*, *Myliusia*, as well as *Cyathella* (Aulocalycidae). According to the parsimony mapping in MacClade, this character evolved in the last common ancestor (LCA) of Hexasterophora and got subsequently lost in the LCA of Dactylocalycidae + Lyssacinosida (followed by a "reversal" in *Cyathella*). Thus, these results suggest that dictyonal skeletons with parallel ray fusion are an autapomorphy of Hexasterophora, and the lyssacine body plan evolved secondarily within Anuncinataria from a dictyonal ground pattern (see further discussion in section Maximum-likelihood ancestral state reconstruction; see also [15]).

Besides the loss of uncinates (but see section Maximum-likelihood ancestral state reconstruction), another potential autapomorphy of Anuncinataria is the ability to produce microscleres with floricoidal tips (paw-shaped distal ends of the secondary rays of hexasters and their derivatives). Such spicules are present in *Myliusia*, *Heterorete* (HMR & MD, unpubl. obs.), *Leucopsacus* (Leucopsacidae), some

Acanthascinae (Rossellidae), and widespread in Euplectellidae (where they are called floricomes; Fig. 3b), but are unknown from sceptrulophorans. However, the scattered nature of this character across Anuncinataria greatly limits its diagnostic/phylogenetic value. Similarly, atrial megascleres (structural spicules of the inner surface layer) dominated by hexactins could be synapomorphic for *Myliusia*, Dactylocalycidae, and Lyssacinosida, but multiple absences in the latter (e.g., in Aulocalycidae) and presence in some sceptrulophorans weaken the usefulness of this character. Morphological character support for the Dactylocalycidae + Lyssacinosida clade is largely restricted to diactin (two-rayed) megascleres, which are very rare in Sceptrulophora (present in only four genera) and were lost twice, in *Acoelocalyx* + *Malacosaccus* (Euplectellidae) and in Aulocalycinae excl. *Aulocalyx* (see also Additional file 14). Finally, support for the sister-group relationship of *Heterorete* and Lychniscosida is largely limited to fused surface networks, a character that also occurs in some sceptrulophorans and is likely rather prone to homoplasy (see also Additional file 14). Clearly, inclusion of sequence data from Lychniscosida and *Myliusia*, as well as increased gene sampling of *Heterorete* (see section Phylogenies inferred from molecular data) are necessary to better resolve the phylogenetic placement of these key taxa and ultimately test the monophyly of Anuncinataria.

Below, we summarize the main findings of our total-evidence analysis concerning the internal relationships of Sceptrulophora and Lyssacinosida. For a more detailed account, the interested reader is referred to the supplementary discussion in Additional file 14.

Sceptrulophora. Relationships within Sceptrulophora are substantially altered compared to those inferred from the morphological data only (Fig. 5), especially concerning the deeper branching order, which appears to be largely driven by the molecular characters. Uncinateridae (*Uncinatera* + *Tretopleura*) is reconstructed as monophyletic, which is supported by the presence of overlapping continuous dictyonal framework rays. Furthermore, these genera are not closely related to Auloplacidae and Aulocalycidae (as in Fig. 5), with which they share several similarities in framework construction, implying that these characters evolved convergently in the three families. Tretodictyidae *s. str.* (see section Phylogenies inferred from molecular data) also includes *Anomochone* and *Sclerothamnopsis*, and the remaining three tretodictyid genera group together in a clade with Uncinateridae. Thus, this phylogeny appears more parsimonious in suggesting a diphyletic instead of a triphyletic (as in Fig. 5) origin of Tretodictyidae. Congruent with the morphology-only analysis, Aphrocallistidae, Auloplacidae, and Farreidae are reconstructed as monophyletic. Furthermore, *Laocoetis* and *Stereochlamis*, the only known extant genera of the paleontologically important Craticulariidae and Cribrospongiidae (cf. [7]) are reconstructed as sister groups, which is supported by the presence of a so-called diplorhysial framework channelization unique to these two families. As in Fig. 6 and 6, Euretidae is clearly polyphyletic, which is not surprising given

that this family constitutes a "waste-bin taxon" for all genera that do not fit into any of the other families (see [17]). Morphological support for many parts of the topology within Sceptrulophora is not clear-cut, especially concerning the deepest nodes. Clearly, many placements of the unsequenced genera in Fig. 9, especially the euretids, can only serve as initial working hypotheses that await to be tested with molecular data.

***Lyssacinosida* s. nov.** Higher-level relationships within Lyssacinosida *s. nov.* are congruent with the molecular phylogenies (Fig. 7, Additional file 23: Figure S9), with Lyssacinosida *inc. sed.* (*Clathrochone*, *Hyaloplacoida*) and Leucopsacidae more closely related to Rossellidae than to Euplectellidae. As discussed in [14], morphological support for this branching order remains unclear. Aulocalycidae is here reconstructed as sister to Leucopsacidae + Rossellidae, in contrast to the ML phylogenies inferred from the molecular supermatrix and the total-evidence matrix restricted to sequenced genera (Fig. 7, Additional file 25: Figure S11), where Aulocalycidae (*Euryplegma*) and Leucopsacidae were reconstructed as sister groups (albeit with weak support). Molecular data from additional aulocalycids will be required to disambiguate between these two hypotheses. However, the latter one is intriguing because Leucopsacidae have choanosomal megascleres exclusively as hexactins, as is the case for all dictyonal taxa as well, whereas the majority of lyssacine hexasterophorans have choanosomal spicules dominated by diactins (see Additional file 14). That is, hexactine choanosomalia are a prerequisite for developing a dictyonal framework, and the evolution of this character in a hypothetical LCA of Leucopsacidae and Aulocalycidae could have provided a pre-adaption that opened the way for the convergent evolution of dictyonal frameworks in the latter. Interestingly, aulocalycid frameworks are characterized by intensive synapticular bridging (see Fig. 2d). Spicule fusion by synapticular bridging is widespread among Lyssacinosida (see Fig. 2h), but otherwise only rarely found (in the "euretids" *Heterorete*, *Tretochone*, and *Pleurochorium*), thus providing some morphological support for inclusion of Aulocalycidae in Lyssacinosida. Besides from that, morphological evidence for this new placement of Aulocalycidae remains scarce, except perhaps for the presence of stauractins and diactins in *Aulocalyx*, spicules that are common in Lyssacinosida but rare in other taxa.

Detailed accounts of morphological support for monophyly of Lyssacinosida, Euplectellidae, *Clathrochone* + *Hyaloplacoida*, Rossellidae, Leucopsacidae, and Aulocalycidae, as well as internal relationships of the latter two taxa, can be found in the first section of the supplementary discussion in Additional file 14.

Euplectellidae. Congruent with Fig. 6, the total-evidence analysis recovered a clade of all genera with the iconical *Euplectella*-like body shape (Fig. 1f), the "venus-flower basket" or VFB clade (which is similar but not identical to Euplectellidae *s. str.* of Mehl [93]). Successive sister groups to the VFB clade include genera with a body shape that can be interpreted as primitive to or derived from a venus-flower basket, so we refer to this larger assemblage as "VFB *sensu lato*". In contrast to Fig. 6, all genera with discoplumicomes (Fig. 3c) group together, so the total-evidence phylogeny is more parsimonious in suggesting only a single origin for this complex type of microsclere. The "discoplumicome clade" contains members of all three currently accepted subfamilies [95], further demonstrating the artificial nature of this division, which is based on mode of attachment to the substrate, a rather homoplastic character. The genera *Walteria* and *Dictyocalyx* (Corbitellinae) together are sister to the discoplumicome clade, but we suspect that this is a misplacement and they are rather related to the VFB *s. l.* clade (see Additional file 14). Sister to all the above groups is a clade containing the majority of Bolosominae (which we call Bolosominae *sensu stricto*), the stalked euplectellids (Fig. 1g). This result is similar to the morphology-only analysis ("core bolosomins" as sister to the remaining euplectellids; Fig. 6) but is more parsimonious in that it includes the new genus from off Hawaii, which is very similar to *Rhabdopectella* in spiculation (MD, pers. obs.), and excludes only the discoplumicome-bearing genera *Saccocalyx* and *Caulocalyx*. We will not yet make any official classificatory changes on the basis of these findings, but we hope that support for this subdivision will solidify with increased sampling of euplectellid genera for molecular phylogenetics.

Rossellidae. The total-evidence topology of Rossellidae differs substantially from that inferred from morphological data alone. In accordance with the molecular results (Fig. 7, Additional file 23: Figure S9), most genera of Rossellidae fall in one of two major clades, Lanuginellinae (*sensu* [83]; i.e., including *Caulophacus* and *Caulophacella*), and a clade with mostly microdiscohexaster (Fig. 3g)-bearing genera (a division that was basically already recognized by Mehl [93]). Only the unsequenced *Vitrollula* and *Aphorme* seem to disrupt this simple picture: *Vitrollula* (with microdiscohexasters) is resolved as sister to Lanuginellinae, and *Aphorme* (without microdiscohexasters) sister to the remaining genera (note that microdiscohexasters are also, likely secondarily, absent in *Bathydorus* and *Trichasterina*). A sister-group relationship of *Vitrollula* and Lanuginellinae is supported by the presence of a significant number of hexactins supplementing the choanosomal diactins (secondarily lost in *Doconesthes* + *Mellonympha*). In contrast, the placement of *Aphorme* finds no obvious support from any morphological characters. Pending resolution of the positions of *Vitrollula* and *Aphorme* with molecular data, it would be tempting to recognize the microdiscohexaster clade as Rossellinae *s. nov.*, because this subfamily is currently purely negatively defined [96]. A subdivision of Rossellidae into Rossellinae *s. nov.* and Lanuginellinae *sensu* [83] would appear to be a natural choice. However, the recent resurrection of Acanthascinae [51] greatly complicates matters because this taxon appears to be an ingroup of Rossellinae *s. nov.* A close relationship between *Acanthascus*, *Rhabdocalyptus*, and *Staurocalyptus* is unambiguously supported by the exclusive presence of discoctasters (Fig. 3e) in these

three taxa, but given the reassignment of subfamily rank to this group [51], a natural classification of Rossellidae that is free of paraphyletic taxa seems to be out of reach for now.

Maximum-likelihood ancestral state reconstruction

For clarity, we present a summary of our conclusions in Fig. 11. Besides from hexactins and syncytial tissue organization, which are the defining autapomorphies of Hexactinellida and were not included in the matrices as they are parsimony-uninformative for ingroup relationships, we inferred that pentactine (five-rayed) megascleres (proportional likelihood [pl] under Mk1/aMk2 = 1.00/0.90) and possibly a dermal (outer tissue layer) skeleton dominated by these spicules (pl = 0.80/0.68) were present in the LCA of Hexactinellida. Pentactine megascleres could also have been the dominant spicule type of the atrial (inner tissue layer) skeleton, but this was only marginally supported (0.54/0.56). For the choanosomal (middle tissue layer) megasclere composition no ancestral type could be found, although hexactins and/or pentactins would be obvious candidates. The presence of uncinates in the LCA of Hexactinellida was not supported (0.23/0.30), suggesting that it is more likely that these spicules evolved convergently in Amphidiscophora and Sceptrulophora. Interestingly, the presence of microhexactins (oxyhexactins; small hexactins with pointed tips and no secondary rays) – the most basic type of microsclere – was also not supported (0.49/0.48). This inference is in line with the observation that these spicules are holactins (proteinaceous axial filaments extending to the ray tips) in Amphidiscophora, but heteractins (axial filaments not extending to the ray tips) in Hexasterophora [2]. Thus, hexasterophoran oxyhexactins could have evolved independently via reduction of secondary rays of hexasters, leaving only a single ray per primary ray. Because other microsclere types (hexasters, amphidiscs) are mutually exclusive in the two subclasses, this raises the possibility that microscleres are not homologous in Amphidiscophora and Hexasterophora.

The ancestral mode of attachment to the substrate was reconstructed as basiphytous (0.82/0.82) for Hexactinellida. However, this result has to be viewed with caution because Hexasterophora, where the vast majority of genera uses this mode, are disproportionately more genus-rich than Amphidiscophora, where this mode never occurs. Thus, in a hypothetical scenario where both subclasses had the same number of genera, the ancestral state would probably be highly ambiguous. Similar arguments can be made for the ancestral hexactinellid body plan reconstructed by the methods employed here: this was marginally supported as dictyonal (0.52/0.59). However, we consider it unlikely that the LCA of Hexactinellida was a dictyonal sponge – we rather suspect that amphidiscophorans retained an ancestral unfused skeleton, because in this group evidence for spicule fusion is entirely lacking [2] and there is no reason to believe that fused skeletons are an ancestral feature of siliceous sponges (Demospongiae + Hexactinellida).

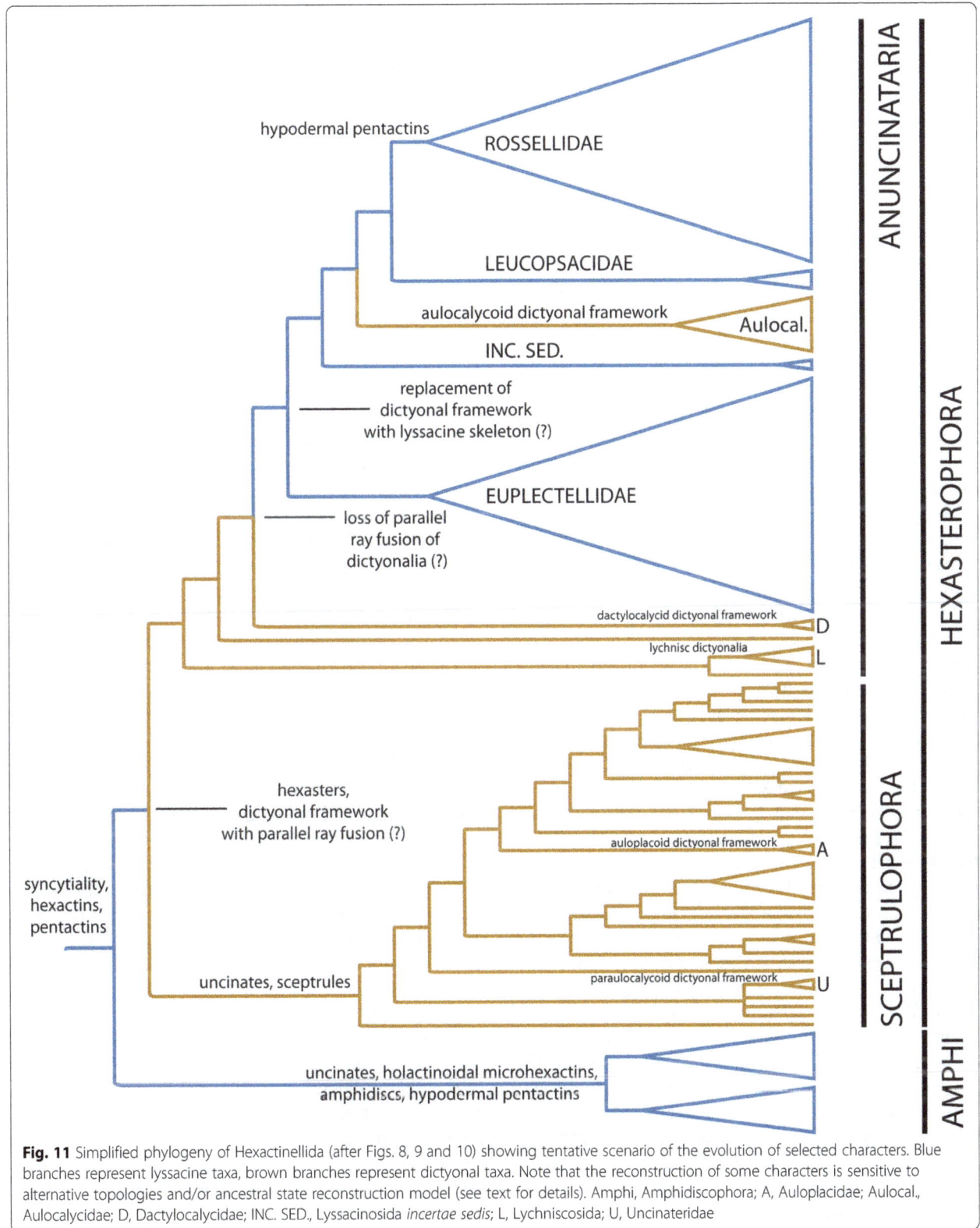

Fig. 11 Simplified phylogeny of Hexactinellida (after Figs. 8, 9 and 10) showing tentative scenario of the evolution of selected characters. Blue branches represent lyssacine taxa, brown branches represent dictyonal taxa. Note that the reconstruction of some characters is sensitive to alternative topologies and/or ancestral state reconstruction model (see text for details). Amphi, Amphidiscophora; A, Auloplacidae; Aulocal., Aulocalycidae; D, Dactylocalycidae; INC. SED., Lyssacinosida *incertae sedis*; L, Lychniscosida; U, Uncinateridae

The LCA of Amphidiscophora was reconstructed as a lyssacine sponge with lophophytous attachment to the substrate (1.00/1.00), dermal skeleton with small megascleres supported by large hypodermal pentactins (1.00/1.00; convergent to Rossellidae), dermal skeleton dominated by pentactins (0.99/0.98), atrial skeleton dominated by pentactins (0.96/0.96), presence of pinular hexactins (0.97/0.93) and pentactins (1.00/1.00; convergent to Lanuginellinae), stauractins (0.89/0.91), uncinates (0.99/1.00), and microscleres as oxyhexactins (0.92/0.92) and amphidiscs of all three size classes (0.99 to 1.00). As these characters occur in almost all genera, this reconstruction is somewhat trivial. Regarding the choanosomal skeletal composition, however, only the presence of hexactins among mixed choanosomalia gained some support under the aMk2 model (0.87) – we hypothesize that the different compositions defining the three families [97] each evolved independently from a simple hexactin-dominated ground state. The aMk2 model analyses also reconstructed additional ancestral states for the LCA of Amphidiscophora that were not supported by the simpler Mk1 model: hypoatrial pentactins (0.69), amphidiscs with additional rays (0.99), oxyoidal (acute-tipped) or clavate (club-tipped) monactin/diactin attachment spicules (0.93), and toothed anchorate attachment spicules (1.00). Regarding the amphidiscs, this result implies that the six-rayed "hexadiscs" found in some genera in addition to the more common two-rayed regular amphidiscs (Fig. 2i) are plesiomorphic remnants and that the latter evolved from the former by ray reduction. The inference about the anchorate attachment spicules is sensitive to the position of *Monorhaphis* and disappears when this genus is placed as sister to Hyalonematidae + Pheronematidae (as in Additional file 27: Figure S14). This was not the case for the oxyoidal attachment spicules – these spicules are unknown from Hyalonematidae, so this result implies that they were secondarily lost in this family.

The LCA of Hexasterophora was reconstructed as a basiphytous (0.98/0.97), dictyonal (0.76/0.84) sponge with parallel ray fusion of dictyonal hexactins (0.66/0.94). As already discussed above, this implies that a lyssacine body plan "re-evolved" within this subclass, followed by independent evolution of dictyonal skeletons in Aulocalycidae. However, this reconstruction was somewhat sensitive to model choice and the topology at the base of Anuncinataria: when the position of *Myliusia* was changed to sister of *Heterorete* + Lychniscosida or to sister of Dactylocalycidae, or when all four taxa were constrained to form a clade, the likelihood of the presence of a dictyonal body plan in the LCA of Hexasterophora dropped below 0.45 under both models (range 0.13–0.43). However, the character "dictyonal bauplan" is an oversimplification (and was thus not used for phylogeny reconstruction), so "parallel ray fusion of dictyonal hexactins" is a more meaningful character to look at. The presence of this construction type in the LCA of Hexasterophora was also not supported under the three alternative topologies when the Mk1 model was used (range 0.26–0.48); however, the aMk2 model supported its presence by pl of 0.75–0.83. These observations demonstrate that a robust resolution of this part of the topology, especially by including sequence data from Lychniscosida and *Myliusia*, will be required to more confidently reconstruct the evolution of non-sceptrulophoran dictyonal frameworks. Furthermore, the distinct construction type

of dactylocalycid frameworks (Fig. 2c) suggest that they might have evolved convergently (see Additional file 14), regardless of phylogenetic considerations. On the other hand, a possible link between dictyonal and lyssacine hexasterophorans are the so-called basidictyonal frameworks occuring in both groups, which are structures of fused spicules involved in the attachment of basiphytous species to the substratum [2]. Interestingly, basidictyonal spicules connect by tip-to-tip fusion [2], which is rare in sceptrulophoran choanosomal frameworks but very common in those of the sister-group of Lyssacinosida, the Dactylocalycidae. Thus, lyssacine hexasterophorans might have evolved by suppression of further development of their basidictyonal spicules into fully-fledged choanosomal dictyonal frameworks, instead re-deploying an ancient genetic program that instructs development of unfused choanosomalia (but see also discussion in [15]). Evolutionary developmental (evo-devo) studies would be of great help in answering these questions, but are likely too difficult to implement in absence of easily manipulatable hexactinellid model systems from different relevant taxa.

With respect to loose spiculation, the LCA of Hexasterophora likely retained pentactine dermal (0.82/0.68) and possibly atrial (0.54/0.56) megascleres from the ground pattern of Hexactinellida. Microscleres of course included hexasters (0.99/0.98), the defining apomorphy of this subclass, most likely in the form of oxy- (Fig. 2j) (0.59/0.71) and disco- (Fig. 3) hexasters (0.92/0.74). Presence of oxyhexactins in the LCA of Hexasterophora was poorly supported (0.48/0.47), suggesting that these spicules might have repeatedly evolved convergently, which appears plausible because under the scenario hypothesized above only loss of some secondary rays is required to evolve these spicules from hexasters.

Comparison of the two different ancestral state reconstruction models further revealed that the Mk1 model frequently supported multiple independent origins of a character towards the tips of a clade, whereas the aMk2 model preferred a single origin at the clade's root, followed by multiple losses. For example, synapticular spicule fusion evolved eight times in Lyssacinosida according to the Mk1 model, whereas the aMk2 model inferred a single origin in the LCA of this group. Further examples include skeletal channelization in Sceptrulophora, choanosomal skeletons dominated by diactins in Lyssacinosida, graphiocomes and floricomes (Fig. 3b, l) in Euplectellidae, and strobiloplumicomes (Fig. 3d) in Lanuginellinae (see also results on Amphidiscophora above). This suggests that the simplifying assumption made by the Mk1 model that gains and losses are equally likely can frequently lead to an inflated estimate of convergent evolution. However, it has to be evaluated on a case-by-case basis whether loss or gain of a particular character is more likely. For instance, floricomes and strobiloplumicomes are quite complex spicules and therefore postulating convergent evolution for this character seems unparsimonious. On the other hand, skeletal channelization of dictyonal frameworks is probably something that is easily evolved, as can be seen for example by the occurrence of channels (epirhyses) in ontogenetically older specimens of the usually unchannelized farreids (e.g., [46]), and also by the multitude of different channelization types (see Additional file 4) that are indicative of convergent evolution. Therefore, although the aMk2 model generally appears to be more realistic by accounting for differences between gain and loss rates, it is important also to conduct analyses under the simpler Mk1 model and compare the results in the light of biological plausibility (see also [98]).

Conclusions

In this study, we have increased the taxon sampling for molecular systematics of Hexactinellida by 15 species, 12 genera, three families, and one order. One major finding was that the order Aulocalycoida is polyphyletic because its two constituent families (Aulocalycidae and Uncinateridae) are resolved as ingroups of Lyssacinosida and Sceptrulophora, respectively. Furthermore, the sceptrule- and uncinate lacking dictyonal genera *Heterorete* (formerly Euretidae) and *Dactylocalyx* (Dactylocalycidae) were resolved as more closely related to Lyssacinosida than to Sceptrulophora, which further demonstrates the artificial nature of Hexactinosida. Consequently, we abolish Aulocalycoida and Hexactinosida, elevate Sceptrulophora from suborder to order, and emend diagnosis and scope of Lyssacinosida to include Aulocalycidae. These updates are timely and bring the Linnean classification of glass sponges in closer agreement with their phylogeny, similar to what was recently proposed for Demospongiae [99].

We further compiled morphological character matrices including all extant genera of Amphidiscophora and Hexasterophora and analyzed these alone and in combination with the molecular data. We compared MP, ML, and Bayesian approaches, as well as "morphology-based phylogenetic binning" [22] and found that MP consistently outperformed the other methods in terms of congruence with well-founded taxonomic and phylogenetic hypotheses. Bayesian analyses performed second best, whereas ML and binning gave largely dubious results. Phylogenies based only on morphological data were partly congruent with the molecular tree (e.g., paraphyly of Hexactinosida, monophyly of many families), but also conflicted in many areas (e.g., monophyletic Aulocalycoida nested within Sceptrulophora). The total-evidence trees were largely congruent with the molecular phylogeny and suggest that the major division of Hexasterophora is not between lyssacine and dictyonal taxa, but instead between taxa with and without sceptrules and uncinates, i.e., between Sceptrulophora and a clade we call Anuncinataria. Besides Lyssacinosida (including Aulocalycidae), Dactylocalycidae, and *Heterorete*, Anuncinataria also includes *Myliusia* (formerly Euretidae) and Lychniscosida, a species-poor relict group that was highly diverse in the Jurassic and Cretaceous. Inclusion of sequence data from the latter two taxa will be crucial to further test the monophyly of Anuncinataria. In general,

placement of the unsequenced genera in our total-evidence phylogeny should not be taken as the last word but as a starting point; these are working hypotheses that need to be further tested by filling the gaps in the molecular dataset. Also, the morphological character matrices should not be viewed as static, but as a resource that is subject to constant revision.

Character mapping and ML ancestral state reconstruction (ASR) on the total-evidence tree allowed us to gain deeper insights into the evolution of skeletal structures in Hexactinellida. Our results suggest that evolution of the dictyonal body plan was more complex than previously thought. Besides from the obvious implication that the dictyonal skeletons of Aulocalycidae evolved convergently from a lyssacine condition and that the peculiar construction types of Uncinateridae and Auloplacidae evolved independently from a more regular type, we found that dictyonal skeletons with parallel ray fusion might have been present in the ground pattern of Hexasterophora and got secondarily lost in the stem lineage of Lyssacinosida. That is, the lyssacine condition in Lyssacinosida might represent a case of evolutionary reversal to an ancestral body plan, the genetic program for which was inherited from the last common ancestor of Hexactinellida. However, this inference was sensitive to ASR model choice and the branching pattern at the base of Anuncinataria, so the possibility that dictyonal frameworks evolved once or multiple times convergently in early-branching anuncinatarians (e.g., Dactylocalycidae) cannot be ruled out until the phylogenetic relationships of these taxa are better resolved.

Concerning loose spiculation, we also found – not unexpectedly – high levels of homoplasy. The degree to which this is due to multiple convergent origins or multiple losses of spicule types depends somewhat on the assumptions of the ASR model used, and has to be evaluated on a case-by-case basis of individual characters. However, it appears that hexactinellids, and sponges in general, are able to retain the genetic instructions to produce certain spicule types over long evolutionary time, even if they are not expressed in the phenotype. For example, discasters (Fig. 3i) and sigmato- or drepanocomes (Fig. 3k) are restricted to Lyssacinosida but within this group are only found in 14 and 12 genera, respectively, scattered across three families. That is, it is the ability to produce these spicules that can be interpreted as an apomorphy of Lyssacinosida, not their actual phenotypic expression. The importance of this phenomenon, which has been called "cryptotypic property" [100], in Hexactinellida was already pointed out by Mehl [93]. Furthermore, Maldonado et al. [101] showed that sponges can be forced to produce spicule types not normally found in a given species just by altering the silica concentration of sea water. Although these phenomena are largely ignored by sponge taxonomists, at least in Hexactinellida the problem of homoplasy seems less severe than in other sponge classes (cf. [99, 102]). Integrating morphology and molecular sequence data has great potential to inform us about the evolution of this fascinating group of animals.

Appendix 1

Table 2 Taxonomic overview of hexactinellid genera

Subclass: Genus	Authority	#spp.	Subfamily	Order: Family
Amphidiscophora				Amphidiscosida
Hyalonema	Gray, 1832	120		Hyalonematidae
Chalaronema	Ijima, 1927	1		Hyalonematidae
Compsocalyx	Schulze, 1904	1		Hyalonematidae
Lophophysema	Schulze, 1900	4		Hyalonematidae
Tabachnickia	Özdikmen, 2010	1		Hyalonematidae
Monorhaphis	Schulze, 1904	1		Monorhaphididae
Pheronema	Leidy, 1868	18		Pheronematidae
Platylistrum	Schulze, 1904	1		Pheronematidae
Poliopogon	Thomson, 1878	6		Pheronematidae
Schulzeviella	Tabachnick, 1990	1		Pheronematidae
Semperella	Gray, 1868	11		Pheronematidae
Sericolophus	Ijima, 1901	5		Pheronematidae
Hexasterophora				Lyssacinosida
Euplectella	Owen, 1841	17	Euplectellinae	Euplectellidae
Acoelocalyx	Topsent, 1910	1	Euplectellinae	Euplectellidae
Chaunangium	Schulze, 1904	1	Euplectellinae	Euplectellidae
Docosaccus	Topsent, 1910	2	Euplectellinae	Euplectellidae
Holascus	Schulze, 1886	15	Euplectellinae	Euplectellidae
Malacosaccus	Schulze, 1886	7	Euplectellinae	Euplectellidae
Placopegma	Schulze, 1895	2	Euplectellinae	Euplectellidae
Bolosoma	Ijima, 1904	8	Bolosominae	Euplectellidae
Amphidiscella	Tabachnick & Lévi, 1997	4	Bolosominae	Euplectellidae
Caulocalyx	Schulze, 1886	1	Bolosominae	Euplectellidae
Hyalostylus	Schulze, 1886	2	Bolosominae	Euplectellidae
Neocaledoniella	Tabachnick & Lévi, 2004	1	Bolosominae	Euplectellidae
Rhabdopectella[a]	Schmidt, 1880	1	Bolosominae	Euplectellidae
Saccocalyx	Schulze, 1896	3	Bolosominae	Euplectellidae
Trachycaulus	Schulze, 1886	1	Bolosominae	Euplectellidae
Vityaziella	Tabachnick & Lévi, 1997	1	Bolosominae	Euplectellidae
Corbitella	Gray, 1867	4	Corbitellinae	Euplectellidae
Atlantisella	Tabachnick, 2002	1	Corbitellinae	Euplectellidae
Dictyaulus	Schulze, 1896	4	Corbitellinae	Euplectellidae
Dictyocalyx	Schulze, 1886	2	Corbitellinae	Euplectellidae
Hertwigia	Schmidt, 1880	1	Corbitellinae	Euplectellidae
Heterotella	Gray, 1867	4	Corbitellinae	Euplectellidae
Ijimaiella	Tabachnick, 2002	1	Corbitellinae	Euplectellidae
Pseudoplectella	Tabachnick, 1990	1	Corbitellinae	Euplectellidae
Regadrella	Schmidt, 1880	8	Corbitellinae	Euplectellidae
Walteria	Schulze, 1886	2	Corbitellinae	Euplectellidae
Leucopsacus	Ijima, 1898	4		Leucopsacidae
Oopsacas	Topsent, 1927	4		Leucopsacidae

Table 2 Taxonomic overview of hexactinellid genera *(Continued)*

Chaunoplectella	Ijima, 1896	1		Leucopsacidae
Rossella	Carter, 1872	20	Rossellinae	Rossellidae
Anoxycalyx	Kirkpatrick, 1907	3	Rossellinae	Rossellidae
Aphorme	Schulze, 1899	1	Rossellinae	Rossellidae
Asconema	Kent, 1870	5	Rossellinae	Rossellidae
Aulosaccus	Ijima, 1896	5	Rossellinae	Rossellidae
Bathydorus	Schulze, 1886	7	Rossellinae	Rossellidae
Crateromorpha	Gray in Carter, 1872	15	Rossellinae	Rossellidae
Hyalascus	Ijima, 1896	9	Rossellinae	Rossellidae
Nodastrella	Dohrmann, Göcke, Reed & Janussen, 2012	2	Rossellinae	Rossellidae
Schaudinnia	Schulze, 1900	1	Rossellinae	Rossellidae
Scyphidium	Schulze, 1900	8	Rossellinae	Rossellidae
Symplectella	Dendy, 1924	1	Rossellinae	Rossellidae
Trichasterina	Schulze, 1900	2	Rossellinae	Rossellidae
Vazella	Gray, 1870	1	Rossellinae	Rossellidae
Vitrollula	Ijima, 1898	1	Rossellinae	Rossellidae
Acanthascus	Schulze, 1886	7	Acanthascinae	Rossellidae
Rhabdocalyptus	Schulze, 1886	18	Acanthascinae	Rossellidae
Staurocalyptus	Ijima, 1897	17	Acanthascinae	Rossellidae
Lanuginella	Schmidt, 1870	1	Lanuginellinae	Rossellidae
Calycosoma	Schulze, 1899	1	Lanuginellinae	Rossellidae
Caulophacella[b]	Lendenfeld, 1915	1	Lanuginellinae	Rossellidae
Caulophacus	Schulze, 1886	26	Lanuginellinae	Rossellidae
Doconesthes	Topsent, 1928	2	Lanuginellinae	Rossellidae
Lanugonychia	Lendenfeld, 1915	1	Lanuginellinae	Rossellidae
Lophocalyx	Schulze, 1887	12	Lanuginellinae	Rossellidae
Mellonympha	Schulze, 1897	1	Lanuginellinae	Rossellidae
Sympagella	Schmidt, 1870	9	Lanuginellinae	Rossellidae
Clathrochone	Tabachnick, 2002	1		*incertae sedis*
Hyaloplacoida	Tabachnick, 1989	1		*incertae sedis*
				Hexactinosida (Sceptrulophora)
Aphrocallistes	Gray, 1858	2		Aphrocallistidae
Heterochone	Ijima, 1927	4		Aphrocallistidae
Auloplax	Schulze, 1904	4		Auloplacidae
Dictyoplax	Reiswig & Dohrmann, 2014	1		Auloplacidae
Laocoetis	Pomel, 1872	1		Craticulariidae
Stereochlamis	Schrammen, 1912	1		Cribrospongiidae
Eurete	Semper, 1868	12	Euretinae	Euretidae
Calyptorete	Okada, 1925	1	Euretinae	Euretidae
Conorete	Ijima, 1927	5	Euretinae	Euretidae
Endorete	Topsent, 1928	1	Euretinae	Euretidae
Gymnorete	Ijima, 1927	3	Euretinae	Euretidae
Heterorete	Dendy, 1916	1	Euretinae	Euretidae
Homoieurete	Reiswig & Kelly, 2011	1	Euretinae	Euretidae

Table 2 Taxonomic overview of hexactinellid genera *(Continued)*

Lefroyella	Thomson, 1878	2	Euretinae	Euretidae
Pararete	Ijima, 1927	7	Euretinae	Euretidae
Pityrete	Topsent, 1928	1	Euretinae	Euretidae
Chonelasma	Schulze, 1886	10	Chonelasmatinae	Euretidae
Bathyxiphus	Schulze, 1899	1	Chonelasmatinae	Euretidae
Myliusia	Gray, 1859	4	Chonelasmatinae	Euretidae
Periphragella	Marshall, 1875	6	Chonelasmatinae	Euretidae
Pinulasma	Reiswig & Stone, 2013	1	Chonelasmatinae	Euretidae
Pleurochorium	Schrammen, 1912	2	Chonelasmatinae	Euretidae
Tretochone	Reid, 1958	1	Chonelasmatinae	Euretidae
Verrucocoeloidea	Reid, 1969	2	Chonelasmatinae	Euretidae
Farrea	Bowerbank, 1862	30		Farreidae
Asceptrulum	Duplessis & Reiswig, 2004	1		Farreidae
Aspidoscopulia	Reiswig, 2002	5		Farreidae
Claviscopulia	Schulze, 1899	1		Farreidae
Lonchiphora	Ijima, 1927	2		Farreidae
Fieldingia	Kent, 1870	2		Fieldingiidae
Tretodictyum	Schulze, 1886	7		Tretodictyidae
Anomochone	Ijima, 1927	3		Tretodictyidae
Cyrtaulon	Schulze, 1886	2		Tretodictyidae
Hexactinella	Carter, 1885	14		Tretodictyidae
Psilocalyx	Ijima, 1927	1		Tretodictyidae
Sclerothamnopsis	Wilson, 1904	2		Tretodictyidae
Sclerothamnus	Marshall, 1875	1		Tretodictyidae
Tretocalyx	Schulze, 1901	1		Tretodictyidae
Sarostegia	Topsent, 1904	1		*incertae sedis*
				Hexactinosida
Dactylocalyx	Stutchbury, 1841	2		Dactylocalycidae
Iphiteon	Bowerbank, 1869	1		Dactylocalycidae
				Aulocalycoida
Aulocalyx	Schulze, 1886	3	Aulocalycinae	Aulocalycidae
Euryplegma	Schulze, 1886	1	Aulocalycinae	Aulocalycidae
Ijimadictyum	Mehl, 1992	1	Aulocalycinae	Aulocalycidae
Indiella	Sautya, Tabachnick & Ingole, 2011	1	Aulocalycinae	Aulocalycidae
Leioplegma	Reiswig & Tsurumi, 1996	1	Aulocalycinae	Aulocalycidae
Rhabdodictyum	Schmidt, 1880	1	Aulocalycinae	Aulocalycidae
Cyathella	Schmidt, 1880	1	Cyathellinae	Aulocalycidae
Uncinatera	Topsent, 1901	1		Uncinateridae
Tretopleura	Ijima, 1927	2		Uncinateridae
				Lychniscosida
Neoaulocystis	Zhuravleva, 1962	4		Aulocystidae
Lychnocystis	Reiswig, 2002	1		Aulocystidae
Scleroplegma	Schmidt, 1880	2		Diapleuridae

[a]Currently classified in Corbitellinae, but see Table 1: footnote c of [3]

[b]This genus was recently reclassified as a subgenus of *Caulophacus* ([83]; following [26]). However, molecular data do not support inclusion of *Caulophacella tenuis* in *Caulophacus* [3, 16]. Therefore, we here retained *Caulophacella* as a terminal taxon

Appendix 2
Changes to Linnean classification, including revised diagnoses
Summary

The subclass Hexasterophora Schmidt, 1870 now contains the following three orders: Sceptrulophora, Lyssacinosida, and Lychniscosida. The former order Hexactinosida is abolished by recognition of Sceptrulophora and Dactylocalycidae as unrelated taxa. The former order Aulocalycoida is abolished by moving its two constituent families to Sceptrulophora and Lyssacinosida, respectively. One family – Dactylocalycidae with the two genera *Dactylocalyx* and *Iphiteon* – and five further genera – *Heterorete*, *Myliusia* (formerly Euretidae), *Deanea*, *Diaretula*, and *Hyalocaulus* (formerly Hexactinosida *inc. sed.*) – cannot presently be assigned to any of the orders and are treated as Hexasterophora *inc. sed.*. Sceptrulophora contains nine families – Aphrocallistidae, Auloplacidae, Craticulariidae, Cribrospongiidae, Euretidae, Farreidae, Fieldingiidae, Tretodictyidae, and Uncinateridae (formerly Aulocalycoida); three genera are treated as Sceptrulophora *inc. sed.*: *Sarostegia* (cf. [17]), *Cyrtaulon* (formerly Tretodictyidae), and *Homoieurete* (formerly Euretidae). Lyssacinosida contains four families: Euplectellidae, Leucopsacidae, Rossellidae, and Aulocalycidae (formerly Aulocalycoida); two genera (*Clathrochone*, *Hyaloplacoida*) are treated as Lyssacinosida *inc. sed.*. Lychniscosida contains two families (Aulocystidae, Diapleuridae).

SCEPTRULOPHORA MEHL, 1992 ord. nov.

Diagnosis (emended from Dohrmann et al. [19]*)*: Dictyonal Hexactinellida with uncinates; sceptrules usually present but can be missing in rare cases. Body shape is highly variable from branching and anastomosing tubes to cup, funnel, or blade forms. Dictyonal skeleton mostly euretoid, but farreoid, auloplacoid, or paraulocalycoid patterns also occur. Channelization of dictyonal framework as epi- and/or aporhyses, schizorhyses, diarhyses, diplorhyses, amararhyses, or absent. Dermalia and atrialia usually pentactins, sometimes hexactins or absent. Microscleres oxy- and/or discohexasters and their derivatives; onycho- and tylo-tipped forms might also occur.

Scope: The following families are included in Sceptrulophora: Aphrocallistidae Gray, 1867, Auloplacidae Schrammen, 1912, Craticulariidae Rauff, 1893, Cribrospongiidae Roemer, 1864 (but see remarks in [19]), Euretidae Zittel, 1877, Farreidae Gray, 1872, Fieldingiidae Tabachnick & Janussen, 2004, Tretodictyidae Schulze, 1886, and Uncinateridae Reiswig, 2002. Uncinateridae (formerly Aulocalycoida Tabachnick & Reiswig, 2000) is here included on the basis of molecular evidence (e.g., Fig. 7) and the presence of uncinates (as well as sceptrules in *Tretopleura* Ijima, 1927).

Remarks: Sceptrules are absent from *Asceptrulum* Duplessis & Reiswig, 2004 (Farreidae) and have not been confirmed so far from the poorly known *Uncinatera* Topsent, 1901 (Uncinateridae), both of which are monospecific genera. However, the presence of uncinates in these species clearly supports their inclusion in Sceptrulophora. The genera *Heterorete* Dendy, 1916 and *Myliusia* Gray, 1859 (formerly Euretidae) have neither uncinates nor sceptrules, and are excluded from Sceptrulophora (see below). Uncinateridae Reiswig, 2002 (*Uncinatera* and *Tretopleura*) have a paraulocalycoid framework construction; the term "auloplacoid" is introduced here to distinguish the skeletal architecture of Auloplacidae Schrammen, 1912 from the prevailing euretoid pattern (see [17]). Elevation of Sceptrulophora from subordinal [19] to ordinal status follows from abolishment of Hexactinosida Schrammen, 1912 (see below). Because the concept of the new order is the same as for the clade name established by Mehl, we prefer to leave the authority with this author.

EURETIDAE ZITTEL, 1877

Diagnosis (emended from Dohrmann et al. [19]*)*: Sceptrulophora with body form either of branching and/or anastomosing tubes, or cup-funnel formed of a ring of tubes, or of a single tube, or of a single-wall funnel with or without lateral oscula extended on marginal tubes, or blade form; dictyonal meshes mainly rectangular or triangular or irregular; meshes usually equal-sided but elongate prismatic mesh series with transverse lamellae developed in some species; dictyonal strands, if developed, orientated longitudinally; with or without dictyonal cortices composed of primary or secondary dictyonalia; dermalia and atrialia are commonly pentactins or pinular hexactins with rays of approximately equal length, or both forms lacking; scopules (which might also be represented by sarules) and uncinates present; microscleres occur as oxyhexasters and/or discohexasters.

Scope: As of May 2016, Euretidae was comprised of 18 genera [5]. Here, we remove the sceptrule- and uncinate-lacking *Heterorete* and *Myliusia* and re-classify them provisionally as Hexasterophora *inc. sed.* because phylogenetic analyses (e.g., Figs. 5, 7 and 10) clearly place them outside of Sceptrulophora but their exact phylogenetic position (and therefore ordinal assignment) remains to be determined (see main text and Additional file 14 for further details). *Homoieurete* Reiswig & Kelly, 2011 is also removed from Euretidae, because molecular evidence (e.g., Fig. 7) strongly suggests that it is unrelated to the other three euretids with available DNA sequence data. Following suggestions of Reiswig and Dohrmann [17], we thus provisionally re-classify *Homoieurete* as Sceptrulophora *inc. sed.*

Remarks: The emended diagnosis reflects the fact that Euretidae no longer contains taxa without sceptrules and uncinates. However, even after removal of the above-mentioned genera, Euretidae clearly constitutes a "waste-bin" taxon that lumps together all sceptrulophoran genera not assignable to any of the other families (see [17]). Our phylogenetic analyses (e.g., Figs. 5 and 9) suggest that these genera are highly interspersed within Sceptrulophora, some being related to each other and others more closely related to different families. If at all, this taxonomic challenge can only be resolved by molecular data. For the time being, we retain Euretidae within the Linnean system while acknowledging its artificial nature.

TRETODICTYIDAE SCHULZE, 1886

Diagnosis (emended from Reiswig and Kelly [44]): Sceptrulophora with body form varying from branching and anastomosing solid cylinders to branching and anasto-mosing tubes to funnel, cup, and irregular globular forms; with three-dimensional, small-meshed, euretoid dictyonal framework several dictyonalia in thickness at the growing edge; primary dictyonal frame consists at least in part of four-sided (square or rectangular) meshes; rays of dictyo-nalia extend only one-mesh in length to the next adjacent dictyonal centrum; longitudinally oriented dictyonal rays aligned and fused side-by-side to form longitudinal strands; schizorhysial channelisation developed by growth of framework in narrow vertical (dermal to atrial) and lon-gitudinal oriented septa bridged by small patches of dic-tyonalia; such growth leaves a confluent system of small gauge channels 1–2 mm wide running mainly longitudin-ally, but connected transversely. Superficial cortices usually not developed but hypersilicification of dermal surfaces with swollen surface nodes occur in three genera; attach-ment of small hexactins to frameworks is rare; spiculation includes strongyloscopules; uncinates of intermediate size with poorly developed brackets and barbs are typical.

Scope: As of May 2016, Tretodictyidae was com-prised of eight genera [5]. Here, we remove *Cyrtaulon* Schulze, 1886 from this family on the basis of molecular evidence (e.g., Fig. 7), and provisionally re-classify it as Sceptrulophora *inc. sed.* Inclusion of *Sclerothamnus* Marshall, 1875, *Sclerothamnopsis* Wil-son, 1904, and *Tretocalyx* Schulze, 1901 is provisional and awaits to be tested with molecular data (see main text and Additional file 14).

Remarks: In the emended diagnosis, we removed "in all but one genus" after "strongyloscopules". The lack of strongyloscopules in *Cyrtaulon* and the presence in-stead of the discohexaster-like "*Cyrtaulon*-spicule" [103] as the only scopule form, as well as presence of large uncinates with fully developed brackets and barbs (see Additional file 14), provide some morphological support for exclusion of *Cyrtaulon* from Tretodictyidae.

LYSSACINOSIDA ZITTEL, 1877

Diagnosis (emended from Definition and Diagnosis of Reiswig [88]): Hexasterophora in which choanosomal

megascleres remain as separate skeletal components, or, where fusion occurs it is by deposition of silica at con-tact points or as synapticula between diactine, tauactine, stauractine, or hexactine megascleres, or by tip-to-ray fusion of hexactine choanosomalia forming longitudinal strands of single continually extended rays with uniaxial connecting beams. Body form is typically a single ovoid, cup or tube bearing a single terminal osculum and deep atrial cavity, but might also be as branching fan or tubes, or tongue-like plate. Attachment to the substrate is either direct or by short peduncle or long stalk and is usually basiphytous with a thin basidictyonal plate of fused hexactins; lophophytous or rhizophytous attach-ment also occurs. Thin-walled forms may have a sieve plate over terminal osculum and a regular series of small parietal oscula; thicker-wall forms may occasionally bifurcate or grow one or more lateral diverticula, each with terminal osculum. Branching in stalks of cup-shaped members is poorly documented as a growth form and may result from secondary settlement. Choanosomal megascleres may be mainly diactins, or unfused or fused hexactins, or a combination of stauractins, tauactins, diactins, rarely pentactins. Dermalia may be large pen-tactins or hexactins unsupported by hypodermalia, or small hexactins, pentactins, stauractins or diactins sup-ported by large pentactine hypodermalia. Atrialia may be either hexactins and/or pentactins and/or stauractins; hypoatrial pentactins may be present. Sceptrules and un-cinates absent. Lateral prostalia may be absent or special diactins or extended hypodermal pentactins or simply the extended distal rays of choanosomal hexactins or pentactins; basalia of lophophytous forms may be mon-actine, diactine or pentactine anchors. Microscleres include single types or combinations of stellate and spherical discohexasters of regular or hemi-form, discocta-sters, discohexactins, floricomes, discoplumicomes, strobi-loplumicomes, sigmatocomes, oxyhexasters of regular and hemi-form, graphiocomes, trichasters, oxyhexactins, ony-choexasters, and onychohexactins.

Scope: Four families are included in Lyssacinosida: Euplectellidae Gray, 1867, Leucopsacidae Ijima, 1903, Rossellidae Schulze, 1885, and Aulocalycidae Ijima, 1927. Two monospecific genera with uncertain family as-signment (Lyssacinosida *inc. sed.*) are also included: *Clathrochone* Tabachnick, 2002 and *Hyaloplacoida* Tabachnick, 1989.

Remarks: Aulocalycidae is here included in Lyssacino-sida on the basis of molecular evidence (e.g., Fig. 7), which is consistent with earlier taxonomic hypotheses (reviewed in [87]) and finds some morphological support from the widespread occurrence of synapticular fusion in this taxon (see main text for discussion). This result is based on sequence data from a single species, *Eury-plegma auriculare* Schulze, 1886; the phylogenetic position of the remaining six genera currently included in Aulocalycidae (see Appendix 1) awaits to be determined with molecular data. The emended diagnosis reflects inclusion of additional characters present in Aulocalycidae. Inclusion of Aulocalycidae in Lyssacinosida requires abolishment of the order Aulocalycoida Tabachnick & Reiswig, 2000. Regarding the *inc. sed.* genera, *Clathrochone* is clearly not a member of any of the other families ([15, 16]; this study); molecular data from *Hyaloplacoida* are required to test the hypothesis that these two genera are sister groups (e.g., Figs. 5 and 10), which would justify erection of a separate family for them [14].

ROSSELLIDAE SCHULZE, 1885

LANUGINELLINAE GRAY, 1872

Diagnosis: Basiphytous, rarely lophophytous, often pedunculate, Rossellidae; dermalia hexactins, pentactins, stauractins, or diactins supported by large hypodermal pentactins; choanosomal spicules diactins, often supplemented by significant amount of hexactins; atrialia pentactins or hexactins often supported by large hypoatrial pentactins; dermal and atrial hexactins and pentactins frequently pinular; prostalia, if present, pentactins or diactins; microscleres include strobiloplumicomes, which may be absent in some species, oxy-, onycho-, or disco-tipped forms (hexasters, hemihexasters, hexactins); microdiscohexasters absent.

Remarks: The revised diagnosis reflects inclusion of *Caulophacus* Schulze, 1886 and *Caulophacella* Lendenfeld, 1915 in the subfamily. These genera were recently transferred from Rossellinae Schulze, 1885 to Lanuginellinae [83] based on molecular evidence [16, 83]. This move is further supported by our phylogenetic analyses of morphological data (e.g., Fig. 6, Additional file 17: Figure S3).

HEXASTEROPHORA *incertae sedis*

Remarks: Our phylogenetic results suggest that Dactylocalycidae Gray, 1867 is more closely related to Lyssacinosida Zittel, 1877 than to Sceptrulophora Mehl, 1992; consequently, order Hexactinosida Schrammen, 1912 (= Sceptrulophora + Dactylocalycidae) is abolished because it is not a monophyletic group. Furthermore, the genera *Heterorete* Dendy, 1916 and *Myliusia* Gray, 1859 also appear to be more closely related to Lyssacinosida than to Sceptrulophora and are consequently removed from Euretidae Zittel, 1877 (see above). Because erecting a Linnean taxon for a clade comprising Lyssacinosida, Dactylocalycidae, *Heterorete*, and *Myliusia* is problematic (see discussion in main text), we here provisionally treat the latter three taxa as Hexasterophora *inc. sed.* until their phylogenetic relationships are more precisely resolved. The genera *Deanea* Bowerbank, 1875, *Diaretula* Schmidt, 1879, and *Hyalocaulus* Marshall & Meyer, 1877 (formerly Hexactinosida *inc. sed.*), which were not included in our phylogenetic analyses, are also reclassified as Hexasterophora *inc. sed.* because their poor documentation does not allow confident assignment to any of the hexasterophoran orders.

DACTYLOCALYCIDAE GRAY, 1867

Diagnosis (emended from Definition and Diagnosis of Reiswig [84]): Dictyonal Hexactinellida with rigid walls composed of branching systems of tubules. Channelization as cavaedia between branching tubules; tubule walls not channelized. Primary framework (tubule walls) is not euretoid in construction; dictyonal polyradial nodes result from tip-to-node and tip-to-tip fusion of dictyonalia. Body form as funnel or cup. Surface spicules as rough club-tipped pentactins and hexactins; sceptrules and uncinates absent. Microscleres include combinations of discohexasters, tylohexasters, hemioxyhexasters to oxyhexactins, and onychohexasters.

Scope: The family currently includes two genera, *Dactylocalyx* Stutchbury, 1841 and *Iphiteon* Bowerbank, 1869.

Remarks: Although monophyly of Dactylocalycidae could not be confirmed with molecular data (e.g., Fig. 7), in the absence of strong evidence for the contrary (see main text for discussion) there are no grounds for abolishing this taxon. The emended diagnosis reflects removal of *Auloplax* Schulze, 1904 [44].

Additional files

Additional file 1: Morphological character matrix of Amphidiscophora in Nexus format. (NEX 2.05 KB)

Additional file 2: Morphological character matrix of Hexasterophora in Nexus format. (NEX 24.1 KB)

Additional file 3: Morphological character matrix of Amphidiscophora with annotated character list. (DOCX 79.8 KB)

Additional file 4: Morphological character matrix of Hexasterophora with annotated character list. (DOCX 258 KB)

Additional file 5: Complete 16S rDNA alignment in Nexus format. (NEX 58.8 KB)

Additional file 6: Complete COI alignment in Nexus format. (NEX 80.8 KB)

Additional file 7: Complete 28S rDNA alignment (incl. secondary structure) in Nexus format. (NEX 116 KB)

Additional file 8: Complete 18S rDNA alignment (incl. secondary structure) in Nexus format. (NEX 116 KB)

Additional file 9: Cleaned concatenated alignment (molecular supermatrix) in Nexus format. (NEX 354 KB)

Additional file 10: Secondary structure for molecular supermatrix (RAxML format). (TXT 4.69 KB)

Additional file 11: Partition information (gene boundaries) for molecular supermatrix (RAxML format). (TXT 80 bytes)

Additional file 12: Combined morphological character matrix of Amphidiscophora and Hexasterophora. (NEX 27.8 KB)

Additional file 13: Combined morphological character matrix of Amphidiscophora and Hexasterophora with additional characters not used for phylogenetic inference, including merged total-evidence MP tree of Amphidiscophora and Hexasterophora. (NEX 33.2 KB)

Additional file 14: Detailed account of the morphology-based and total-evidence phylogenies. (DOCX 172 KB)

Additional file 15: Figure S1. Phylogeny of Amphidiscophora inferred from the morphological data matrix with MrBayes. 50% majority rule consensus tree from 9000 post-burnin samples. Average standard deviation of split frequencies between two independent runs was 0.003873. Bayesian posterior probabilities <1.00 shown on branches. H, Hyalonematidae; M, Monorhaphididae; P, Pheronematidae. Scale bar, expected number of character replacements per character. (JPG 222 KB)

Additional file 16: Figure S2. Phylogeny of Amphidiscophora inferred from the morphological data matrix with RAxML. Bootstrap values >50% shown on branches (based on 1000 pseudoreplicates). H, Hyalonematidae; M, Monorhaphididae; P, Pheronematidae. Scale bar, expected number of character replacements per character. (JPG 179 KB)

Additional file 17: Figure S3. Phylogeny of Hexasterophora inferred from the morphological data matrix with MrBayes. 50% majority rule consensus tree from 9000 post-burnin samples. Average standard deviation of split frequencies between two independent runs was 0.026249. Bayesian posterior probabilities <1.00 shown on branches. Current family assignment given after genus names. Subfamilies indicated with letters: A = Acanthasci-nae (Rossellidae)/Aulocalycinae (Aulocalycidae), B = Bolosominae (Euplectelli-dae), C = Corbitellinae (Euplectellidae)/Chonelasmatinae (Euretidae)/Cyathellinae (Aulocalycidae), E = Euplectellinae (Euplectellidae)/Euretinae (Euretidae), L = Lanuginellinae (Rossellidae), R = Rossellinae (Rossellidae). Scale bar, expected number of character replacements per character. (JPG 714 KB)

Additional file 18: Figure S4. Phylogeny of Hexasterophora inferred from the morphological data matrix with RAxML. Bootstrap values >50% shown on branches (based on 650 pseudoreplicates). Current family assignment given after genus names. Subfamilies indicated with letters: A = Acanthascinae (Rossellidae)/Aulocalycinae (Aulocalycidae), B = Bolosominae (Euplectellidae), C = Corbitellinae (Euplectellidae)/Chonelasmatinae (Euretidae)/Cyathellinae (Aulocalycidae), E = Euplectellinae (Euplectellidae)/Euretinae (Euretidae), L = Lanuginellinae (Rossellidae), R = Rossellinae (Rossellidae). Scale bar, expected number of character replacements per character. (JPG 645 KB)

Additional file 19: Figure S5. Mitochondrial 16S rDNA phylogeny of Hexactinellida inferred with RAxML. Bootstrap values >50% shown on branches (based on 450 pseudoreplicates). Newly sampled species highlighted in bold. *, sequence data from mitochondrial genome sequencing projects [72, 73]. Scale bar, expected number of substitutions per site. (JPG 523 KB)

Additional file 20: Figure S6. Mitochondrial COI phylogeny of Hexactinellida inferred with RAxML from nucleotide alignment. Bootstrap values >50% shown on branches (based on 450 pseudoreplicates). Newly sampled species highlighted in bold. *, sequence data from mitochondrial genome sequencing projects [72, 73]. Scale bar, expected number of substitutions per site. (JPG 543 KB)

Additional file 21: Figure S7. 28S rDNA phylogeny of Hexactinellida inferred with RAxML. Bootstrap values >50% shown on branches (based on 600 pseudoreplicates). Newly sampled species highlighted in bold. Scale bar, expected number of substitutions per site. (JPG 514 KB)

Additional file 22: Figure S8. 18S rDNA phylogeny of Hexactinellida inferred with RAxML. Bootstrap values >50% shown on branches (based on 500 pseudoreplicates). Newly sampled species highlighted in bold. Scale bar, expected number of substitutions per site. (JPG 368 KB)

Additional file 23: Figure S9. Phylogeny of Hexactinellida inferred from concatenated molecular markers with MrBayes. 50% majority rule consensus tree from 25,000 post-burnin samples. Average standard deviation of split frequencies between two independent runs was 0.003858. Bayesian posterior probabilities <1.00 shown on branches. Newly sampled species highlighted in bold. *, 16S and

COI sequence data from mitochondrial genome sequencing projects [72, 73]. Scale bar, expected number of substitutions per site. (JPG 615 KB)

Additional file 24: Figure S10. Phylogeny of Hexasterophora inferred with RAxML from reduced taxon set (one species per genus) of concatenated molecular markers. Bootstrap values >50% shown on branches (based on 450 pseudoreplicates). Newly sampled genera highlighted in bold. *, 16S and COI sequence data from mitochondrial genome sequencing project [72]. Scale bar, expected number of substitutions per site. (JPG 843 KB)

Additional file 25: Figure S11. Phylogeny of Hexasterophora inferred with RAxML from concatenated molecular and morphological data, including only genera with sequence data. Bootstrap values >50% shown on branches (based on 450 pseudoreplicates). Newly sampled genera highlighted in bold. *, 16S and COI sequence data from mitochondrial genome sequencing project [72]. Scale bar, expected number of substitutions/character replacements per site/character. (JPG 378 KB)

Additional file 26: Figure S12. Phylogeny of Hexasterophora inferred with MrBayes from concatenated molecular and morphological data, including only genera with sequence data. Fifty percent majority rule consensus tree from 45,000 post-burnin samples. Average standard deviation of split frequencies between two independent runs was 0.002564. Bayesian posterior probabilities <1.00 shown on branches. Newly sampled genera highlighted in bold. *, 16S and COI sequence data from mitochondrial genome sequencing project [72]. Scale bar, expected number of substitutions/character replacements per site/character.(JPG 422 KB)

Additional file 27: Figure S14. Phylogeny of Amphidiscophora inferred with MrBayes from concatenated molecular and morphological data, including all genera. Genera with sequence data highlighted in bold and connected with thick branches. Fifty percent majority rule consensus tree from 45,000 post-burnin samples. Average standard deviation of split frequencies between two independent runs was 0.003744. Bayesian posterior probabilities <1.00 shown on branches. H, Hyalonematidae; M, Monorhaphididae; P, Pheronematidae. Scale bar, expected number of substitutions/character replacements per site/character. (JPG 152 KB)

Additional file 28: Figure S13. Phylogeny of Amphidiscophora inferred with RAxML from concatenated molecular and morphological data, including all genera. Genera with sequence data highlighted in bold and connected with thick branches. Bootstrap values >50% shown on branches (based on 1000 pseudoreplicates). H, Hyalonematidae; M, Monorhaphididae; P, Pheronematidae. Scale bar, expected number of substitutions/character replacements per site/character. (JPG 162 KB)

Additional file 29: Figure S15. Placement of Amphidiscophora genera without sequence data (taxon names preceded by "QUERY__") on the molecular backbone phylogeny using weighted morphology-based phylogenetic binning [22] as implemented in RAxML. H, Hyalonematidae; M, Monorhaphididae; P, Pheronematidae. (JPG 330 KB)

Additional file 30: Figure S16. Phylogeny of Hexasterophora inferred with RAxML from concatenated molecular and morphological data, including all genera. Genera with sequence data highlighted in bold and connected with thick branches. Bootstrap values >50% shown on branches (based on 550 pseudoreplicates). Current family assignment given after genus names: Aphr = Aphrocallistidae, Auloca = Aulocalycidae, Aulocy = Aulocystidae, Aulop = Auloplacidae, Crat = Craticulariidae, Crib = Cribrospongiidae, Dact = Dactylocalycidae, Diap = Diapleuridae, Eupl = Euplectellidae, Eure = Euretidae, Farr = Farreidae, Fiel = Fieldingiidae, Leuc = Leucopsacidae, Lyss inc. sed. = Lyssacinosida incertae sedis, Ross = Rossellidae, Scep inc. sed. = Sceptrulophora incertae sedis, Tret = Tretodictyidae, Unci = Uncinateridae. Subfamilies indicated with letters: A = Acanthascinae (Rossellidae)/Aulocalycinae (Aulocalycidae), B = Bolosominae (Euplectellidae), C = Corbitellinae (Euplectellidae)/Chonelasmatinae (Euretidae)/Cyathellinae (Aulocalycidae), E = Euplectellinae (Euplectellidae)/Euretinae (Euretidae), L = Lanuginellinae (Rossellidae), R = Rossellinae (Rossellidae). Scale bar, expected number of substitutions/character replacements per site/character. (JPG 648 KB)

Additional file 31: Figure S17. Phylogeny of Hexasterophora inferred with MrBayes from concatenated molecular and morphological data, including all genera. Genera with sequence data highlighted in bold and connected with thick branches. Fifty percent majority rule consensus tree from 45,000 post-burnin samples. Average standard deviation of split frequencies between two independent runs was 0.043509. Bayesian posterior probabilities <1.00 shown on branches. Current family assignment given after genus names: Aphr = Aphrocallistidae, Auloca = Aulocalycidae, Aulocy = Aulo-cystidae, Aulop = Auloplacidae, Crat = Craticulariidae, Crib = Cribrospongiidae, Dact = Dactylocalycidae, Diap = Diapleuridae, Eupl = Euplectellidae, Eure = Eur- etidae, Farr = Farreidae, Fiel = Fieldingiidae, Leuc = Leucopsacidae, Lyss inc. sed. = Lyssacinosida incertae sedis, Ross = Rossellidae, Scep inc. sed. = Sceptrulo- phora incertae sedis, Tret = Tretodictyidae, Unci = Uncinateridae. Subfamilies indicated with letters: A = Acanthascinae (Rossellidae)/Aulocalycinae (Aulocalycidae), B = Bolosominae (Euplectellidae), C = Corbitellinae (Euplectelli- dae)/Chonelasmatinae (Euretidae)/Cyathellinae (Aulocalycidae), E = Euplectelli- nae (Euplectellidae)/Euretinae (Euretidae), L = Lanuginellinae (Rossellidae), R = Rossellinae (Rossellidae). Scale bar, expected number of substitutions/character replacements per site/character. (JPG 637 KB)

Additional file 32: Figure S18. Placement of Hexasterophora genera without sequence data (taxon names preceded by "QUERY__") on the molecular backbone phylogeny (Additional file 24: Figure S10) using weighted morphology-based phylogenetic binning [22] as implemented in RAxML. Current family assignment given after genus names. Subfamilies indicated with letters: A = Acanthascinae (Rossellidae)/Aulocalycinae (Aulocalycidae), B = Bolosominae (Euplectellidae), C = Corbitellinae (Euplectellidae)/Chonelasmati- nae (Euretidae)/Cyathellinae (Aulocalycidae), E = Euplectellinae (Euplectellidae)/ Euretinae (Euretidae), L = Lanuginellinae (Rossellidae), R = Rossellinae (Rosselli- dae). (JPG 5.21 MB)

Abbreviations

aMk2: Asymmetrical Mk 2-parameter model; ASR: Ancestral state reconstruction; bp: Base pairs; BS: Bootstrap support; CI: Consistency index; COI: Cytochrome oxidase subunit I; G4: Four-rate category gamma correction for among-site rate variation; GTR: General-time-reversible; inc. sed.: Incertae sedis; LCA: Last common ancestor; LD clade: Clade containing Lyssacinosida and Dactylocalycidae; MCMC: Markov Chain Monte Carlo; Mk: Markov-k model; Mk1: 1-parameter Mk model; ML: Maximum likelihood; MP: Maximum parsimony; MPT: Most parsimonious tree; MRC: Majority-rule consensus tree; mt: Mitochondrial; PCR: Polymerase chain reaction; pl: Proportional likelihood; PP: Posterior probability; RC: Rescaled consistency index; rDNA: Ribosomal DNA; RI: Retention index; s. nov.: Sensu novo; s.l.: Sensu lato; s.s.: Sensu stricto; TL: Treelength; VFB clade: Venus-flower-basket clade

Acknowledgements

MD and HMR thank the organizers of the August 2011 PorToL Integrative Taxonomy Workshop (Christina Diaz, Bob Thacker, Shirley Pomponi, Malcolm and April Hill, Joe Lopez, Dennis Lavrov, Allen Collins, and Niamh Redmond) for their invitation to participate and funding to attend. We are indebted to Shirley Pomponi and John Reed for access to HBOI specimens, Rob van Soest for providing Cyrtaulon and additional Dactylocalyx specimens, and Philippe Bouchet (MNH Paris) for access to Laocoetis material. We thank Daniela Henkel for the morphological character matrix used in [23], and Dennis Lavrov and Karri Haen for providing mtDNA sequences before public release. Dorte Janussen and an anonymous reviewer are thanked for critical comments on an earlier version of the manuscript. MD acknowledges Gert Wörheide (LMU) for providing work space and computational resources. The Willi Hennig Society is acknowledged for making TNT available free of charge.

Funding

MD was supported by a Smithsonian Institution Postdoctoral Fellowship and by the Deutsche Forschungsgemeinschaft (DFG grants DO 1742/1-1, 2). Additional funding for molecular work was provided by a US National Science Foundation Porifera Tree of Life grant to Allen Collins (Smithsonian Institution). Collection of the Hawaii sponges was funded by a grant from the Papahanaumokuakea Marine National Monument (administered by the National Marine Sanctuaries Program). MK was supported by New Zealand Foundation for Research, Science and Technology (C01X0224, SFAS033), Ministry of Fisheries (ZBD2004-01), NOAA Satellite Operations Facility (NRAM053), CSIRO Wealth from Oceans Flagship, and Coasts and Oceans Research Programme 2 Marine Biological Resources: Discovery and definition of the marine biota of New Zealand (2014/2015 and 2015/2016 SCI). JNAH thanks the Marine Barcoding of Life Initiative (MarBol), funded by the Alfred P. Sloan Foundation and the GeoBio-CenterLMU, for financing the subsampling and extraction of sponge specimens within the custodianship of the Queensland Museum, and the Taxonomic Research Informatics Network (TRIN) and Atlas of Living Australia (ALA) project.

Authors' contributions

MD conceived and designed the study, performed molecular work, compiled morphological data sets, analyzed the data, interpreted results, and wrote the paper. HMR prepared Figs. 2 and 3. CK and MK provided underwater photographs used in Fig. 1. CK, MK, AP, JNAH, and HMR contributed specimens and provided intellectual input and critical comments on the manuscript. All authors read and approved the final manuscript.

Competing interests

The authors declare that they have no competing interests.

Author details

[1]Department of Earth & Environmental Sciences, Palaeontology & Geobiology, Molecular Geo- & Palaeobiology Lab, Ludwig-Maximilians-University Munich, Richard-Wagner-Str. 10, 80333 Munich, Germany. [2]Hawaii Undersea Research Laboratory, University of Hawaii at Manoa, 1000 Pope Rd, MSB 229, Honolulu 96822, HI, USA. [3]Coasts and Oceans National Centre, National Institute of Water and Atmospheric Research (NIWA) Ltd, Private Bag 99940, Newmarket, Auckland 1149, New Zealand. [4]Institute of Paleobiology, Polish Academy of Sciences, ul. Twarda 51/55, 00-818 Warszawa, Poland. [5]Biodiversity & Geosciences Program, Queensland Museum, South Brisbane, QLD 4101, Australia. [6]Eskitis Institute for Drug Discovery, Griffith University, Nathan, QLD 4111, Australia. [7]Natural History Section, Royal British Columbia Museum, 675 Belleville Street, Victoria, BC V8W 9W2, Canada. [8]Department of Biology, University of Victoria, 3800 Finnerty Road, Victoria, BC V8P 4H9, Canada.

References

1. Hooper JNA, van Soest RWM. Systema Porifera. A guide to the classification of sponges. New York: Plenum; 2002.
2. Leys SP, Mackie GO, Reiswig HM. The biology of glass sponges. Adv Mar Biol. 2007;52:1–145.
3. Dohrmann M, Janussen D, Reitner J, Collins AG, Wörheide G. Phylogeny and evolution of glass sponges (Porifera, Hexactinellida). Syst Biol. 2008;57:388–405.
4. Philippe H, Derelle R, Lopez P, Pick K, Borchiellini C, Boury-Esnault N, Vacelet J, Renard E, Houliston E, Quéinnec E, Da Silva C, Wincker P, Le Guyader H, Leys S, Jackson DJ, Schreiber F, Erpenbeck D, Morgenstern B, Wörheide G, Manuel M. Phylogenomics revives traditional views on deep animal relationships. Curr Biol. 2009;19:706–12.
5. van Soest RWM, Boury-Esnault N, Hooper JNA, Rützler K, de Voogd NJ, Alvarez de Glasby B, Hajdu E, Pisera AB, Manconi R, Schönberg C, Klautau M, Picton B, Kelly M, Vacelet J, Dohrmann M, Díaz C, Cárdenas P, Carballo JL. World Porifera Database. 2016. http://www.marinespecies.org/porifera.
6. Leinfelder RR, Krautter M, Laternser R, Nose M, Schmid DU, Schweigert G, Werner W, Keupp H, Brugger H, Herrmann K, Rehfeld-Kiefer U, Schroeder JH, Reinhold C, Koch R, Zeiss A, Schweizer V, Christmann H, Menges G, Luterbacher H. The origin of Jurassic reefs: current research developments and results. Facies. 1994;31:1–56.
7. Krautter M. Fossil Hexactinellida: an overview. In: Hooper JNA, van Soest RWM, editors. Systema Porifera. A guide to the classification of sponges. New York: Plenum; 2002. p. 1211–23.
8. Pisera A. Palaeontology of sponges – a review. Can J Zoolog. 2006;84:242–61.
9. Maldonado M, Aguilar R, Bannister RJ, Bell JJ, Conway KW, Dayton PK, Díaz C, Gutt J, Kelly M, Kenchington ELR, Leys SP, Pomponi SA, Rapp HT, Rützler K, Tendal OS, Vacelet J, Young CM. Sponge grounds as key marine habitats: a synthetic review of types, structure, functional roles, and conservation concerns. In: Rossi S, editor. Marine animal forests. Switzerland: Springer International Publishing; 2016. p. 1–39.

10. Sundar VC, Yablon AD, Grazul JL, Ilan M, Aizenberg J. Fibre-optical features of a glass sponge. Nature. 2003;424:899–900.

11. Aizenberg J, Sundar VC, Yablon AD, Weaver JC, Chen G. Biological glass fibres: correlation between optical and structural properties. Proc Natl Acad Sci U S A. 2004;101:3358–63.

12. Monn MA, Weaver JC, Zhang T, Aizenberg J, Kesari H. New functional insights into the internal architecture of the laminated anchor spicules of Euplectella aspergillum. Proc Natl Acad Sci U S A. 2015;112:4976–81.

13. Cárdenas P, Pérez T, Boury-Esnault N. Sponge systematics facing new challenges. Adv Mar Biol. 2012;61:79–209.

14. Wörheide G, Dohrmann M, Erpenbeck D, Larroux C, Maldonado M, Voigt O, Borchiellini C, Lavrov DV. Deep phylogeny and evolution of sponges (Phylum Porifera). Adv Mar Biol. 2012;61:1–78.

15. Dohrmann M, Collins AG, Wörheide G. New insights into the phylogeny of glass sponges (Porifera, Hexactinellida): monophyly of Lyssacinosida and Euplectellinae, and the phylogenetic position of Euretidae. Mol Phylogenet Evol. 2009;52:257–62.

16. Dohrmann M, Haen KM, Lavrov DV, Wörheide G. Molecular phylogeny of glass sponges (Porifera, Hexactinellida): increased taxon sampling and inclusion of the mitochondrial protein-coding gene, cytochrome oxidase subunit I. Hydrobiologia. 2012;687:11–20.

17. Reiswig HM, Dohrmann M. Three new species of glass sponges (Porifera: Hexactinellida) from the West Indies, and molecular phylogenetics of Euretidae and Auloplacidae (Sceptrulophora). Zool J Linn Soc. 2014;171:233–53.

18. Dohrmann M, Göcke C, Reed J, Janussen D. Integrative taxonomy justifies a new genus, Nodastrella gen. nov., for North Atlantic "Rossella" species (Porifera: Hexactinellida: Rossellidae). Zootaxa. 2012;3383:1–13.

19. Dohrmann M, Göcke C, Janussen D, Reitner J, Lüter C, Wörheide G. Systematics and spicule evolution in dictyonal sponges (Hexactinellida: Sceptrulophora) with description of two new species. Zool J Linn Soc. 2011;163:1003–25.

20. Schulze FE. Über den Bau und das System der Hexactinelliden. Phys Abh königl preuß Akad Wiss Berlin (Phys-Math Cl). 1886;1:1–97

21. Schulze FE. Amerikanische Hexactinelliden nach dem Materiale der Albatross-Expedition. Jena: Fischer; 1899

22. Berger SA, Stamatakis A. Accuracy of morphology-based phylogenetic fossil placement under maximum likelihood. Proc AICCSA. 2010;2010:1–8.

23. Henkel D, Borkenhagen K, Janussen D. Phylogeny of the Hexactinellida: phylogenetic reconstruction of the subclass Hexasterophora based on morphological characters. J Mar Biol Ass UK. 2015;95:1365–9.

24. Janussen D, Reiswig HM. Re-description of Cyathella lutea SCHMIDT and formation of a new subfamily Cyathellinae (Hexactinellida, Aulocalycoida, Aulocalycidae). Senckenb Biol. 2003;82:1–10.

25. Duplessis K, Reiswig HM. Three new species and a new genus of Farreidae (Porifera: Hexactinellida: Hexactinosida). Proc Biol Soc Wash. 2004;117:199–212.

26. Janussen D, Tabachnick KR, Tendal OS. Deep-sea Hexactinellida (Porifera) of the Weddell Sea. Deep-Sea Res Pt II. 2004;51:1857–82.

27. Tabachnick KR, Janussen D. Description of a new species and subspecies of Fieldingia, erection of a new family Fieldingidae, and a new order Fieldingida (Porifera; Hexactinellida; Hexasterophora). Boll Mus Ist Biol Univ Genova. 2004;68:623–37.

28. Tabachnick KR, Lévi C, Richer de Forges B. Lyssacinosa du Pacifique sud-ouest (Porifera: Hexactinellida). In: Marshall B, editor. Tropical Deep-Sea Benthos 23, vol. 191. 2004. p. 11–71. Mém. Mus. Natl. Hist. Nat.

29. Menshenina LL, Tabachnick KR. Revision of Pleurochorium annandalei (Porifera, Hexactinellida). Boll Mus Ist Biol Univ Genova. 2004;68:463–75.

30. Lopes DA, Hajdu E, Reiswig HM. Redescription of two Hexactinosida (Porifera, Hexactinellida) from the southwestern Atlantic, collected by Programme REVIZEE. Zootaxa. 2005;1066:43–56.

31. Lopes DA, Hajdu E, Reiswig HM. Taxonomy of Euretidae (Porifera, Hexactinellida, Hexactinosida) of Campos Basin, southwestern Atlantic, with a description of a new species. Mar Biol Res. 2007;3:243–55.

32. Menshenina LL, Tabachnick KR, Janussen D. Revision of the subgenus Neopsacas (Hexactinellida, Rossellidae, Crateromorpha) with the description of new species and subspecies. Zootaxa. 2007;1436:55–68.

33. Menshenina LL, Tabachnick KR, Lopes DA, Hajdu E. Revision of Calycosoma Schulze, 1899 and finding of Lophocalyx Schulze, 1887 (six new species) in the Atlantic Ocean (Hexactinellida, Rossellidae). In: Custódio MR, Lôbo-Hajdu G, Hajdu E, Muricy G, editors. Porifera research: biodiversity, innovation & sustainability. Rio de Janeiro: Museu Nacional; 2007. p. 449 65.

34. Tabachnick KR, Menshenina LL. Revision of the genus Asconema (Porifera: Hexactinellida: Rossellidae). J Mar Biol Ass UK. 2007;87:1403–29.

35. Reiswig HM, Dohrmann M, Pomponi SA, Wörheide G. Two new tretodictyids (Hexactinellida: Hexactinosida: Tretodictyidae) from the coasts of North America. Zootaxa. 2008;1721:53–64.

36. Tabachnick KR, Collins AG. Glass sponges (Porifera, Hexactinellida) of the northern Mid-Atlantic Ridge. Mar Biol Res. 2008;4:25–47.

37. Tabachnick KR, Janussen D, Menshenina LL. New Australian Hexactinellida (Porifera) with a revision of Euplectella aspergillum. Zootaxa. 2008;1866:7–68.

38. Janussen D, Reiswig HM. Hexactinellida (Porifera) from the ANDEEP III expedition to the Weddell Sea, Antarctica. Zootaxa. 2009;2136:1–20.

39. Tabachnick KR, Menshenina LL, Lopes DA, Hajdu E. Two new Hyalonema species (Hyalonematidae: Amphidiscosida) from eastern and south-eastern Brazil, and further Hexactinellida (Porifera) collected from seamounts off south-eastern Brazil by the RV 'Marion Dufresne' MD55 expedition. J Mar Biol Ass UK. 2009;89:1243–50.

40. Reiswig HM. A new species of Tretodictyum (Porifera: Hexactinellida: Tretodictyidae) from off Cocos Island, tropical eastern Pacific Ocean. Proc Biol Soc Wash. 2010;123:242–50.

41. Sautya S, Tabachnick KR, Ingole B. First record of Hyalascus (Hexactinellida: Rossellidae) from the Indian Ocean, with description of a new species from a volcanic seamount in the Andaman Sea. Zootaxa. 2010;2667:64–8.

42. Göcke C, Janussen D. ANT XXIV/2 (SYSTCO) Hexactinellida (Porifera) and bathymetric traits of Antarctic glass sponges (incorporating ANDEEP-material); including an emendation of the rediscovered genus Lonchiphora. Deep Sea Res Pt II. 2011;58:2013–21.

43. Lopes DA, Hajdu E, Reiswig HM. Taxonomy of Farrea (Porifera, Hexactinellida, Hexactinosida) from the southwestern Atlantic, with description of a new species and a discussion on the recognition of subspecies in Porifera. Can J Zoolog. 2011;89:169–89.

44. Reiswig HM, Kelly M. The marine fauna of New Zealand: Hexasterophoran glass sponges of New Zealand (Porifera: Hexactinellida: Hexasterophora): Orders Hexactinosida, Aulocalycoida and Lychniscosida. NIWA Biodiv Mem. 2011;124:1–176.

45. Sautya S, Tabachnick KR, Ingole B. A new genus and species of deep-sea glass sponge (Porifera, Hexactinellida, Aulocalycidae) from the Indian Ocean. ZooKeys. 2011;136:13–21.

46. Tabachnick KR, Menshenina LL, Pisera A, Ehrlich H. Revision of Aspidoscopulia Reiswig, 2002 (Porifera: Hexactinellida: Farreidae) with description of two new species. Zootaxa. 2011;2883:1–22.

47. Göcke C, Janussen D. Hexactinellida of the genus Rossella, of ANT XXIV/ 2 (SYSTCO I) Expedition — Antarctic Eastern Weddell Sea. Zootaxa. 2013;3692:102–22.

48. Kahn AS, Geller JB, Reiswig HM, Smith Jr KL. Bathydorus laniger and Docosaccus maculatus (Lyssacinosida; Hexactinellida): two new species of glass sponge from the abyssal eastern North Pacific Ocean. Zootaxa. 2013;3646:386–400.

49. Lopes DA, Tabachnick KR. New data on glass sponges (Porifera, Hexactinellida) of the northern Mid-Atlantic Ridge. Part 1. Farreidae. Mar Biol Res. 2013;9:462–8.

50. Murillo FJ, Tabachnick KR, Menshenina LL. Glass sponges off the Newfoundland (Northwest Atlantic): description of a new species of Dictyaulus (Porifera: Hexactinellida: Euplectellidae). J Mar Biol. 2013;2013: 438485.

51. Reiswig HM, Stone RP. New glass sponges (Porifera: Hexactinellida) from deep waters of the central Aleutian Islands, Alaska. Zootaxa. 2013;3628:1–64.

52. Tabachnick KR, Menshenina LL. New data on glass sponges (Porifera, Hexactinellida) of the northern Mid-Atlantic Ridge. Part 2. Aphrocallistidae, Euretidae, Euplectellidae and Rossellidae (with descriptions of two new species of Sympagella). Mar Biol Res. 2013;9:469–87.

53. Reiswig HM. Six new species of glass sponges (Porifera: Hexactinellida) from the north-eastern Pacific Ocean. J Mar Biol Ass UK. 2014;94:267–84.

54. Gong L, Li X, Qiu J-W. Two new species of Hexactinellida (Porifera) from the South China Sea. Zootaxa. 2015;4034:182–92.

55. Reiswig HM. First Lanuginellinae (Porifera, Hexactinellida, Rossellidae) from the NE Pacific and first species of Doconesthes from the Pacific Ocean. Zootaxa. 2015;3920:572–8.

56. Maddison WP, Maddison DR. Mesquite: a modular system for evolutionary analysis. Available at http://mesquiteproject.org/mesquite/mesquite.html. 2011. Accessed 3 Feb 2017.

57. Goloboff PA, Farris JS, Nixon KC. TNT, a free program for phylogenetic analysis. Cladistics. 2008;24:774–86.

58. Goloboff PA, Carpenter JM, Arias JS, Esquivel DRM. Weighting against homoplasy improves phylogenetic analysis of morphological data sets. Cladistics. 2008;24:758–73.

59. Maddison WP, Maddison DR. MacClade. Sunderland: Sinauer Associates; 2002.

60. Wright AM, Hillis DM. Bayesian analysis using a simple likelihood model outperforms parsimony for estimation of phylogeny from discrete morphological data. PLoS One. 2014;9:e109210.

61. O'Reilly JE, Puttick MN, Parry L, Tanner AR, Tarver JE, Fleming J, Pisani D, Donoghue PCJ. Bayesian methods outperform parsimony but at the expense of precision in the estimation of phylogeny from discrete morphological data. Biol Lett. 2016;12:20160081.

62. Ronquist F, Teslenko M, van der Mark P, Ayres DL, Darling A, Höhna S, Larget B, Liu L, Suchard MA, Huelsenbeck JP. MrBayes 3.2: efficient Bayesian phylogenetic inference and model choice across a large model space. Syst Biol. 2012;61:539–42.

63. Lewis PO. A likelihood approach to estimating phylogeny from discrete morphological character data. Syst Biol. 2001;50:913–25.

64. Yang Z. Maximum likelihood phylogenetic estimation from DNA sequences with variable rates over sites: approximate methods. J Mol Evol. 1994;39:306–14.

65. Altekar G, Dwarkadas S, Huelsenbeck JP, Ronquist F. Parallel metropolis coupled Markov chain Monte Carlo for Bayesian phylogenetic inference. Bioinformatics. 2004;20:407–15.

66. Rambaut A, Suchard M, Drummond AJ. Tracer v1.6. Available from http://tree.bio.ed.ac.uk/software/tracer/. 2013. Accessed 3 Feb 2017.

67. Stamatakis A. RAxML version 8: a tool for phylogenetic analysis and post-analysis of large phylogenies. Bioinformatics. 2014;30:1312–3.

68. Stamatakis A, Hoover P, Rougemont J. A rapid bootstrap algorithm for the RAxML web servers. Syst Biol. 2008;57:758–71.

69. Pattengale ND, Alipour M, Bininda-Emonds ORP, Moret BME, Stamatakis A. How many bootstrap replicates are necessary? J Comput Biol. 2010;17:337–54.

70. Dohrmann M. Symplectella rowi (Porifera: Lyssacinosida) is a rossellid, not a euplectellid. J Mar Biol Ass UK. 2016;96:291–5.

71. Tautz D, Renz M. An optimized freeze-squeeze method for the recovery of DNA fragments from agarose gel. Anal Biochem. 1983;132:14–9.

72. Haen KM, Pett W, Lavrov DV. Eight new mtDNA sequences of glass sponges reveal an extensive usage of +1 frameshifting in mitochondrial translation. Gene. 2014;535:336–44.

73. Zhang Y, Sun J, Li X, Qiu J-W. The mitochondrial genome of the deep-sea glass sponge Lophophysema eversa (Porifera, Hexacinellida, Hyalonematidae). Mitochondr DNA. 2016;27:1273–4.

74. Larkin MA, Blackshields G, Brown NP, Chenna R, McGettigan PA, McWilliam H, Valentin F, Wallace IM, Wilm A, Lopez R, Thompson JD, Gibson TJ, Higgins DG. ClustalW and ClustalX version 2.0. Bioinformatics. 2007;23:2947–8.

75. Lanave C, Preparata G, Saccone C, Serio G. A new method for calculating evolutionary substitution rates. J Mol Evol. 1984;20:86–93.

76. Savill NJ, Hoyle DC, Higgs PG. RNA sequence evolution with secondary structure constraints: comparison of substitution rate models using maximum-likelihood methods. Genetics. 2001;157:399–411.

77. Gouy M, Guindon S, Gascuel O. SeaView version 4: a multiplatform graphical user interface for sequence alignment and phylogenetic tree building. Mol Biol Evol. 2010;27:221–4.

78. Schöniger M, von Haeseler A. A stochastic model for the evolution of autocorrelated DNA sequences. Mol Phylogenet Evol. 1994;3:240–7.

79. Voigt O, Erpenbeck D, Wörheide G. Molecular phylogenetic application of rDNA in early diverging Metazoa: first comparative analysis and phylogenetic application of complete SSU rRNA secondary structures in Porifera. BMC Evol Biol. 2008;8:69.

80. Felsenstein J. Confidence limits on phylogenies: an approach using the bootstrap. Evolution. 1985;39:783–91.

81. Berger SA, Stamatakis A, Lücking R. Morphology-based phylogenetic binning of the lichen genera Graphis and Allographa (Ascomycota: Graphidaceae) using molecular site weight calibration. Taxon. 2011;60:1450–7.

82. Duchêne S, Lanfear R. Phylogenetic uncertainty can bias the number of evolutionary transitions estimated from ancestral state reconstruction methods. J Exp Zool (Mol Dev Evol). 2015;324:517–24.

83. Boury-Esnault N, Vacelet J, Reiswig HM, Fourt M, Aguilar R, Chevaldonné P. Mediterranean hexactinellid sponges, with the description of a new Sympagella species (Porifera, Hexactinellida). J Mar Biol Ass UK. 2015;95:1353–64.

84. Reiswig HM. Family Dactylocalycidae Gray, 1867. In: Hooper JNA, van Soest RWM, editors. Systema Porifera. A guide to the classification of sponges. New York: Plenum; 2002. p. 1293–300.

85. Lévi C. Spongiaires des zones bathyale, abyssale et hadale. Galathea Report Scientific Results of The Danish Deep-Sea Expedition Round the World, 1950–52. 1964;7:63–112.

86. Reid REH. Mesozoic and Cenozoic hexactinellid sponges: Lychniscosa and Hexactinosa. In: Kaesler RL, editor. Treatise on invertebrate paleontology. Part E (revised). Porifera, vol. 3. Boulder: The Geological Society of America; 2004. p. 449–512.

87. Reiswig HM. Family Aulocalycidae Ijima, 1927. In: Hooper JNA, van Soest RWM, editors. Systema Porifera. A guide to the classification of sponges. New York: Plenum; 2002. p. 1362–71.

88. Reiswig HM. Order Lyssacinosida Zittel, 1877. In: Hooper JNA, van Soest RWM, editors. Systema Porifera. A guide to the classification of sponges. New York: Plenum; 2002. p. 1387.

89. Sanderson MJ, McMahon MM, Steel M. Terraces in phylogenetic tree space. Science. 2011;448:448–50.

90. Sanderson MJ, McMahon MM, Stamatakis A, Zwickl DJ, Steel M. Impacts of terraces on phylogenetic inference. Syst Biol. 2015;64:709–26.

91. Felsenstein J. Cases in which parsimony or compatibility methods will be positively misleading. Syst Zool. 1978;27:401–10.

92. Tabachnick KR, Menshenina LL. An approach to the phylogenetic reconstruction of Amphidiscophora (Porifera: Hexactinellida). Mem Qld Mus. 1999;44:607–15.

93. Mehl D. Die Entwicklung der Hexactinellida seit dem Mesozoikum. Paläobiologie, Phylogenie und Evolutionsökologie. Berl Geowiss Abh E. 1992;2:1–164.

94. Schulze FE. Report on the Hexactinellida collected by H.M.S. Challenger during the years 1873–1876. Zoology. 1887;21:1–514

95. Tabachnick KR. Family Euplectellidae Gray, 1867. In: Hooper JNA, van Soest RWM, editors. Systema Porifera. A guide to the classification of sponges. New York: Plenum; 2002. p. 1388–434.

96. Tabachnick KR. Family Rossellidae Schulze, 1885. In: Hooper JNA, van Soest RWM, editors. Systema Porifera. A guide to the classification of sponges. New York: Plenum; 2002. p. 1441–505.

97. Reiswig HM. Order Amphidiscosida Schrammen, 1924. In: Hooper JNA, van Soest RWM, editors. Systema Porifera. A guide to the classification of sponges. New York: Plenum; 2002. p. 1231.

98. Griffith OW, Blackburn DG, Brandley MC, van Dyke JU, Whittington CM, Thompson MB. Ancestral state reconstructions require biological evidence to test evolutionary hypotheses: a case study examining the evolution of reproductive mode in squamate reptiles. J Exp Zool (Mol Dev Evol). 2015;324:493–503.

99. Morrow C, Cárdenas P. Proposal for a revised classification of the Demospongiae (Porifera). Frontiers Zool. 2015;12:7.

100. Osche G. Über latente Potenzen und ihre Rolle im Evolutionsgeschehen – Ein Beitrag zur Theorie der Pluripotenzphaenomens. Zool Anz. 1965;174:411–40.

101. Maldonado M, Carmona MC, Uriz MJ, Cruzado A. Decline in Mesozoic reef-building sponges explained by silicon limitation. Nature. 1999;401:785–8.

102. Voigt O, Wülfing E, Wörheide G. Molecular phylogenetic evaluation of classification and scenarios of character evolution in calcareous sponges (Porifera, class Calcarea). PLoS One. 2012;7:e33417.

103. Reiswig HM. Family Tretodictyidae Schulze, 1886. In: Hooper JNA, van Soest RWM, editors. Systema Porifera. A guide to the classification of sponges. New York: Plenum; 2002. p. 1341–54.

2

Muscle development in the shark *Scyliorhinus canicula*: implications for the evolution of the gnathostome head and paired appendage musculature

Janine M. Ziermann[1*], Renata Freitas[2,3] and Rui Diogo[4]

Abstract

Background: The origin of jawed vertebrates was marked by profound reconfigurations of the skeleton and muscles of the head and by the acquisition of two sets of paired appendages. Extant cartilaginous fish retained numerous plesiomorphic characters of jawed vertebrates, which include several aspects of their musculature. Therefore, myogenic studies on sharks are essential in yielding clues on the developmental processes involved in the origin of the muscular anatomy.

Results: Here we provide a detailed description of the development of specific muscular units integrating the cephalic and appendicular musculature of the shark model, *Scyliorhinus canicula*. In addition, we analyze the muscle development across gnathostomes by comparing the developmental onset of muscle groups in distinct taxa. Our data reveal that appendicular myogenesis occurs earlier in the pectoral than in the pelvic appendages. Additionally, the pectoral musculature includes muscles that have their primordial developmental origin in the head. This culminates in a tight muscular connection between the pectoral girdle and the cranium, which founds no parallel in the pelvic fins. Moreover, we identified a lateral to ventral pattern of formation of the cephalic muscles, that has been equally documented in osteichthyans but, in contrast with these gnathostomes, the hyoid muscles develop earlier than mandibular muscle in *S. canicula*.

Conclusion: Our analyses reveal considerable differences in the formation of the pectoral and pelvic musculatures in *S. canicula*, reinforcing the idea that head tissues have contributed to the formation of the pectoral appendages in the common ancestor of extant gnathostomes. In addition, temporal differences in the formation of some cranial muscles between chondrichthyans and osteichthyans might support the hypothesis that the similarity between the musculature of the mandibular arch and of the other pharyngeal arches represents a derived feature of jawed vertebrates.

Keywords: Muscles, Cranial, *Cucullaris*, Head, Pectoral, Pelvic, Fin, Shark, Limb

Background

The origin of jawed vertebrates (gnathostomes) is undoubtedly one of the major events in the history of life, as it drastically changed the feeding modes among vertebrates [1]. The origin of jaws, paired fins, and the cephalic and appendicular musculature was probably of chief importance for the transition from suspension feeding to predation in this particular vertebrate lineage [1]. These novel morphological features may also have contributed to the vast radiation of gnathostomes, which make up more than 99.9% of all living vertebrates [2]. Chondrichthyans such as the sharks are considered to have morphological characteristics that retained various plesiomorphic gnathostome traits [3, 4]. These cartilaginous fishes have been around for over 400 million years and are, therefore, amongst the oldest surviving vertebrate groups [5]. Although they have unique features that evolved in ways distinct from other fishes,

* Correspondence: jziermann@yahoo.de
[1]Department of Anatomy, Howard University College of Medicine, 520 W St NW, Washington, DC 20059, USA
Full list of author information is available at the end of the article

they possess, for example, the most plesiomorphic paired fin structure of modern vertebrates [5]. Therefore, sharks are also relevant for discussions on the development and evolution of not only fish muscles but also of the muscles of vertebrates as a whole [6]. Moreover, striking similarities were detected between the musculature of chondrichthyans and placoderms, which are fossil representatives of the most basal gnathostomes [7, 8]. This makes them ideal extant models to study the evolution of paired (i.e., pectoral and pelvic) appendage musculature, providing a unique opportunity to investigate the developmental processes involved in the formation of these tissues during early evolution of gnathostomes [9–11]. Sharks, in addition, belong to the living sister group of osteichthyans (bony fish + tetrapods), where developmental studies are mostly performed, which makes them fundamental models for phylogenetic sampling (e.g., [10–12]). Data from cyclostomes - which are the only extant representatives of agnathans - such as the lamprey are also extremely relevant to understand the origin and early evolution of gnathostome morphology [13–20].

Comparative analyses have provided descriptions of the musculature of the head, neck, and locomotory appendages across various vertebrate lineages [18, 21–26]. However, little information is available on how and when each of the specific muscles develop in organisms that may have retained plesiomorphic gnathostome features, as with the shark. Currie and colleagues used shark models to investigate how the mechanisms that generate appendicular muscles evolved [12, 27]. These authors confirmed the observations obtained in studies carried out at the end of the nineteenth century showing that, in sharks, the appendicular musculature is formed by epithelial somitic extensions that penetrate the fin buds during development. However, their work was not focused on providing a detailed description on the development of specific muscles, such as those that connect the pectoral appendages to the skull. The only studies providing details on the development of individual muscles of sharks were published several decades ago (e.g., cephalic musculature reviewed by Edgeworth [23]), but they lack validation with novel methodological approaches. Therefore, detailed studies on shark muscle development are required to further explore potential ancient developmental processes involved in the formation of the cephalic and appendicular musculature in gnathostomes.

Several studies suggest a striking conservation of the developmental patterning of cephalic muscles, which is particularly well documented in amphibians [28–31]. These include the observation that these muscles tend to differentiate from the anterior to the posterior. For example, mandibular and hyoid muscles normally appear earlier than the muscles of the branchial (i.e., the most

posterior pharyngeal) arches. Their development also tends to follow a lateral to medial direction. For instance, lateral muscles of one arch tend to differentiate earlier than the more medially and ventral muscles of the same arch. In addition, these muscles normally develop from their region of origin towards their region of insertion [28, 32, 33]. It remains unknown, however, if this temporal and spatial sequence of developmental events represents the plesiomorphic feature, present in the common ancestor of all gnathostomes. Diogo and colleagues suggested that, in general, the developmental order of appearance of the cephalic muscles of amphibians [28, 29] and zebrafish [34] parallels the evolutionary order of appearance. This is also the case for the cephalic muscles in the head, neck, and limb muscles of primates [35]. Developmental studies on sharks are crucial to investigate if such patterns are also seen in chondrichthyans, exploring the conservation of these developmental patterns within gnathostomes, and whether there is a parallelism between ontogeny and phylogeny in vertebrate muscle development in general.

Within the broader analysis of cephalic muscle development in vertebrates, special attention has been given to the puzzling muscle cucullaris, which is deeply related to one of the most crucial evolutionary events during vertebrate evolutionary history: the evolution of the neck [36]. Recently, Ziermann and colleagues proposed that neck evolution was a long, stepwise macroevolutionary event, involving a stage in which an undivided cucullaris was connected to the branchial arches and the pectoral girdles [18], followed by its subdivision into the levatores arcuum branchialium attaching to the branchial arches and the protractor pectoralis attaching to the pectoral girdle, in osteichthyans [18]. Subsequently, there was further differentiation of the protractor pectoralis into various muscles (e.g., trapezius, sternocleidomastoideus) that took place during the evolution of tetrapods, where the head became further separated from the trunk [18]. Remarkably, recent analyses of the musculature in placoderm fossils suggest that the cucullaris might not have attached to the pectoral girdle in at least some of the members of this extinct group [36]. The comparison of the osteichthyans developmental data with the information obtained via the detailed analyses of the development of the cucullaris in chondrichthyans may offer additional information to discuss the ancestral condition of the cucullaris and thus the evolution of the neck within gnathostomes.

Additionally, comprehensive myological analyses in shark embryos may also help elucidate how paired appendage musculature was acquired within the gnathostome lineage. Two influential hypotheses were proposed during the late nineteenth century to explain the origin of two sets of paired appendages (pectoral and pelvic) in vertebrates: the gill - arch theory and the lateral fin-fold theory. The gill-arch theory proposes that pectoral and pelvic appendages evolved from modified gill arches and the pelvic appendages secondarily migrated caudally [37], whereas the lateral fin-fold theory suggests that pectoral and pelvic fins derive from a hypothetical bilateral continuous embryonic finfold. Both theories are consistent with the hypothesis that pectoral and pelvic appendages are serial homologous [38–40]. However, within these two theories the fin-fold theory, which lacks paleontological evidences [41, 42], excludes the contribution of head tissues to the formation of pectoral appendages. When molecular analyzes became available, the involvement of a common set of molecular mechanisms activated within a continuous dorsal/ventral field of competence to form appendages ("competent stripe", [13]) during the development of not only paired but also unpaired appendages were consistent with this idea that all these appendages do share similar developmental mechanisms [9–11, 43–46]. However, Gillis and colleagues, have shown that there are remarkable similarities in the developmental mechanisms operating during the ontogeny of the branchial arches and pectoral fin development [47, 48], thus reigniting discussions on Gegenbaur's hypothesis.

It is worth noting that the theories regarding the origin of two sets of paired appendages in gnathostomes mainly target the initial developmental components of fins, which are the fin mesenchyme and the endoskeleton that differentiates within and from it. However, dismissed from these theories are additional components essential for fin/limb function and which probably reinforced their adaptive rate, such as muscles, nerves, or blood vessels. Comparative myogenic studies performed by Diogo and colleagues, integrated with data from other authors and fields, suggest that the musculature of pectoral (fore-) and pelvic (hind-) appendages are particularly different in the proximal (girdle) region of these appendages [22, 49, 50]. These data question the existence of a common serial homologue musculature in pectoral and pelvic appendages, and indicate that what makes the pectoral and pelvic appendages so unique, and so remarkably similar in derived gnathostomes such as tetrapods, might in fact be the result of derived co-option [51]. To gain insight into these questions, it is crucial to comparatively evaluate muscle development in the pectoral and pelvic appendages in animals retaining a plesiomorphic fin musculature within gnathostomes, the sharks, and evaluate the contribution of cranial muscles to appendicular muscles during their formation.

Therefore, to discuss the broader developmental and evolutionary issues mentioned above, we present a detailed timeline of the development of both the cephalic and paired appendicular muscles in a shark species,

Scyliorhinus canicula. We identified heterochronic events during the development of the cephalic muscles of the shark as compared to the developmental pattern reported for most osteichthyans. That is, in our analyses of the shark the hyoid muscles develop earlier than mandibular muscles. This pattern contrasts with the observations in most osteichthyans where usually the mandibular muscles develop before the hyoid muscles or both groups develop simultaneously. In addition, we found that, although the development of the pectoral and pelvic appendicular muscles share similarities, there are significant differences concerning the timing of their formation. Moreover, tight muscular connections, involving several muscular units, develop between the pectoral girdle and the cranium of sharks, which finds no parallel during myogenesis of the pelvic fins. Our results highlight the importance to trace the distinct evolutionary processes analyzing different tissues individually and making use of model organisms at key phylogenetic positions.

Methods
Collection and staging of embryos
Scyliorhinus canicula (L. 1758) eggs were collected from the Menai Strait (North Wales). Embryos were isolated from egg cases and dissected from the yolk sac in ice-cold phosphate-buffered saline (PBS). Specimens were then staged according to Ballard et al. [52], before being fixed and processed as described below.

SEM and histology
For the scanning electron microscopy (SEM), specimens were fixed in 1% glutaraldehyde, then treated with 1% osmium tetroxide, dehydrated in a graded ethanol series, and transferred to acetone. Subsequently the specimens were critical-point dried, mounted onto carbon discs, sputter-coated with gold particles and visualized in a Jeol JSM-T300 Scanning Electron Microscope. For histology, embryos were fixed in 4% paraformaldehyde, dehydrated in a graded ethanol series, washed in Xylene, and embedded in paraffin. The resultant microtome sections (10 µm) were stained using Mallory's Triple Stain.

Whole-mount immunochemistry
S. canicula muscle development was characterized in the embryonic time comprising stages 23 to 32 using immunochemistry with antibody against Myosin Heavy Chain (MyHC; A4–1025, DSHB), a marker of muscle differentiation [12], following previously established and described protocols [53]. We analyzed one embryo per stage for muscle development using immunochemistry, as our previous muscle developmental studies have indicated that the intraspecific variability concerning the timing of muscle development depends also on sampling density and here the stages were clearly separated [54].

Regarding immunochemistry, the specimens were fixed in 4% PFA, then washed in PBS with 1% triton (PBT-1) for 3 h, incubated in 0.25% trypsin for 2–5 min, rinsed in PBT-1, and immersed in pre-cooled acetone for 10 min. After a brief rinse in PBT-1, the embryos were placed in blocking solution containing 10% goat serum (GS), 1% dimethyl sulfoxide, and 5% H_2O_2 in PBT-1, overnight. The MyHC antibody was used in a concentration of 1:10 and was diluted in PBT-1 containing 10% GS. Goat anti-mouse IgG secondary antibody, HRP (Thermofisher), was used at a concentration of 1:500 diluted in PBT-1 with 1% GS. Embryos were then washed in 1% GS in PBT-1, followed by PBS, and then incubated in 0.5 mg/ml diaminobenzidine (DAB). The reaction was developed by transferring embryos to fresh DAB activated with 0.003% H_2O_2.

Muscle characterizations

One side of the embryos was dissected with micro-dissection tools under a dissection microscope to analyze the development of deep muscles. The specimens were photographed at a dissecting microscope (Nikon SMZ-2B) equipped with a Nikon DS Fi1 5 Megapixel Color Camera Head. Myological terminology used in the present paper follows that proposed by Diogo and Abdala [21] and updated by Ziermann et al. [18] and Diogo and Ziermann [22] for adult sharks.

Results

Pectoral fin development and muscle differentiation

The development of the cephalic and appendicular musculature of *S. canicula* is summarized in Table 1 and shown in Figs. 1, 2, 3, 4, 5 and 6. Muscle projections are detected extending from the myotome towards the pectoral fin fields between **stages 26** and **27** (Fig. 1d-e; j-k). MyHC staining indicates that muscle projections invade the pectoral fin territory between **stages 28** and **29** (Fig. 1l-m). The former shows the development of the dorsal muscle *adductor superficialis* and the ventral muscle *abductor superficialis* (Figs. 1l-m and 3c-d). At **stage 30**, muscle projections are detected throughout the fin, both dorsally and ventrally (Fig. 1f). MyHC staining further indicates that the *abductor superficialis*, which connects the girdle to the fin, is now prominent at this stage (Fig. 1n). MyHC staining also highlights the rostroventral development of the *pterygialis cranialis* at **stage 28** (Fig. 4c). Between **stages 31** and **32**, the *adductor superficialis* and *abductor superficialis* pursue development (Figs. 1g-h, o-p; 6d), and the *pterygialis cranialis* continues to expand and differentiate without major changes relative to the previous stages (Figs. 5d, e and 6c). Finally, at **stage 34 and** prior to hatching, the appendicular skeleton appears strongly associated to the musculature both in the girdle and in the pectoral fins (Fig. 1i).

The first signs of pectoral fin development are detectable at **stage 19**, budding out from the lateral plate mesoderm and positioned approximately between somites 6 to 16 (Fig. 1a). By **stage 23**, the pectoral fin buds are visible lateral to the yolk stalk [53] (Fig. 1b). However, the apical ectodermal ridge (AER), a crucial signaling center during fin/limb development, is undetectable up to **stage 24** (Fig. 1c). Then, the pectoral fin buds reshape dorsoventrally acquiring a disc-like structure capped distally by the AER (Fig. 1c). This distal structure is rapidly converted into an apical ectodermal fold [53] (AEF) by **stage 25** (Fig. 1d).

Pelvic fin development and muscle differentiation

The pelvic fins start to bud out from the lateral plate mesoderm laterally to the cloaca region at **stage 25** and their development is close to completion prior to hatching [9] (Fig. 2a-f). As for the pectoral appendages, the formation of the AER seems to be transient, rapidly giving rise to an AEF, at **stage 26** (Fig. 2b). As seen in the pectoral appendages, MyHC staining and histology suggest that muscle projections extend ventrally towards the base of the pelvic fins between **stage 27** and **stage 28** (Fig. 2c, g). At **stage 30** an undivided *abductor* and the *adductor superficialis* can be identified (Fig. 2d, h). However, pelvic fins show an even fainter staining on both the dorsal and ventral sides indicating just the initial development of the *adductor superficialis* and *abductor*, respectively (compare Fig. 1n with Fig. 2h). The latter is easier to see at **stage 31** (Fig. 2e, i). These muscles then expand along the proximodistal axis of the pelvic fins during stage 31 (Fig. 2e) reaching the margin of the terminal finfold by **stage 32** (Fig. 6d). While formation of the endoskeletal elements continues at **stage 33**, the distinction between a proximal abductor and a distal abductor, as well as the presence of a separate muscle protractor of the pelvic fin as described in adult sharks [22], cannot be made. As these are superficial structures, it is likely that they are not differentiated yet in any of the stages analyzed by us, which is plausible due to the fact that the pelvic fins start to form later than the pectoral fins.

Cranial development and muscle differentiation

The first signs of eye and nasal pit development are detected in *S. canicula* at **stage 17** and **21**, respectively (data not shown). As early as **stage 19**, features such as the otic vesicles and the curvatures delineating the forebrain, the midbrain, and the hindbrain are visible [10]. At **stage 23**, the ganglia of cranial nerves V and VII are clearly observable through the skin (Fig. 3a). Furthermore, six pharyngeal arches are present as anlagen and the proximity to the posteroventrally looping heart is clearly visible [52] (Fig. 3a). The first pharyngeal arch is

Table 1 Myogenic cephalic and pectoral/pelvic fin development in *Scyliorhinus canicula*. Stages according to Ballard et al. [52]. Terminology and adult characteristics follows Ziermann et al. [18]. x = present (independent on status of differentiation); o = absent (or not stained); (number) = number of repetitive muscles (usually corresponding to branchial arches and counting from anterior to posterior). Orange box = dorsal constrictor of mandibular arch. Yellow box = *constrictor hyoideus*. Green box = only one abductor could be found in the *S. canicula* stages investigated here. (?) could not be observed because of overlying muscles. *Adult condition is from *Squalus acanthias* [18]

Muscle/Stage	23	24	26	27	28	29	30	31	32	Adult*
Extraocular muscles				o	x	x	x	x	x	x
Mandibular muscles										
intermandibularis			o	x	x	x	x	x	x	x
adductor mandibulae A2				o	x	x	x	x	x	x
preorbitalis (labial muscle)				o	x	x	x	x	x	x
levator arcus palatini				o	x	x	x	x	x	x
spiracularis				o		x	x	x	x	x
Hyoid muscles										
interhyoideus			o	x	x	x	x	x	x	x
adductor arcus palatini	o	x	x	x	x	x	x	x	x	x
constrictor hyoideus dorsalis	o	x	x	x	x	x	x	x	x	x
True branchial muscles										
cucullaris					o	x	x	x	x	x
constrictores branchiales	o	(3)	(3)	(3)	(4)	(4)	(4)	(4)	(4)	(4)*
adductores branchiales									o	(5)*
interarcuales laterales									o	(4)*
interbranchiales									o	(4)*
Ventral branchial muscles										
coracobranchiales					o	(2)	(3)	(4)	(?)	(5)*
Epibranchial muscles										
interpharyngobranchiales									o	(3)*
subspinalis									o	x
Hypobranchial muscles										
coracomandibularis				o	x	x	x	x	x	x
sternohyoideus					o	x	x	x	x	x
coracoarcualis				o	x	x	x	x	x	x
Pectoral fin										
pterygialis cranialis				o	x	x	x	x	x	x
abductor superficialis				o	x	x	x	x	x	x
adductor superficialis				o	x	x	x	x	x	x
Pelvic fin										
protractor									o	x
adductor superficialis					o	x	x	x	x	x
abductor proximalis					o					x
abductor distalis					o	x	x	x	x	x

the mandibular arch, the second is the hyoid arch, and the following four are branchial arches I-IV (i.e., pharyngeal arches III-VI). The first three pharyngeal pouches are open, with the first one being the spiracle (C1, hyomandibular or spiracular cleft). Only the myotomes and the cardiac wall show MyHC staining at this stage. The most anterior myotome has only its ventral half stained and is superior to the posterior half of the fourth pharyngeal arch.

By **stage 24** the posterior branchial arches are better defined compared to the previous stage and the fourth pharyngeal cleft is visible. The hyoid arch and branchial

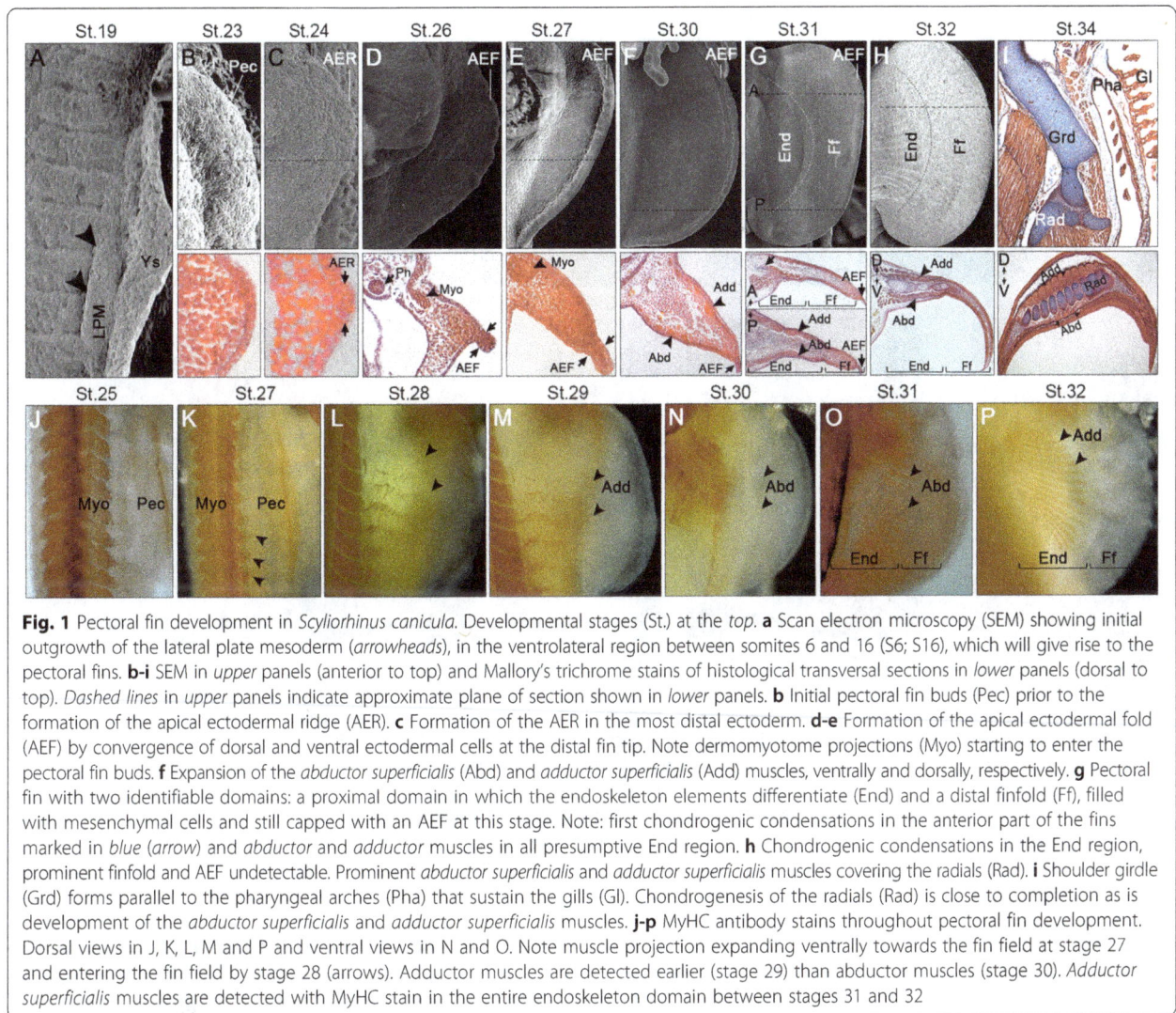

Fig. 1 Pectoral fin development in *Scyliorhinus canicula*. Developmental stages (St.) at the *top*. **a** Scan electron microscopy (SEM) showing initial outgrowth of the lateral plate mesoderm (*arrowheads*), in the ventrolateral region between somites 6 and 16 (S6; S16), which will give rise to the pectoral fins. **b-i** SEM in *upper* panels (anterior to top) and Mallory's trichrome stains of histological transversal sections in *lower* panels (dorsal to top). *Dashed lines* in upper panels indicate approximate plane of section shown in *lower* panels. **b** Initial pectoral fin buds (Pec) prior to the formation of the apical ectodermal ridge (AER). **c** Formation of the AER in the most distal ectoderm. **d-e** Formation of the apical ectodermal fold (AEF) by convergence of dorsal and ventral ectodermal cells at the distal fin tip. Note dermomyotome projections (Myo) starting to enter the pectoral fin buds. **f** Expansion of the *abductor superficialis* (Abd) and *adductor superficialis* (Add) muscles, ventrally and dorsally, respectively. **g** Pectoral fin with two identifiable domains: a proximal domain in which the endoskeleton elements differentiate (End) and a distal finfold (Ff), filled with mesenchymal cells and still capped with an AEF at this stage. Note: first chondrogenic condensations in the anterior part of the fins marked in *blue* (arrow) and *abductor* and *adductor* muscles in all presumptive End region. **h** Chondrogenic condensations in the End region, prominent finfold and AEF undetectable. Prominent *abductor superficialis* and *adductor superficialis* muscles covering the radials (Rad). **i** Shoulder girdle (Grd) forms parallel to the pharyngeal arches (Pha) that sustain the gills (Gl). Chondrogenesis of the radials (Rad) is close to completion as is development of the *abductor superficialis* and *adductor superficialis* muscles. **j-p** MyHC antibody stains throughout pectoral fin development. Dorsal views in J, K, L, M and P and ventral views in N and O. Note muscle projection expanding ventrally towards the fin field at stage 27 and entering the fin field by stage 28 (arrows). Adductor muscles are detected earlier (stage 29) than abductor muscles (stage 30). *Adductor superficialis* muscles are detected with MyHC stain in the entire endoskeleton domain between stages 31 and 32

arches I and II show the first external gill buds that reach from their arch posteriorly over the adjacent cleft. The gill buds decrease in size from anterior to posterior (Fig. 3b). Dorsally between branchial arches I and II is a small anteroposteriorly orientated muscle, which is the ventral portion of an anterior developing myotome (Fig. 3b). Anteriorly on the hyoid arch and branchial arches I-III, faint MyHC staining can be detected indicating the initial differentiation of the dorsal constrictor muscles (*constrictor hyoideus* of the hyoid arch, and *constrictores branchiales I to III* of the branchial arches) (Fig. 3b). The staining decreases from anterior to posterior, which is undetectable in the mandibular arch. By **stage 26**, the fifth pharyngeal cleft is visible and the first myotome, which lies dorsally between branchial arches III and IV, is larger and more visible than in previous stages (Fig. 3c). The external gill branches on the hyoid arch and branchial arches I and II expand and the first gills appear on branchial arch III.

In **stage 27** the ventral portion of the most anterior differentiating myotome becomes visible dorsally between branchial arches I and II (Fig. 3d). The mandibular arch also shows an elongated thin muscle anlage anteroventrally located; this is the *intermandibularis* anlage, which is only separated by a small gap from the developing *interhyoideus* anlage of the hyoid arch (Fig. 3e). This *interhyoideus* anlage is continuous with the ventral portion of the *constrictor hyoideus*. The muscle anlagen of the dorsal constrictors extend and are more clearly visible (Fig. 3d, e): the hyoid arch includes the *constrictor hyoideus* anlage, while the branchial arches include the *constrictores branchiales I-III* (*constrictores branchiales superficialis*). These muscles are mainly dorsoventrally orientated, the *constrictor hyoideus* being the longest and stretching almost the entire length of the hyoid arch, and the *constrictor branchialis III* being the shortest, covering only half of the dorsoventral extension of the third branchial arch (Fig. 3d). The staining of the

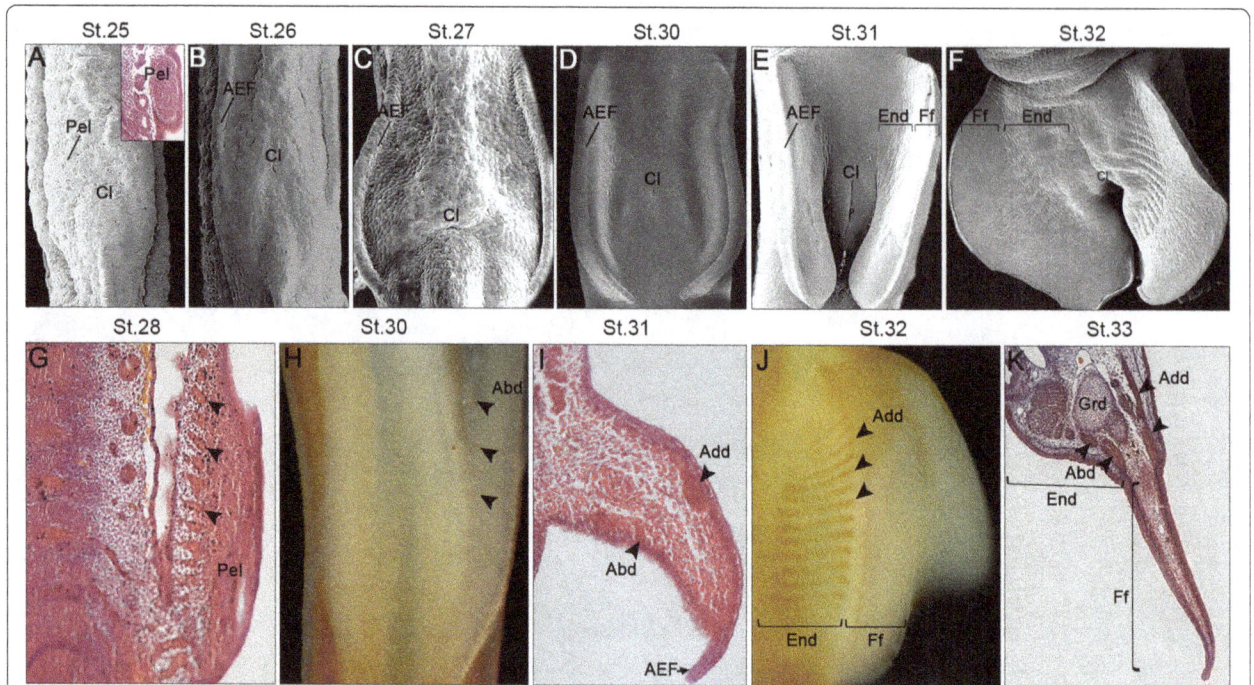

Fig. 2 Pelvic fin development in *Scyliorhinus canicula*. Developmental stages (St.) at the *top*. **a** Scan electron microscopy (SEM) and Mallory's trichrome stains of histological transversal sections showing initial pelvic fin outgrowth (Pel), laterally to the prospective cloaca region (Cl). **b-f** SEMs showing the progression of pelvic fin development (ventral views). Note formation of the apical ectodermal fold (AEF) at stage 26 and separation of a proximal domain in which the endoskeleton differentiates (End) and a distal domain, the finfold (Ff) by stage 32. **g-k** Muscle development between stages 28 and 33 shown with Mallory's trichrome stains (**g**, **i**, **k**) and MyHC antibody stain (H,J). **g** Muscle projections entering the fin territory (*arrows*). **h** Abductor muscles observed ventrally (Abd). **i** Abductor and adductor superficialis (Add) muscles detected ventrally and dorsally, respectively. **j** Dorsal view showing adductor muscles detected from the proximal part of the fin to the distal border of the endoskeleton domain (End), but not in the finfold (Ff). **k** Chondrogenesis is close to completion, with the pelvic girdle (Grd) clearly detected proximally and prominent abductor and adductor muscles

constrictors decreases from anterior (hyoid arch) to posterior (Fig. 3d) indicating the later development of more posterior muscles.

Major differences are then observed at **stage 28** (Fig. 4a-c). External gills appear on the mandibular arch and all posterior external gills are lengthened. Dorsal extraocular muscles are visible, the mandibular arch muscle anlage is expanded, and, from a lateral view, the *constrictor dorsalis* of this arch is visible just rostral to the spiracle (Fig. 4a). The most rostral portion of the mandibular muscles is the developing *preorbitalis*, caudally adjacent to the anlage of the *adductor mandibulae A2*; and caudomedially there is the *intermandibularis* anlage (Fig. 4b). At this stage, all mandibular arch muscles derive originally from a single elongated anlage that then separates during later stages into several regions (dorsal, middle and ventral), which then gives rise to one or more muscles. The hyoid arch muscle anlage also differentiates into three portions (Fig. 4a, b): the most rostro-dorsal one is the anlage of the *adductor arcus palatini*, the large lateral one is the anlage of the *constrictor hyoideus dorsalis*, and the most ventral portion is the *interhyoideus* anlage. The *adductor arcus palatini* develops from the most

rostrodorsal portion of the *constrictor hyoideus* primordium; this rostrodorsal portion also gives rise to the *constrictor hyoideus dorsalis*. The *constrictores branchiales I-IV* stretch almost the entire dorsoventral length of their respective arches (Fig. 4a). The ventral branchial muscles *coracobranchiales* and the hypobranchial muscles *coracomandibularis* and *coracoarcualis* are distinguishable from all the other cephalic muscles because they develop from their region of insertion (i.e., from the mandible and branchial arches, respectively; compare Figs. 4b, f, 5b, d for *coracomandibularis* and Figs. 4f, 5b, d, 6c for *coracoarcualis*), while the mandibular, hyoid and branchial muscles develop from their region of origin, as do most cephalic muscles of other vertebrates (Fig. 4b; compare for examples Figs. 3e, 4f, 5d, 6c for the development of the *intermandibularis* and *interhyoideus*).

At **stage 29** the ventral extraocular muscles become visible (Fig. 4d). The differentiation of the mandibular muscles is now also noticeable from a lateral view (Fig. 4d). The dorsal constrictor separates into a dorsal part that extends posteriorly, the *spiracularis*, and a lateral portion, extending inferiorly and medially, the *levator arcus palatini* (Fig. 4d). The *preorbitalis, adductor*

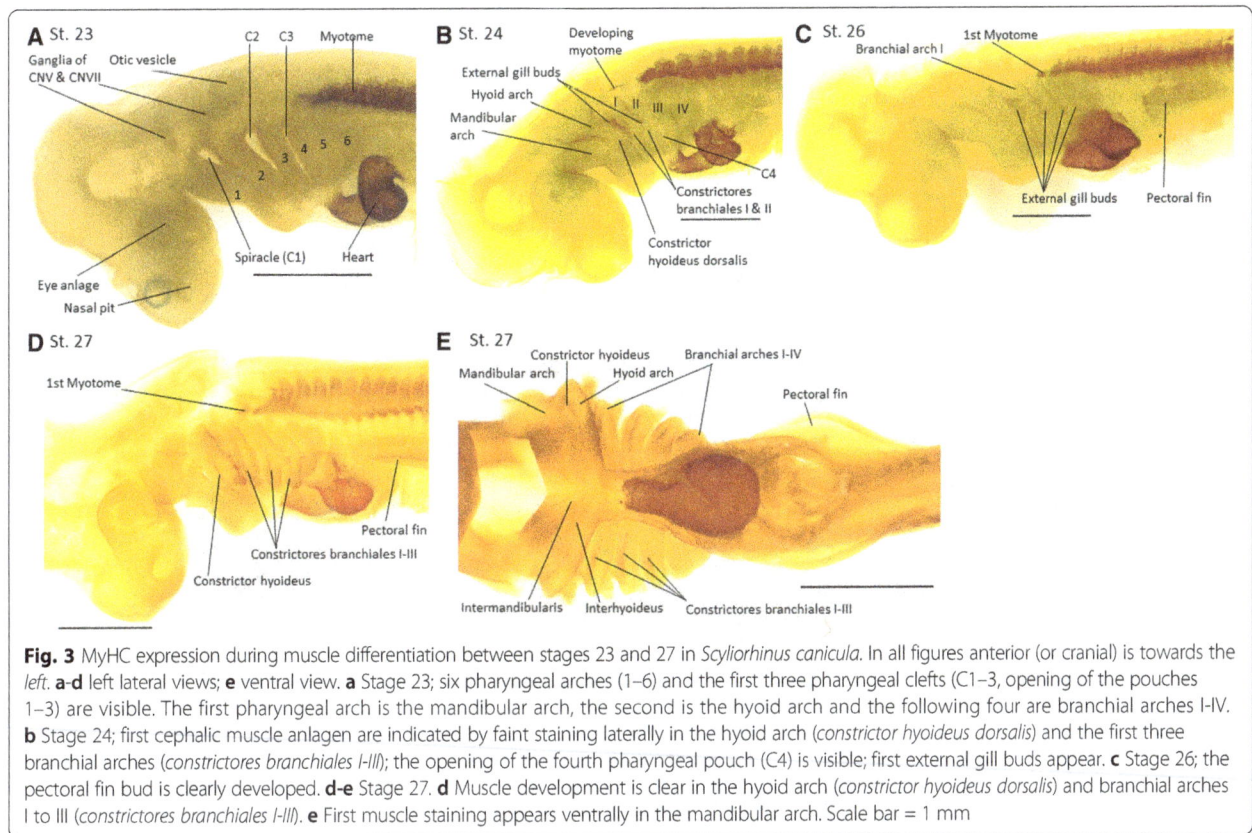

Fig. 3 MyHC expression during muscle differentiation between stages 23 and 27 in *Scyliorhinus canicula*. In all figures anterior (or cranial) is towards the *left*. **a–d** left lateral views; **e** ventral view. **a** Stage 23; six pharyngeal arches (1–6) and the first three pharyngeal clefts (C1–3, opening of the pouches 1–3) are visible. The first pharyngeal arch is the mandibular arch, the second is the hyoid arch and the following four are branchial arches I–IV. **b** Stage 24; first cephalic muscle anlagen are indicated by faint staining laterally in the hyoid arch (*constrictor hyoideus dorsalis*) and the first three branchial arches (*constrictores branchiales I–III*); the opening of the fourth pharyngeal pouch (C4) is visible; first external gill buds appear. **c** Stage 26; the pectoral fin bud is clearly developed. **d–e** Stage 27. **d** Muscle development is clear in the hyoid arch (*constrictor hyoideus dorsalis*) and branchial arches I to III (*constrictores branchiales I–III*). **e** First muscle staining appears ventrally in the mandibular arch. Scale bar = 1 mm

mandibulae A2 and *intermandibularis* can be distinguished (Fig. 4d, f). The hyoid muscles *adductor arcus palatini* and *constrictor hyoideus dorsalis* start to separate, while ventrally the *interhyoideus* muscle grows (Fig. 4d, f). Within the branchial muscles only four *constrictores branchiales* – extending from lateral to ventral regions – are clearly visible (Fig. 4e, f). Ventrally the *coracobranchiales I–II* appear as faint stains at the base of the branchial arches I–II (Fig. 4f). Dorsal to branchial arches I and II, there is a faint anteroposteriorly orientated muscle staining, which indicates the initial differentiation of the *cucullaris* (Fig. 4e). The hypobranchial muscles *coracomandibularis* and *coracoarcualis* are now more distinguishable than in stage 28, and the hypobranchial muscle *sternohyoideus* is faintly visible just rostral to the *coracoarcualis* (Fig. 4f).

In **stage 30** the muscles described in stage 29 become more clearly separated (Fig. 5a), especially in ventral view (Fig. 5b). The main contrast to the former stage is that the *cucullaris* is now plainly visible as a thin band-like structure extending dorsally from branchial arch I to branchial arch IV (Fig. 5a). Ventrally the *sternohyoideus* becomes further distinguishable from the *coracoarcualis* (Fig. 5b). The *coracobranchiales I–III* can be seen ventrally to the respective branchial arches (Fig. 5b). In **stage 31** the lateral muscles of all arches continue to

grow and differentiate without major changes relative to the previous stage (Fig. 5c), the *intermandibularis* and *interhyoideus* come into contact, and four *coracobranchiales* are now visible (Fig. 5d).

In **stage 32** all muscles are further differentiated (Fig. 6) and the head now appears, in a ventral view, almost completely covered by muscles (Fig. 6c). The *intermandibularis* and *interhyoideus* reach the ventral midline. The *coracobranchiales* are completely covered by the *constrictores branchiales*, and staining of deeper muscles is not visible. The *cucullaris* spans the entire length dorsally to the branchial arches (Fig. 6a, b). The cartilages are not clearly distinguishable because they are almost translucent at this stage with this methodology, but there was a faint attachment of the *cucullaris* onto the caudal branchial arch IV. Immediately caudal to the last branchial arch, the fibers of this muscle extend lateroventrally towards the pectoral girdle (Fig. 6a, b).

We observed neither the deeper true branchial muscles (*adductores branchiales*, *interbranchiales* and *interarcuales laterales*) nor the epibranchial muscles (*interpharyngobranchiales* and *subspinalis*); in adult sharks these deeper muscles are superficially covered by the *constrictores branchiales*, and attach onto branchial arches [18] (Table 1). That these muscles/bundles could not be seen can be explained by either the fact that they

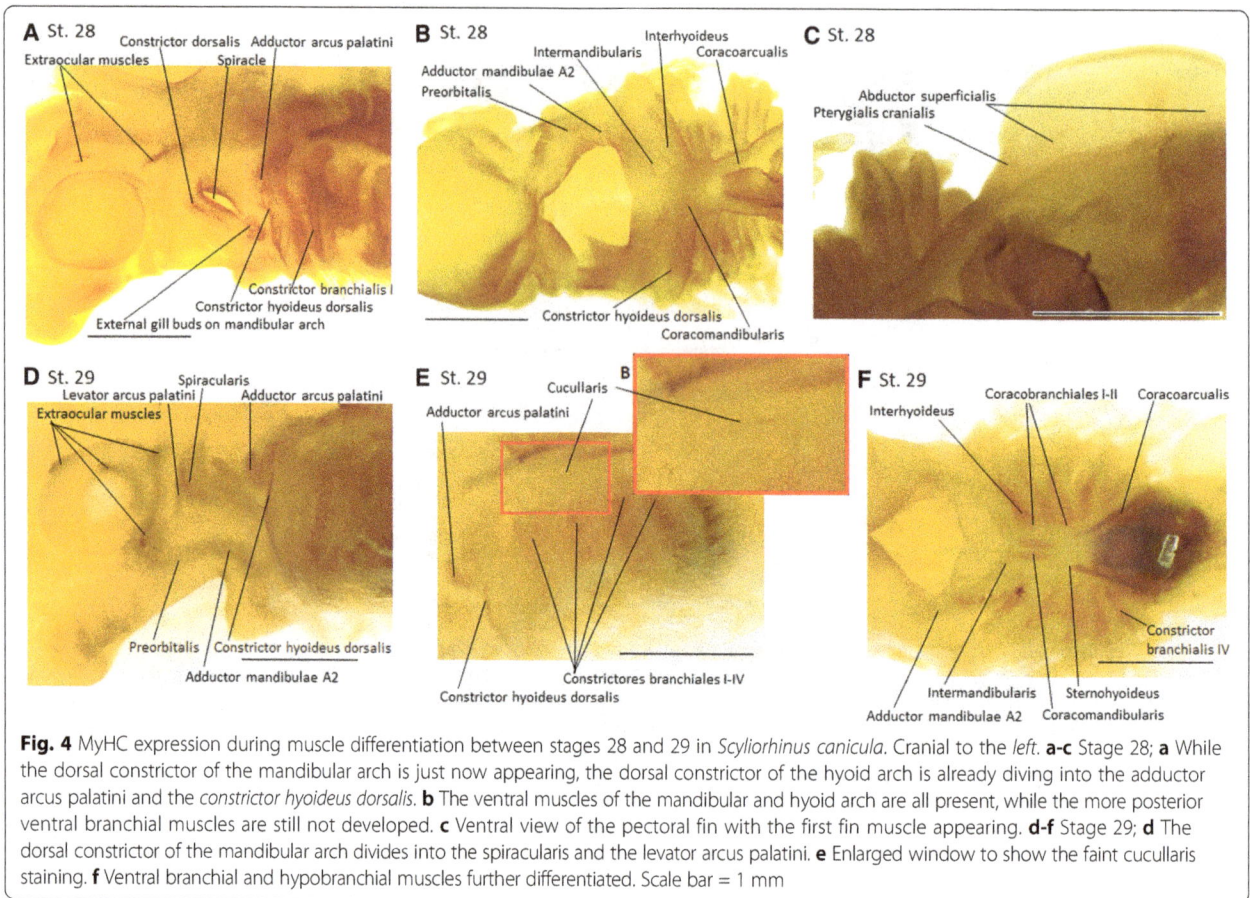

Fig. 4 MyHC expression during muscle differentiation between stages 28 and 29 in *Scyliorhinus canicula*. Cranial to the *left*. **a-c** Stage 28; **a** While the dorsal constrictor of the mandibular arch is just now appearing, the dorsal constrictor of the hyoid arch is already diving into the adductor arcus palatini and the *constrictor hyoideus dorsalis*. **b** The ventral muscles of the mandibular and hyoid arch are all present, while the more posterior ventral branchial muscles are still not developed. **c** Ventral view of the pectoral fin with the first fin muscle appearing. **d-f** Stage 29; **d** The dorsal constrictor of the mandibular arch divides into the spiracularis and the levator arcus palatini. **e** Enlarged window to show the faint cucullaris staining. **f** Ventral branchial and hypobranchial muscles further differentiated. Scale bar = 1 mm

are not yet developed/differentiated or – more likely taking into account that all the other, more superficial, muscles of the branchial region are already seen – the penetration of the antibody was not deep enough. In fact, based on our studies of muscle development in other fishes and in tetrapods (cited above), it is very likely that most, or all, of these muscles started to differentiate in the oldest shark specimen(s) analyzed by us.

Thus, in summary, the anlagen (primordia) of mandibular arch muscles appear after the anlagen of hyoid arch muscles. The first branchial muscle anlagen can be observed simultaneously with the anlagen of the hyoid arch muscles, with the staining fainting from anterior to posterior (Fig. 3d). Still, analyzing the detailed appearance of muscles we could observe that muscles develop following an anterior to posterior direction, from the hyoid to the branchial arches and from outside to inside, i.e., lateral muscles develop before ventral muscles (except within the mandibular arch), and superficial muscles develop before deep muscles. Most muscles develop from their region of origin to their region of insertion (compare for examples Figs. 3e, 4f, 5d, 6c for the development of the *intermandibularis* and *interhyoideus*),

except for the ventral branchial muscles *coracobranchiales* and the hypobranchial muscles *coracomandibularis* and *coracoarcualis*, which develop from their region of insertion to their region of origin (compare Figs. 4b, f, 5b, d for *coracomandibularis* and Figs. 4f, 5b, d, 6c for *coracoarcualis*).

Discussion

On the origin of the pectoral and pelvic musculatures

In chondrichthyan and osteichthyan fishes the pectoral appendages are invariably described as developing before the pelvic appendages, in contrast to the condition found in most tetrapods, where they develop relatively simultaneously [55, 56]. In *S. canicula* the pelvic fin indeed starts to develop later than the pectoral one: the first signs of a pectoral fin outgrowth were identified as early as stage 19, while pelvic fin development was only detected at stage 25 (Fig. 1). Moreover, as previously suggested [9], all events characterizing fish fin development, such as formation of a transient AER, conversion of this structure into an AEF, outgrowth and differentiation of endoskeleton elements, occur earlier in the pectoral fins than in the pelvic fins. Here we show that appendicular myogenesis also occurs earlier in the

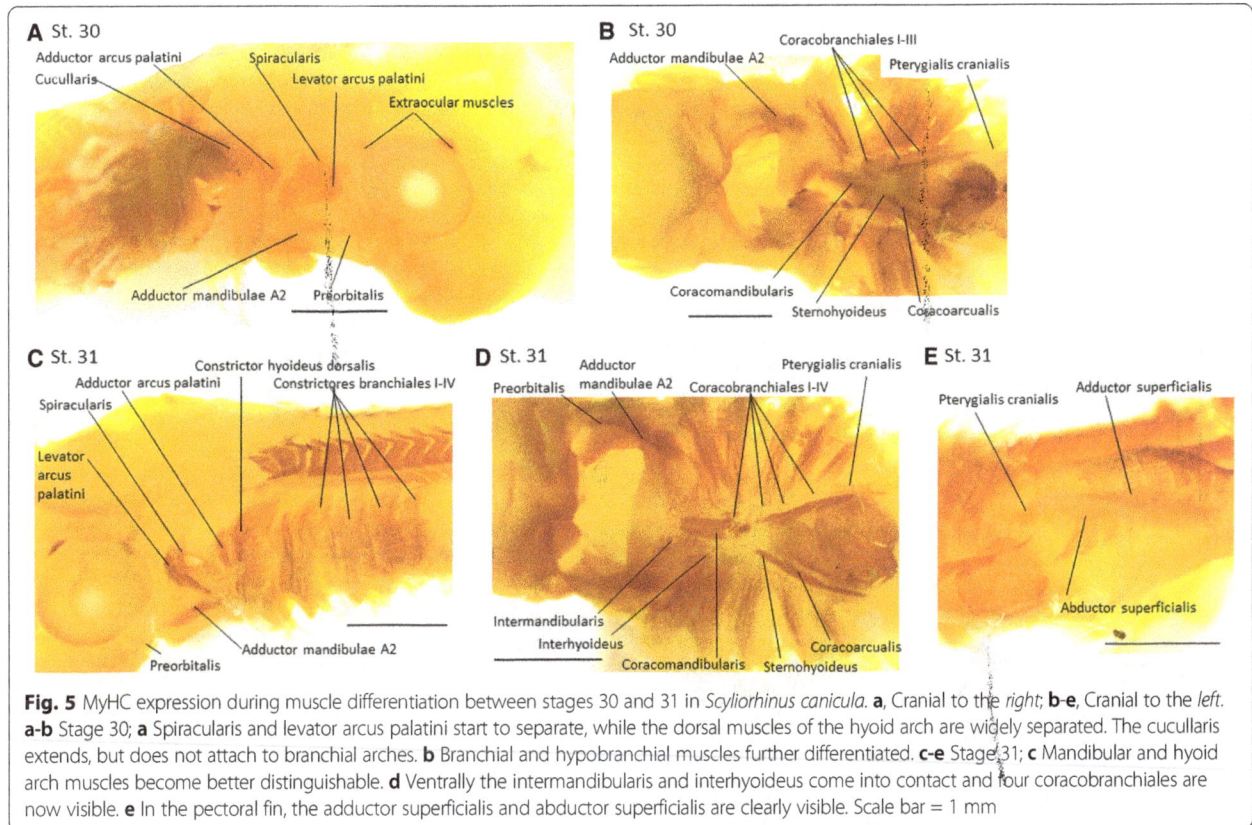

Fig. 5 MyHC expression during muscle differentiation between stages 30 and 31 in *Scyliorhinus canicula*. **a**, Cranial to the *right*; **b-e**, Cranial to the *left*. **a-b** Stage 30; **a** Spiracularis and levator arcus palatini start to separate, while the dorsal muscles of the hyoid arch are widely separated. The cucullaris extends, but does not attach to branchial arches. **b** Branchial and hypobranchial muscles further differentiated. **c-e** Stage 31; **c** Mandibular and hyoid arch muscles become better distinguishable. **d** Ventrally the intermandibularis and interhyoideus come into contact and four coracobranchiales are now visible. **e** In the pectoral fin, the adductor superficialis and abductor superficialis are clearly visible. Scale bar = 1 mm

pectoral fins (stage 28) than in the pelvic fins (stage 29) in *S. canicula*. Interestingly, the formation of the *abductor* and *adductor* muscles in the pectoral fins occurred simultaneously with the formation of the preaxial muscle, the *pterygialis cranialis* (Table 1). All these muscles seem to have been present in the last common ancestor (LCA) of the crown-group Gnathostomata [6, 22, 57]. Therefore, the non-simultaneous development of pectoral and pelvic musculature, which is commonly observed in osteichthyan fishes, may reflect the ancestral developmental process in the gnathostome lineage.

In adult chondrichthyans the pectoral *abductor* and *adductor* and the pelvic *adductor* have superficial and deep bundles, but the pelvic *abductor* has instead proximal and distal bundles, demonstrating that the significant anatomical differences between the pectoral and pelvic appendages of sharks concern not only hard tissues, but also soft tissues such as muscles [22, 49, 58]. Our results demonstrate that during early developmental stages the muscles of the pectoral and pelvic fins are more similar to each other than in adulthood. While developing, there are mainly two major undivided muscles in each fin, *abductor* and *adductor*, except for a preaxial muscle present only in the pectoral fin (the *pterygialis cranialis*). Thus, during late development, most likely after hatching, pelvic muscles undergo further elaboration becoming rather

distinct from pectoral muscles. Interestingly, while we could observe the formation of the *abductor* and *adductor* muscles in the pelvic fins up to stage 34, we could not detect the development of the *protractor* even in the oldest stages, which suggests that considerable development of the pelvic musculature occurs, in fact, after hatching.

Apart from the differences between the adult *abductors* of the pectoral and pelvic appendages, and between the time of appearance of the musculature of each of these appendages, there is another major difference between the musculature of these appendages in *S. canicula*: the presence of several muscles connecting the pectoral girdle to cranial elements, such as the *coracomandibularis*, *coracoarcualis*, *coracobranchiales*, and the *cucullaris*, which all develop from the head region to the pectoral girdle region and which have no corresponding muscles in the pelvic appendage. This latter difference stresses the point that there are major functional and evolutionary reasons for the spatial correlation of the pectoral girdle with the skull in early gnathostomes: the internal branchial chamber seems to restrict the development of the pectoral girdle more anteriorly, which forms a protection for the pericardial cavity and an insertion for the pectoral fins [58].

In fact, studies in chondrichthyans have shown that the formation of branchial arches in sharks and the

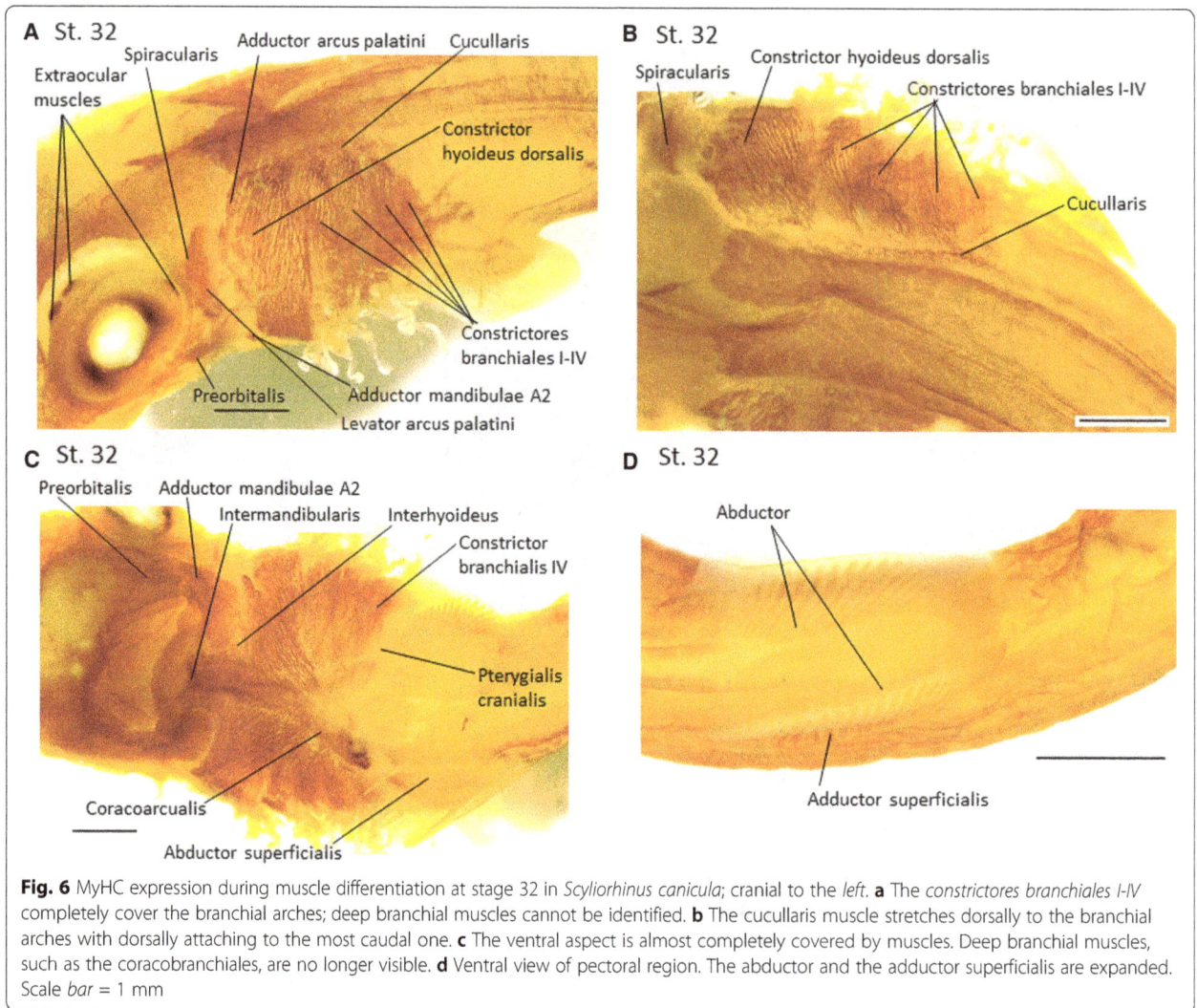

Fig. 6 MyHC expression during muscle differentiation at stage 32 in *Scyliorhinus canicula*; cranial to the *left*. **a** The *constrictores branchiales I-IV* completely cover the branchial arches; deep branchial muscles cannot be identified. **b** The cucullaris muscle stretches dorsally to the branchial arches with dorsally attaching to the most caudal one. **c** The ventral aspect is almost completely covered by muscles. Deep branchial muscles, such as the coracobranchiales, are no longer visible. **d** Ventral view of pectoral region. The abductor and the adductor superficialis are expanded. Scale *bar* = 1 mm

tetrapod forelimb share strikingly similar developmental mechanisms [47, 48], somewhat consistent with the view that the branchial arches and the pectoral appendage might be highly related evolutionarily/developmentally see also [59]. For example, *sonic hedgehog (Shh)* is crucial to establish the anteroposterior polarity in both the outgrowing fin–/limb-bud and the developing gill arch [48]. Other studies analyzed the body wall formation in lampreys as compared to gnathostomes, which is relevant as the gnathostome paired appendages start as outgrowths of body wall somatopleure [60]. The somatopleure is a tissue containing somatic lateral plate mesoderm and overlying ectoderm [60]. Lampreys are cyclostomes, i.e., vertebrates without jaw and paired fins. Tulenko and colleagues [60] suggest that the somatopleure is eliminated in lampreys while the lateral plate mesoderm is separated from the ectoderm and isolated to the coelomic linings during myotome extension. One way to interpret those data is that the somatopleure may

have originally persisted close to the gills, established a pectoral fin, and afterwards, spread posteriorly to the pelvic level [60]. This model has similarities to variations of the gill arch hypothesis [59]. Interestingly, a recent study identified a *Tbx5* fin enhancer, CNS12, in the noncoding region downstream of *Tbx5* locus [61]. The enhancer CNS12 was suggested to have driven the reporter gene expression in the lateral plate mesoderm posterior to the heart – a region where vertebrates with pectoral appendages show an apomorphic *Tbx5* expression pattern [61]. In the cephalochordate amphioxus *Tbx4/5* is expressed in the pharyngeal and posterior mesoderm together with cardiac genes and is relevant for the development of a noncentralized heart [62]. Other marker genes for vertebrate head and trunk mesoderm are also expressed in overlapping domains in amphioxus dorsal mesoderm, what indicates that the mesoderm is not yet differentiated along the craniocaudal axis [62]. These data thus support the hypothesis that the mesoderm of the posterior head region, the heart, and pectoral appendages might have originated from a common ancestral region. This scenario might be an example for deep homology, in which structures evolve by the modification of pre-existing genetic regulatory circuits established in early metazoans [63]. In fact, it was recently shown that the pharyngeal (head) muscles and the myocardium are developmentally and evolutionary more linked to each other than previously thought and that the so-called cardiopharyngeal field was likely present in the last common ancestor or of Olfactores (tunicates + vertebrates) [64].

The observations regarding the development of the *cucullaris* in *S. canicula* reinforce the idea of an ancestral close association between the head and pectoral girdle musculature. We showed that the anlage of the *cucullaris* clearly appears in the dorsal region of the branchial arches, without any connection between it and the anterior somites in early development (Fig. 4e). This further supports the hypothesis that the *cucullaris* is a true branchial muscle, as defended by classical authors such as Edgeworth [23] and in more recent developmental and molecular works [65]. *Tbx1* mutant mice, for example, have no *trapezius* or *sternocleidomastoideus*, which are derivatives of the *cucullaris*, and no branchial muscles, while somite-derived limb muscles are unaffected [66]. Further evidence was provided by a fate map study in *Ambystoma mexicanum* where it was shown that the lateral plate mesoderm contributes to posterior *branchial arch levators* and to the *cucullaris*, what led the authors to suggest that this mesoderm should be regarded as posterior cranial mesoderm [67].

There are further indications consistent with the idea that the musculature connecting the head to the pectoral girdle could have derived from, or co-opted, similar developmental mechanisms than those used by the musculature of the posterior pharyngeal arches. In fact, there are three groups of muscles that connect these appendages to head structures other than the *cucullaris*: the hypobranchial muscles *coracomandibularis* and *coracoarcualis* attach the pectoral girdle to the mandible and (mostly) to the ceratohyal, respectively, while the ventral branchial muscles *coracobranchiales* connect it to the branchial arches. The muscles *coracomandibularis* and *coracoarcualis* cannot be used to support a similarity between the pectoral and posterior pharyngeal arch musculatures, because they are hypobranchial muscles derived from somites, and not branchiomeric head muscles, as are most of the muscles that connect the posterior pharyngeal arches to other cranial structures. However, the presence of the true branchial muscles *coracobranchiales* connecting the pectoral girdle to the branchial arches, exactly as numerous branchial muscles connect the branchial arches to each other, might

constitute an argument consistent with the idea of a deep association between the pectoral girdle musculature and the branchial arch musculature. Particularly because the pelvic musculature has of course no muscles at all connecting it to the head, and thus to any branchial arch. These data contradict the hypothesis that pectoral and pelvic appendages and associated soft tissues are strictly serial homologous because this does not refer merely to the different topological position of the pectoral vs. pelvic appendages. Instead, this refers to completely different types of tissues, derived from completely types of primordia, being part of each of these two types of appendages. That is, the pectoral appendage includes/is related to branchial muscles that are derived from the cardiopharyngeal field, while no such muscles are related/part of the pelvic appendage, which exclusively includes muscles derived from somites.

Authors have recently suggested that in some placoderm fossils the pectoral and pelvic appendages seem to be more similar than previously thought [36]. However, these studies do not include reconstructions of appendicular soft tissues such as muscles, which are crucial to discuss the similarity vs. dissimilarity of the pelvic and pectoral appendages as a whole. Anatomical and developmental studies performed on extant animals considered to have retained plesiomorphic musculature of gnathostomes [7], like chondrichthyans, show several aspects of dissimilarity between these appendages [18, 22]. Thus, strict serial homology does not seem to explain the origin of all the tissues that constitute/attach onto the pectoral and pelvic appendages. Further information on the cephalic and appendicular musculature of basal gnathostome lineages, such as the placoderms, would be ideal to infer to which extent chondrichthyans are plesiomorphic for this specific trait. We favor a scenario in which the different components of the pectoral and pelvic appendages may have arisen from distinct evolutionary processes leading to the integration of homoplastic structures. Moreover, the morphological evolution of the pectoral and pelvic appendages may have been conditioned by the position where they develop along the body axis, which result in distinct muscular phenotypes, including the crucial difference of the strong musculature connecting the pectoral girdle and the cranium.

Myogenic progression during muscle development
Previous studies have suggested that various vertebrate groups share a temporal and spatial myogenic progression during the development of the cephalic muscles: from lateral/superficial to ventral/medial (outside-in), from origin to insertion, and from anterior to posterior [28, 32, 33]. Additionally, cephalic muscle differentiation seems to be tightly correlated with the development of cephalic cartilages, which was formerly described by

various authors [68–71]. However, it was not previously addressed whether a similar myogenic progression is also detected in chondrichthyans. Our data reveal that, in *S. canicula,* the lateral muscles of one arch differentiate before the ventral muscles of the same arch. Thus, cephalic muscle development occurs following a lateral to ventral myogenic progression, which resemble the process described in osteichthyans such as zebrafish, lungfish, amphibians, and birds [28–33, 70, 72]. This pattern is clearer in the branchial arches than in the mandibular and hyoid arches, with the mandibular arch being the only exception: the ventral *intermandibularis* develops before the other muscles of the first arch, which are more lateral.

In both the head and paired appendages of *S. canicula*, muscles normally develop from their region of origin to their region of insertion, as was previously reported for the cephalic musculature of other osteichthyans [28, 29, 32, 33, 70]. The only exceptions, within the cephalic muscles analyzed by us, are the *coracomandibularis, coracoarcualis* and the *coracobranchiales*, which developed in the head region from their adult region of insertion (mandible, ceratohyal, and branchial arches, respectively) and then extend posteriorly during development towards their adult region of origin (pectoral girdle). Only a few other exceptions to this origin-insertion myogenic progression were formerly described [72]. Muscles with attachments on these cartilages remain without other attachments until the formation of the cartilages that lie in the adult region of origin of these muscles (e.g., otic capsule, pterygoid bone). This indicates that head muscle development depends on the underlying skeletal development [71], and this is probably why we see such a pattern in the *coracomandibularis, coracoarcualis* and *coracobranchiales* of *S. canicula*, as the coracoid develops later than Meckel's cartilage, ceratohyal, and the branchial cartilages.

As most previous studies describing an anterior to posterior myogenic progression of the cephalic muscles of various non-chondrichthyan taxa (Table 2) differ in their methodology, we used fiber development as a criterion to access and compare order of development, and we grouped all muscles of the same pharyngeal arch into a single group. By doing this, one can consistently compare the data obtained for each taxon, and detect their developmental progression. All embryonic/larval amphibians shown in Table 2 develop their mandibular and hyoid arch muscles simultaneously and in most species (13 out of 20) these muscles also develop simultaneously with the first branchial arch muscles. In the amniote groups Aves and Theria the mandibular arch muscles clearly develop earlier than the hyoid arch muscles, which even develop later than the muscles of branchial arch I in Theria (Table 2). In *S. canicula* the mandibular muscles, in contrast, develop after the hyoid muscles,

while within the hyoid and branchial muscles the normal anteroposterior myogenic progression of muscle differentiation takes place. Furthermore, other branchial muscles develop following an anterior to posterior myogenic progression with the muscles associated to the last arch developing latest.

However, one should note that in other fishes there are also exceptions to the anteroposterior myogenic progression (Table 2), which makes it difficult to infer whether this pattern is even the most commonly found in non-tetrapod vertebrates. For instance, two studies of the Australian lungfish (*Neoceratodus forsteri*) found minor differences in the developmental pattern of cephalic muscles [32, 73] and, while other developmental studies of lungfishes exist, none of them mentions the timing to clarify this pattern. In the zebrafish, the mandibular arch muscles develop before the hyoid arch muscles [70] and a recent study of the Longnose Gar (*Lepisosteus osseus;* Actinopterygii) describes a simultaneous development of mandibular and hyoid muscles [74]. The *Polypterus senegalus* belongs to the Polypteriformes, which is the most basal extant actinopterygian family [75], and was also described as developing the hyoid muscles before other cranial muscles [76].

What can be inferred from the developmental studies on gnathostome muscle development, summarized here, is that the hyoid arch muscles develop and differentiate before the branchial arch muscles in non-amniote vertebrates, as also described in the results presented here. The order of appearance of the mandibular arch muscles in vertebrates seems to be more variable (Table 2). Unfortunately, no study of agnathans explicitly states the order of development of each cephalic muscle, which is required in the future, to investigate which pattern is plesiomorphic, and to discuss its implications for our understanding of the evolution of the musculature in vertebrates and gnathostomes.

Associations between ontogeny and phylogeny

Our previous works have indicated that in zebrafish and salamanders there is generally a parallelism between the order in which each cephalic muscle develops and the order in which each muscle was acquired during evolution ('phylo-ontogenetic' parallelism), barring only a few exceptions [28, 34]. A major problem with inferring a parallelism between the developmental order of appearance of muscles in sharks and the order in which the muscles appeared in evolution is that most of the muscles found in sharks and other gnathostomes are not present in any non-gnathostome extant taxon. Furthermore, there are insufficient detailed muscle reconstructions in fossils representing the transitions from agnathans to gnathostomes, making it difficult to infer the evolutionary order of appearance of the shark muscles.

Table 2 Relative order of cephalic muscle development in selected vertebrates (based on first appearance of myofibers). Sources of developmental description are shown in the right column. Muscles in the same box develop simultaneously. However, it should be mentioned that if M, H, B, appear simultaneously, it is almost always because only the most anterior one or two arches develop simultaneous with the M and H. **M** – Mandibular arch muscles, **H** – Hyoid arch muscles, **B** – Branchial arch muscles, **Hy** – Hypobranchial arch muscles, **L** – Laryngeal muscles, **A** – extrinsic ocular muscles

Species/Order of appearance	1	2	3	4	5	6	Source
Non-tetrapods							
Scyliorhinus canicula	HB	MAHy					THIS STUDY
Polypterus senegalus	H	MA	Hy	B			Noda et al. [76]
Danio rerio	MAHy	H	B				Schilling and Kimmel [70]
Neoceratodus forsteri	MH	BHy	A				Ericsson et al. [73]
N. forsteri	H	M	B	Hy	L	A	Ziermann [32]
Tetrapods							
Urodela							
Ambystoma mexicanum	MHBHy	L	A				Ziermann [32]
A. mexicanum	MHB						Ericsson and Olsson [33]
Ichthyotriton (Mesotriton) alpestris	MH	BHy	A				Ziermann [32]
Lissotriton vulgaris	MHHy	BL	A				Ziermann [32]
Necturus maculosus	MHBHy						Platt [78]
Anura							
Ascaphus truei	MHBHy	A	L				Ziermann [32]
Xenopus laevis	MHA	BHy	L				Ziermann and Olsson [31]
Hymenochirus boettgeri	MH	BHyL	A				Ziermann [32]
Discoglossus galganoi	MH	BHy	LA				Ziermann [32]
Discoglossus pictus	MHBHyL	A					Ziermann [32]
D. pictus	MHB						Schlosser and Roth [79]
Bombina orientalis	MH	B					Ziermann [32]
Bombina variegata	MH	B	L	HyA			Ziermann [32]
Pelodytes punctatus	MHB	L	Hy	A			Ziermann [32]
Pelobates fuscus	MHB	HyL	A				Ziermann [32]
Hyla cinerea	MHB	HyL	A				Ziermann [32]
Lepidobatrachus laevis	MHBHyL	A					Ziermann [32]
Bufo brongersmai	MH	BHy L	A				Ziermann [32]
Bufo speciosus	MHB	L	Hy				Ziermann [32]
Phrynomerus bifasciatus	MHB	HyA	L				Ziermann [32]
Kaloula pulchra	MHB	HyL	A				Ziermann [32]
Eleutherodactylus coqui	MHBHy						Schlosser and Roth [80]
Aves							
Gallus domesticus	A	M	H	B			Noden et al. [81]
Coturnix coturnix	M	HB					McClearn and Noden [72]
Theria							
Monodelphis domestica	Hy	MB	HL	A			Smith [82]
Mus musculus	Hy	M	B	H			Kaufman and Kaufman [83]

However, given the phylogenetic position of sharks and their plesiomorphic muscular structure among gnathostomes [36], our observation that mandibular muscles develop after hyoid muscles implies that this might have been the ancestral process in the LCA of the crown-group Gnathostomata. In line with this idea, the simultaneous development of these two muscular units observed in osteichthyans may reflect derivation from the ancestral process. Interestingly, Miyashita [77] recently proposed that the ancestral mandibular arch was distinct from the pharyngeal arches, and only became secondarily similar to those structures during the evolutionary process that culminated with the origin of gnathostomes. Thus, mandibular and hyoid structures may have arisen from independent developmental processes, which then converge becoming increasingly similar over time. The patterns of cephalic muscle development observed in S. canicula, and particularly the fact that the mandibular muscles develop later than those of more posterior arches, breaking the seemingly stable anteroposterior myogenic progression seen in these latter arches in most osteichthyan clades, can thus provide important insights for further studies on the origin and early evolution of the gnathostome jaws.

Conclusions

1. Our dissections and analysis of muscles in S. canicula are consistent with idea that there is an anatomical/functional association between the musculature associated with the pectoral girdle and that associated with the posterior branchial arches in the crown-group Gnathostomata. This contradicts the view that the pectoral and pelvic appendages are strict serial homologues in these animals. Instead, we favor a scenario in which the pectoral girdle musculature may have arisen from a nonhomologous process to the one involved in the origin of the pelvic musculature.

2. In both the head and paired appendages of S. canicula, muscles normally develop from their region of origin to their region of insertion. The only exceptions within all the cephalic muscles are the coracomandibularis, coracoarcualis, and the coracobranchiales, which develop from their adult region of insertion (mandible, ceratohyal, and branchial arches, respectively), and then extend posteriorly towards their adult region of origin (pectoral girdle). Furthermore, during cephalic muscle development, a lateral to ventral pattern can be observed, with the mandibular arch being the only one where there is an exception with a ventral muscle developing before the lateral ones. In S. canicula the mandibular arch muscles develop later than the hyoid muscles, while among the hyoid and branchial muscles one can observe an anteroposterior myogenic progression. Even with the exceptions described here, cranial muscle development appears to be highly conserved in gnathostomes.

3. In the chondrichthyan species analyzed here, the mandibular muscles develop later than the hyoid muscles as was also described for P. senegalus [76], which is a member of the most basal extant actinopterygian group Polypteriformes [75]. In contrast, in most osteichthyans the mandibular muscles develop at the same time, or even earlier, then the hyoid muscles. A parallelism between ontogeny and phylogeny could be established if future studies provide further evidence consistent with Miyashita's recent idea [77] that the mandibular arch was originally not integrated with or was not similar to the ancestral pharyngeal arches, and only became secondarily integrated with/similar to them in the transitions that lead to the LCA of crown-group Gnathostomata.

Acknowledgements

The experiments were conducted by RF in the Martin J. Cohn Lab (University of Florida) and we are grateful for his contribution providing insight and expertise that greatly assisted the research. We also acknowledge Raul Guizzo, as well as two anonymous reviewers, for carefully reviewing a previous version of this paper, as well as numerous colleagues for discussions on the subjects presented here.

Funding

RF was funded by FCT – Fundação para a Ciência e a Tecnologia (FCT), Portugal. JMZ and RD were funded by the Howard University College of Medicine. Funding sources did not influence the design of the study, the collection, analysis, or interpretation of data, or the writing of the manuscript.

Authors' contributions

RD and JMZ designed the study. RF conducted the experiments and JMZ dissected specimens in RD's Lab. JMZ and RF analyzed the specimen regarding muscle development. RF provided Figs. 1 and 2, and JMZ Figs. 4, 5 and 6. JMZ, RF, and RD discussed the data and wrote the manuscript. All authors read and approved the final manuscript.

Competing interests

The authors declare that they have no competing interests.

Author details

[1]Department of Anatomy, Howard University College of Medicine, 520 W St NW, Washington, DC 20059, USA. [2]IBMC—Institute for Molecular and Cell Biology, Oporto, Portugal. [3]I3S, Institute for Innovation and Health Research, University of Oporto, Oporto, Portugal. [4]Department of Anatomy, Howard University College of Medicine, Washington, DC 20059, USA.

References

1. Mallatt J. The origin of the vertebrate jaw: neoclassical ideas versus newer, development-based ideas. Zoo Sci. 2008;25:990–8.
2. Coates MI. Palaeontology: beyond the age of fishes. Nature. 2009;458:413–4.
3. Janvier P. Early vertebrates: Oxford University Press; 1996.
4. Brazeau MD. The braincase and jaws of a Devonian 'acanthodian'and modern gnathostome origins. Nature. 2009;457:305–8.
5. Cole NJ, Currie PD. Insights from sharks: evolutionary and developmental models of fin development. Dev Dyn. 2007;236:2421–31.
6. Diogo R, Johnston P, Molnar JL, Esteve-Altava B. Characteristic tetrapod musculoskeletal limb phenotype emerged more than 400 MYA in basal lobe-finned fishes. Sci Rep. 2016;6:37592.
7. Johanson Z. Placoderm branchial and hypobranchial muscles and origins in jawed vertebrates. J Vert Paleo. 2003;23:735–49.
8. Carr RK, Johanson Z, Ritchie A. The phyllolepid placoderm Cowralepis mclachlani: insights into the evolution of feeding mechanisms in jawed vertebrates. J Morph. 2009;270:775–804.
9. Freitas R, Zhang G, Cohn MJ. Biphasic Hoxd gene expression in shark paired fins reveals an ancient origin of the distal limb domain. PLoS One. 2007;2:e754.
10. Freitas R, Zhang G, Cohn MJ. Evidence that mechanisms of fin development evolved in the midline of early vertebrates. Nature. 2006;422:1033–7.
11. Tanaka M, Münsterberg A, Anderson WG, Prescott AR, Hazon N, Tickle C. Fin development in a cartilaginous fish and the origin of vertebrate limbs. Nature. 2002;416:527–31.
12. Neyt C, Jagla K, Thisse C, Thisse B, Haines L, Currie P. Evolutionary origins of vertebrate appendicular muscle. Nature. 2000;408:82–6.
13. Yonei-Tamura S, Abe G, Tanaka Y, Anno H, Noro M, Ide H, et al. Competent stripes for diverse positions of limbs/fins in gnathostome embryos. Evol Dev. 2008;10:737 45.

14. Kuratani S. Evolutionary developmental studies of cyclostomes and the origin of the vertebrate neck. Dev Growth Diff. 2008;50:S189–94.

15. Kuratani S, Horigome N, Hirano S. Developmental morphology of the head mesoderm and reevaluation of segmental theories of the vertebrate head: evidence from embryos of an agnathan vertebrate, Lampetra japonica. Dev Biol. 1999;210:381–400.

16. Kuratani S, Nobusada Y, Horigome N, Shigetani Y. Embryology of the lamprey and evolution of the vertebrate jaw: insights from molecular and developmental perspectives. Phil Tr Royal Soc B. 2001;356:1615–32.

17. Kuratani S, Murakami Y, Nobusada Y, Kusakabe R, Hirona S. Developmental fate of the mandibular mesoderm in the lamprey, Lethenteron japonicum: comparative morphology and development of the Gnathostome jaw with special reference to the nature of the trabecula cranii. J Exp Zool Part B. 2004;302B:458–68.

18. Ziermann JM, Miyashita T, Diogo R. Cephalic muscles of cyclostomes (hagfishes and lampreys) and Chondrichthyes (sharks, rays and holocephalans): comparative anatomy and early evolution of the vertebrate head muscles. Zool J Lin Soc - London. 2014;172:771–802.

19. Cerny R, Cattell M, Sauka-Spengler T, Bronner-Fraser M, Yu F, Medeiros DM. Evidence for the prepattern/cooption model of vertebrate jaw evolution. PNAS. 2010;107:17262–7.

20. Medeiros DM, Crump JG. New perspectives on pharyngeal dorsoventral patterning in development and evolution of the vertebrate jaw. Dev Biol. 2012;371:121–35.

21. Diogo R, Abdala V. Muscles of vertebrates - comparative anatomy, evolution, homologies and development. New Hampshire: CRC Press; Science Publisher, Enfield; 2010.

22. Diogo R, Ziermann JM. Muscles of chondrichthyan paired appendages: comparison with osteichthyans, deconstruction of the fore–hindlimb serial homology dogma, and new insights on the evolution of the vertebrate neck. Anat Rec. 2015;298:513–30.

23. Edgeworth FH. The cranial muscles of vertebrates. London: Cambridge at the University Press; 1935.

24. Diogo R, Abdala V, Lonergan N, Wood BA. From fish to modern humans - comparative anatomy, homologies and evolution of the head and neck musculature. J Anat. 2008;213:391–424.

25. Diogo R, Abdala V, Aziz M, Lonergan N, Wood B. From fish to modern humans - comparative anatomy, homologies and evolution of the pectoral and forelimb musculature. J Anat. 2009;214:694–716.

26. Diogo R, Ziermann JM. Development, metamorphosis, morphology, and diversity: the evolution of chordate muscles and the origin of vertebrates. Dev Dyn. 2015;244:1046–57.

27. Cole NJ, Hall TE, Don EK, Berger S, Boisvert CA, Neyt C, et al. Development and evolution of the muscles of the pelvic fin. PLoS Biol. 2011;9:e1001168.

28. Ziermann JM, Diogo R. Cranial muscle development in the model organism Ambystoma mexicanum: implications for tetrapod and vertebrate comparative and evolutionary morphology and notes on ontogeny and phylogeny. Anat Rec. 2013;296:1031–48.

29. Ziermann JM, Diogo R. Cranial muscle development in frogs with different developmental modes: direct development vs. biphasic development. J Morph. 2014;275:398–413.

30. Ziermann JM, Mitgutsch C, Olsson L. Analyzing developmental sequences with Parsimov – a case study of cranial muscle development in anuran larvae. J Exp Zool Part B. 2014;322B:584–604.

31. Ziermann JM, Ölsson L. Patterns of spatial and temporal cranial muscle development in the African clawed frog, Xenopus laevis (Anura: Pipidaé). J Morph. 2007;268:791–804.

32. Ziermann JM. Evolutionäre Entwicklung larvaler Cranialmuskulatur der Anura und der Einfluss von Sequenzheterochronien, Biologisch-Pharmazeutischen Fakultät. Jena: Institut für Spezielle Zoologie und Evolutionsbiologie mit Phyletischem Museum, Friedrich Schiller University Jena; 2008. p. 347.

33. Ericsson R, Olsson L. Patterns of spatial and temporal visceral arch muscle development in the Mexican axolotl (Ambystoma mexicanum). J Morph. 2004;261:131–40.

34. Diogo R, Hinits Y, Hughes SM. Development of mandibular, hyoid and hypobranchial muscles in the zebrafish: homologies and evolution of these muscles within bony fishes and tetrapods. BMC Dev Biol. 2008;8:1–22.

35. Diogo R, Molnar JL, Smith TD. The anatomy and ontogeny of the head, neck, pectoral, and upper limb muscles of Lemur catta and Propithecus

Coquereli (primates): discussion on the parallelism between ontogeny and phylogeny and implications for evolutionary and developmental biology. Anat Rec. 2014;297:1435–53.

36. Trinajstic K, Sanchez S, Dupret V, Tafforeau P, Long J, Young G, et al. Fossil musculature of the most primitive jawed vertebrates. Science. 2013;341:160–4.

37. Gegenbaur C. Elements of comparative anatomy, Macmillan and Company, 1878.

38. Vicq-d'Azyr F. Parallèle des os qui composent les extrémités. Mém Acad Sci. 1774;1774:519–57.

39. Oken L. Lehrbuch der Naturphilosophie, Schultheß, 1843.

40. Owen R. On the nature of limbs: a discourse delivered on Friday, February 9, at an evening meeting of the Royal Institution of great Britain, John Van Voorst, 1849.

41. Kerr JG. Text-book of embryology. Vol. II: vertebrata with the exception of mammalia. London: Macmillan and Co. Limited; 1919.

42. Coates MI. The origin of vertebrate limbs. Development. 1994;1994:169–80.

43. Ruvinsky I, Gibson-Brown JJ. Genetic and developmental bases of serial homology in vertebrate limb evolution. Development. 2000;127:5233–44.

44. Young NM, Hallgrimsson B. Serial homology and the evolution of mammalian limb covariation structure. Evolution. 2005;59:2691–704.

45. Abbasi AA. Evolution of vertebrate appendicular structures: insight from genetic and palaeontological data. Dev Dyn. 2011;240:1005–16.

46. Dahn RD, Davis MC, Pappano WN, Shubin NH. Sonic hedgehog function in chondrichthyan fins and the evolution of appendage patterning. Nature. 2007;445:311–4.

47. Gillis JA, Dahn RD, Shubin NH. Shared developmental mechanisms pattern the vertebrate gill arch and paired fin skeletons. PNAS. 2009;106:5720–4.

48. Gillis JA, Hall BK. A shared role for sonic hedgehog signalling in patterning chondrichthyan gill arch appendages and tetrapod limbs. Development. 2016;143:1313–7.

49. Diogo R, Linde-Medina M, Abdala V, Ashley-Ross MA. New, puzzling insights from comparative myological studies on the old and unsolved forelimb/hindlimb enigma. Biol Rev. 2013;88:196–214.

50. Diogo R, Ziermann JM. Development of fore- and hindlimb muscles in frogs: morphogenesis, homeotic transformations, digit reduction, and the forelimb-hindlimb enigma. J Exp Zool Part B. 2014;322B:86–105.

51. Diogo R, Molnar J. Comparative anatomy, evolution, and homologies of Tetrapod Hindlimb muscles, comparison with forelimb muscles, and deconstruction of the forelimb-Hindlimb serial homology hypothesis. Anat Rec. 2014;297:1047–75.

52. Ballard WW, Mellinger J, Lechenault H. A series of normal stages for development of Scyliorhinus canicula, the lesser spotted dogfish (Chondrichthyes: Scyliorhinidae). J Exp Zool. 1993;267:318–36.

53. Freitas R, Cohn MJ. Analysis of EphA4 in the lesser spotted catshark identifies a primitive gnathostome expression pattern and reveals co-option during evolution of shark-specific morphology. Dev Genes Evol. 2004;214:466–72.

54. de Jong IML, Colbert MW, Witte F, Richardson MK. Polymorphism in developmental timing: intraspecific heterochrony in a Lake Victoria cichlid. Evol Dev. 2009;11:625–35.

55. Bininda-Emonds ORP, Jeffery JE, Sánchez-Villagra MR, Hanken J, Colbert M, Pieau C, et al. Forelimb-hindlimb developmental timing across tetrapods. BMC Evol Biol. 2007;7:1–7. doi:10.1186/1471-2148-7-182.

56. Yamanoue Y, Setiamarga D, Matsuura K. Pelvic fins in teleosts: structure, function and evolution. J Fish Biol. 2010;77:1173–208.

57. Molnar J, Johnston P, Esteve-Altava B, Diogo R. Musculoskeletal anatomy of the pelvic fin in Polypterus and the phylogenetic distribution and homology of pre- and postaxial pelvic appendicular muscles. J Anat. 2017;230:532–41.

58. Coates MI, Cohn MJ. Fins, limbs, and tails: outgrowths and axial patterning in vertebrate evolution. BioEssays. 1998;20:371–81.

59. Coates MI. The evolution of paired fins. Theory Biosci. 2003;122:266–87.

60. Tulenko FJ, McCauley DW, MacKenzie EL, Mazan S, Kuratani S, Sugahara F, et al. Body wall development in lamprey and a new perspective on the origin of vertebrate paired fins. PNAS. 2013;110:11899–904.

61. Adachi N, Robinson M, Goolsbee A, Shubin NH. Regulatory evolution of Tbx5 and the origin of paired appendages. Proc Natl Acad Sci U S A. 2016; 113(36):10115–20.

62. Pascual-Anaya J, Albuixech-Crespo B, Somorjai IML, Carmona R, Oisi Y, Álvarez S, et al. The evolutionary origins of chordate hematopoiesis and vertebrate endothelia. Dev Biol. 2013;375:182–92.

63. Shubin N, Tabin C, Carroll S. Deep homology and the origins of evolutionary novelty. Nature. 2009;457:818–23.

64. Diogo R, Kelly RG, Christiaen L, Levine M, Ziermann JM, Molnar JL, et al. A new heart for a new head in vertebrate cardiopharyngeal evolution. Nature. 2015;520:466–73.

65. Lescroart F, Hamou W, Francou A, Théveniau-Ruissy M, Kelly RG, Buckingham M. Clonal analysis reveals a common origin between nonsomite-derived neck muscles and heart myocardium. PNAS. 2015;112:1446–51.

66. Theis S, Patel K, Valasek P, Otto A, Pu Q, Harel I, et al. The occipital lateral plate mesoderm is a novel source for vertebrate neck musculature. Development. 2010;137:2961–71.

67. Sefton EM, Bhullar B-AS, Mohaddes Z, Hanken J. Evolution of the head-trunk interface in tetrapod vertebrates. elife. 2016;5:e09972.

68. Hacker A, Guthrie S. A distinct developmental program for the cranial paraxial mesoderm in the chick embryo. Development. 1998;125:3461–72.

69. Noden DM. The embryonic origins of avian cephalic and cervical muscles and associated connective tissues. Am J Anat. 1983;168:257–76.

70. Schilling TF, Kimmel CB. Musculoskeletal patterning in the pharyngeal segments of the zebrafish embryo. Development. 1997;124:2945–60.

71. Noden DM. The role of the neural crest in patterning of avian cranial skeletal, connective, and muscle tissues. Dev Biol. 1983;96:144–65.

72. McClearn D, Noden DM. Ontogeny of architectural complexity in embryonic quail visceral arch muscles. Am J Anat. 1988;183:277–93.

73. Ericsson R, Joss J, and Olsson L. Early head development in the Australian lungfish, *Neoceratodus forsteri*. In: Jørgensen JM, Joss J, editors. Enfield: The Biology of Lungfishes, Science publisher, CRC Press, Tailor and Francis; 2010. pp. 149ff.

74. Konstantinidis P, Warth P, Naumann B, Metscher B, Hilton EJ, Olsson L. The developmental pattern of the musculature associated with the mandibular and hyoid arches in the Longnose gar, *Lepisosteus osseus* (Actinopterygii, Ginglymodi, Lepisosteiformes). Copeia. 2015;103:920–32.

75. Inoue JG, Miya M, Tsukamoto K, Nishida M. Basal actinopterygian relationships: a mitogenomic perspective on the phylogeny of the "ancient fish". Mol Phylogenet Evol. 2003;26:110–20.

76. Noda M, Miyake T, Okabe M. Development of cranial muscles in the actinopterygian fish Senegal bichir, Polypterus senegalus Cuvier. J Morph. 1829:2017.

77. Miyashita T. Fishing for jaws in early vertebrate evolution: a new hypothesis of mandibular confinement. Biol Rev. 2015; doi:10.1111/brv.12187.

78. Platt JB. The development of the cartilaginous skull and of the branchial and hypoglossal musculature in *Necturus*. Morphol Jb. 1898;25:377–463.

79. Schlosser G, Roth G. Evolution of nerve development in frogs I: the development of the peripheral nervous system in *Discoglossus pictus* (Discoglossidae). Brain Behav Evol. 1997;50:62–93.

80. Schlosser G, Roth G. Evolution of nerve development in frogs II: modified development of the peripheral nervous system in the direct-developing frog *Eleutherodactylus coqui* (Leptodactylidae). Brain Behav Evol. 1997;50:94–128.

81. Noden DM, Marcucio R, Borycki A-G, Emerson CP, JR. Differentiation of avian craniofacial muscles: 1. Patterns of early regulatory Gene expression and myosin heavy chain synthesis. Dev Dyn. 1999;216:96–112.

82. Smith KK. Development of craniofacial musculature in *Monodelphis domestica* (Marsupialia, Didelphidae). J Morph. 1994;222:149–73.

83. Kaufman MH, and Kaufman MH. The atlas of mouse development, Academic press San Diego. 1992.

3

How animals distribute themselves in space: variable energy landscapes

Juan F. Masello[1][*] [iD], Akiko Kato[2], Julia Sommerfeld[1], Thomas Mattern[1] and Petra Quillfeldt[1]

Abstract

Background: Foraging efficiency determines whether animals will be able to raise healthy broods, maintain their own condition, avoid predators and ultimately increase their fitness. Using accelerometers and GPS loggers, features of the habitat and the way animals deal with variable conditions can be translated into energetic costs of movement, which, in turn, can be translated to energy landscapes. We investigated energy landscapes in Gentoo Penguins *Pygoscelis papua* from two colonies at New Island, Falkland/Malvinas Islands.

Results: In our study, the marine areas used by the penguins, parameters of dive depth and the proportion of pelagic and benthic dives varied both between years and colonies. As a consequence, the energy landscapes also varied between the years, and we discuss how this was related to differences in food availability, which were also reflected in differences in carbon and nitrogen stable isotope values and isotopic niche metrics. In the second year, the energy landscape was characterized by lower foraging costs per energy gain, and breeding success was also higher in this year. Additionally, an area around three South American Fur Seal *Arctocephalus australis* colonies was never used.

Conclusions: These results confirm that energy landscapes vary in time and that the seabirds forage in areas of the energy landscapes that result in minimized energetic costs. Thus, our results support the view of energy landscapes and fear of predation as mechanisms underlying animal foraging behaviour. Furthermore, we show that energy landscapes are useful in linking energy gain and variable energy costs of foraging to breeding success.

Keywords: Energetic costs, Energy landscape, Foraging effort, Foraging strategy, Landscape of fear, Ecological mechanism, Movement ecology, Non-lethal effects of predation, Tri-axial acceleration, Variable costs of foraging

Background

Animals do not distribute themselves randomly. An extensive literature on wild animal movements and habitat use shows that some locations are highly used, while other nearby locations are avoided [1–6]. Understanding the behavioural decisions that makes a place a foraging 'hot-spot' as compared to a corridor or even a no-go area will be crucial for securing safe spaces for wild animals facing expanding human influence [7] and climate change [8]. Optimal foraging theory [9, 10] predicts that animals will select patches abundant in resources where the gain per unit cost is high. Any unnecessarily extensive movements might increase the risk of predation, and thus, predator avoidance also influences the movements of many animals [5, 11, 12].

In addition to the description of the movement of organisms (e.g. [13]), it is important to consider movements in the context of ecological factors [5, 14–16]. Foraging costs have usually been investigated in terms of time, energy gained or energy consumed [17–19]. However, even minor landscape features may directly affect animal movements by imposing considerable energy barriers on travel [7]. Likewise, the degree of variation in the landscape will account for variable energy cost of movements [20], which can be translated into an energy landscape for animals foraging in it [21, 22]. Consequently, in landscapes where resources are not distributed in a way that resembles the energy landscape, animals will forage in areas of the energy landscape that result in minimized costs and maximised net energetic gain [21]. This prediction has been supported by studies that investigated foraging movements through energy landscapes using animal-attached devices to derive the energetic costs of foraging

* Correspondence: juan.f.masello@bio.uni-giessen.de
[1]Department of Animal Ecology & Systematics, Justus Liebig University Giessen, Heinrich-Buff-Ring 26, D-35392 Giessen, Germany
Full list of author information is available at the end of the article

[7, 21–24]. In marine environments or "seascapes", oceanographic conditions and currents vary over time related to oceanographic cycles and climate change [25–30], resulting in changes in food availability and distribution and thus, in energy landscapes. Such temporal changes of energy landscapes (between or within years) and their consequences on animal behaviour have not been investigated to date. Filling such a gap in our knowledge is particularly relevant in the context of climate change.

Seabirds have evolved a multitude of foraging strategies in order to successfully prey on marine food, such as species-specific preferences of prey or the use of open-ocean versus coastal habitats [16, 31, 32]. During the breeding season, seabirds are central-place foragers, exploiting resources within a given range around their colonies or nests [18, 33, 34]. In a previous study, we investigated simultaneous ecological segregation among species and colonies of a diving seabird assemblage, sharing a sector of the south-western Atlantic Ocean during the breeding season [5]. In that study, we deployed GPS-temperature-depth (GPS-TD) loggers on Gentoo, Rockhopper, and Magellanic penguins (*Pygoscelis papua*, *Eudyptes chrysocome*, *Spheniscus magellanicus*), and Imperial Shags (*Phalacrocorax atriceps*) breeding at New Island, Falkland / Malvinas Islands, during the breeding season. Because the studied seabird colonies at New Island were much closer to each other (2–7 km) than the average foraging range of the species (9–27 km), we expected large overlaps among the foraging areas. However, we found little, if any, overlap due to strong spatial and temporal segregation [5]. Particularly striking, we observed strong differences in foraging areas, diving depth, time of foraging and prey choice among birds of the same species, breeding in different colonies at the same island [5]. We concluded that the observed differences were most likely caused by optimal foraging of individuals in relation to habitat differences on a local scale, leading to a complex pattern of interactions with environmental covariates, combined with avoidance of predation [5]. Such a flexible foraging strategy was also observed in Gentoo Penguins from Antarctica, where differences were found among years [29, 35]. Flexible foraging habits would provide a buffer against changes in prey availability [29].

In the present study, we investigated the mechanisms behind the flexible foraging strategies in Gentoo Penguins. During two different years, using two colonies of Gentoo Penguins that previously showed strong spatial and temporal segregation [5], and GPS and tri-axial acceleration data for the calculation of energetic costs of movement [21], we aim to show that 1) energy landscapes vary in time (e.g. between breeding seasons) resembling the interaction between foraging effort and prey availability, 2) the

seabirds will forage in areas of the energy landscapes that result in minimized energetic costs, 3) as central-place foragers are constraint in the area where they can forage, temporal changes in the energy landscape and associated changes in energy costs of foraging will affect the breeding success.

Methods

Study site and species

The study was conducted at New Island Nature Reserve in the Falkland Islands / Islas Malvinas, south-western Atlantic Ocean [36, 37]. At the continental slope, the Falkland Current generates a strong upwelling of productive Sub-Antarctic superficial water ([37] and references therein). This area of increased productivity attracts many seabird species, 13 of which breed in colonies distributed over New I. [38]. Among them is the Gentoo Penguin, which we investigated in two breeding colonies: one at the North End (51° 41.402′ S 61° 15.003′ W), and one at the South End (51° 44.677′ S 61°17.683′ W; Fig. 1) of New Island.

In a previous study, we found complete spatial segregation between these two colonies of Gentoo Penguins, regardless of their proximity (7 km apart), during the studied breeding season (chick guard 2008) [5]. Our study also showed that Gentoo Penguins started foraging very close (from 0.4 to 2.5 km) to the breeding colonies [5]. Gentoo Penguins have been found to be neritic foragers during the breeding season and among the main avian benthic consumers of the sub-Antarctic area, their diet varying greatly between locations and in time [32, 39]. Miller et al. [29] and Handley et al. [40] found that the prey of Gentoo Penguins comprised mainly benthic prey but regularly included pelagic prey. An earlier study of Gentoo Penguins at New I. [41, 42] was in line with these findings, as the diet comprised mainly lobster krill (*Munida gregaria*; 56%), followed by both benthic and pelagic fish (main items: *Micromesistius* sp., Nototheniidae and Perciformes; 34%) and squid (mainly *Gonatus antarcticus*; 9%) in 1986/87. For the North End colony at New I., Clausen et al. [43] found that Gentoo Penguins foraged mainly on pelagic prey (*Sprattus fuegensis*). In East Falkland, the principal prey items during chick guard were rock cod *Patagonotothen* spp. (78% in 2012), and Patagonian longfin squid (*Loligo gahi*) (7% in 2012) [40].

Instrumentation and fieldwork procedures

GPS-temperature-depth (GPS-TD; earth & Ocean Technologies, Kiel, Germany) and micro tri-axial accelerometer (Axy; TechoSmArt Europe, Rome, Italy) loggers were simultaneously deployed on 32 Gentoo Penguins from the South End and North End colonies during chick guard (December) in 2013 and 2014 (Table 1). We were not allowed to work on the North End colony

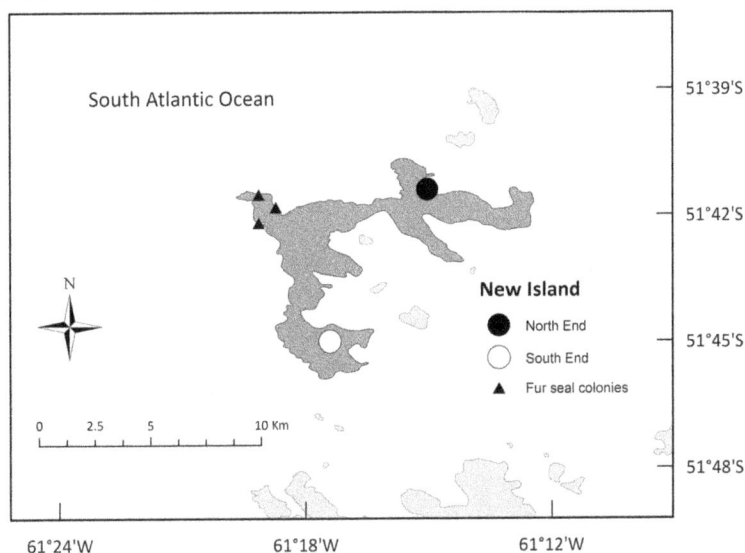

Fig. 1 Location of the Gentoo Penguin *Pygoscelis papua* colonies studied. New Island (in *dark grey*) is located in the Falkland Islands/Islas Malvinas, Southwestern Atlantic. South American Fur seal *Arctocephalus australis* colonies are indicated with black triangles. See bathymetric map in Additional File 1: Figure S1

during 2013 due to the activities of a film crew. No loggers were deployed in days of bad weather conditions in order to ensure an effective protection of the chicks and the adult birds. Birds were captured mostly by hand, in the vicinity of their nests, with the occasional help of a hook attached to a rod [44]. Chicks were also captured to protect them from predators like Brown Skuas *Catharacta antarctica* and Striated Caracaras *Phalcoboenus australis* during the handling of the adult. Handling time was kept to a minimum, mostly below 15 min and always

Table 1 Parameters of foraging trips used for the calculations of energy landscapes. The data correspond to Gentoo Penguin *Pygoscelis papua* breeding at New Island (Falkland/Malvinas Is.), during chick guard (December) in 2013 and 2014. Only the first foraging trip of each individual was included in the calculations in order to avoid individuals with more than one trip having more weight in the analyses

	2013	2014	
	South End	South End	North End
Individuals tagged	16	8	8
Number of complete data sets obtained (first foraging trips)	13	4	6
Trip length [km]	131.1 ± 59.1 (67.0–281.7)	92.7 ± 64.7 (24.1–169.4)	56.9 ± 13.7 (33.8–75.5)
t-test between seasons	t = 1.113	P = 0.283	
Mann-Whitney Rank Test between colonies		T = 24.000	P = 0.749
Maximum distance from colony [km]	69.1 ± 9.8 (51.3–87.6)	49.6 ± 33.0 (13.8–89.3)	33.3 ± 17.3 (15.3–60.1)
Mann-Whitney Rank Test between seasons	T = 26.000	P = 0.282	
t-test between colonies		t = −1.038	P = 0.330
Trip duration [min]	1811.5 ± 754.4 (770.6–2965.1)	1636.6 ± 1162.8 (320.7–3066.6)	1183.0 ± 353.4 (798.2–1650.8)
t-test between seasons	t = 0.129	P = 0.725	
Mann-Whitney Rank Test between colonies		T = 25.000	P = 0.610
Start time of foraging (local time)	07:14:53 ± 06:14:24 (02:12:13–19:16:23)	15:07:12 ± 04:50:53 (08:10:50–18:38:50)	10:22:05 ± 08:05:17 (02:47:49–20:26:48)
Mann-Whitney Rank Test between seasons	T = 54.000	P = **0.048**	
t-test between colonies		t = 1.044	P = 0.327

Sample sizes vary with respect to deployments, as not all parameters could be calculated for all individuals, mainly due to some batteries running out before the finalization of an ongoing trip. Statistically significant values are marked bold

below 20 min. Extreme care was taken to minimize stress to the captured birds, with the head covered during handling in order to minimize the risk of adults regurgitating. During this procedure no great signs of stress were apparent: none of the birds regurgitated. The attachment of the loggers on the adult penguin was carried out using adhesive Tesa® 4651 tape as described by Wilson et al. [45]. Both loggers (GPS-TD: 75 to 145 g; Axy: 19 g) represent a maximum of 2.5% of the adult body mass (mean 6459 ± 172 g, n = 16) [5]. In a previous study [46], we showed that handling and short-term logger attachments like the ones here carried out showed limited effect on the behaviour and physiology of the birds. Other studies have also found no negative effects of similar GPS-loggers in the foraging behaviour or the breeding success of the birds [47–50]. GPS-TD loggers recorded detailed position (longitude, latitude; sampling interval: 5 min), dive depth (resolution: 3.5 cm; sampling interval: 1 s), and time of day. While at sea, GPS functionality was pressure controlled so as to attempt to obtain a GPS fix upon resurfacing from dives. The Axy loggers recorded acceleration (sampling interval: 50 Hz) measured in three directions (x, y, z, i.e. surge, sway, heave) (e.g. [51]). After the deployment procedure and immediately before the release of the adult bird, chicks were returned to the nest. The adults were released some 20 m from their nests. All birds returned to their nests and attended their chicks shortly after being released.

The birds were recaptured in the vicinity of their nests after 2 to 12 days (median: 5 d) of logger deployment. All birds were recaptured and loggers recovered except in one case. Despite intensive efforts, we were not able to recapture one bird tagged in the South End colony in December 2014. It may be possible that the penguin abandoned the nest or that it was predated, as several Southern Sea Lions *Otaria flavescens* were intensively hunting at the penguin landing place during the deployment period. We observed several cases of Gentoo Penguin predation by sea lions while waiting for our tagged birds to return to the colony. Surprisingly, the two chicks belonging to the nest with the missing penguin developed normally, suggesting that they were adequately provisioned by the remaining parent. In any case, the unrecovered device was lost, at latest, during the natural moulting period (shortly after the breeding season) preventing any long-term consequences for the bird.

After logger recovery, the penguins were released as described above. All birds returned to their nests and attended their chicks shortly after being released except in one case. In this instance, the adult penguin took longer than usual to return to its nest and two Striated Caracaras predated the two chicks. No other cases of nest desertion were recorded and all chicks survived at least until the starting of the crèche period, a time when we were not able to identify individual chicks anymore.

Spatial and temporal data

From 32 deployments in this study, we obtained 23 complete sets of tri-axial acceleration and GPS data, comprising location, time, and dive depth, which we used in the following analyses (Tables 1 and 2). Failures to produce complete data sets were due to 1) three GPS-TD loggers fully damaged by salt water reaching the electronic components, 2) two broken GPS antennas, and 3) four batteries that were unexpectedly depleted before the end of the first foraging trips. In 2013, seven Axy loggers were damaged by salt water but the data could be recovered. In 2014, all Axy loggers were recovered without any damage, as the logger coating was purposely reinforced by TechoSmArt and, additionally, the units were placed inside a tightly closed finger of a lab glove and then inside a heat-shrink tubing before deployment.

As in previous studies (e.g. [5]), we defined foraging trips from the time when the birds departed from the colony to the sea until returning to the colony. Bathymetry data were obtained from the global sea floor topography from satellite altimetry and ship depth soundings (Global Topography; Additional file 1: Figure S1) [52] available at [53]. Positional data obtained from GPS-TD-loggers were used to plot and analyse the trips performed by the birds in ArcGIS 9.3 (ESRI, Redlands, USA). Trip length was calculated as the total cumulative linear distance between all positional fixes along the foraging trip, outside of the colony. For each trip, the maximum distance from the colony was calculated as the linear grand circle distance between the furthest point of the plotted trip and the geographical coordinates of the departure colony, determined by GPS. Trip duration was determined as the time elapsed between departure and return from the colony. Foraging dives were identified using purpose-written software in Matlab (The Mathworks Inc., Nattick, USA) and purpose-written script for IGOR Pro 6.3.7.2 (WaveMetrics, Lake Oswego, USA). Following Mattern et al. (2007) dive events could only be accepted when depths >3 m were reached. The bottom phase was defined as a period of the dive with little vertical undulation following a steady descent and before a steady ascent back to the surface [50, 54]. The maximum depth (in m) reached during a dive event (hereafter event maximum depth), and the number of dive events during a particular foraging trip were also calculated (Table 2). For each dive, we calculated a geographical position either by using the half way point between GPS fixes recorded immediately before and after the dive, or by calculating the relative position along a linear interpolated line between the last fix obtained and before the first fix

Table 2 Dive parameters used for the calculations of energy landscapes corresponding to Gentoo Penguin *Pygoscelis papua*. The study was conducted on penguins breeding at New Island (Falkland/Malvinas Is.), during chick guard (December) in 2013 and 2014. Only the first foraging trip of each individual was included in the calculations in order to avoid individuals with more than one trip having more weight in the analyses. For sample sizes see Table 1. For means, ranges are given in brackets, while for medians 75 and 25% quartiles are given

	2013	2014	
	South End	South End	North End
Maximum dive depth [m]	188.3	178.2	156.3
Mean number of dives per foraging trip (MND)	298 (176–674)	265 (81–648)	280 (192–343)
Mann-Whitney Rank Test between seasons	T = 50.000	P = 0.405	
Mann-Whitney Rank Test between colonies		T = 36.000	P = 0.445
Mean dive duration (DD), benthic dives [s]	166 (112–215)	175 (145–244)	180 (125–213)
Mann-Whitney Rank Test between seasons	T = 96	**P = 0.002**	
Mann-Whitney Rank Test between colonies		t = 0.367	P = 0.721
Mean dive duration (DD), pelagic dives [s]	109 (87–158)	118 (112–140)	123 (108–146)
Mann-Whitney Rank Test between seasons	t = −1.610	P = 0.126	
Mann-Whitney Rank Test between colonies		t = −0.409	P = 0.690
Median dive event maximum depth [m]	21.9 (8.0–97.1)	45.1 (14.2–93.2)	45.2 (16.0–91.2)
Mann-Whitney Rank Test between seasons	T = 54,929,247.5	**P < 0.001**	
Mann-Whitney Rank Test between colonies		T = 34,821,241.0	P = 0.985
Median dive depth of pelagic dives [m]	15.8 (6.3–77.0)	12.7 (5.8–41.2)	21.1 (9.2–48.9)
Mann-Whitney Rank Test between seasons	T = 1,203,123.5	**P < 0.001**	
Mann-Whitney Rank Test between colonies		T = 834,201.5	**P < 0.001**
Mean proportion of benthic dives (pBD) [%]	24 (10–40)	54 (22–72)	48 (30–76)
Mean proportion of pelagic dives (pPD) [%]	76 (61–90)	46 (27–78)	52 (24–70)
t-test between seasons	t = −3.828	**P = 0.002**	
t-test between colonies		t = −0.426	P = 0.678
Minimum benthic bottom time (mBBT) [s]	2	3	2

Statistically significant values are marked bold

after the dive occurred based on the time the dive occurred relative to these fixes.

As Gentoo Penguins were found to take both benthic and pelagic prey at the Falkland Islands [5, 40], the foraging dives performed by the individuals were split in benthic and pelagic ones for further analyses. This was done by calculating an index of benthic diving behaviour developed by Tremblay and Cherel [54]. This method assumes that benthic divers dive serially to a specific depth, and therefore consecutive dives reach the same depth zone. These are called intra-depth zone (IDZ) dives [54]. As in previous studies, the IDZ was defined as the depth ± 10% of the maximum depth reached by the preceding dive [16, 55]. During the current study, Gentoo Penguins performed a varying proportion of benthic and pelagic dives, which was taken into account in the following analyses (Table 2). As the inspection of histograms showed that the data for pelagic dives was left shifted, the median dive depth per colony per year was used for further calculations involving pelagic dives (Table 2; see Additional file 1: Figure S2; see also

'Calculation of energy'). The geographical location of benthic and pelagic dives was checked in order to detect any potential bias in the distribution of the data. Benthic and pelagic dives were distributed evenly in the same depth areas of the ocean around New I. (see Additional file 1: Figures S3, S4). We also calculated the mean number of dives performed during the foraging trips (Table 1). In all calculations, only the first foraging trip of each individual was included in order to avoid individuals with more than one trip having more weight in the data. In a previous study [5], we found that the Gentoo Penguin from New I. showed no sexual differences in foraging behaviour parameters. Therefore, in this study, we pooled the data of males and females.

The nonparametric fixed kernel density estimator was used to determine the 20, 40, 60 and 80% density contour areas (estimated foraging range) [56] of dive locations (i.e. GPS position at the onset of a dive event). Kernel densities indicate the places in a foraging trip where birds spent most of their time [56]. Hawth's Analysis Tools [57] in ArcGIS 9.3 were used to estimate a fixed kernel density using the quartic approximation of a true Gaussian kernel function [57]. GPS data-points at the colonies were excluded in order to avoid an overestimation of their importance.

When normality and equal variance tests passed (all $P > 0.05$), we used t-tests implemented in R to test for differences between colonies and seasons on the calculated trip and dive parameters (Tables 1 and 2) [58]. In cases where normality and equality of variance were not satisfied ($P < 0.05$), we used Mann-Whitney rank sum tests in order to investigate differences.

Calculation of energy

Using a purpose-written script for IGOR Pro 6.3.7.2 (WaveMetrics, Lake Oswego, USA) and tri-axial acceleration data from Axy accelerometers, we calculated the Overall Dynamic Body Acceleration (ODBA) for all first foraging trips and individuals. ODBA is a linear proxy for metabolic energy that can be further converted into energy expenditure (e.g. [51, 59–63] but see [64]). ODBA (expressed as gravitational force g) was calculated as described in Wilson et al. [21]. We used the sum of the absolute values of dynamic acceleration from each of the three spatial axes (i.e. surge, sway, and heave) after subtracting the static acceleration (= smoothed acceleration) from the raw acceleration values [21]:

$$ODBA = |Ax| + |Ay| + |Az| \qquad (1)$$

A_x, A_y and A_z are the derived dynamic accelerations at any point in time corresponding to the three orthogonal axes.

The sum of ODBA during dives was related to the maximum dive depth (see Additional file 1: Figures S5–S10). However, a general additive model (GAM; see Additional file 1: Table S1) revealed that this relationship differed between studied years, colonies, and between benthic and pelagic dives. Thus, the regressions with the best fit were determined for the different combination of years, colonies and dive types in SigmaPlot 10.0 (Systat Software, San Jose, USA; see Additional file 1: Table S2; Figures S5–S10). We used the regressions between the sum of ODBA during the dive of the deployed penguins and the maximum dive depth (see Additional file 1: Table S2), together with the bathymetric data points from the Global Topography [52] to calculate benthic ODBAs for a grid of the marine area around New I. (approximately 100 km around the island; $n = 26,196$) separate for each colony and season. For the pelagic ODBAs, we used the corresponding regressions (see Additional file 1: Table S2) and the median dive depth per colony per year (Table 1; see 'Analyses of spatial and temporal data' for method validation).

The distance between each point in the marine area grid around New I. for which bathymetric data were available (see Additional file 1: Figure S1) and the Gentoo Penguin breeding colonies on New I. was calculated with the Hawth's Analysis Tools [57] in ArcGIS 9.3. Using this distance and the mean swimming speed previously calculated for Gentoo Penguins (2.3 m s^{-1}) [65], we were able to calculate the travel time needed for the birds to reach each of the 26,196 locations around New I. for which bathymetric data were available. The travel time (TT, in s), and their minimum metabolic cost of transportation (16.1 W kg^{-1}) [65], allowed us subsequently to calculate the minimum cost of travelling (CT, in J kg^{-1}) to each location:

$$CT = TT * 16.1W \ kg^{-1} \qquad (2)$$

Recent research demonstrated a linear relationship between ODBA and metabolic rate in all species examined to date (summarised in [21]; but see also [66]). Halsey et al. [61] investigated the relationship between the rate of oxygen consumption V_o (in ml min^{-1}; an indirect measure of energy expenditure) and ODBA for 10 different species including Magellanic and Rockhopper penguins. The robust results obtained ($R^2 = 0.99$) allowed Halsey et al. [61] to propose a relationships between the species mean body mass (BM) and both the slope and intercept of the predictive relationships for all 10 species (including the two penguin species; $P < 0.001$ in all cases): intercept, y = 2.75 * BM$^{0.73}$ slope y = 3.52 * BM$^{0.94}$. Thus, following Halsey et al. [61], we first calculated:

$$V_o = 10.78 + ODBA * 20.45 \qquad (3)$$

Although some inter-species variation can be observed in the analysis by Halsey et al. [61], the relationship for both penguin species is quite similar, allowing us to safely estimate a relationship between V_o and ODBA in Gentoo Penguins using the calculation method proposed by these authors.

In order to convert the uptake of 1 l of oxygen into energy expenditure we used the mean value of the oxidative catabolism of lipids, glucose and protein provided by Heldmaier et al. [67] (20 kJ), such that 1 ml O_2/min equals 0.333 J s^{-1}. To derive the mass-specific power (MP, in J kg^{-1} s^{-1}) [21], the energy expenditure was divided by the mean weight of Gentoo Penguins (6.5 kg) [5]:

$$MP = V_o * 0.333/6.5kg \qquad (4)$$

The MP (4) can be calculated for each bathymetric data point in the grid of the marine area around New I. separately for benthic dives (MP$_{benthic}$, based on bathymetric depth) and pelagic dives (MP$_{pelagic}$, based on the median dive depth during pelagic dives).

Subsequently, we calculated the MP for the grid of the marine area around New I. (see Additional file 1: Figure S1) and for both colonies and years, based on the mean number of dives per foraging trip (MND) and mean dive duration (DD, duration in s of the dive event; Table 2), assuming a gradient of bottom depths from 3 m to the maximum depth (= bathymetric depth), for benthic and pelagic dives as follows:

$$MP_{MND\ benthic} = DD_{benthic} * \left(MP_{benthic\ (3\ m\ depth)} + MP_{benthic}\right) * MND/2 * pBD \tag{5}$$

$$MP_{MND\ pelagic} = DD_{pelagic} * \left(MP_{pelagic\ (3\ m\ depth)} + MP_{pelagic}\right) * MND/2 * pPD \tag{6}$$

where pBD is the mean proportion of benthic dives and pPD the mean proportion of pelagic dives (Table 2). These parameters together with previous calculations of the cost of travelling (CT), allowed us to calculate the total cost of foraging (TCF, in J kg^{-1}) as:

$$TCF = MP_{MND\ benthic} + MP_{MND\ pelagic} + CT * 2 \tag{7}$$

In order to build energy landscapes that also take into account the energy gained during foraging, we calculated bottom times (duration in s of bottom dive phase) and minimum benthic bottom times (mBBT; Table 1). The bottom times from the first foraging trip of each individual showed a relationship with maximum dive depth. This relationship also differed between studied years, colonies and between benthic and pelagic dives (GAM; see Additional file 1: Table S3). The regressions with the best fit were again determined for the different combination of years, colonies and dive types in SigmaPlot 10.0 (Systat Software, San Jose, USA; see Additional file 1: Table S4; Figures S11–S16). The regressions between bottom time and maximum dive depth (see Additional file 1: Table S4), together with bathymetric data [52] allowed us to calculate the sum of benthic bottom time (BBT) for each bathymetric point (see Additional file 1: Figure S1), separately for each colony and year. The minimum benthic bottom time for each colony and year is shown in Table 1. For pelagic bottom times (PBT), we used the corresponding regressions (see Additional file 1: Table S4) and the median dive depth per colony per year (Table 2; see 'Analyses of spatial and temporal data' for method validation). For the calculation of the total bottom time (TBT, in s), we took into account that the birds start diving close to the colony (as also found in [5]) and increase dive depth while gaining distance. A mean is calculated and the mean multiplied per the mean number of dives:

$$TBT = (mBBT + BBT)/2 * MND * pBD + PBT * MND * pPD \tag{8}$$

Finally, dividing TCF (7) by TBT (8) we were able to calculated the total relative cost (TRC, in J kg^{-1} s^{-1}) as the total cost of foraging (TCF; diving plus commuting) relative to the total bottom time (TBT). Using TRC values calculated for the grid of the marine area around New I. for which bathymetric data was available (n = 26,196; see Additional file 1: Figure S1), we constructed the energy landscape by applying the Inverse Distance Weighted (IDW) interpolation in ArcGIS 9.3 to the resulting data grid. The IDW interpolation was chosen as 1) a large set of sample values was available, and 2) the sample data points represented the minimum and maximum values in our surface [68]. Thus, the energy landscapes here presented are based on the bathymetry of the area and the total cost of foraging (diving plus commuting) relative to the bottom time (in J kg^{-1} s^{-1}), and take into account the different proportion of benthic and pelagic dives carried out by the penguins in each studied colony and year.

Stable isotope niche analysis

We analysed carbon (δ^{13}C) and nitrogen (δ^{15}N) stable isotope values of chick feather samples as a marker of breeding season foraging ecology. Feathers were sampled when the chicks were around 2 months old (February), ensuring that the feathers were grown during the time of deployment of the loggers (December). Twenty feathers were analysed from each colony and year except for the North End colony in 2014, for which we analysed 18 samples. Carbon and nitrogen isotope analyses were carried out on 0.65–0.75 mg sample aliquots, weighed into tin cups. Carbon and nitrogen isotope ratios were measured simultaneously by continuous-flow isotope ratio mass spectrometry (CF-IRMS) at the UC Davis Stable Isotope Facility, using a PDZ Europa ANCA-GSL elemental analyser interfaced to a PDZ Europa 20–20 isotope ratio mass spectrometer (Sercon Ltd., Cheshire, UK). Laboratory standard measurements have been previously calibrated against NIST Standard Reference Materials indicated a standard deviation is 0.2‰ for ^{13}C and 0.3‰ for ^{15}N. Stable isotope ratios were expressed in δ notation as parts per thousand (‰) deviation from the international standards V-PeeDee Belemnite for δ^{13}C and to atmospheric N$_2$ for δ^{15}N.

The isotopic niches of birds from the two colonies were calculated using SIBER (Stable Isotope Bayesian Ellipses in R) [69]. In this analysis, the location of the centroid (LOC) indicates where the niche is centred in isotope space. A Bayesian approach based on multivariate ellipse metrics was used to calculate the standard ellipse area (SEA), which represents the core isotope niche width as described by Jackson et al. (2011). To describe the spread of the data points, parameters proposed by Layman et al. [70] were calculated. As proxies of intra-population trophic diversity, the mean distance to centroid (CD) and the mean nearest-neighbour distance (NND) are given. Information on the trophic length of the community is given as the $\delta^{15}N$ range (NR), and an estimate of the diversity of basal resources is provided by the $\delta^{13}C$ range (CR).

Results

The marine areas used by Gentoo Penguins varied among years, and so did the degree of spatial segregation between colonies (Fig. 2). This was most evident when kernel densities were considered (Fig. 3). In 2013, birds from the South End colony performed the longest trips, which took them furthest away from the colonies and which were more extended in time (Table 1). However, most trip parameters did not differ significantly between colonies or between years due to large inter-individual variability (Table 1).

The mean number of dives per foraging trip was similar for both colonies and years (Table 2). Birds from the South End colony carried out more pelagic dives in 2013, while the proportion of pelagic and benthic dives was almost equal for both colonies in 2014 (Table 2). The maximum dive depth was achieved by a bird from the South End colony in 2013 (Table 2). However, the median of the event maximum depth was largest in 2014 and showed no differences between colonies (Table 2). The deepest pelagic dives corresponded to birds from the North End colony (Table 2).

Gentoo Penguins preferentially used the areas of the energy landscape that resulted in lower foraging costs per bottom time gain, mostly below 225 J kg^{-1} s^{-1} in 2013 and below 175 J kg^{-1} s^{-1} in 2014 (Fig. 4). There was no evident relationship between the foraging areas used by the Gentoo Penguins and depth or distance to the colony (Fig. 3). The selection of the foraging areas varied noticeably in space (Figs. 2 and 3) and water depth (Fig. 3, Additional file 1: Figure S1), but in all cases implied minimal power requirements compared with other parts of the landscape accessible to the penguins around the colony (Fig. 4).

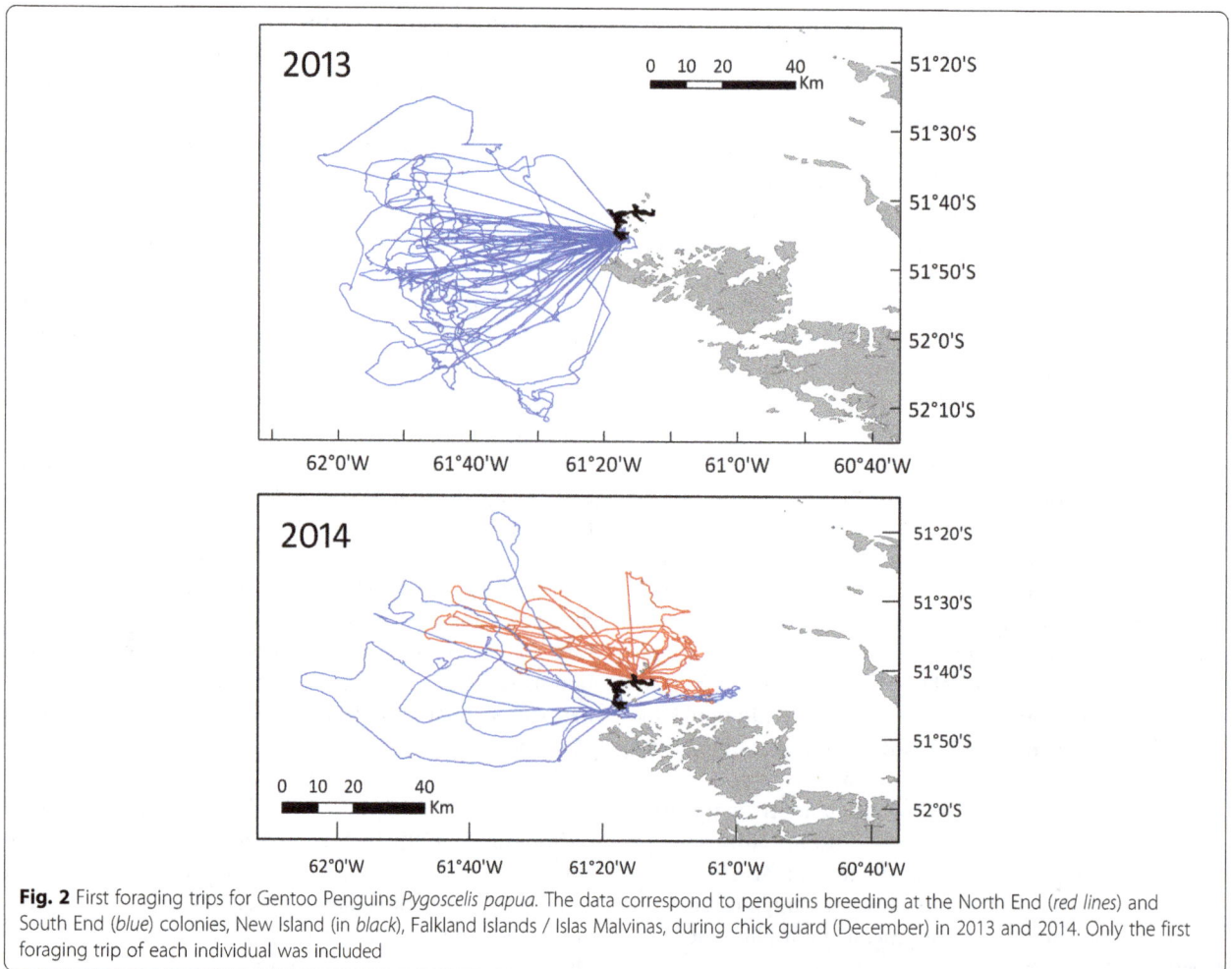

Fig. 2 First foraging trips for Gentoo Penguins *Pygoscelis papua*. The data correspond to penguins breeding at the North End (*red lines*) and South End (*blue*) colonies, New Island (in *black*), Falkland Islands / Islas Malvinas, during chick guard (December) in 2013 and 2014. Only the first foraging trip of each individual was included

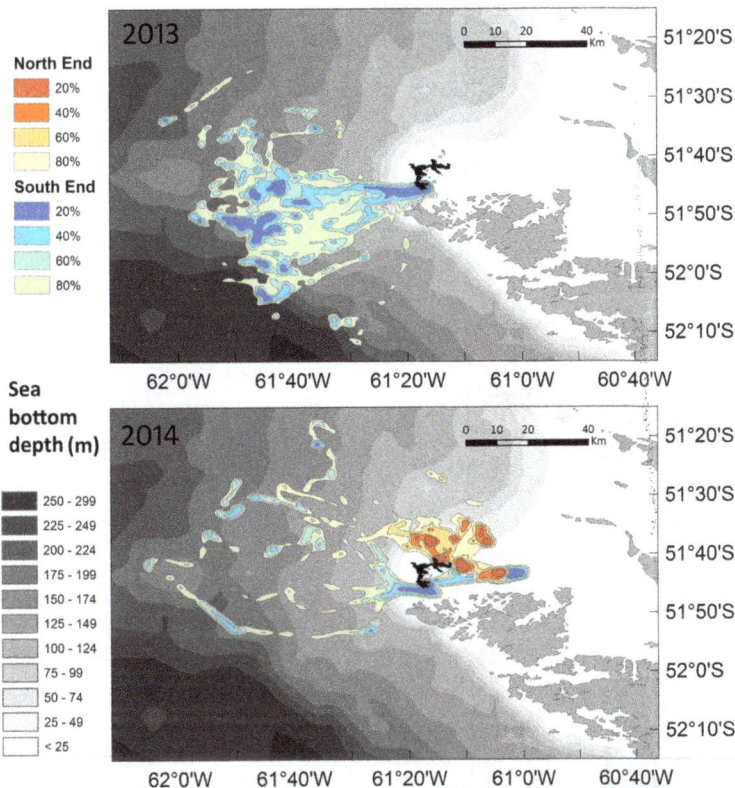

Fig. 3 Kernel density distribution (20, 40, 60, and 80%) of dive locations. Kernel density distribution shows the places where the Gentoo Penguins *Pygoscelis papua* spent most of their foraging time, for birds breeding at the North End (*shades of red*) and South End (*shades of blue*) colonies, New Island (in *black*), Falkland Islands / Islas Malvinas. Depth zones (in m) are based on data from the Global Topography (Smith & Sandwell 1997) and an IDW interpolation in ArcGIS 9.3. Only dives performed during the first foraging trip of each individual were included

The energy landscapes varied strongly in time (i.e. between the 2 years), but no obvious differences were observed between the energy landscapes calculated for the two colonies in 2014 (Fig. 4). We compared the foraging costs per bottom time gain extracted from the energy landscapes and corresponding to the locations where actual dive events were carried out (distribution pattern shown in Fig. 5). When comparing the means for each deployed penguin, the highest mean foraging costs per bottom time gain was observed for the South End colony in 2013 (mean ± SD, 2013: 163.7 ± 9.7, 2014: 107.8 ± 22.2, J kg^{-1} s^{-1}; t = 7.790, d.f. = 17, P < 0.001). No differences in foraging costs per bottom time gain were observed between the colonies in 2014 (South End: 107.8 ± 22.2, North End: 106.7 ± 13.8, J kg^{-1} s^{-1}; t = 0.109, d.f. = 11, P = 0.915).

At the beginning of the fieldwork (December, i.e. late incubation and early chick-feeding), we counted all active nests at the colonies. The North End colony consisted of 2378 nests in 2013 and 2073 nests in 2014. The South End colony contained 2044 nests in 2013 and 2072 nests in 2014. During the crèche period (mid-January), the colonies were revisited to count the number of chicks as a measure of breeding success.

The North End colony contained 1352 chicks in 2013 and 3172 in 2014. In the South End colony we counted 2458 chicks in 2013 and 2171 chicks in 2014. However, the South End colony was affected by an outbreak of avian pox in January 2015, which affected the numbers corresponding to the second season of this study (December 2014 to February 2015). Despite this disease, the overall breeding success was higher in 2014 (1.29 chick per nest) than in 2013 (0.86 chicks per nest).

Stable isotope niche analysis

The SIBER analyses corresponding to Gentoo Penguin chick feathers revealed differences between the years (Fig. 6, Table 3). In 2014, we measured lower δ^{13}C (GLM, effect of site: F = 5.66, P = 0.020, effect of year: F = 26.68, P < 0.001) and higher δ^{15}N isotope values (GLM, effect of site: F = 0.37, P = 0.544, effect of year: F = 14.92, P < 0.001). All niche metrics (Table 3) were larger in 2013 than in 2014, indicating a higher variability in the feeding ecology among individuals. Furthermore, the South End colony (which was represented by the birds carrying data loggers) had the highest niche metrics among all four groups (Table 3).

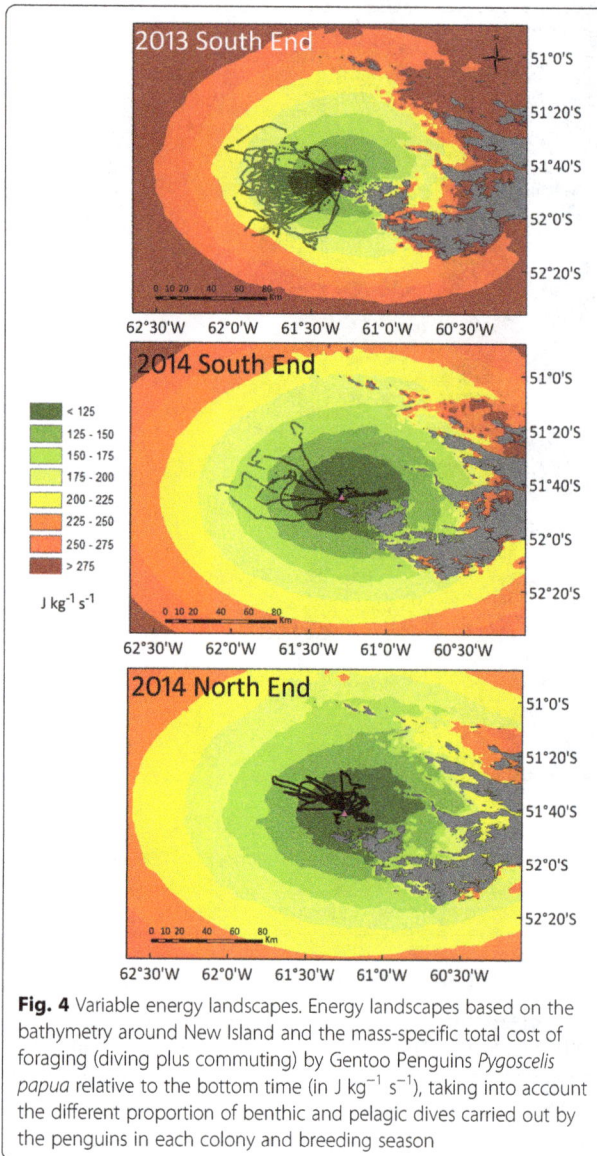

Fig. 4 Variable energy landscapes. Energy landscapes based on the bathymetry around New Island and the mass-specific total cost of foraging (diving plus commuting) by Gentoo Penguins *Pygoscelis papua* relative to the bottom time (in J kg^{-1} s^{-1}), taking into account the different proportion of benthic and pelagic dives carried out by the penguins in each colony and breeding season

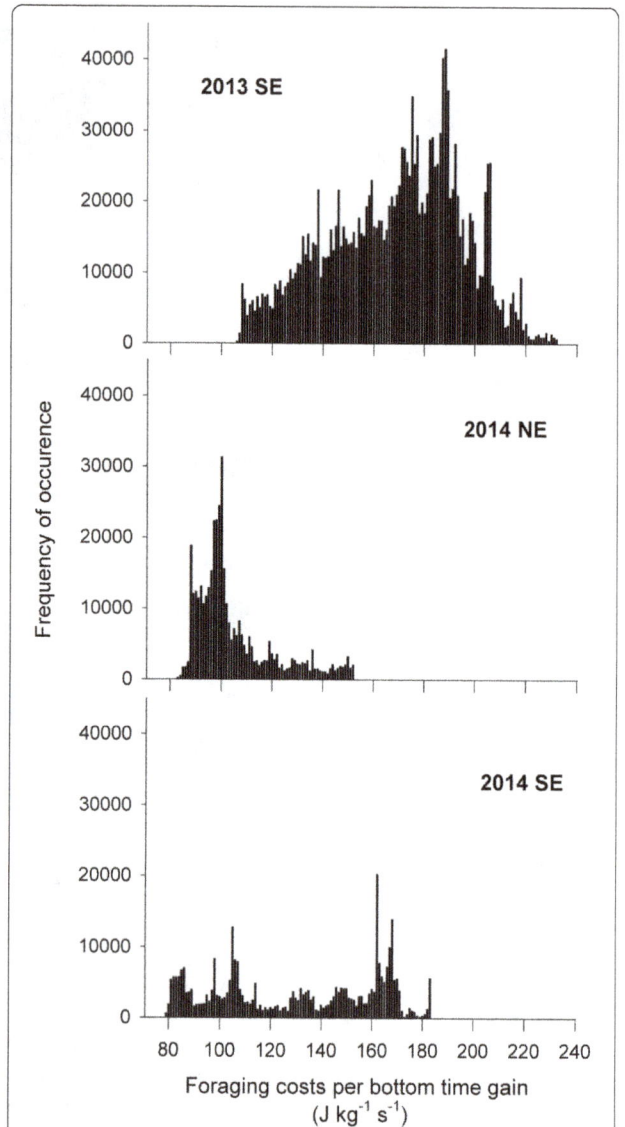

Fig. 5 Frequencies of foraging costs per bottom time gain. Data are shown in J kg^{-1} s^{-1}, for each colony and breeding season of Gentoo Penguins *Pygoscelis papua* breeding at the North End and South End colonies, New Island, Falkland Islands / Islas Malvinas, during chick guard (December) in 2013 and 2014

Discussion

The costs associated to movements are frequently determined by the landscapes through which animals move [7, 12, 21]. Hence, the energy landscape approach to movement ecology predicts that individuals will modulate different foraging parameters in order to maximize net energy gain during foraging avoiding costly areas [21, 22, 24].

As in previous studies of animal movement, Gentoo penguins in our study consistently foraged in areas of the energy landscape that resulted in lower foraging costs. However, the results of the present study show that, in line with our prediction, the energy landscape changed temporally, namely between the two seasons. During the first season, in December 2013, Gentoo penguins experienced an energy landscape with increased

foraging costs around New Island when compared to the second season, in December 2014. Despite these higher costs, Gentoo Penguins in 2013 travelled further (albeit not statistically significantly), and foraged most of the time in more costly areas of the energy landscape than in 2014 (Fig. 5). The breeding success data were in line with this: in a situation of higher energy expenditure (2013), the breeding success was low (0.86 chicks/nest), compared to a situation of lower energy expenditure (2014: 1.29 chicks/nest).

Variation in energy landscapes over time may be due to changes in the landscapes that make the movements of the animal more challenging [21]. In marine environments

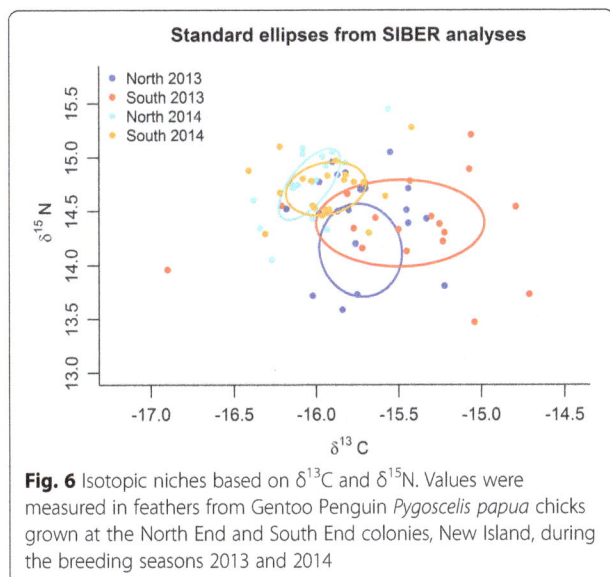

Fig. 6 Isotopic niches based on $\delta^{13}C$ and $\delta^{15}N$. Values were measured in feathers from Gentoo Penguin *Pygoscelis papua* chicks grown at the North End and South End colonies, New Island, during the breeding seasons 2013 and 2014

or "seascapes", the energy landscapes may vary in time following changing oceanographic conditions or as a consequence of fluctuating food availability. In the Falkland Islands, the total catches of rock cod and Patagonian longfin squid, the two main items in the diet of Gentoo Penguin during guard [40], were lower in 2013 (32,436 and 40,168 t respectively) than in 2014 (56,686 and 48,702 t respectively) [71]. The Falkland Islands fisheries statistics thus suggested lower food availability during 2013 compared with 2014 [71], which was reflected in the more expensive energy landscape. This is also in agreement with the lower chlorophyll *a* concentrations observed in the area southwest of the Falkland Islands in 2013 (see Additional file 1: Figure S17, A) with respect to 2014 (see Additional file 1: Figure S17, B; Giovanni Ocean Color Time-Series, National Aeronautics and Space Administration, USA). Also during 2013, we observed a predominance of lobster krill remains in the scats of Gentoo Penguins breeding on New Island. Previous studies showed that lobster krill is a less preferred item in the diet of Gentoo Penguins at the Falkland Islands both during guard and crèche formation [39, 40]. The reduced availability of preferred prey and the generally lower ocean productivity may have forced the Gentoo Penguins from New Island to prey on a less preferred prey in 2013. Additionally, $\delta^{15}N$ was lower in 2013, suggesting lower trophic level prey (e.g. lobster krill), and all $\delta^{13}C$ and $\delta^{15}N$ niche metrics were larger in 2013 than in 2014 (Table 3), suggesting a higher variability in the feeding ecology among individuals.

A high degree of plasticity in foraging behaviour and diet was also reported for Gentoo Penguins both from Antarctica and Sub-Antarctic islands as a buffer against changes in prey availability [29, 35, 72]. Similarly, in our study of Imperial Shags at New Island, we also observed such plasticity in the diet, diving, and foraging behaviour over time [16]. In the case of Imperial Shags, pelagic dives dominated in poorer years in terms of breeding success. In our present study, Gentoo Penguins performed a significantly higher proportion of pelagic dives during 2013 (Table 2), probably preying on the pelagic phase of lobster krill [73]. This switch to a predominantly pelagic foraging strategy in 2013 could be interpreted as a strategy to overcome a more expensive energy landscape. In years when food availability makes benthic foraging altogether too costly, birds could switch to a more cost-effective pelagic strategy.

The balance between energy gain and variable energy costs of foraging will directly affect the survival and reproduction of individuals in a particular landscape [22, 24]. It follows that in the context of natural selection individuals that move efficiently to areas of the best energy gain per energy expenditure will increase their fitness, leading to the evolution of a variety of energy-saving mechanisms [22]. However, this could be a too simplistic approach, as movement can also depend on other factors in addition to the availability of prey, like the probability of being predated [11, 12, 21, 74].

Table 3 Isotopic niche metrics of Gentoo Penguins *Pygoscelis papua*. Parameters are based on carbon ($\delta^{13}C$) and nitrogen ($\delta^{15}N$) stable isotopes of chick feather samples as a marker of breeding season foraging ecology from two colonies at New Island and two breeding seasons calculated with the SIAR package. *SE* South End colony, *NE* North End colony

Symbol	Explanation	NE, 2013	SE, 2013	NE, 2014	SE, 2014
		n = 20	*n* = 20	*n* = 18	*n* = 20
LOC	Location of centroid (mean $\delta^{13}C$, mean $\delta^{15}N$)	−15.73, 14.45	−15.49, 14.39	−16.04, 14.75	−15.94, 14.71
SEA	Area of the standard ellipse (isotope niche width)	0.34	0.63	0.15	0.19
SEAc	as above, corrected for sample size	0.36	0.67	0.16	0.20
NR	trophic length (range in $\delta^{15}N$)	1.46	1.74	1.40	0.99
CR	diversity of basal resources (range in $\delta^{13}C$)	0.96	2.19	0.82	0.99
CD	niche width 2 (Mean distance to centroid)	0.43	0.52	0.31	0.29
NND	mean Nearest Neighbour Distance	0.16	0.26	0.13	0.15

Gentoo Penguins from New Island did not forage in all areas of the energy landscape with lower foraging costs. An area with the lowest foraging costs i.e. < 125 J kg^{-1} s^{-1} located to the north-west of New Island was avoided in both years of this study and also during a previous study (Figs. 2, 3 and 4) [5]. This area surrounds three South American Fur Seal *Arctocephalus australis* colonies (Fig. 1). According to the landscape of fear approach to movement ecology [12, 75], the spatial and temporal use of the landscapes would be driven by the fear of being killed (risk of predation). Our results are also in line with the landscape of fear approach, adding to a number of studies showing the importance of non-lethal effects of predation on seabird foraging behaviour (e.g. [11]). Moreover, the foraging movements observed during this study provide further support to the complementarity of the energy and fear landscape paradigms proposed by Gallagher et al. [12], as a way of better understanding the mechanistic basis of movement ecology.

Conclusions

This study clearly illustrates that in order to adequately understand the mechanistic basis of movement ecology it is necessary to consider a variety of factors and complementary approaches. A complementary approach looking at the energy gain and variable energy costs of foraging (energy landscapes) and the non-lethal effects of predation (landscape of fear) that also considers the fluctuations in food availability and/or the spatial and temporal changes of the landscapes will certainly help us understanding the complex decisions made by wild animals during foraging. Energy landscapes are also useful in linking energy gain and variable energy costs of foraging to breeding success. Thus, long term studies of the energy landscapes experienced by populations of wild animals could also help understanding demographic changes and their consequences for conservation. Moreover, investigating energy landscapes over time may become a useful tool for the identification of key areas for conservation spatial planning.

Additional file

Additional file 1: Table S1. GAM investigating the sum of ODBAas a function of maximum dive depth. **Table S2**. Relationship between the sum of ODBA and maximum dive depth. **Table S3**. GAM investigating the bottom time as a function of event maximum depth. **Table S4**. Relationship between bottom time and event maximum depth. **Figure S1**. Depth zones. **Figure S2**. Distribution of depth during benthic (A) and pelagic (B) dives. **Figure S3**. Distribution in different depths of benthic (A) and pelagic (B) dives. **Figure S4**. Benthic dives example. **Figure S5**.Sum of ODBA versus maximum dive depth for benthic dives (South End, 2013). **Figure S6**. Sum of ODBA versus maximum dive depth for pelagic dives (South End, 2013). **Figure S7**. Sum of ODBA versus maximum dive depth for benthic dives (South End, 2014). **Figure S8**. Sum of ODBA versus maximum dive depth for pelagic dives (South End, 2014). **Figure S9**. Sum of ODBA versus maximum dive depth for benthic dives (North End, 2014). **Figure S10**. Sum of ODBA versus maximum dive

depth for pelagic dives (North End, 2014). **Figure S11**. Bottom time versus event maximum depth for benthic dives (South End, 2013). **Figure S12**. Bottom time versus event maximum depth for pelagic dives (South End, 2013). **Figure S13**. Bottom time versus event maximum depth for benthic dives (South End, 2014). **Figure S14**. Bottom time versus event maximum depth for pelagic dives (South End, 2014). **Figure S15**. Bottom time versus event maximum depth for benthic dives (North End, 2014). **Figure S16**. Bottom time versus event maximum depth for pelagic dives (North End, 2014). **Figure S17**. Chlorophyll a concentration. (DOCX 2492 kb)

Abbreviations
Axy: Micro tri-axial accelerometer; BBT: Benthic bottom time; CD: Distance to centroid; CF-IRMS: Continuous-flow isotope ratio mass spectrometry; CR: Carbon range; CT: Cost of travelling; DD: Dive duration; GAM: Generalized additive models; GLM: General linear model; GPS-TD logger: General position system-temperature-depth logger; IDW: Inverse distance weighted; LOC: Location of centroid; mBBT: Minimum benthic bottom times; MND: Mean number of dives per foraging trip; MP: Mass-specific power; NND: Nearest-neighbour distance; NR: Nitrogen range; ODBA: Overall dynamic body acceleration; pBD: Proportion of benthic dives; PBT: Pelagic bottom times; pPD: Proportion of pelagic dives; SEA: Standard ellipse area; SIBER: Stable isotope bayesian ellipses; TBT: Total bottom time; TCF: Total cost of foraging; TRC: Total relative cost; TT: Travel time

Acknowledgments
We are grateful to the New Island Conservation Trust for permission to work on the island and for providing logistic support. Martin Wikelski (Max Planck Institute for Ornithology, Radolfzell, Germany) provided the GPS data loggers. We would like to thank Ian, Maria and Georgina Strange, Melanie Marx, Cristoph Kaula, Jessica Winter, Richard Phillips and Pauline Sackett (British Antarctic Survey, UK), Sylvia Kuhn (Max Planck Institute for Ornithology), Paul Brickle (South Atlantic Environmental Research Institute, Stanley, Falkland Islands), Mikako Saito, Leiv Poncet, and Tatiana de Mendonça Pinto Alves, for their contributions to the fieldwork, lab work, and logistics.

Funding
The study and JFM received financial support from the German Research Foundation (Deutsche Forschungsgemeinschaft, DFG, grant MA2574/5^{-1}).

Authors' contributions
JFM and PQ conceived and designed the study. JFM generated the field data. AK and TM developed purpose-written software and script. TM and JS contributed to the calculation of dive and acceleration parameters respectively. JFM and PQ analysed and interpreted the tracking, acceleration and dive data. JFM drafted the manuscript. All authors revised, and approved the final version of the manuscript for publication.

Author details
[1]Department of Animal Ecology & Systematics, Justus Liebig University Giessen, Heinrich-Buff-Ring 26, D-35392 Giessen, Germany. [2]Centre d'Etudes Biologiques de Chizé, UMR7372 CNRS-Université La Rochelle, 79360 Villiers en Bois, France.

References
1. Wolf JBW, Kauermann G, Trillmich F. Males in the shade: habitat use and sexual segregation in the Galapagos sea lion (*Zalophus californianus wollebaeki*). Behav Ecol Sociobiol. 2005;59:293–302.
2. Nathan R, Getz WM, Revilla E, Holyoak M, Kadmon R, Saltz D, Smouse PE. A movement ecology paradigm for unifying organismal movement research. Proc Natl Acad Sci U S A. 2008;105:19052–9.
3. Revilla E, Wiegand T. Movement ecology special feature: individual movement behavior, matrix heterogeneity, and the dynamics of spatially structured populations. Proc Natl Acad Sci U S A. 2008;105:19120–5.
4. Ballard G, Dugger KM, Nur N, Ainley DG. Foraging strategies of Adelie penguins: adjusting body condition to cope with environmental variability. Mar Ecol Prog Ser. 2010;405:287–302.

5. Masello JF, Mundry R, Poisbleau M, Demongin L, Voigt CC, Wikelski M, Quillfeldt P. Diving seabirds share foraging space and time within and among species. Ecosphere. 2010;1:19.11–28.

6. Wilson RP. Resource partitioning and niche hyper-volume overlap in free-living Pygoscelid penguins. Funct Ecol. 2010;24:646–57.

7. Wall J, Douglas-Hamilton I, Vollrath F. Elephants avoid costly mountaineering. Curr Biol. 2006;16:527–9.

8. Wilcove DS, Wikelski M. Going, Going, Gone: Is Animal Migration Disappearing? PLoS Biol. 2008;6:1361–4.

9. MacArthur RH, Pianka ER. On the optimal use of a patchy environment. Am Nat. 1966;100:603–10.

10. Schoener TW. Theory of feeding strategies. Annu Rev Ecol Syst. 1971;2:369–404.

11. Riou S, Hamer KC. Predation risk and reproductive effort: impacts of moonlight on food provisioning and chick growth in Manx shearwaters. Anim Behav. 2008;76:1743–8.

12. Gallagher AJ, Creel S, Wilson RP, Cooke SJ. Energy landscapes and the landscape of fear. Trends Ecol Evol. 2017;32:88–96.

13. Holyoak M, Casagrandi R, Nathan R, Revilla E, Spiegel O. Trends and missing parts in the study of movement ecology. Proc Natl Acad Sci U S A. 2008;105:19060–5.

14. Holland RA, Wikelski M, Kümmeth F, Bosque C. The secret life of oilbirds: new insights into the movement ecology of a unique avian frugivore. PLoS One. 2009;4:8264.8261–6.

15. Roshier DA, Doerr VAJ, Doerr ED. Animal movement in dynamic landscapes: interaction between behavioural strategies and resource distributions. Oecologia. 2008;156:465–77.

16. Quillfeldt P, Schroff S, van Noordwijk HJ, Michalik A, Ludynia K, Masello JF. Flexible foraging behavior of a sexually dimorphic seabird: large males do not always dive deep. Mar Ecol Prog Ser. 2011;428:271–87.

17. Langman VA, Roberts TJ, Black J, Maloiy GMO, Heglund NC, Weber J-M, Kram R, Taylor CR. Moving cheaply: energetics of walking in the African elephant. J Exp Biol. 1995;198:629–32.

18. Elliott KH, Davoren GK, Gaston AJ. Increasing energy expenditure for a deep-diving bird alters time allocation during the dive cycle. Anim Behav. 2008;75:1311–7.

19. Ballance LT, Ainley DG, Ballard G, Barton K. An energetic correlate between colony size and foraging effort in seabirds, an example of the Adelie penguin Pygoscelis adeliae. J Avian Biol. 2009;40:279–88.

20. Rubenson J, Henry HT, Dimoulas PM, Marsh RL. The cost of running uphill: linking organismal and muscle energy use in guinea fowl (Numida meleagris). J Exp Biol. 2006;209:2395–408.

21. Wilson RP, Quintana F, Hobson VJ. Construction of energy landscapes can clarify the movement and distribution of foraging animals. Proc R Soc B. 2012;279:975–80.

22. Shepard ELC, Wilson RP, Rees WG, Grundy E, Lambertucci SA, Simon BV. Energy landscapes shape animal movement ecology. Am Nat. 2013;182:298–312.

23. Brownscombe JW, Gutowsky LF, Danylchuk AJ, Cooke SJ. Foraging behaviour and activity of a marine benthivorous fish estimated using tri-axial accelerometer biologgers. Mar Ecol Prog Ser. 2014;505:241–51.

24. Mosser AA, Avgar T, Brown GS, Walker CS, Fryxell JM. Towards an energetic landscape: broad-scale accelerometry in woodland caribou. J Anim Ecol. 2014;83:916–22.

25. Boersma PD. Breeding patterns of Galapagos penguins as an indicator of oceanographic conditions. Science. 1978;200:1481–3.

26. Ballance LT, Pitman RL, Fiedler PC. Oceanographic influences on seabirds and cetaceans of the eastern tropical Pacific: a review. Prog Oceanogr. 2006;69:360–90.

27. Quillfeldt P, Strange I, Masello JF. Sea surface temperatures, variable food supply and behavioural buffering capacity in thin-billed prions Pachyptila belcheri: breeding success, provisioning and chick begging. J Avian Biol. 2007;38:298–308.

28. Quillfeldt P, Masello JF, McGill RAR, Adams M, Furness RW. Moving polewards in winter: a recent change in migratory strategy. Front Zool. 2010;7:15.11.

29. Miller AK, Karnovsky NJ, Trivelpiece WZ. Flexible foraging strategies of gentoo penguins Pygoscelis papua over 5 years in the south Shetland Islands, Antarctica. Mar Biol. 2009;156:2527–37.

30. Fort J, Cherel Y, Harding AMA, Welcker J, Jakubas D, Steen H, Karnovsky NJ, Gremillet D. Geographic and seasonal variability in the isotopic niche of little auks. Mar Ecol Prog Ser. 2010;414:293–302.

31. Shealer DA. Foraging behavior and food of seabirds. In: Schreiber EA, Burger J, editors. Biology of marine birds. Boca Raton, Florida: CRC Press; 2002. p. 1–722.

32. Lescroël A, Bost C-A. Foraging under contrasting oceanographic conditions: the gentoo penguin at Kerguelen archipelago. Mar Ecol Prog Ser. 2005;302:245–61.

33. Aronson RB, Givnish TJ. Optimal central-place foragers: a comparison with null hypotheses. Ecology. 1983;64:395–9.

34. Ropert-Coudert Y, Gremillet D, Kato A, Ryan PG, Naito Y, Le Maho Y. A fine-scale time budget of cape gannets provides insights into the foraging strategies of coastal seabirds. Anim Behav. 2004;67:985–92.

35. Hinke JT, Salwicka K, Trivelpiece SG, Watters GM, Trivelpiece WZ. Divergent responses of Pygoscelis penguins reveal a common environmental driver. Oecologia. 2007;153:845–55.

36. Agnew DJ. Critical aspects of the Falkland Islands pelagic ecosystem: distribution, spawning and migration of pelagic animals in relation to oil exploration. Aquat Conserv Mar Freshwat Ecosyst. 2002;12:39–50.

37. Arkhipkin A, Brickle P, Laptikhovsky V. The use of island water dynamics by spawning red cod, Salilota australis (Pisces: Moridae) on the Patagonian shelf (Southwest Atlantic). Fish Res. 2010;105:156–62.

38. Strange I, Catry P, Strange G, Quillfeldt P. New Island, Falkland Islands. A South Atlantic wildlife sanctuary for conservation management. New Island Conservation Trust: Stanley; 2007.

39. Clausen AP, Pütz K. Recent trends in diet composition and productivity of gentoo, magellanic and rockhopper penguins in the Falkland Islands. Aquat Conserv Mar Freshwat Ecosyst. 2002;12:51–61.

40. Handley JM, Baylis AM, Brickle P, Pistorius P. Temporal variation in the diet of gentoo penguins at the Falkland Islands. Polar Biol. 2016;39:283–96.

41. Thompson KR: An assessment of the potential for competition between seabirds and fisheries in the Falkland Islands. In Falkland Islands Foundation project report. Brighton: Falkland Islands Foundation; 1989.

42. Thompson KR. Predation on Gonatus antarcticus by Falkland Islands seabirds. Antarct Sci. 1994;6:269–74.

43. Clausen AP, Arkhipkin AI, Laptikhovsky VV, Huin N. What is out there: diversity in feeding of gentoo penguins (Pygoscelis papua) around the Falkland Islands (Southwest Atlantic). Polar Biol. 2005;28:652–62.

44. Pütz K, Ingham RJ, Smith JG. Foraging movements of magellanic penguins Spheniscus magellanicus during the breeding season in the Falkland Islands. Aquat Conserv Mar Freshwat Ecosyst. 2002;12:75–87.

45. Wilson RP, Pütz K, Peters G, Culik B, Scolaro JA, Charassin J-B, Ropert-Coudert Y. Long-term attachment of transmitting and recording devices to penguins and other seabirds. Wildl Soc Bull. 1997;25:101–6.

46. Ludynia K, Dehnhard N, Poisbleau M, Demongin L, Masello JF, Quillfeldt P. Evaluating the impact of handling and logger attachment on foraging parameters and physiology in southern Rockhopper penguins. PLoS One. 2012;7:e50429.50421–11.

47. Grémillet D, Dell'Omo G, Ryan PG, Peters G, Ropert-Coudert Y, Weeks SJ. Offshore diplomacy, or how seabirds mitigate intra-specific competition: a case study based on GPS tracking of cape gannets from neighbouring colonies. Mar Ecol Prog Ser. 2004;268:265–79.

48. Garthe S, Montevecchi WA, Chapdelaine G, Rail JF, Hedd A. Contrasting foraging tactics by northern gannets (Sula bassana) breeding in different oceanographic domains with different prey fields. Mar Biol. 2007;151:687–94.

49. Garthe S, Montevecchi WA, Davoren GK. Flight destinations and foraging behaviour of northern gannets (Sula bassana) preying on a small forage fish in a low-Arctic ecosystem. Deep-Sea Research II. 2007;54:311–20.

50. Mattern T, Ellenberg U, Houston DM, Davis LS. Consistent foraging routes and benthic foraging behaviour in yellow-eyed penguins. Mar Ecol Prog Ser. 2007;343:295–306.

51. Wilson RP, Shepard ELC, Liebsch N. Prying into the intimate details of animal lives: use of a daily diary on animals. Endanger Species Res. 2008;4:123–37.

52. Smith WH, Sandwell DT. Global sea floor topography from satellite altimetry and ship depth soundings. Science. 1997;277:1956–62.

53. Global Topography Scripps Institution of Oceanography, University of California San Diego, La Jolla, USA. 2017. http://topex.ucsd.edu/WWW_html/mar_topo.html. Accessed 3 July 2017.

54. Tremblay Y, Cherel Y. Benthic and pelagic dives: a new foraging behaviour in rockhopper penguins. Mar Ecol ProgSer. 2000;204:257–67.

55. Tremblay Y, Cook TR, Cherel Y. Time budget and diving behaviour of chick-rearing Crozet shags. Can J Zool. 2005;83:971–82.

56. Wood AG, Naef-Daenzer B, Prince PA, Croxall JP. Quantifying habitat use in satellite-tracked pelagic seabirds: application of kernel estimation to albatross location. J Avian Biol. 2000;31:278–86.

57. Beyer HL: Hawth's analysis Tools for ArcGIS. 2004. http://www.spatialecology.com/htools. Accessed 3 July 2017.

58. R Development Core Team. R: a language and environment for statistical computing. Vienna: R Foundation for Statistical Computing; 2016.

59. Wilson RP, White CR, Quintana F, Halsey LG, Liebsch N, Martin GR, Butler PJ. Moving towards acceleration for estimates of activity-specific metabolic rate in free-living animals: the case of the cormorant. J Anim Ecol. 2006;75:1081–90.

60. Halsey LG, Green JA, Wilson RP, Frappell PB. Accelerometry to estimate energy expenditure during activity: best practice with data loggers. Physiol Biochem Zool. 2009;82:396–404.

61. Halsey LG, Shepard ELC, Quintana F, Gómez Laich A, Green JA, Wilson RP. The relationship between oxygen consumption and body acceleration in a range of species. Comparative Biochemistry and Physiology A. 2009;152:197–202.

62. Shepard ELC, Wilson RP, Quintana F, Gómez Laich A, Forman DW. Pushed for time or saving on fuel: fine-scale energy budgets shed light on currencies in a diving bird. Proc R Soc B. 2009;276:3149–55.

63. Gleiss AC, Wilson RP, Shepard EL. Making overall dynamic body acceleration work: on the theory of acceleration as a proxy for energy expenditure. Methods Ecol Evol. 2011;2:23–33.

64. Halsey LG. Relationships grow with time: a note of caution about energy expenditure-proxy correlations, focussing on accelerometry as an example. Funct Ecol. 2017;31:1176–83.

65. Culik B, Wilson R, Dannfeld R, Adelung D, Spairani H, Coria NRC. Pygoscelid penguins in a swim canal. Polar Biol. 1991;11:277–82.

66. Elliott KH. Measurement of flying and diving metabolic rate in wild animals: review and recommendations. Comp Biochem Physiol A Mol Integr Physiol. 2016;202:63–77.

67. Heldmaier G, Neuweiler G, Rössler W. Vergleichende Tierphysiologie. Heidelberg: Springer; 2013.

68. Johnston K, Ver Hoef JM, Krivoruchko K, Lucas N: Using ArcGIS geostatistical analyst. Redlands: ESRI; 2001.

69. Jackson AL, Inger R, Parnell AC, Bearhop S. Comparing isotopic niche widths among and within communities: SIBER – stable isotope Bayesian ellipses in R. J Anim Ecol. 2011;80:595–602.

70. Layman CA, Arrington DA, Montaña CG, Post DM. Can stable isotope ratios provide for community-wide measures of trophic structure? Ecology. 2007;88:42–8.

71. Government FI: Fisheries department fisheries statistics, volume 20, 2015. pp. 1–94. Stanley: FIG Fisheries Department; 2016:1-94.

72. Carpenter-Kling T, Handley JM, Green DB, Reisinger RR, Makhado AB, Crawford RJM, Pistorius PA. A novel foraging strategy in gentoo penguins breeding at sub-Antarctic Marion Island. Mar Biol. 2017;164:33.

73. Williams BG. The pelagic and benthic phases of post-metamorphic *Munida gregaria* (Fabricius) (Decapoda, Anomura). J Exp Mar Biol Ecol. 1980;42:125–41.

74. Avgar T, Mosser A, Brown GS, Fryxell JM. Environmental and individual drivers of animal movement patterns across a wide geographical gradient. J Anim Ecol. 2013;82:96–106.

75. Laundré JW, Hernández L, Altendorf KB. Wolves, elk, and bison: reestablishing the "landscape of fear" in Yellowstone National Park, USA. Can J Zool. 2001;79:1401–9.

4

Gene expression profiling of whole blood cells supports a more efficient mitochondrial respiration in hypoxia-challenged gilthead sea bream (*Sparus aurata*)

Juan Antonio Martos-Sitcha[1], Azucena Bermejo-Nogales[1,2], Josep Alvar Calduch-Giner[1] and Jaume Pérez-Sánchez[1*] (iD)

Abstract

Background: Acclimation to abiotic challenges, including decreases in O_2 availability, requires physiological and anatomical phenotyping to accommodate the organism to the environmental conditions. The retention of a nucleus and functional mitochondria in mature fish red blood cells makes blood a promising tissue to analyse the transcriptome and metabolic responses of hypoxia-challenged fish in an integrative and non-invasive manner.

Methods: Juvenile gilthead sea bream (*Sparus aurata*) were reared at 20–21 °C under normoxic conditions (> 85% O_2 saturation) followed by exposure to a gradual decrease in water O_2 concentration to 3.0 ppm (41–42% O_2 saturation) for 24 h or 1.3 ppm (18–19% O_2 saturation) for up to 4 h. Blood samples were collected at three different sampling points for haematological, biochemical and transcriptomic analysis.

Results: Blood physiological hallmarks remained almost unaltered at 3.0 ppm, but the haematocrit and circulating levels of haemoglobin, glucose and lactate were consistently increased when fish were maintained below the limiting oxygen saturation at 1.3 ppm. These findings were concurrent with an increase in total plasma antioxidant activity and plasma cortisol levels, whereas the opposite trend was observed for growth-promoting factors, such as insulin-like growth factor I. Additionally, gene expression profiling of whole blood cells revealed changes in upstream master regulators of mitochondria (*pgcβ* and *nrf1*), antioxidant enzymes (*gpx1*, *gst3*, and *sod2*), outer and inner membrane translocases (*tom70*, *tom22*, *tim44*, *tim10*, and *tim9*), components of the mitochondrial dynamics system (*mfn2*, *miffb*, *miro1a*, and *miro2*), apoptotic factors (*aifm1*), uncoupling proteins (*ucp2*) and oxidative enzymes of fatty acid β-oxidation (*acca2*, *ech*, and *hadh*), the tricarboxylic acid cycle (*cs*) and the oxidative phosphorylation pathway. The overall response is an extensive reduction in gene expression of almost all respiratory chain enzyme subunits of the five complexes, although mitochondrial-encoded catalytic subunits and nuclear-encoded regulatory subunits of Complex IV were primarily increased in hypoxic fish.

Conclusions: Our results demonstrate the re-adjustment of mitochondrial machinery at transcriptional level to cope with a decreased basal metabolic rate, consistent with a low risk of oxidative stress, diminished aerobic ATP production and higher O_2-carrying capacity. Taken together, these results suggest that whole blood cells can be used as a highly informative target tissue of metabolic condition.

Keywords: Blood transcriptomics, Hypoxia, Limiting oxygen saturation, Mitochondrial activity, Oxphos, *Sparus aurata*

* Correspondence: jaime.perez.sanchez@csic.es
[1]Nutrigenomics and Fish Growth Endocrinology Group, Institute of
Aquaculture Torre de la Sal, Consejo Superior de Investigaciones Científicas
(IATS-CSIC), Ribera de Cabanes, E-12595 Castellón, Spain
Full list of author information is available at the end of the article

Background

Among the abiotic factors, dissolved oxygen (O_2) is particularly important as the major limiting factor of fish aerobic metabolism [1, 2]. When regulatory mechanisms are no longer sufficient to maintain the O_2 consumption rate (MO_2), further reductions in MO_2 occur at a certain level of O_2 saturation [3]. This threshold is termed the limiting oxygen saturation (LOS) in fed fish able to maintain a routine metabolic rate, and according to the oxystatic control theory of feed intake, fish adjust their feed intake to meet dietary O_2 demands [4]. Therefore, changes in LOS, produced by fluctuations in O_2 solubility associated with variations in water temperature, should be considered and regulated to ensure a non-compromised physiological function and guarantee the welfare of farmed fish fed high or low O_2-demanding diets [5, 6]. This regulation is mediated through O_2 sensors that trigger anaerobic metabolic rates to compensate for the decreasing aerobic ATP production [7, 8]. For this purpose, eukaryotic cells switch from mitochondrial oxidative phosphorylation (OXPHOS) to the less efficient anaerobic glycolytic pathway, which induces stress and lactic acidosis (reviewed in [9]). The hallmarks of human muscle adaptation to hypoxia are a decrease in muscle oxidative capacity concomitant with a decrease in aerobic work capacity [10, 11]. In this regard, hypo-metabolic states should be considered as part of the adaptive response to hypoxia instead of a negative result in hypoxia-tolerant individuals [12] since this metabolic depression prevents the accumulation of toxic by-products from anaerobic metabolism [13].

In fish, microarray gene expression profiling of liver and skeletal muscle demonstrated that metabolic suppression is a key adaptive strategy in the hypoxic goby fish, *Gillichthys mirabilis*, to drive energy resources from growth towards metabolic processes that are essential for hypoxia survival [14]. However, in *Fundulus grandis*, both cardiac and hepatic tissues displayed increases in the gene expression of different enzyme subunits of the OXPHOS pathway in response to short-term hypoxic exposure [15]. Similarly, confounding results have been reported in European sea bass (*Dicentrarchus labrax*), as early life exposure to moderate hypoxia has long-lasting detrimental effects on growth performance with no improvement of hypoxia tolerance in juvenile fish despite the enhanced expression of glycolytic enzymes, which are target genes of hypoxia-inducible factors [16]. Whether this response is tissue- or fish species-specific remains unclear. Importantly, the red blood cells (RBC) of fish and almost all amphibians, reptiles and birds retain a nucleus and functional mitochondria [17]. These RBCs present new research opportunities, and previous research attempts have demonstrated that the expression of mitochondrial uncoupling proteins is highly regulated

by hypoxia stimuli in the whole blood cells of gilthead sea bream (*Sparus aurata*) [8]. Certainly, fish microarray meta-analysis revealed that mitochondria are particularly sensitive to cellular stress triggered by a wide range of nutritional and environmental stress stimuli [18]. Hence, a PCR-array containing 88 mitochondrial-related markers has been useful to examine changes in hepatic and muscle metabolism in response to short-term fasting [19] or aquaculture stressors that mimic thermal stress and daily operational farming activities in gilthead sea bream [20]. There is little information on blood transcriptomics, and the aim of the present study was to provide new insights into the regulation and adaptive responses of hypoxic metabolism in fish, combining non-invasive transcriptional approaches based on mitochondrial markers with conventional measures of blood haematology and biochemistry. This type of approach is crucial to determine whether samples collected without sacrificing animals provide a reliable measure of mitochondrial functioning and energy metabolism at the level of the whole organism.

Methods

Animal care

Juvenile gilthead sea bream of Atlantic origin (Ferme Marine du Douhet, Bordeaux, France) were reared from early life stages at the indoor experimental facilities of the Institute of Aquaculture Torre de la Sal (IATS-CSIC, Castellón, Spain) under natural photoperiod and temperature conditions at our latitude (40°5′N; 0°10′E). Seawater was pumped ashore (open system) and filtered through a 10-μm filter. The O_2 content of water effluents under standard conditions remained consistently higher than 85% saturation, and unionised ammonia under both control and experimental conditions remained below toxic levels (<0.02 mg/L). For sampling, the fish were anaesthetised using 3-aminobenzoic acid ethyl ester (100 mg/L), and blood was drawn from caudal vessels using EDTA-treated syringes.

Experimental setup and sampling

Juvenile fish of 230–260 g body weight were distributed in 500-L tanks (16 fish per tank) allocated in a recirculatory system equipped with physical and biological filters and programmable temperature. The water temperature was maintained at 20–21 °C. Fish were fed daily to visual satiety using a commercial diet (INICIO Forte 824/EFICO Forte 824; BioMar, Palencia, Spain), and all fish were fasted during the hypoxia challenges. The water conditions for the control fish (normoxic fish) remained unchanged, whereas hypoxic fish experienced a gradual decrease in the water O_2 concentration until reaching i) 3.0 ppm (41–42% O_2 saturation; moderate hypoxia, H1) for 24 h or ii) 1.3 ppm (18–19% O_2 saturation; severe hypoxia, H2) for up to 4 h in two different hypoxic tests (Fig. 1). Both low dissolved O_2 levels tested were obtained by the cessation of normal aeration in the tank, achieving an accurate balance between the consumption rates of the animals and the supply of clean and oxygenated water by means of an electrovalve within the established O_2 steady-state condition.

In each test, normoxic or hypoxia-challenged fish were sampled at three different sampling points after decreasing the water O_2 concentration (8 fish per time and condition): i) H1: T0, T1 (24 h), T2 (48 h), and ii) H2: T0, T1 (7 h), T2 (11 h). One blood aliquot (150 μL) was directly collected into a microtube containing 500 μL of stabilising lysis solution (REAL Total RNA Spin Blood Kit, Durviz, Valencia, Spain) and stored at −80 °C until total RNA extraction. Other aliquots were processed for haematocrit (Hc) and haemoglobin (Hb) determinations.

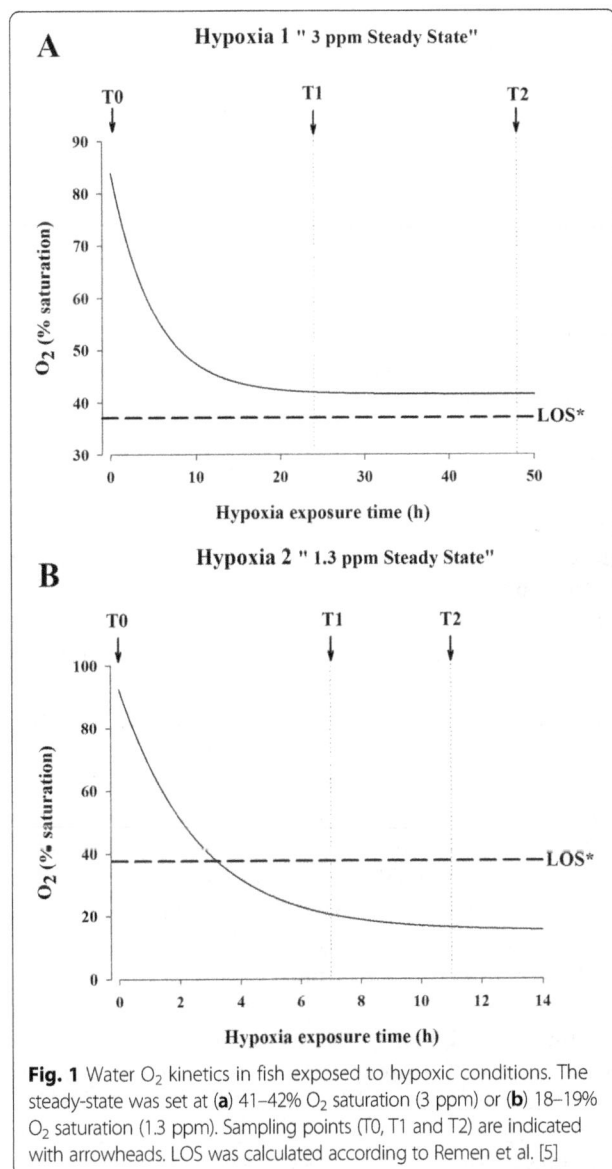

Fig. 1 Water O_2 kinetics in fish exposed to hypoxic conditions. The steady-state was set at (**a**) 41–42% O_2 saturation (3 ppm) or (**b**) 18–19% O_2 saturation (1.3 ppm). Sampling points (T0, T1 and T2) are indicated with arrowheads. LOS was calculated according to Remen et al. [5]

The remaining blood was centrifuged at 3000 × g for 20 min at 4 °C, and the plasma samples were frozen and stored at −20 °C until biochemical and hormonal analyses were performed.

Blood biochemistry and hormonal parameters
Hc was measured using heparinised capillary tubes centrifuged at 1500 × g for 30 min in a Sigma 1–14 centrifuge (Sigma, Osterode am Harz, Germany). The Hb concentration was assessed using a Hemocue Hb 201+ (Hemocue, Ängelholm, Sweden). Plasma glucose was analysed using the glucose oxidase method (Thermo Electron, Louisville, CO, USA). Blood lactate was measured in deproteinised samples (perchloric acid 8%) using an enzymatic method based on the use of lactate dehydrogenase (Instruchemie, Delfzijl, The Netherlands). Total antioxidant capacity in plasma samples was measured using a commercial kit (Cayman Chemical, Ann Arbor, MI, USA) adapted to 96-well microplates. This assay relies on the ability of the antioxidants in the samples to inhibit the oxidation of ABTS (2,2′-azino-di-[3-ethylbenzothiazoline sulphonate]) to the ABTS radical cation by metamyoglobin, a derivatised form of myoglobin. The capacity of the sample to prevent ABTS oxidation was compared with that of Trolox (water-soluble tocopherol analogue) and quantified as mM Trolox equivalents. Plasma cortisol levels were analysed using an EIA kit (Kit RE52061 m IBL, International GmbH, Germany). The limit of detection of the assay was 2.46 ng/mL with intra- and inter-assay coefficients of variation lower that 3% and 5%, respectively. Plasma insulin-like growth factors (Igf) were extracted using acid-ethanol cryoprecipitation [21], and the concentration was measured using a generic fish Igf-I RIA validated for Mediterranean perciform fish [22]. The sensitivity and midrange of the assay were 0.05 and 0.7–0.8 ng/mL, respectively.

Gene expression analysis
Total RNA from total blood cells was extracted using the REAL Total RNA Spin Blood Kit (Durviz) including a DNase step. The RNA yield was >2.5 μg, with absorbance measures ($A_{260/280}$) of 1.9–2.1. The cDNA was synthesised using the High-Capacity cDNA Archive Kit (Applied Biosystems, Foster City, CA, USA) with random decamers and 500 ng of total RNA in a final volume of 100 μL. Reverse transcription (RT) reactions were incubated for 10 min at 25 °C and 2 h at 37 °C. Negative control reactions were run without the RT enzyme. qPCR was performed using an Eppendorf Mastercycler Ep Realplex Real-Time Detection System (Eppendorf, Wesseling-Berzdorf, Germany). Diluted RT reactions were conveniently used for qPCR assays in 25 μL volume in combination with a SYBR Green Master

Mix (Bio-Rad, Hercules, CA, USA) and specific primers at a final concentration of 0.9 μM (Additional file 1: Table S1). The 96-well PCR-array layout was designed for the simultaneous profiling of a panel of 85 mitochondrial genes under uniform cycling conditions and associated with different biological processes, such as molecular chaperones (7), antioxidant defence (8), transcription factors (5), outer and inner membrane translocation (8), mitochondrial dynamics and apoptosis (10), fatty acid oxidation and the tricarboxylic acid cycle (5), OXPHOS (41) and respiration uncoupling (1). The programme used for PCR amplification included an initial denaturation step at 95 °C for 3 min, followed by 40 cycles of denaturation for 15 s at 95 °C and annealing/extension for 60 s at 60 °C. All the pipetting operations were conducted using an EpMotion 5070 Liquid Handling Robot (Eppendorf, Hamburg, Germany) to improve data reproducibility. The efficiency of PCRs (>92%) was assessed, and the specificity of the reactions was verified through an analysis of melting curves (ramping rates of 0.5 °C/10 s over a temperature range of 55–95 °C) and linearity of serial dilutions of the RT reactions (>0.99). Fluorescence data acquired during the extension phase were normalised using the delta-delta C_T method [23]. A range of potential housekeeping genes (*β-actin*, *cox4a*, *elongation factor 1*, *α-tubulin* and *18S rRNA*) was initially tested for gene expression stability using Genorm software. The most stable gene in relation to different experimental conditions (normoxia and hypoxia) was *cox4a* (M score = 0.31); therefore, this gene was used as the housekeeping gene in the normalisation procedure. For multi-gene analysis, the data on gene expression were in reference to the expression level of *sod1* obtained in normoxic fish, for which a value of 1 was arbitrarily assigned (Table 1).

This manuscript follows the ZFIN Zebrafish Nomenclature Guidelines for gene and protein names and symbols (https://wiki.zfin.org/display/general/ZFIN+Zebrafish+Nomenclature+Guidelines).

Statistical analysis
The data on biochemical and hormonal parameters were analysed using two-way analysis of variance (ANOVA), followed by the Holm-Sidak test. The data on gene expression were analysed using Student's *t* test. The significance level was set at $P < 0.05$. All analyses were performed using SigmaPlot Version 13 for Windows.

Results
Hypoxic effects on blood haematology and biochemistry
Over the course of the first hypoxia test (H1, 41–42% O_2 saturation), measurements of haematological parameters

and plasma glucose and lactate levels remained unaltered in both normoxic (>85% O_2 saturation) and hypoxia-challenged fish (Fig. 2a, c, e and g, respectively). In contrast, these parameters significantly increased in fish exposed to severe hypoxia (H2, 18–19% O_2 saturation) (Fig. 2b, d, f and h). The same trend was observed for total plasma antioxidant activity and plasma cortisol levels (Fig. 3a, b), although the cortisol increase was more pronounced at the last sampling point. The opposite regulation was observed for circulating Igf-I, although a statistically significant effect was observed at the last sampling point (Fig. 3c). No variations in all the parameters studied were observed in fish maintained under normoxic conditions in sub-experiment H2.

Hypoxic effects in whole blood cell gene expression profiling
Based on the results of hormonal and metabolic parameters, gene expression profiling of whole blood cells was restricted to the last sampling point of the severe hypoxia experiment (H2). The relative gene expression and fold-changes (FC) of mitochondrial-related genes are summarised in Table 1. For easier interpretation and visualisation of the results, the FC of differentially expressed genes is indicated using square symbols in red (up-regulated) or green (down-regulated). With the exception of *pgc1α*, all the genes included in the array were detected in all samples analysed. Among these genes, 41 out of 84 were differentially expressed, and the overall response involved repressed expression in response to severe hypoxia. This response was mediated by antioxidant enzymes (*gpx1*, *gst3*, and *sod2*), the transcription factor *nrf1*, outer and inner membrane translocases (*tom70*, *tom22*, *tim44*, *tim10*, and *tim9*), markers of mitochondrial dynamics and apoptosis (*mfn2*, *miffb*, *miro1a*, *miro2*, and *aifm1*), fatty acid β-oxidation (*acaa2* and *hadh*), tricarboxylic acid cycle (*cs*), respiration uncoupling (*ucp2*) and respiratory enzyme subunits of Complex I (*ndufa3*, *ndufa4*, *ndufa7*, *ndufb5*, and *ndufs7*), Complex II (*sdha*, *sdhaf1*, and *sdhaf2*), Complex III (*uqcrc1*, *uqcrc2*, and *uqcrh*) and Complex V (*atp5c1*, *atp5g1*, *atp5l*, and *atpaf2*), encoded by either mitochondrial or nuclear DNA. The nuclear-encoded assembly factors of Complex IV (*sco1*, *surf1*, and *cox15*) were also significantly down-regulated, but the opposite trend was observed for catalytic (*coxi*) and regulatory (*cox5a2* and *cox8b*) enzyme subunits of mitochondrial or nuclear origin, respectively. This up-regulation was also observed for the transcription factor *pgc1ß* and the outer membrane translocase *tom34*. The molecular chaperones were the only factors that did not significantly change under hypoxic conditions, although the overall trend was a down-regulation in hypoxic fish.

Table 1 Relative gene expression of mitochondrial-related genes in total blood cells

Gene name	Symbol	Relative expression Control	Hypoxia	FC
Molecular chaperones				
10 kDa heat shock protein	hsp10	0.40 ± 0.03	0.30 ± 0.07	-1.34
40 kDa heat shock protein DnaJ (Hsp40) homolog, subfamily A, member 3A	dnaja3a	0.17 ± 0.02	0.12 ± 0.02	-1.49
Iron-sulfur cluster co-chaperone protein HscB	dnajc20	0.06 ± 0.01	0.05 ± 0.01	-1.26
60 kDa heat shock protein	hsp60	0.19 ± 0.02	0.15 ± 0.03	-1.25
75 kDa Glucose-regulated protein	grp-75	0.30 ± 0.03	0.23 ± 0.03	-1.34
Derlin-1	der-1	0.19 ± 0.02	0.23 ± 0.04	1.19
Glucose-regulated protein, 170 kDa	grp-170	0.36 ± 0.06	0.27 ± 0.04	-1.35
Antioxidant enzymes				
Catalase	cat	4.85 ± 0.52	3.94 ± 0.38	-1.23
Glutathione peroxidase 1	gpx1	18.26 ± 1.33	13.90 ± 1.25*	-1.31
Glutathione reductase	gr	0.38 ± 0.06	0.26 ± 0.03	-1.47
Glutathione S-transferase 3	gst3	0.17 ± 0.06	0.02 ± 0.01*	-9.68
Peroxiredoxin 3	prdx3	0.23 ± 0.04	0.21 ± 0.03	-1.14
Peroxiredoxin 5	prdx5	0.34 ± 0.04	0.26 ± 0.03	-1.30
Superoxide dismutase [Cu-Zn]	sod1	1.08 ± 0.08	1.12 ± 0.11	1.05
Superoxide dismutase [Mn]	sod2	0.55 ± 0.05	0.28 ± 0.05**	-1.97
Transcription factors				
GA-binding protein alpha chain	gabpa	0.31 ± 0.06	0.18 ± 0.02	1.66
Mitochondrial transcription factor A	mt-tfa	0.13 ± 0.02	0.09 ± 0.02	-1.52
Nuclear respiratory factor 1	nrf1	0.22 ± 0.05	0.10 ± 0.01*	-2.21
Proliferator-activated receptor gamma coactivator 1 alpha	pgc1α	nd	nd	-
Proliferator-activated receptor gamma coactivator 1 beta	pgc1β	0.28 ± 0.09	0.80 ± 0.14*	2.85
Outer membrane translocases (TOM complex)				
Mitochondrial import receptor subunit Tom70	tom70	0.65 ± 0.04	0.36 ± 0.06**	-1.79
Mitochondrial import receptor subunit Tom34	tom34	0.55 ± 0.02	0.76 ± 0.10*	1.39
Mitochondrial import receptor subunit Tom22	tom22	0.23 ± 0.03	0.09 ± 0.02**	-2.47
Inner membrane translocases (TIM complex)				
Mitochondrial import inner membrane translocase subunit 44	tim44	0.16 ± 0.05	0.05 ± 0.02*	-3.62
Mitochondrial import inner membrane translocase subunit 23	tim23	0.28 ± 0.02	0.28 ± 0.04	1.02
Mitochondrial import inner membrane translocase subunit Tim8A	tim8A	0.83 ± 0.06	1.14 ± 0.22	1.38
Mitochondrial import inner membrane translocase subunit Tim10	tim10	0.19 ± 0.06	0.06 ± 0.02*	-3.20
Mitochondrial import inner membrane translocase subunit Tim9	tim9	0.13 ± 0.03	0.05 ± 0.01*	-2.55
Mitochondrial dynamics and apoptosis				
Mitochondrial fission 1 protein	fis1	0.63 ± 0.04	0.72 ± 0.09	1.14
Mitofusin 1	mfn1	0.06 ± 0.01	0.08 ± 0.02	1.37
Mitofusin 2	mfn2	0.21 ± 0.02	0.12 ± 0.01*	-1.72
Mitochondrial fission factor homolog B	miffb	0.35 ± 0.04	0.19 ± 0.03*	-1.82
Mitochondrial Rho GTPase 1	miro1a	0.12 ± 0.04	0.03 ± 0.01*	-3.70
Mitochondrial Rho GTPase 2	miro2	0.31 ± 0.05	0.12 ± 0.02*	-2.69
Apoptosis-related protein 1	aifm1	0.23 ± 0.02	0.15 ± 0.03**	-1.53
Apoptosis-related protein 3	aifm3	0.28 ± 0.05	0.24 ± 0.05	-1.15
Apoptosis regulator BAX	bax	0.33 ± 0.04	0.31 ± 0.02	-1.06
Bcl-2-like protein 1	bclx	0.65 ± 0.08	0.52 ± 0.07	-1.25
FA oxidation & TCA				
3-ketoacyl-CoA thiolase	acaa2	0.43 ± 0.06	0.18 ± 0.03*	-2.38
Carnitine palmitoyltransferase 1A	cpt1a	0.53 ± 0.07	0.33 ± 0.05	-1.59
Enoyl-CoA hydratase	ech	0.16 ± 0.01	0.17 ± 0.03	1.07
Hydroxyacyl-CoA dehydrogenase	hadh	0.82 ± 0.03	0.53 ± 0.08**	-1.56
Citrate synthase	cs	0.62 ± 0.03	0.47 ± 0.05**	-1.32
OXPHOS (Complex I)				
NADH-ubiquinone oxidoreductase chain 2	nd2	4.95 ± 0.60	5.07 ± 1.49	1.02
NADH-ubiquinone oxidoreductase chain 5	nd5	0.98 ± 0.26	0.80 ± 0.21	-1.22
NADH dehydrogenase [ubiquinone] 1 alpha subcomplex subunit 1	ndufa1	0.83 ± 0.06	0.77 ± 0.10	-1.08
NADH dehydrogenase [ubiquinone] 1 alpha subcomplex subunit 3	ndufa3	0.81 ± 0.06	0.56 ± 0.07*	-1.46
NADH dehydrogenase [ubiquinone] 1 alpha subcomplex subunit 4	ndufa4	1.10 ± 0.07	0.78 ± 0.04**	-1.40
NADH dehydrogenase [ubiquinone] 1 alpha subcomplex subunit 7	ndufa7	0.21 ± 0.02	0.12 ± 0.01**	-1.69
NADH dehydrogenase [ubiquinone] 1 beta subcomplex subunit 5	ndufb5	0.52 ± 0.02	0.32 ± 0.05**	-1.59
NADH dehydrogenase [ubiquinone] 1 beta subcomplex subunit 10	ndufb10	0.59 ± 0.03	0.75 ± 0.11	1.28
NADH dehydrogenase iron-sulfur protein 2	ndufs2	0.54 ± 0.04	0.44 ± 0.04	-1.23
NADH dehydrogenase iron-sulfur protein 7	ndufs7	0.60 ± 0.04	0.42 ± 0.05**	-1.42
NADH dehydrogenase (ubiquinone) 1 alpha subcomplex, assembly factor 2	ndufaf2	0.46 ± 0.05	0.32 ± 0.05	-1.43
OXPHOS (Complex II)				
Succinate dehydrogenase [ubiquinone] flavoprotein subunit	sdha	0.27 ± 0.04	0.13 ± 0.02*	-2.10
Succinate dehydrogenase cytochrome b560 subunit	sdhc	0.16 ± 0.02	0.19 ± 0.03	1.20
Succinate dehydrogenase [ubiquinone] cytochrome b small subunit B	sdhd	0.13 ± 0.02	0.17 ± 0.01	1.29
Succinate dehydrogenase assembly factor 1	sdhaf1	0.20 ± 0.06	0.07 ± 0.02*	-2.99
Succinate dehydrogenase assembly factor 2	sdhaf2	0.19 ± 0.04	0.07 ± 0.02*	-2.61
OXPHOS (Complex III)				
Cytochrome b	cyb	3.81 ± 0.42	5.08 ± 1.37	1.33
Cytochrome b-c1 complex subunit Rieske	uqcrfs1	0.12 ± 0.03	0.06 ± 0.01	-2.01
Cytochrome b-c1 complex subunit 1	uqcrc1	0.31 ± 0.06	0.14 ± 0.02*	-2.14
Cytochrome b-c1 complex subunit 2	uqcrc2	0.64 ± 0.06	0.33 ± 0.04**	-1.96
Cytochrome b-c1 complex subunit 6	uqcrh	0.27 ± 0.05	0.16 ± 0.02*	-1.72
Cytochrome b-c1 complex subunit 8	uqcrq	2.21 ± 0.26	2.09 ± 0.32	-1.06
Cytochrome b-c1 complex subunit 9	uqcr10	0.46 ± 0.02	0.40 ± 0.05	-1.15
Ubiquinol-cytochrome c reductase complex chaperone CBP3 homolog	uqcc	0.17 ± 0.02	0.11 ± 0.02	-1.50
OXPHOS (Complex IV)				
Cytochrome c oxidase subunit I	coxi	2.52 ± 0.37	4.69 ± 1.34*	1.86
Cytochrome c oxidase subunit II	coxii	0.96 ± 0.09	1.28 ± 0.31	1.33
Cytochrome c oxidase subunit III	coxiii	1.44 ± 0.18	1.76 ± 0.46	1.22
Cytochrome c oxidase subunit 4 isoform 1	cox4a	0.59 ± 0.00	0.59 ± 0.00	1.00
Cytochrome c oxidase subunit 5A, mitochondrial-like isoform 2	cox5a2	0.04 ± 0.01	0.09 ± 0.02**	2.48
Cytochrome c oxidase subunit 6A isoform 2	cox6a2	0.79 ± 0.07	1.09 ± 0.22	1.38
Cytochrome c oxidase subunit 6C-1	cox6c1	0.44 ± 0.06	0.35 ± 0.03	-1.25
Cytochrome c oxidase subunit 7B	cox7b	0.68 ± 0.09	0.78 ± 0.08	1.14
Cytochrome c oxidase subunit 8B	cox8b	1.02 ± 0.10	1.39 ± 0.16*	1.37
SCO1 protein homolog, mitochondrial	sco1	0.19 ± 0.06	0.05 ± 0.02*	-3.83
Surfeit locus protein 1	surf1	0.22 ± 0.04	0.09 ± 0.01*	-2.47
Cytochrome c oxidase assembly protein COX15 homolog	cox15	0.15 ± 0.02	0.10 ± 0.01**	-1.48
OXPHOS (Complex V)				
ATP synthase subunit gamma	atp5c1	0.36 ± 0.04	0.19 ± 0.04*	-1.83
ATP synthase subunit beta	atp5b	0.40 ± 0.05	0.29 ± 0.04	-1.39
ATP synthase lipid-binding protein	atp5g1	0.61 ± 0.06	0.36 ± 0.06*	-1.68
ATP synthase subunit g	atp5l	0.64 ± 0.06	0.42 ± 0.05*	-1.53
Mitochondrial F1 complex assembly factor 2	atpaf2	0.04 ± 0.01	0.01 ± 0.00*	-6.02
Respiration uncoupling				
Uncoupling protein 2	ucp2	0.12 ± 0.05	0.02 ± 0.02*	-6.26

Gilthead sea breams were exposed to normoxic (oxygen saturation > 85%) and hypoxic (1.3 ppm, oxygen saturation = 18–19%) conditions. Data are presented as the mean ± SEM (n = 7–8). Statistically significant differences between normoxic and hypoxic fish are indicated (*P < 0.05, **P < 0.01; Student's t test). *nd*: non-detected. Gene names of mitochondrial-encoded catalytic subunits of the OXPHOS pathway are highlighted in bold and italicised. Gene names of nuclear-encoded catalytic subunits of the OXPHOS pathway are highlighted in bold. Gene names of nuclear-encoded regulatory subunits are presented in normal font. Gene names of nuclear-encoded assembly factors are italicised. Square symbols are used for up- (red) and down-regulated genes (green)

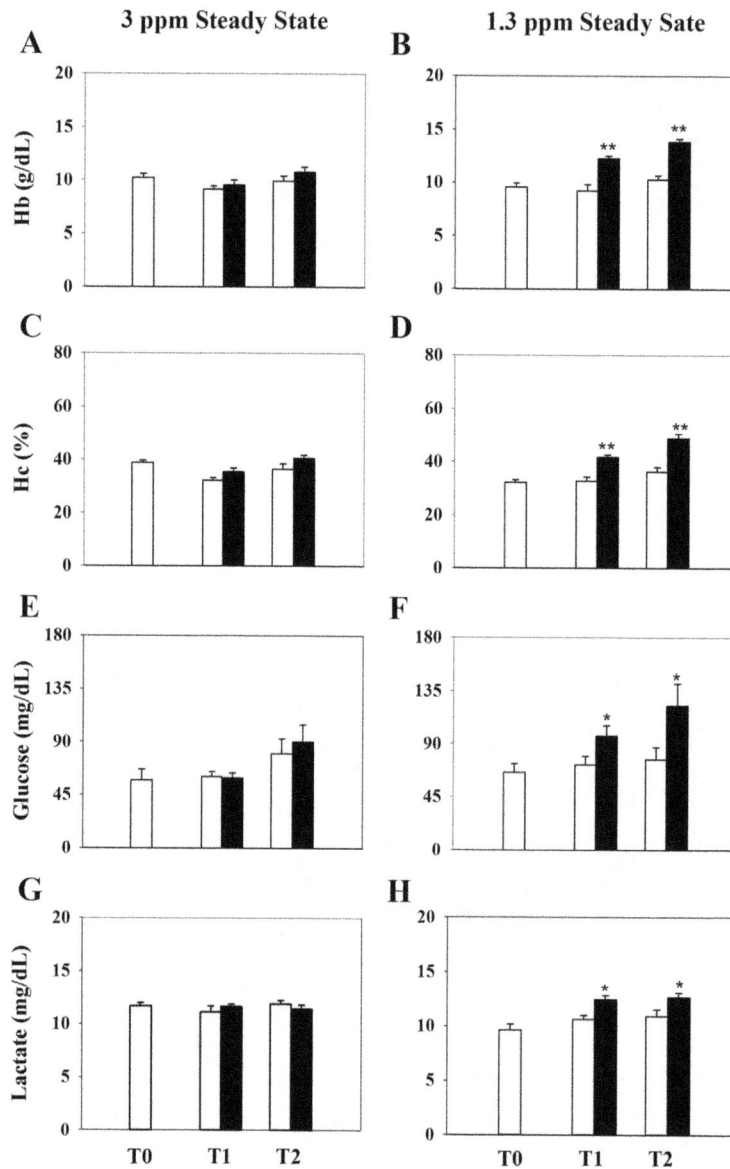

Fig. 2 Effects of normoxia (*white bars*) and hypoxia (*black bars*) on blood haematology and biochemistry. Hypoxia levels were set above (**a, c, e, g**) or below (**b, d, f, h**) the LOS. Data are presented as the mean ± SEM (n = 7–8). Statistically significant differences between normoxic and hypoxic fish are indicated (*$P < 0.05$, **$P < 0.01$; two-way analysis of variance (ANOVA) followed by the Holm-Sidak test)

Discussion

Studies in gilthead sea bream have indicated that the response of this species to a progressive decline in O_2 concentration is to reduce its swimming activity, indicative of an increasing metabolic stress and/or a coping strategy to prolong survival time when hypoxia cannot be avoided [5]. In the same study, the threshold level of LOS determined in 400-g fish varied from 17% O_2 saturation at 12 °C and 36% O_2 saturation at 20 °C. These O_2 concentrations can be implemented in aquaculture as a lower limit for acceptable decreases in O_2 concentration with respect to the physiological function and welfare of farmed gilthead sea bream. Therefore, as further explained below, it

is not surprising that data on blood biochemistry and haematology in fish exposed to O_2 concentrations above the LOS did not significantly vary after 24 h of hypoxia challenge. In contrast, a consistent response, exacerbated over time, was observed for blood parameters measured a few hours after exposure to O_2 concentrations below the LOS. In this case, the gene expression profile of whole blood cells was analysed, and the molecular signatures of hypoxic fish revealed important changes consistent with reduced but more efficient aerobic ATP production.

Living organisms are characterised by continuous switching between resting and active states, which includes long resting periods with low ATP production

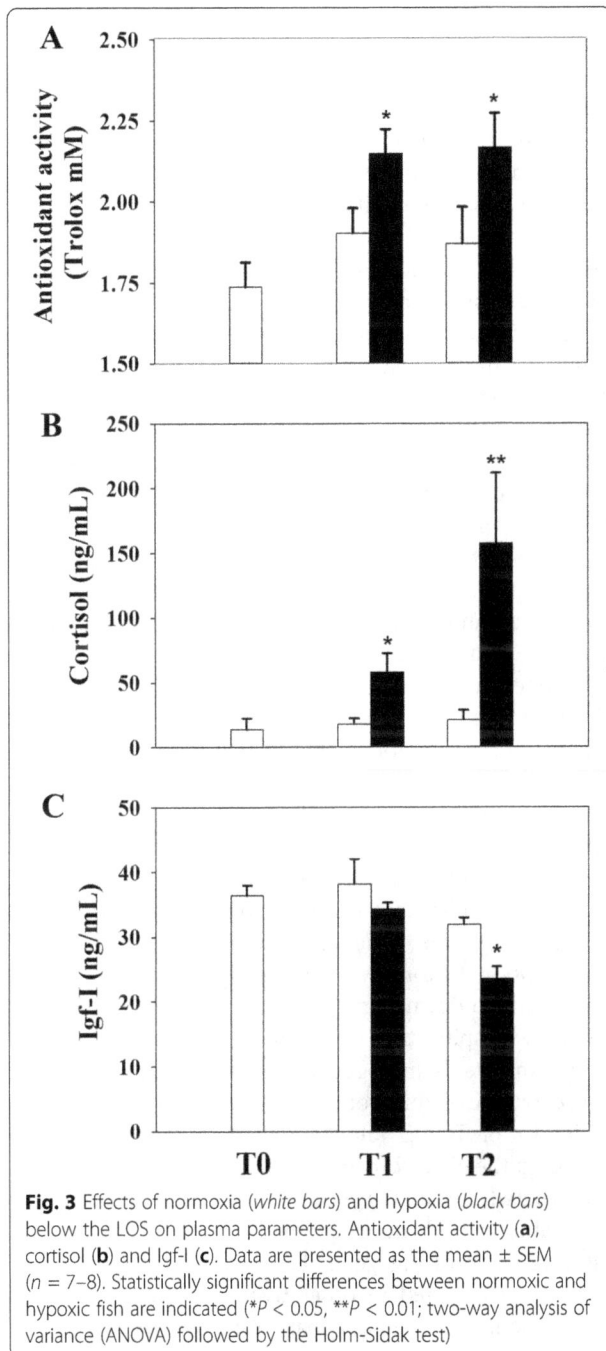

Fig. 3 Effects of normoxia (*white bars*) and hypoxia (*black bars*) below the LOS on plasma parameters. Antioxidant activity (**a**), cortisol (**b**) and Igf-I (**c**). Data are presented as the mean ± SEM (n = 7–8). Statistically significant differences between normoxic and hypoxic fish are indicated (*$P < 0.05$, **$P < 0.01$; two-way analysis of variance (ANOVA) followed by the Holm-Sidak test)

present study showed that all measured haematological and biochemical parameters remained mostly unaltered in fish maintained at 20–21 °C and 41–42% O_2 saturation. In contrast, a pronounced increase in Hc, Hb and plasma glucose and lactate levels was reported after exposure to severe hypoxia (18–19% O_2 saturation) for 4 h under steady-state conditions. Indeed, this rapid response could reflect an increase in blood O_2-carrying capacity [27] associated in the short term with erythrocyte release from a storage organ or with a reduction in plasma volume rather than the formation of new Hb [28]. Consistent with [8], this finding likely reflects metabolic changes mediated by O_2 sensors that drive the shift of the redox cellular status of NADH to a more reduced form with a rapid recycling of NAD^+ to NADH. Certainly, hypoxic situations must improve and adjust the metabolic and O_2-carrying capacities of challenged fish to cope and reach internal homeostasis [29]. The trigger observed in plasma antioxidant capacity after acute and severe hypoxia demonstrates a general decrease in metabolic rates that also reflects the aerobic/anaerobic shift of metabolism [25, 30, 31].

The increase in plasma cortisol levels observed after severe hypoxia indicates a stressful scenario in the experimental model used in the present study. Other common features of hypoxic and stress conditions include a decrease in plasma Igf-I levels and concomitant growth inhibition [32, 33]. In this sense, a characteristic response in challenged gilthead sea bream produced by crowding, and presumably also through hypoxia, is the overall downregulated expression of hepatic *igfs* and growth hormone receptors [34]. Studies in rodents support the involvement of the Gh/Igf system in the regulation of key antioxidant enzymes, ROS production and scavenging as well mitochondrial biogenesis and activity [35, 36]. However, thus far, the precise mechanisms underlying these Gh/Igf-mediated effects remain unexplored in fish. Moreover, confounding results have been reported for the aerobic/anaerobic shift during hypoxia exposure, although studies in the euryoxic mudsucker *Gillichthys mirabilis* showed a tissue-specific gene regulation resulting in suppressed protein synthesis in skeletal muscle and enhanced anaerobic ATP production in the liver tissue [14]. Similarly, zebra fish (*Danio rerio*) embryos survive during severe hypoxia (0–5% O_2 saturation) through changes in the gene and protein expression of master regulators of O_2 homeostasis, such as the hypoxia-inducible factor 1 alpha (*hif*-1α/Hif-1α) [37–40]. Additionally, long-term adaptive responses in the gene expression of several pathways related to cell architecture, cell division and energy metabolism have been underlined in the gills of adult hypoxic fish [41]. In our experimental model, this hypothesis was perfectly consistent with hypoxic-mediated effects on mitochondrial-related markers of blood cells (see below).

[24]. Similarly, the response to hypoxia has two main aspects characterised by defence and rescue, where the early defence stage is achieved by reducing energy needs (hypo-metabolic state) and the dependence on aerobic metabolism [25]. In the case of gilthead sea bream, the antioxidant defences in fish fed diets supplemented with methionine and white tea were insufficient to avoid oxidative stress under moderate hypoxia induced by 40% O_2 saturation at 22–23 °C [26]. However, LOS increases with decreasing temperature [6], and the results of the

Most mitochondrial proteins are encoded by nuclear DNA; thus, a healthy metabolic mitochondria phenotype is highly dependent on the protein import system, which involves two assembly complexes: the translocases of the outer membrane (TOM complex) and the translocases of the inner membrane (TIM complex) (see [42, 43] for review). Thus, as demonstrated in mammalian cells [44], the TOM/TIM complex is highly inducible and regulated at both transcriptional and post-transcriptional levels under conditions of chronic stress or energy deficit to ensure the maintenance of adequate mitochondrial protein import rates. Similarly, juveniles of gilthead sea bream exhibit a clear up-regulation in the gene expression of hepatic protein subunits of the TOM/TIM complex in response to aerobic energy stimuli after exposure to cyclic decreases in water temperature [20]. Conversely, the present study demonstrated that severe hypoxia induced a pronounced down-regulation of *tom70* and *tom22* subunits in whole blood cells concurrent with decreases in mRNAs encoding protein subunits of TIM23 (*tim44*) and TIM22 (*tim10* and *tim9*) complexes. In addition, co-expression analyses revealed the up-regulation of *tom34*, which acts as a co-chaperone of the Hsp70/Hsp90 complex, inhibiting mitochondrial protein translocation when expressed in excess [45]. Taken together, these findings suggest in hypoxic fish an orchestration of the TOM/TIM complex that could enable adjustments in mitochondrial protein translocation to reduce plasma oxidative capacity and the risk of oxidative stress, a feature that is consistent with the down-regulated expression of markers of ROS production and scavenging, including *ucp2*, mitochondrial superoxide dismutase (*sod2*), enzymes of the glutathione system (*gpx1* and *gst3*) and enzymes of fatty acid β-oxidation and TCA (*acaa2*, *hadh*, and *cs*). Importantly, the same trend was observed for mitochondrial (*hsp10*, *dnaja3a*, *dnajc20*, *hsp60*, and *grp-75*) and endoplasmic reticulum (*grp-170*) molecular chaperones, suggesting that proper protein folding was primarily assured in the blood cells of gilthead sea bream under the depressed metabolism induced by hypoxia exposure. Similarly, severe hypoxia did not induce the gene expression of heat shock proteins in rainbow trout (*Oncorhynchus mykiss*) RBCs cultured in vitro when the hypoxia challenge was not accompanied by a heat shock treatment [46].

Mitochondrial dynamics is an essential process that adapts mitochondria morphology to the bioenergetics requirements of the cell (see [47] for review). The mechanism of this biological process involves the balance of two opposing procedures (fusion and fission), but it is also greatly affected by the "railways" used by the mitochondria to move inside the cells. The functionality of these organelles favours the redistribution of mitochondria within the cell to ensure high oxidative capacity

under conditions of high energy demand, enabling the removal of dysfunctional or damaged mitochondria. This mechanism is highly conserved from yeast to mammals [48], and the molecular identity of major components of the fusion (*mfn1* and *mfn2*) and fission (*fis1* and *miffb*) system, as well as those of the MIRO system (*miro1a* and *miro2*) has been characterised in gilthead sea bream and uploaded to public database repositories [20]. Nevertheless, experimental evidences demonstrated that the gene expression of some of these effectors is highly induced in response to aerobic stimuli after cold-water exposure. In contrast, in the present study, severe hypoxia significantly repressed the expression of most components of this biological process (*mfn2*, *miffb*, *miro1a*, and *miro2*), including the well-known mitochondrial apoptotic factor *aifm1*. Consistently, the knockout of the transcriptional regulator *pgc1ß* is associated with a selective reduction in the expression in mice [49]. The lack of Pgc1ß also impaired the thermogenic response of adipose tissue and hepatic lipid metabolism in response to high fat dietary loads [50]. Therefore, Pgc1ß is essential for proper metabolic tuning in stress situations, contributing to the maintenance of the basal expression of mitochondrial and metabolic-related genes. However, in the present experimental model, the opposite regulation was observed for *pgc1ß* and *mfn2*, suggesting that the up-regulated expression of *pgc1ß* was more a consequence than the cause of the overall repressed expression of mitochondria-related genes. This notion was supported by the observation that the mitochondrial transcription factor *nrf1*, another target gene of *pgc1ß* [50], was also down-regulated in hypoxia-challenged fish. Notably, despite the overlapping gene expression of *pgc1ß* and its homologue *pgc1α*, the compensation of Pgc1α or Pgc1ß functions was not completely observed in Pgc1α or Pgc1ß knockout rodents [51–53]. In the case of gilthead sea bream blood cells, this effect is more exacerbated because *pgc1α* mRNAs were almost undetectable in both normoxic and hypoxic fish, although the expression of this gene at noticeable levels has previously been reported in other tissues of this fish species [20]. Whether this effect is part of the evolutionary pressure to select the conservation of functional mitochondria in the nucleated RBCs of non-mammalian vertebrates remains to be established [17].

The ultimate effector for coping with changes in energy needs and aerobic ATP production is the regulation of the OXPHOS pathway, which comprises five enzyme complexes (I-V) with catalytic enzymatic subunits encoded by both nuclear or mitochondrial DNA, whereas the enzyme subunits with regulatory or assembly properties are strictly of nuclear origin [19]. Changes in the enzymatic activities of the OXPHOS pathway have been studied for many years both in mammals and fish (e.g., [54–56]). Little is known at the molecular level, although gene expression profiling of liver, skeletal muscle and cardiac muscle tissues revealed that both the direction and magnitude of change is highly dependent on the metabolic capabilities of each tissue [8]. Thus far, the molecular fingerprinting of the OXPHOS pathway remains primarily unexplored in blood cells, and this is the first to address the specific regulation of this pathway in response to environmental stressors, evidenced by the general depletion of several components of Complexes I, II, III and V in response to severe hypoxic stimuli. Assembly factors of Complex IV (*sco1*, *surf1* and *cox15*) were also down-regulated in the present experimental model. These enzyme subunits play an important role in energy production, and mutations or defects in these molecules produce adverse effects in the appropriate function of the OXPHOS pathway in mammals [57–61]. However, this observation contrasted with the overall overexpression of catalytic and regulatory subunits of Complex IV, which was statistically significant for the catalytic *coxi* and the regulatory *cox5a2* and *cox8b* subunits. CoxI protein is encoded by mitochondrial DNA and represents one of the largest subunits of Complex IV, which contains the bimetallic centre where O_2 binds and is reduced to H_2O [62, 63]. In addition, the observed increase in the gene expression of Cox5a and Cox8 family subunits highlights their importance during the completion of the holocomplex monomer, which contains the functional structure of the cytochrome *c* binding site (see [64] for review). Therefore, we hypothesised that the net

effect should be a reduced mitochondrial ATP production due to the overall suppression of mRNAs encoding the enzyme subunits of Complexes I, II, III and IV, although the opposite regulation of the catalytic/regulatory components of Complex IV should be accompanied by subsequent mechanisms that allow a better exploitation of available oxygen in the most energetically favorable way. Modifications in mitochondrial properties also occur in other vertebrates, and the hypo-metabolic steady-state observed in overwintering frogs (*Rana temporaria*) occurred during hypoxic submergence by increases in mitochondrial O_2 affinity and a reduction in resting (state 4) and active (state 3) respiration rates in mitochondria isolated from skeletal muscle [65]. Similarly, early studies in the freshwater European eel (*Anguilla anguilla*) suggest that the efficiency of OXPHOS is increased after acclimation to high hydrostatic pressure, decreasing the enzymatic activity of Complex II in red muscle, whereas that of Complex IV is significantly increased [66]. This situation would enable a reduction in the electron leak and the optimisation of the respiratory chain. Similarly, more recent studies in gilthead sea bream have revealed that the gene expression ratio of the enzyme subunits of Complexes I and IV is altered in heart and liver tissue during the recovery state after severe hypoxia exposure [67]. Thus, as reviewed by [68], it is now evident that variations in the mitochondrial efficiency of ATP production exist among individuals, populations and environments, and even within the same individual over time. This spatial and temporal variability in mitochondrial machinery adds an

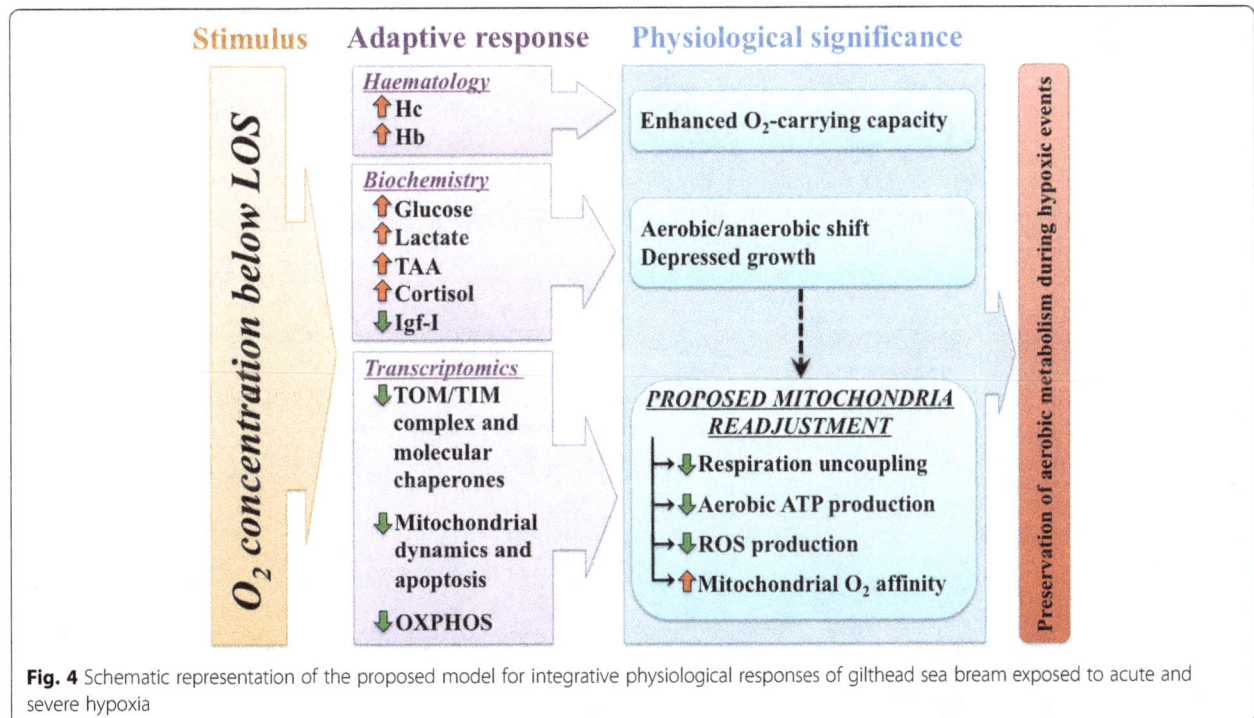

Fig. 4 Schematic representation of the proposed model for integrative physiological responses of gilthead sea bream exposed to acute and severe hypoxia

additional layer of complexity to the regulation of energy metabolism, and the maintenance of aerobic metabolism is becoming recognised as a primary hypoxia survival strategy in most organisms, including fish [69]. Even so, the usage of transcriptomic analysis with other experimental approaches related to mitochondrial activity and respiration would be necessary for the better understanding about the proposed re-adjustment of mitochondrial function in hypoxia-challenged fish.

Conclusions and future perspectives

As summarized in Fig. 4, the integrated data on blood haematology, biochemistry and transcriptomics in response to water O_2 concentrations below the LOS highlighted an enhanced O_2-carrying capacity as a result of higher Hc and Hb concentrations in response to strong hypoxic stimuli. Changes in plasma antioxidant capacity, as well as hormone and metabolite levels supported reduced energy needs and also reflected an aerobic/anaerobic shift. These results were further confirmed by gene expression profiling of a wide representation of mitochondrial-related markers, including antioxidant enzymes and molecular chaperones, effectors of mitochondrial dynamics and apoptosis, and key components of the respiratory chain, suggesting that the mitochondrial bioenergetics of fish blood cells are finely adjusted at the transcriptional level through changes in water O_2 concentrations. The induced gene expression profiles of catalytic and regulatory enzyme subunits of Complex IV should be considered an adaptive process to ensure reduced but more efficient aerobic ATP production consistent with reduced respiration uncoupling, as suggested by the decreased expression of $ucp2$. These results indicate that the gilthead sea bream is a highly euryoxic fish. Further studies are underway to determine the resilience of gilthead sea bream to high rearing densities and low O_2 concentrations, exploring the potential benefits of hypoxic preconditioning for improving the aerobic scope and swimming metabolic activity of farmed fish.

Acknowledgements
The authors are grateful to M.A. González for her excellent technical assistance with sampling and PCR analyses.

Funding
This work was financially supported by a grant from the European Commission of the European Union under the Horizon 2020 research infrastructure project AQUAEXCEL2020 (652831). Additional funding was provided by Generalitat Valenciana (PROMETEOII/2014/085). JAMS received a Postdoctoral Research Fellowship (Juan de la Cierva-Formación, Reference FJCI-2014-20,161) from MINECO.

Authors' contributions
JACG and JPS conceived and designed the study. ABN, JACG and JAMS performed the experimental procedures. JAMS, JACG and JPS analysed and interpreted the data. JAMS and JPS drafted the original manuscript. All authors reviewed, edited and approved the final manuscript.

Competing interests
The authors declare that they have no competing interests.

Author details
[1]Nutrigenomics and Fish Growth Endocrinology Group, Institute of Aquaculture Torre de la Sal, Consejo Superior de Investigaciones Científicas (IATS-CSIC), Ribera de Cabanes, E-12595 Castellón, Spain. [2]Present address: Endocrine Disruption and Toxicity of Contaminants, Department of Environment, INIA, Madrid, Spain.

References
1. Fry FEJ. Effects of the environment on animal activity. Univ Tor Stud Biol Ser. 1947;68:5–62.
2. Fry FEJ. The effect of environmental factors on the physiology of fish. In: Hroar WS, Randall DJ, editors. Fish physiology. Environmental relations and behavior 6. New York: Academic Press, Elsevier; 1971. p. 1–87.
3. Pörtner HO, Grieshaber MK. Critical PO2 (s) in oxyconforming and oxyregulating animals gas exchange, metabolic rate and the mode of energy production. In: Bicudo JEPW, editor. The vertebrate gas transport cascade adaptations to environment and mode of life. Boca Raton: CRC Press; 1993. p. 330–57.
4. Saravanan S, Geurden I, Figueiredo-Silva AC, Kaushik SJ, Haidar MN, Verreth JA, Schrama JW. Control of voluntary feed intake in fish: a role for dietary oxygen demand in Nile tilapia (Oreochromis niloticus) fed diets with different macronutrient profiles. Br J Nutr. 2012;108:1519–29.
5. Remen M, Nederlof MAJ, Folkedal O, Thorsheim G, Sitjà-Bobadilla A, Pérez-Sánchez J, Oppedal F, Olsen RE. Effect of temperature on the metabolism, behavior and oxygen requirements of Sparus aurata. Aquacult Env Interac. 2015;7:115–23.
6. Remen M, Sievers M, Torgersen T, Oppedal F. The oxygen threshold for maximal feed intake of Atlantic salmon post-smolts is highly temperature-dependent. Aquaculture. 2016;464:582–92.
7. Lushchak VI, Bagnyukova TV. Effects of different environmental oxygen levels on free radical processes in fish. Comp Biochem Physiol B-Biochem Mol Biol. 2006;144:283–9.
8. Bermejo-Nogales A, Calduch-Giner JA, Pérez-Sánchez J. Tissue-specific gene expression and functional regulation of uncoupling protein 2 (UCP2) by hypoxia and nutrient availability in gilthead sea bream (Sparus aurata): implications on the physiological significance of UCP1-3 variants. Fish Physiol Biochem. 2014;40:751–62.
9. Khacho M, Tarabay M, Patten D, Khacho P, MacLaurin JG, Guadagno J, Bergeron R, Cregan SP, Harper M-E, Park DS, Slack RS. Acidosis overrides oxygen deprivation to maintain mitochondrial function and cell survival. Nat Commun. 2014;5:3550.
10. Hoppeler H, Vogt M. Muscle tissue adaptations to hypoxia. J Exp Biol. 2001; 204:3133–9.
11. Murray AJ. Metabolic adaptation of skeletal muscle to high altitude hypoxia: how new technologies could resolve the controversies. Genome Med. 2009; 1:117.
12. Gamboa JL, Andrade FH. Muscle endurance and mitochondrial function after chronic normobaric hypoxia: contrast of respiratory and limb muscles. Pflug Arch Eur J Physiol. 2012;463:327–38.
13. Donohoe PH, West TG, Boutilier RG. Respiratory, metabolic and acid–base correlates of aerobic metabolic rate reduction in overwintering frogs. Am J Physiol-Reg I. 1998;43:R704–10.
14. Gracey AY, Troll JV, Somero GN. Hypoxia-induced gene expression profiling in the euryoxic fish Gillichthys mirabilis. Proc Natl Acad Sci U S A. 2001;98: 1993–8.
15. Everett MV, Antal CE, Crawford DL. The effect of short-term hypoxic exposure on metabolic gene expression. J Exp Zool Part A. 2012;317:9–23.
16. Vanderplancke G, Claireaux G, Quazuguel P, Madec L, Ferraresso S, Sévère A, Zambonino-Infante JL, Mazurais D. Hypoxic episode during the larval period has long-term effects on European sea bass juveniles (Dicentrarchus labrax). Mar Biol. 2015;162:367–76.
17. Stier A, Bize P, Schull Q, Zoll J, Singh F, Geny B, Gros F, Royer C, Massemin S, Criscuolo F. Avian erythrocytes have functional mitochondria, opening novel perspectives for birds as animal models in the study of ageing. Front Zool. 2013;10:33.

18. Calduch-Giner JA, Echasseriau Y, Crespo D, Baron D, Planas JV, Prunet P, Pérez-Sánchez J. Transcriptional assessment by microarray analysis and large-scale meta-analysis of the metabolic capacity of cardiac and skeletal muscle tissues to cope with reduced nutrient availability in gilthead sea bream (Sparus aurata L.). Mar Biotechnol. 2014;16:423–35.

19. Bermejo-Nogales A, Calduch-Giner JA, Pérez-Sánchez J. Unraveling the molecular signatures of oxidative phosphorylation to cope with the nutritially changing metabolic capabilities of liver and muscle tissues in farmed fish. PLoS One. 2015;10(4):e0122889.

20. Bermejo-Nogales A, Nederlof M, Benedito-Palos L, Ballester-Lozano GF, Folkedal O, Olsen RE, Sitjà-Bobadilla A, Pérez-Sánchez J. Metabolic and transcriptional responses of gilthead sea bream (Sparus aurata) to environmental stress: new insights in fish mitochondrial phenotyping. Gen Comp Endocrinol. 2014a;205:305–15.

21. Shimizu M, Swanson P, Fukada H, Hara A, Dickhoff WW. Comparison of extraction methods and assay validation for salmon insulin-like growth factor-I using commercially available components. Gen Comp Endocrinol. 2000;119:26–36.

22. Vega-Rubín de Celis S, Gómez-Requeni P, Pérez-Sánchez J. Production and characterization of recombinantly derived peptides and antibodies for accurate determinations of somatolactin, growth hormone and insulin-like growth factor-I in European sea bass (Dicentrarchus labrax). Gen Comp Endocrinol. 2004;139:266–77.

23. Livak KJ, Schmittgen TD. Analysis of relative gene expression data using real-time quantitative PCR and the 2$^{-\Delta\Delta Ct}$ method. Methods. 2001;25:402–8.

24. Kadenbach B, Ramzan R, Wen L, Vogt S. New extension of the Mitchell theory for oxidative phosphorylation in mitochondria of living organism. BBA-Gen Subj. 1800;2010:205–12.

25. Hochachka PW, Buck LT, Doll CJ, Land SC. Unifying theory of hypoxia tolerance: molecular/metabolic defense and rescue mechanisms for surviving oxygen lack. Proc Natl Acad Sci U S A. 1996;93:9493–8.

26. Pérez-Jiménez A, Peres H, Rubio VC, Oliva-Teles A. The effect of hypoxia on intermediary metabolism and oxidative status in gilthead sea bream (Sparus aurata) fed on diets supplemented with methionine and white tea. Comp Biochem Physiol C-Toxicol Pharmacol. 2012;155:506–16.

27. Wood SC, Johansen K. Adaptation to hypoxia by increased HbO$_2$ affinity and decreased red cell ATP concentration. Nat New Biol. 1972;237:278–9.

28. Soivio A, Nikinmaa M, Westman K. The blood oxygen binding properties of hypoxic Salmo gairdneri. J Comp Physiol. 1980;136(1):83–7.

29. Storey KB. Regulation of hypometabolism: insights into epigenetic controls. J Exp Biol. 2015;218(1):150–9.

30. Dalla Via J, Van den Thillart G, Cattani O, Cortesi P. Behavioural responses and biochemical correlates in Solea solea to gradual hypoxic exposure.

31. Virani NA, Rees BB. Oxygen consumption, blood lactate and inter-individual variation in the gulf killifish, Fundulus grandis, during hypoxia and recovery. Comp Biochem Physiol A-Mol Integr Physiol. 2000;126:397–405.

32. Mommsen TP, Vijayan MM, Moon TW. Cortisol in teleosts: dynamics, mechanisms of action, and metabolic regulation. Rev Fish Biol Fish.

33. Dyer AR, Upton Z, Stone D, Thomas PM, Soole KL, Higgs N, Quinn K, Carragher JF. Development and validation of a radioimmunoassay for fish insulin-like growth factor I (IGF-I) and the effect of aquaculture related stressors on circulating IGF-I levels. Gen Comp Endocrinol. 2004;135:268–75.

34. Saera-Vila A, Calduch-Giner JA, Prunet P, Pérez-Sanchez J. Dynamics of liver GH/IGF axis and selected stress markers in juvenile gilthead sea bream (Sparus aurata) exposed to acute confinement: differential stress response of growth hormone receptors. Comp Biochem Physiol A-Mol Integr Physiol. 2009;154:197–203.

35. Brown-Borg HM, Rakoczy SG, Romanick MA, Kennedy MA. Effects of growth hormone and insulin like growth factor-1 on hepatocyte antioxidative enzymes. Exp Biol Med. 2002;227:94–104.

36. Brown-Borg HM, Rakoczy SG. Growth hormone administration to long-living dwarf mice alters multiple components of the antioxidative defense system. Mech Aging Dev. 2003;124:1013–24.

37. Ton C, Stamatiou D, Liew CC. Gene expression profile of zebrafish exposed to hypoxia during development. Physiol Genomics. 2003;13:97–106.

38. Woods IG, Imam FB. Transcriptome analysis of severe hypoxic stress during development in zebrafish. Genomics data. 2015;6:83–8.

39. Robertson CE, Wright PA, Köblitz L, Bernier NJ. Hypoxia-inducible factor-1 mediates adaptive developmental plasticity of hypoxia tolerance in zebrafish, Danio rerio. P Royal Soc Lond B Bio. 2014;281(1786):20140637.

40. Köblitz L, Fiechtner B, Baus K, Lussnig R, Pelster B. Developmental expression and hypoxic induction of hypoxia inducible transcription factors in the zebrafish. PLoS One. 2015;10(6):e0128938.

41. van der Meer DL, van den Thillart GE, Witte F, de Bakker MA, Besser J, Richardson MK, Spaink HP, Leito JTD, Bagowski CP. Gene expression profiling of the long-term adaptive response to hypoxia in the gills of adult zebrafish. Am J Physiol-Reg I. 2005;289:1512–9.

42. Pfanner N, Meijer M. Mitochondrial biogenesis: the tom and Tim machine. Curr Biol. 2010;7:100–3.

43. Smits P, Smeitink J, van den Heuvel L. Mitochondrial translation and beyond: processes implicated in combined oxidative phosphorylation deficiencies. J Biomed Biotechnol. 2010;2010:737385.

44. Ljubicic V, Joseph AM, Saleem A, Uguccioni G, Collu-Marchese M, Lai RY, Nguyen LM-D, Hood D. A. Transcriptional and post-transcriptional regulation of mitochondrial biogenesis in skeletal muscle: effects of exercise and aging. BBA-Gen Subj. 2010;1800:223–34.

45. Faou P, Hoogenraad NJ. Tom34: a cytosolic cochaperone of the Hsp90/Hsp70 protein complex involved in mitochondrial protein import. BBA-Mol Cell Res. 2012;1823:348–57.

46. Currie S, Tufts BL, Moyes CD. Influence of bioenergetic stress on heat shock protein gene expression in nucleated red blood cells of fish. Am J of Physiol-Reg I. 1999;276:990–6.

47. Ferree A, Shirihai O. Mitochondrial dynamics: the intersection of form and function. In: Kadenbach B, editor. Mitochondrial oxidative Phosphorylation. New York: Springer; 2012. p. 13–40.

48. Anesti V, Scorrano L. The relationship between mitochondrial shape and function and the cytoskeleton. BBA–Bioenerg. 2006;1757:692–9.

49. Liesa M, Borda-d'Água B, Medina-Gómez G, Lelliott CJ, Paz JC, Rojo M, Palacín M, Vidal-Puig A, Zorzano A. Mitochondrial fusion is increased by the nuclear coactivator PGC-1β. PLoS One. 2008;3(10):e3613.

50. Patti ME, Butte AJ, Crunkhorn S, Cusi K, Berria R, Kashyap S, Miyazaki Y, Kohane I, Costello M, Saccone R, Landaker EJ, Goldfine AB, Mun E, DeFronzo R, Finlayson J, Kahn CR, Mandarino LJ. Coordinated reduction of genes of oxidative metabolism in humans with insulin resistance and diabetes: potential role of PGC1 and NRF1. P Nat Acad Sci USA. 2003;100:8466–71.

51. Arany Z, He H, Lin J, Hoyer K, Handschin C, Toka O, Ahmad F, Matsui T, Chin S, Wu P-H, Rybkin II, Shelton JM, Manieri M, Cinti S, Schoen FJ, Bassel-Duby R, Rosenzweig A, Ingwall JS, Spiegelman BM. Transcriptional coactivator PGC-1α controls the energy state and contractile function of cardiac muscle. Cell Metab 2005;1:259–271.

52. Leone TC, Lehman JJ, Finck BN, Schaeffer PJ, Wende AR, Boudina S, Courtois M, Wozniak DF, Sambandam N, Bernal-Mizrachi C, Chen Z, Holloszy JO, Medeiros DM, Schmidt RE, Saffitz JE, Abel ED, Semenkovich CF, Kelly DP. PGC-1α deficiency causes multi-system energy metabolic derangements: muscle dysfunction, abnormal weight control and hepatic steatosis. PLoS Biol. 2005;3(4):e101.

53. Lelliott CJ, Medina-Gomez G, Petrovic N, Kis A, Feldmann HM, Bjursell M, Parker N, Curtis K, Campbell M, Hu P, Zhang D, Litwin SE, Zaha VG, Fountain KT, Boudina S, Jimenez-Linan M, Blount M, Lopez M, Meirhaeghe A, Bohlooly-Y M, Storlien L, Strömstedt M, Snaith M, Oresic M, Abel ED, Cannon B, Vidal-Puig A. Ablation of PGC-1b results in defective mitochondrial activity, thermogenesis, hepatic function, and cardiac performance. PLoS Biol. 2006; 4(11):e369.

54. Holloszy JO. Biochemical adaptations in muscle effects of exercise on mitochondrial oxygen uptake and respiratory enzyme activity in skeletal muscle. J Biol Chem. 1967;242:2278–82.

55. Zerbetto E, Vergani L, Dabbeni-Sala F. Quantification of muscle mitochondrial oxidative phosphorylation enzymes via histochemical staining of blue native polyacrylamide gels. Electrophoresis. 1997;18:2059–64.

56. Guderley H. Metabolic responses to low temperature in fish muscle. Biol Rev. 2004;79:409–27.

57. Zhu Z, Yao J, Johns T, Fu K, De Bie I, Macmillan C, Cuthbert AP, Newbold RF, Wang J, Chevrette M, Brown GK, Brown RM, Shoubridge EA. SURF1, encoding a factor involved in the biogenesis of cytochrome c oxidase, is mutated in Leigh syndrome. Nat Genet. 1998;20:337–43.

58. Antonicka H, Mattman A, Carlson CG, Glerum DM, Hoffbuhr KC, Leary SC, Kennaway NG, Shoubridge EA. Mutations in COX15 produce a defect in the mitochondrial heme biosynthetic pathway, causing early-onset fatal hypertrophic cardiomyopathy. Am J Hum Genet. 2003;72:101–14.

59. Williams SL, Valnot I, Rustin P, Taanman JW. Cytochrome c oxidase subassemblies in fibroblast cultures from patients carrying mutations in COX10, SCO1, or SURF1. J Biol Chem. 2004;279:7462–9.

60. Stiburek L, Vesela K, Hansikova H, Pecina P, Tesarova M, Cerna L, Houstek J, Zeman J. Tissue-specific cytochrome c oxidase assembly defects due to mutations in SCO2 and SURF1. Biochem J. 2005;392:625–32.

61. Smith D, Gray J, Mitchell L, Antholine WE, Hosler JP. Assembly of cytochrome-c oxidase in the absence of assembly protein Surf1p leads to loss of the active site heme. J Biol Chem. 2005;280:17652–6.

62. García-Horsman JA, Barquera B, Rumbley J, Ma J, Gennis RB. The superfamily of heme-copper respiratory oxidases. J Bacteriol. 1994;176(18):5587.

63. Lenka N, Vijayasarathy C, Mullick J, Avadhani NG. Structural organization and transcription regulation of nuclear genes encoding the mammalian cytochrome c oxidase complex. Prog Nucl Res Molec Biol. 1998;61:309–44.

64. Ghezzi D, Zeviani M. Assembly factors of human mitochondrial respiratory chain complexes: physiology and Pathophysiology. In: Kadenbach B, editor. Mitochondrial oxidative Phosphorylation. New York: Springer; 2012. p. 65–106.

65. St-Pierre J, Brand MD, Boutilier RG. The effect of metabolic depression on proton leak rate in mitochondria from hibernating frogs. J Exp Biol. 2000; 203:1469–76.

66. Theron M, Guerrero F, Sebert P. Improvement in the efficiency of oxidative phosphorylation in the freshwater eel acclimated to 10.1 MPa hydrostatic pressure. J Exp Biol. 2000;203:3019–23.

67. Magnoni L, Martos-Sitcha JA, Queiroz A, Calduch-Giner JA, Magalhães Gonçalves JF, Rocha CMR, Abreu HT, Schrama JW, Ozorio ROA, Pérez-Sánchez J. Dietary supplementation of heat-treated *Gracilaria* and *Ulva* seaweeds enhanced acute hypoxia tolerance in gilthead Seabream (*Sparus aurata*). Biol Open. 2017;6(6):897–908. doi:10.1242/bio.024299.

68. Salin K, Auer SK, Rey B, Selman C, Metcalfe NB. Variation in the link between oxygen consumption and ATP production, and its relevance for animal performance. Proc R Soc B. 2015;282:20151028.

69. Rogers NJ, Urbina MA, Reardon EE, McKenzie DJ, Wilson RW. A new analysis of hypoxia tolerance in fishes using a database of critical oxygen level (Pcrit). Conserv Physiol. 2016;4(1):cow012.

5

Comparative genomics analyses of alpha-keratins reveal insights into evolutionary adaptation of marine mammals

Xiaohui Sun, Zepeng Zhang, Yingying Sun, Jing Li, Shixia Xu[*] and Guang Yang[*]

Abstract

Background: Diversity of hair in marine mammals was suggested as an evolutionary innovation to adapt aquatic environment, yet its genetic basis remained poorly explored. We scanned α-keratin genes, one major structural components of hair, in 16 genomes of mammalian species, including seven cetaceans, two pinnipeds, polar bear, manatee and five terrestrial species.

Results: Extensive gene loss and high pseudogenization rate of α-keratin genes were identified in cetaceans when compared to terrestrial artiodactylans (average number of α-keratins 37.29 vs. 58.33; pseudogenization rate 29.89% vs. 8.00%), especially of hair follicle-specific keratin genes (average pseudogenization rate in cetaceans of 43.88% relative to 3.80% artiodactylian average). Compared to toothed whale, the much more number of intact functional α-keratin genes was examined in the baleen whale that had specific keratinized baleen. In contrast, the number of keratin genes in pinnipeds, polar bear and manatee were comparable to those of their respective terrestrial relatives. Additionally, four keratin genes (K39, K9, K42, and K74) were found to be pseudogenes or lost uniquely in cetaceans and manatees.

Conclusions: Species-specific evolution of α-keratin gene family identified in the marine mammals might be responsible for their different hair characteristics. Increased gene loss and pseudogenization rate identified in cetacean lineages was likely to contribute to hair-less phenotype to adaptation for complete aquatic environment. However, the fully aquatic manatee still remained the comparable number of intact genes to its terrestrial relative, probably due to its perioral bristles and bristle-like hairs on the oral disk. By contrast, similar evolution pattern of α-keratin gene repertoire in the pinnipeds, polar bear and their terrestrial relatives was likely due to abundant hair to keep warm when they went ashore. Interestingly, some keratin genes were exclusively lost in cetaceans and manatees, likely as a result of convergent hair-loss phenotype to inhabit completely aquatic environment in both groups.

Keywords: Marine mammals, Hair, α-keratin, Gene loss, Pseudogenization rate

Background

Marine mammals are specific groups with evolutionary histories that their separate terrestrial relatives returned to the ocean on separate occasions and adapted to living all or part of their life in the aquatic environment. The living groups of marine mammals are generally including the following five groups: pinnipeds (seals, sea lions, fur seals, and walruses), cetaceans (whales, dolphins, and porpoises), sea otters, sirenians (dugongs and manatees), and polar bears [1]. Despite their independent evolutionary origins, these clades of the marine mammals have developed a series of adaptations to full or part aquatic environments, including morphological (e.g. streamlined shape, paddle-like limbs and feet) and physiological features adaptations (such as superb diving skill, echolocation, the thickened blubber, and etc.) [2].

Hair emergence is one of the major innovations in the mammalian evolution. Hair plays a key role in protection from mechanical insults, facilitated homeothermy, sense the immediate surrounding, sexual dimorphism,

* Correspondence: xushixia78@163.com; gyang@njnu.edu.cn
Jiangsu Key Laboratory for Biodiversity and Biotechnology, College of Life Sciences, Nanjing Normal University, Nanjing 210023, China

attraction of mates and etc. [3, 4]. Similar to other mammals, marine mammals are characterized by the presence of hair coat but with different evolution pattern among them. In order to reduce drag in the water, cetaceans and sirenians lack of hair coat because both are completely aquatic though cetaceans have body hair temporarily at fetuses [5, 6]. In contrast, the walrus has lost much of their hair (fur) and are characterized by thick layers of blubber under the skin to keep warm due to spending considerable time on land [5]. The Weddell seal grows a thin fur coat around its whole body except for small areas around the flippers [5]. The polar bear's fur consists of a dense layer of 'under fur' and an outer layer of guard hairs that appear white to tan but are actually transparent. The guard hair is 5–15 cm long over most of the bear's body [7]. The sea otter, however, has an exceptionally thick and dense coat of fur although almost time living in the sea [8]. However, the genetic bases of hair evolution in these clades of marine mammals remain poorly explored.

Hair is a kind of strongly keratinized tissue and is mainly composed of alpha-keratins and keratin associated-proteins (KRTAPs) [9, 10], which are encoded by a large number ssof multigene families and arranged in clusters on chromosomes [9, 10]. Alpha-keratins and KRTAPs are fibre-reinforced structures consisting of intermediate filaments embedded in an amorphous protein matrix which can provide support for the stability and rigidity of epidermal cells and tissue morphology [11]. The KRTAPs, as well known KAPs, are unique to mammals. According to the amino acid composition, KRTAP was divided into two major groups: high/ultrahigh cysteine (HS) and high glycine-tyrosine (HGT) that are essential for the formation of rigid and resistant hair shafts with α-keratins [9, 12]. By contrast, the α-keratin gene family was divided into two categories, i.e. epithelial keratin genes and hair keratin genes. A total of 54 functional keratin genes were identified in the human genome, which can be divided into 28 type I and 26 type II genes [10, 11]. The type I keratins consist of 17 epithelial and 11 hair keratins while the type II members comprise 20 epithelial and 6 hair keratins. In the 37 epithelial keratins, nine are specifically expressed in the hair follicle. Thus, the nine genes and 17 hair keratins are collectively referred to as hair follicle-specific keratins, which are reported to control the growth and variation of hair [10]. Previous studies characterized KRTAP gene repertoire in 22 mammals and found that gene family repertoire expansion, contraction, and pseudogenization were related to hair diversities in mammals [13]. In addition, an increase rate of hair keratin genes loss and pseudogenization were examined in two cetacean species when compared with fur terrestrial relatives, which was suggested to be associated with cetacean hairless phenotype [14]. However, previous studies only determined in one or two cetacean species could not provide comprehensive insight into genetic basis of hair diversities in marine mammals. Thus, more marine mammals were added in the present study, including seven cetacean species, two pinniped species, polar bear as well as one manatee. We first scanned the α-keratin repertoire from the genomes of the 11 marine mammals, and compared with that of their terrestrial relatives to test if α-keratin gene diversities are related to their hair features. Moreover, we tested whether there was common genes loss along with pseudogenization in the full aquatic cetaceans and sirenians as response for convergent hair-loss phenotype. Specially, we also tested if baleen whales retained much more α-keratin genes than toothed whales due to its specific keratinized baleen. From all the analyses, we expected to address the underlying genetic basis of the different hair phenotype in marine mammals.

Methods
α-keratin and KRTAP genes scanning and identification
We first identified α-keratin gene repertoires in the cow (*Bos taurus*) (coverage 7×,Btau_4.6.1) from UCSC Genome Browser website (http://genome.ucsc.edu/) taking all the known α-keratin gene sequences of human [14] as queries using BLASTN and TBLASTN algorithm [15]. All exon/intron junctions follow the cannonical AG/GT rule of splicing [16]. Meanwhile, we combined the online websites GENEWISE prediction (http://www.ebi.ac.uk/Tools/psa/genewise/) to verifying the accuracy of the sequence. The putative α-keratin genes in the cow were then taken as queries to explore the α-keratin multigene family in the genome of 11 marine mammals, including bowhead whale (*Balaena mysticetus*, coverage 150×, http://www.bowhead-whale.org/downloads/), minke whale (*B. acutorostrata*, coverage 92×, BalAcu1.0, NCBI), sperm whale (*Physeter macrocephalus*, coverage 75×, Physeter_macrocephalus_2.0.2, NCBI), Yangtze River dolphin (*Lipotes vexillifer*, Lipotes_vexillifer_v1, coverage 115×, NCBI), killer whale (*Orcinus orca*, coverage 200×, Oorc_1.1, NCBI), bottlenose dolphin (*Tursiops truncatus*, coverage 30×, Ttru_1.4, UCSC), Yangtze finless porpoise (*Neophocaena asiaeorientalis asiaeorientalis*, unpublished data), Weddell seal (*Leptonychotes weddellii*, coverage 82×, LepWed_1.0, NCBI), Florida manatee (*Trichechus manatus latirostris*, coverage 150×, TriManLat_1.0, NCBI), Polar bear (*Ursus maritimus*, coverage 101×, UrsMar_1.0, NCBI), Pacific walrus (*Odobenus rosmarus*, coverage 200×, Oros_1.0, NCBI, Additional file 1). The orthologous α-keratin genes were also scanned from the genome of their respective terrestrial relatives, such as sheep (*Ovis aries*, coverage 142×, Oar_v3.1, NCBI), alpaca (*Vicugna pacos*, coverage 22×, Vicugna_pacos-2.0.1, NCBI), African savanna elephant (*Loxodonta africana*, coverage 7×, Loxafr3.0, NCBI), giant panda (*Ailuropoda melanoleuca*, coverage 60×, AilMel_1.0, NCBI) (Additional file 1). Type I and type II keratins have

their unique flanking sequences, which can ensures maximum searching these genes in mammalian genomes. In addition, in order to obtain as much as possible of the keratin repertoire, all the newly annotated gene sequences were taken as queries in blast searches against their own genomes. Finally, all genes were checked to test whether their best hit was a α-keratin gene from NCBI genome database http://www.ncbi.nlm.nih.gov/blast/Blast.cgi using BLASTN algorithm [17]. All the identified α-keratin genes were separated into three categories: intact genes, incomplete genes (abbreviated to ic and pseudogenes (abbreviated to p), according to amino acid alignment and blast results. All the keratins were named according to the revised keratin nomenclature of human [10].

In addition, the published human [13] KRTAP genes were as queries to search unannotated genomic sequences of KRTAP genes in seal, walrus and manatee by using BLASTN algorithm [15]. All KRTAP genes have single exon in mammals. The homology fragment more than 30 bp was retrieved in the present study in order to scan the KRTAP gene repertoire as fully as possible. Like α-keratin genes, all the newly annotated KRTAP gene sequences were used as queries in blast searches against the own genomes. In the end all KRTAP genes were checked whether their best hit was a KRTAP gene from NCBI genome database.

Phylogenetic inference for α-keratin genes classification

First, all nucleotide sequences were aligned using MUSCLE in MEGA6.05 [18] and checked by eye. Second, to classify type I and II keratin genes to their respective family, we reconstructed the phylogenetic tree under the GTR + G+ I model using Maximum likelihood (ML) by RAxML [19] and Bayesian inference (BI) by MrBayes 3.2.3 [20]. The ML tree evaluated the best tree for each cluster, and supported for the nodes obtained with 1000 replications. For the Bayesian analyses, two simultaneous independent runs were performed for 40×10^6 iterations of a Markov ChainMonte Carlo algorithm, with six simultaneous chains, sampling every 1000 generations. In case of misplacement in phylogenetic trees, we will recheck their sequence for correction. Based on the sequence homology and phylogenetic relationship, we divided α-keratin genes into different subfamilies. Finally, we mapped α-keratin gene organization of each species on their genomes.

Gene conversion and recombination detection

To detect whether the gene conversion and recombination presented among keratin gene family that grouped together in the same species rather than with the members of the same family, we used the RDP4 software [21] to detect gene conversion and recombination events using RDP, Geneconv, Bootscan, MaxChi and Chimaera with 1000 permutations and cut off p value of 0.001.

Likelihood analysis of gene gain and loss

To estimate the average gene gain/loss rate and to identify gene families that have undergone significant size changes, we applied CAFÉ v3.0 [22], a tool for the statistical analysis of the evolution of the gene family size. CAFÉ v3.0 can be estimated gene gain and loss rate among α-keratin gene family, calculated ancestral states of gene family sizes for each node in the phylogenetic tree and identified the average expansion or contraction gene family on each branch. For the CAFÉ analysis, we reconstructed the ultrametric tree as a starting tree for Bayesian inference of tree topologies and node ages using Markov chain Monte Carlo (MCMC) in BEAST v1.8.1 [23]. We ran six independent MCMC, 3×10^8 steps long under individual gene models previously selected for reconstruction. We then checked for the convergence and stationarity using Tracer v1.6 [23]. Finally, we extracted the maximum clade credibility tree for combined tree sets using TreeAnnotator v1.8.1 [23].

Results and discussion
α-keratin gene family related to hair features

The advent of mainstream sequencing technologies has allowed the whole genomes of many mammals to be assembled, which provides possibility to explore the genetic basis of adaptation to their specific lifestyles at genome level. In the present study, we scanned the α-keratin repertoire from the genomes of 11 marine mammals (including seven cetaceans, two pinnipeds, polar bear and manatee) and compared with their terrestrial relatives (Table 1) to explore the genetic basis of hair diversity in marine mammals as response for their adaptations to aquatic lifestyle with different degrees. The genomic regions containing keratin gene families are reported to be conservative with special flanking sequences of two clusters, which facilitate to obtain the complete keratin repertoire in mammalian genomes. For example, type I keratins are flanked by SMARCE1 and EIF1 genes while type II keratins are flanked by FAIM2 and EIF4B genes (Fig. 1). However, we have not retrieved the FAIM2 gene in the 5′ flanking regions of five species, including Weddell seal, giant panda, bottlenosed dolphin, sperm whale, and bowhead whale (Fig. 1). Actually, the FAIM2 gene was identified in such five genomes but located in the different chromosomes with α-keratins. Similar phenomenon was also found in the EIF1and EIF4B that was not found in the flanking region of Weddell seal and bottlenose dolphin, respectively (Fig. 1). A total of 780 α-keratin genes were identified in the 16 mammalian species examined in our study, of which 383 belong to type I keratins, 397 belong to type II keratins.

Table 1 Numbers of α-keratin genes present in 16 mammalian species

Species	Total genes	Type I genes	Type II genes	Hair keratins	Hair follicle-specific epithelial keratins	Hair follicle-specific keratins
Weddell seal	54(1)[9]	28[5]	26(1)[4]	16[3]	9[1]	25[4]
Pacific walrus	58(2)[2]	29	29(2)[2]	18(1)	9	27(1)
Giant panda	58(2)[6]	30[2]	28(2)[4]	18(1)[2]	9[1]	27(1)[3]
Polar Bear	58(4)[9]	29[4]	29(4)[5]	19(2)[1]	9[1]	28(2)[2]
Bottlenosed dolphin	36(8)[7]	18(3)[4]	18(5)[3]	12(3)[3]	2(1)[1]	14(4)[4]
Killer whale	36(10)[2]	18(4)[2]	18(6)	11(3)	3 (2)[1]	14(5)[1]
Yangtze finless porpoise	36(16)[3]	18(8)[1]	18(8)[2]	12(9)[1]	3(3)	15(12)[1]
Yangtze River dolphin	34(8)	17(4)	17(4)	9(1)	3(3)	12(4)
Sperm whale	32(6)[3]	16(4)[2]	16(2)[1]	8(1)	1	9(1)
Minke whale	45(13)	19(4)	26(9)	13(5)	4(2)	17(7)
Bowhead whale	42(17)[7]	20(8)[3]	22(9)[4]	13(7)[2]	4(3)[1]	17(10)[3]
Cow	61(4)[2]	29(1)	32(3)[2]	18(1)[1]	9	27(1)[1]
Sheep	60(5)[3]	29	31(5)[3]	18[2]	9	27[2]
Alpaca	54(5)[7]	28(2)[3]	26(3)[4]	16(2)[4]	9	25(2)[4]
African savanna elephant	60(5)[3]	29(2)	31(3)[3]	18(1)[3]	9	27(1)[3]
Florida manatee	56(6)[4]	26(1)[2]	30(5)[2]	17(1)[1]	9(1)	26(2)[1]

Number of pesudogene and incomplete genes are represented in parenthesis and brackets, respectively
Hair keratins and hair follicle-specific epithelial keratins are collectively referred to as hair follicle-specific keratin

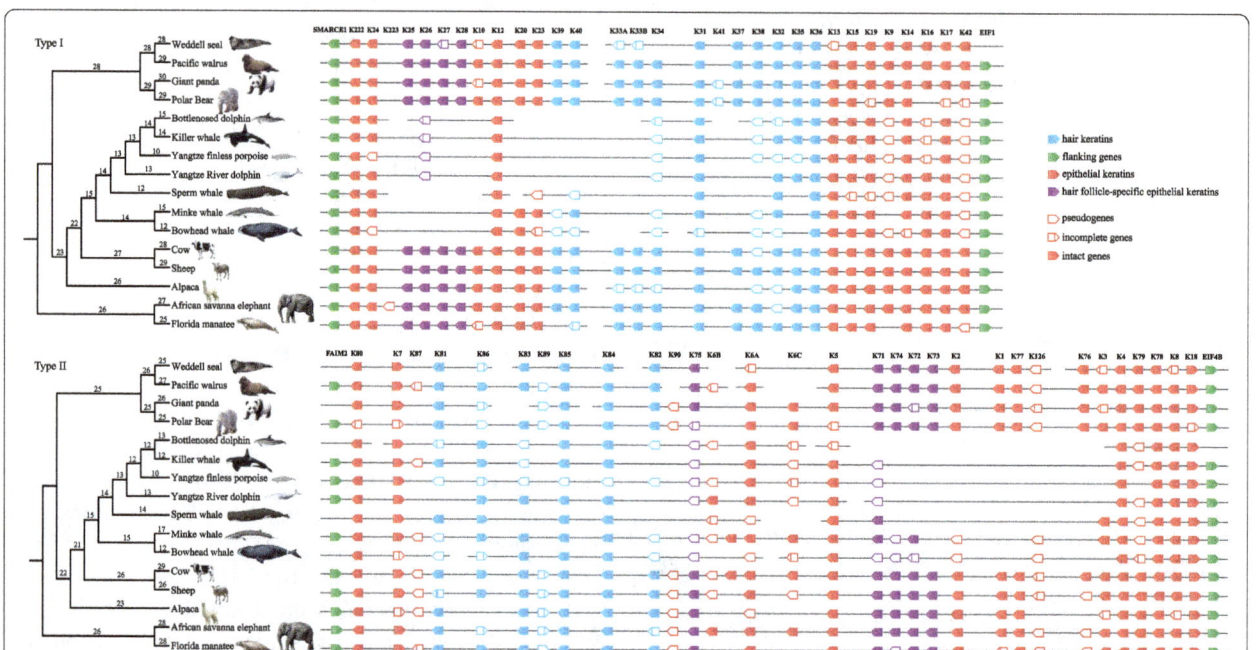

Fig. 1 Genomic organization of α-keratin genes in 16 mammalian species according to phylogenetic tree, genomic position and blast results. The α-keratin genes name was based on that of the human. Connecting lines indicate α-keratin genes on the same chromosome/genomic scaffold. Direction of the arrows of each α-keratin gene indicate the direction in the contig. Filled figures: intact genes; empty figures: pseudogenes; empty figures with a vertical line: incomplete genes. In addition, all genes are divided into different colors: flanking genes (*green*); hair keratins (*blue*); epithelial keratins (*red*); hair follicle-specific epithelial keratins (*purple*). The numbers show the. Ancestral states of gene family sizes for each node in the phylogenetic tree

In contrast with the terrestrial mammals, the marine mammals have lost some α-keratins and are characterized with increased pseudogenization rate, likely related with their hair diversities in response to their living environment. Average of 44.27 α-keratin genes observed in the marine mammals has slightly differences to the terrestrial mammals which has averagely 58.8 α-keratins. Moreover, the pseudogenization rate of marine mammals (18.69%) is nearly three times of that of their terrestrial relatives (7.17%). In the case of the marine mammals, however, different evolution pattern of α-keratins among them might be responsible for their different hair characteristics. We found an average number of α-keratin genes in cetaceans of 37.29, which was significantly lower than that of artiodactylian species with an average number of 58.33. The number of α-keratin genes ranged from 32 in the sperm whale to 45 in the minke whale, while in artiodactylans it ranged from 54 in the alpaca to 61 in the cow (Table 1). The higher rate of pseudogenization rate was identified in cetacean species (average 29.89% compared to artiodactylans' average of 8.00%). The cetacean species are nearly completely absent of hair coat with only a few bristle in order to reduce drag of swimming for adaptation to completely aquatic environment. There is evidence of an apparent reduction in the α-keratin gene repertoire and increased rate of pseudogenization in cetaceans, which may be associated with their hairless phenotype. For the fully aquatic manatee, body hairs are greatly reduced but with perioral bristles and bristle-like hairs on the oral disk that have specialized sensory and feeding function [24]. Accordingly, the manatee still remained the comparable number (56 in manatee vs. 60 in elephant) and rate of pseudogenization (8.33% in manatee vs. 10.71% in elephant) to its terrestrial relative, i.e. elephant (Table 1). By contrast, gene repertoires in the pinnipeds (56) and polar bear (58) were similar to that of panda (58) (Table 1). In addition, we detected 49 and 45 intact genes in pinnipeds and polar bear, comparable to that of the panda with 50 ones. Similar evolution pattern of α-keratin gene repertoire in the pinnipeds, polar bear and their terrestrial relatives is likely due to abundant hair to keep warm when they went ashore to mate, give birth, molt or escape from predators.

Further, we found that the gene gain and loss of α-keratin was consistent with the hair diversification of marine mammals. we employed the software CAFÉ to estimate the gene gain and loss of α-keratin typeI and typeII at each ancestral node in the phylogenetic tree. It was noted that loss of 6.5 genes in the ~53–56 million years since the cetacean ancestor split from the cow/sheep clade. By contrast, the gain of five genes in the common ancestor of the cow/sheep. However, average one gene gain or lose in other marine mammals and their terrestrial relatives.

Hair follicle-specific keratin genes consisted of 9 hair follicle-specific epithelial keratin genes (K71-K75, K25-K28) and 17 hair keratins (K31-K40, K81-K86) expressed in hair follicle and hair which could directly affect hair growth [25]. A large number of diseases of hair were caused by mutations in hair follicle-specific keratin genes. For example, a mutation in K75 leads to loose anagen hair syndrome, which is characterized by easily pluckable hair [26]. In addition, the K71 mutation will make almost all types of mouse hair abnormal and easily fall off [27]. Subsequently, increased gene loss and pseudogenization rate were found to focus on hair follicle-specific keratin genes in cetacean lineages (Fig. 1). Only an average of 14 hair follicle-specific keratins were identified in cetaceans but with an average of 26.33 in artiodactylans. Most importantly, the averaged pseudogenized rate in cetaceans was 43.88%, 11 times higher than that in artiodactylans (3.80%). In summary, the unique gene loss and/or loss of their function in hair follicle-specific keratin genes might contribute to the hairless phenotype.

Species-specific evolution of α-keratin gene repertoire
The α-keratin gene family was further classified according to the phylogenetic relationship reconstructed using ML and Bayesian approaches. We found that almost all of the α-keratin genes of 16 species could be clustered into respective clades according to each subfamily, which can be regarded as orthologous genes in phylogenetic tree (Figs. 2 and 3). However, some genes within one specie grouped together rather than with the members of the same family in other species, such as among the genes of K31, K33A, K33B and K34 in type I, as well as the K6A, K6B, K6C, K81, K83, K87 and K86 in the type II (Figs. 2 and 3). This phenomenon suggested that such members within a repetitive family did not evolve independently of each other but under concerted evolution [13], which was also found in other mammals such as horse, dog, human, etc. [14]. We further used the RDP4 to test whether the gene conversion and recombination resulted in these 19 gene groups supposed to be under concerted evolution. The result showed that only three gene group supposed to be under concerted evolution was due to significant level of gene conversion and recombination (Additional file 2).

Species-specific evolution of α-keratin gene family was identified in the marine mammals, including differences in the total number of genes, functional genes, as well as pseudogenes. Nine keratin subfamilies were completely absent in cetaceans when compared to its terrestrial relative of artiodactylans, such as six type I genes (K10, K25, K27, K28, K33A, K33B) and three type II genes (K1, K73, K77) (Fig. 1). Most importantly, the cetacean species were also exclusively lost of 10 keratin genes (type I: K10, K25, K27, K28, K33A, K33B, K37; type II:

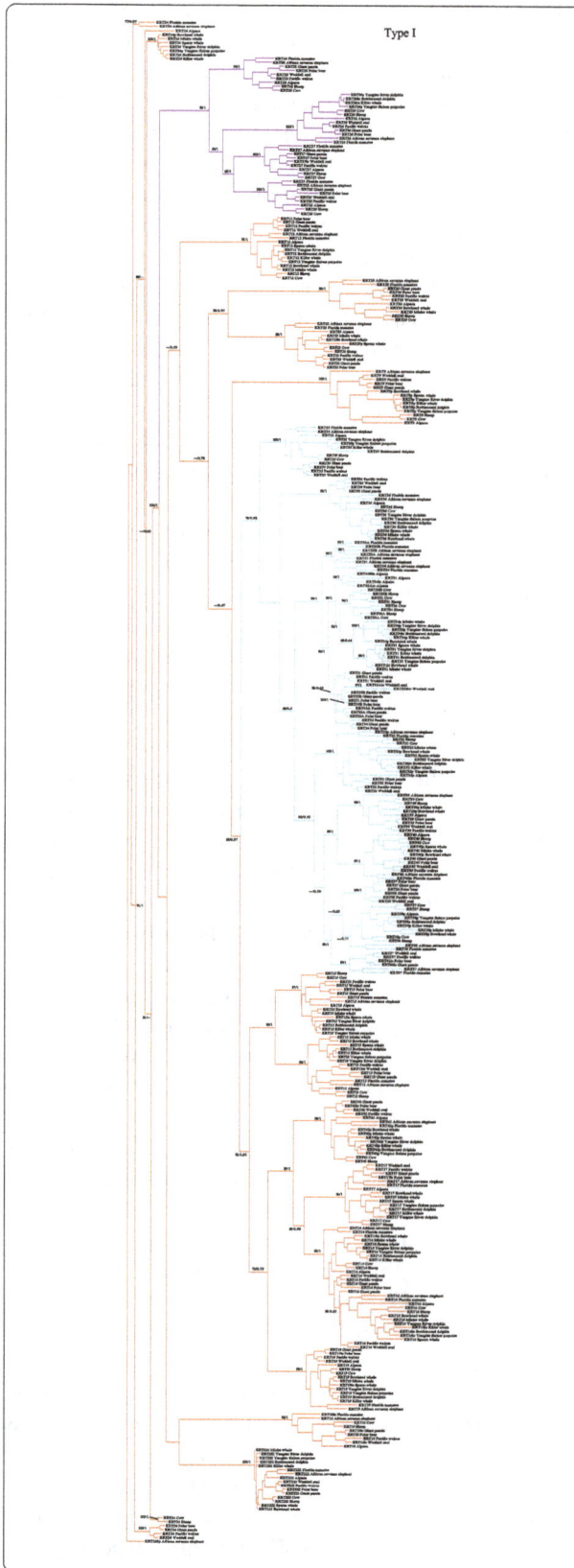

Fig. 2 Phylogenetic tree of type I keratins. Maximum likelihood phylogram describing phylogenetic relationships among the type I keratin genes. Numbers above the nodes correspond to maximum likelihood bootstrap support values, and numbers below the nodes correspond to Bayesian posterior probabilities. Branches in *blue* indicated hair-type keratins, *purple* indicated hair follicle-specific epithelial keratins and *red* indicated epithelial keratins

K1, K73, K77) when compared with other marine mammals (Fig. 1). Besides these lost genes, we also examined 12 pseudogenes in cetaceans (type I: K9, K26, K34, K38, K39, K42; type II: K2, K6B, K74, K75, K79, K82, Fig. 1). Previous study has reported that deficient for K6B (epithelial keratins) may contributed to the absence of hair and nail in mice [28]. By contrast, mutations in K1 and K10 were shown to be associated with bullous congenital ichthyosiform erythroderma (BCIE) [29]. Collectively, cetaceans were exclusively lost K6B, K1 and K10, which is likely to contribute to its hair-loss phenotype to adapt to the completely aquatic environment. For complete aquatic manatee, however, we found that only four keratin subfamilies, such as K9, K6C, K6B, K39, were uniquely lost when compared with its terrestrial relatives, the elephant (Fig. 1).

When only cetaceans were considered, we found the average number of α-keratin genes in baleen whales was 43.5, slightly higher than that in toothed whales (averaged 34.8). Specially, pseudogenization rate of hair follicle-specific keratin genes was higher in toothed whales (42.86%) than in baleen whales (35.29%). The much higher number of intact functional keratin genes in the baleen whales may be associated with the presence of keratinized baleens.

Sharing keratin genes loss and pseudogenes in cetaceans and manatee as response for convergent hair-less phenotype

Marine mammals are relatively independent in evolution but with similar phenotype changes and associated physiological features to adaptation to aquatic lifestyle. Remarkably, both cetaceans and sirenians are fully aquatic and therefore encompass convergent hair-less phenotype to adapt to aquatic environment, since the lack of fur improves their hydrodynamic and subaquatic movements. Cetaceans have apparent reduction of the α-keratin gene repertoire and increased rate of pseudogenization while the manatees have a comparable number of α-keratin and pseudogenization rate when compared to their respective terrestrial relatives. Importantly, both groups were found to share common gene loss in the four subfamilies, i.e. K39, K9, K42, K74 that play a key role in hair development, likely to be responsible for hair-less phenotype. For example, it has been reported that heterozygous mutations in K74 could

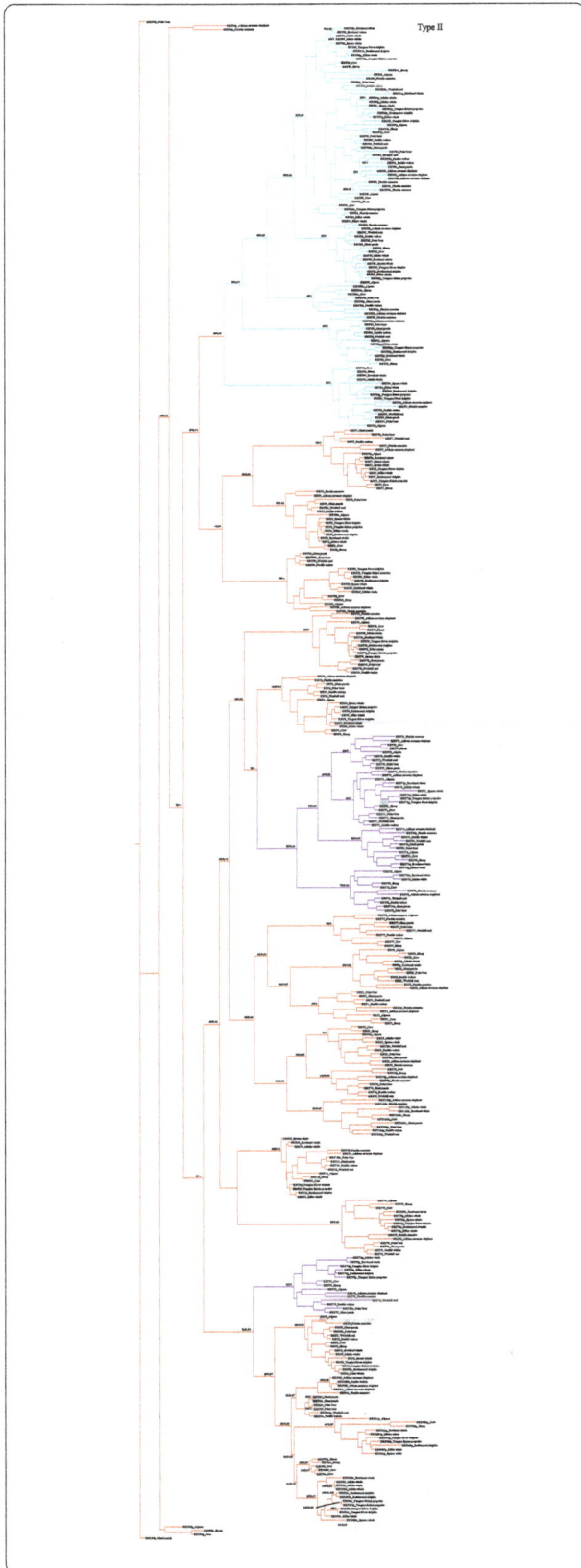

Fig. 3 Phylogenetic tree of type II keratins. Maximum likelihood phylogram describing phylogenetic relationships among the type II keratin genes. Numbers above the nodes correspond to maximum likelihood bootstrap support values, and numbers below the nodes correspond to Bayesian posterior probabilities. Branches in *blue* indicated hair-type keratins, *purple* indicated hair follicle-specific epithelial keratins and *red* indicated epithelial keratins

cause autosomal dominant woolly hair, and/or hypotrichosis simplex [30] whereas mutations in K9 gene caused epidermolytic palmoplantar keratoderma, leading to manifests cytolysis and epidermal thickening [31]. In addition, K39 as a member of the hair-keratins, was identified to play a key role in late hair differentiation [32]. By contrast, when another components of hair, KRTAP, are considered, a significant gene number reduction (81 in manatee vs. 112 in elephant) and increased pseudogenization rate (41.98% in manatees vs. 20% in elephant) were detected in manatee. Similarly, it has been recently reported that dolphins only have 35 KRTAP genes (compared to 145 genes in the cow) but with highest pseudogenization rate (74% relative to the 19% of mammalian average) [13]. However, we have detected average of 65 functional KRTAP genes in the pinnipeds that have abundant hair, which were comparable to that of terrestrial relatives, such as panda with 64 KRTAP genes [13]. Collectively, a reduced number of functional KRTAP genes, high percentage of KRTAP pseudogenes, and common keratin gene loss in both cetaceans and manatees might be responsible for their convergent hair-less phenotype.

Conclusion

In this study, we provide a comprehensive characterization of α-keratin genes among marine mammals and shed light on the mechanisms involved in the evolution of this gene family. Our results show an apparent reduction in the α-keratin gene repertoire and increased rate of pseudogenization in cetaceans when compared to terrestrial counterparts, which may be associated with their hairless phenotype. In contrast, the evolution pattern of α-keratin genes in pinnipeds was comparable to that of their respective terrestrial relatives, which is well matched with its fur coat. Interestingly, we found a reduced number of functional KRTAP genes, high percentage of KRTAP pseudogenes, and common keratin gene loss in both cetaceans and manatees, which might be as a result of convergent hair-loss phenotype to inhabit completely aquatic environment. Species-specific evolution of α-keratin gene repertoire identified in marine mammals was likely to contribute to their hair diversities phenotype although their life or part inhabiting in aquatic environment. Our study first indicated the genetic basis of hair diversities of marine mammals.

Abbreviations

BCIE: Bullous congenital ichthyosiform erythroderma; EIF1: Eukaryotic translation initiation factor 1; EIF4B: Eukaryotic translation initiation factor 4B; FAIM2: Fas apoptotic inhibitory molecule 2; HGT: High glycine-tyrosine; HS: High/ultrahigh cysteine; KRTAPs: Keratin associated-proteins; ML: Maximum likelihood; SMARCE1: SWI/SNF related, matrix associated, actin dependent regulator of chromatin, subfamily e, member 1

Acknowledgements

We thank members of the Jiangsu Key Laboratory for Biodiversity and Biotechnology, Nanjing Normal University, for their contributions to this paper. A special thank-you is also due from Zhengfei Wang and Professor Qi Wu for their technical support.

Funding

This work was supported by the National Natural Science Foundation of China (NSFC, grant number 31570379 to S X), the National Science Fund for Distinguished Young Scholars to GY (grant number 31325025), and the State Key Program of National Natural Science of China to GY (grant number 31630071); the National Key Programme of Research and Development, Ministry of Science and Technology (Grant number 2016YFC0503200 to GY and SX), the Priority Academic Program Development of Jiangsu Higher Education Institutions to GY. and S X, the Natural Science Foundation of Jiangsu Province of China (grant number BK20141449) to SX.

Authors' contributions

SX and GY conceived the project and designed the experiments. XS, ZZ, YS, and JL performed the molecular evolution analysis. XS wrote the manuscript, SX and GY improve the manuscript. All authors read and approved the final manuscript.

Competing interests

The authors declare that they have no competing interests.

References

1. Uhen MD. Evolution of marine mammals: back to the sea after 300 million years. Anat Rec (Hoboken). 2007;290:514–22.
2. Berta A, Sumich JL, Kovacs KM. Marine mammals: evolutionary biology. Cambridge: Academic Press. 2005.
3. Bergman J. Why mammal body hair is an evolutionary enigma. CRS Quarterly. 2004;40:240–3.
4. Maderson PFA. Mammalian skin evolution: a reevaluation. Exp Dermatol. 2003;12:233–6.
5. Berta A. Pinnipedia, overview. J Zool. 2002;83:1525–31.
6. Drake SE, Crish SD, George JC, Stimmelmayr R, Thewissen JG. Sensory hairs in the bowhead whale, Balaena Mysticetus (Cetacea, Mammalia). Anat Rec (Hoboken). 2015;298:1327–35.
7. Yonenaga A. Future man-made fiber. High-performance and specialty fibers. Japan: Springer; 2016. p. 435–51.
8. Riedman M, Estes JA. The sea otter (Enhydra Lutris): behavior, ecology, and natural history. Biol Reprod. 1990;90:1–126.
9. Shimomura Y, Ito M. Human hair keratin-associated proteins. J Investig Dermatol Symp Proc. 2005;10:230–3.
10. Moll R, Divo M, Langbein L. The human keratins: biology and pathology. Histochem Cell Biol. 2008;129:705–33.
11. Schweizer J, Bowden PE, Coulombe PA, Langbein L, Lane EB, Magin TM, Maltais L, Omary MB, Parry DA, Rogers MA, Wright MW. New consensus nomenclature for mammalian keratins. J Cell Biol 2006;174:169-174.
12. Wu DD, Irwin DM, Zhang YP. Molecular evolution of the keratin associated protein gene family in mammals, role in the evolution of mammalian hair. BMC Evol Biol. 2008;8:241.
13. Khan I, Maldonado E, Vasconcelos V, O'Brien SJ, Johnson WE, Antunes A. Mammalian keratin associated proteins (KRTAPs) subgenomes: disentangling hair diversity and adaptation to terrestrial and aquatic environments. BMC Genomics. 2014;15:779.
14. Nery MF, Arroyo JI, Opazo JC. Increased rate of hair keratin gene loss in the cetacean lineage. BMC Genomics. 2014;15:869.
15. Altschul SF, Madden TL, Schäffer AA, Zhang JH, Zhang Z, Miller W, Lipman DJ. Gapped BLAST and PSI-BLAST: a new generation of protein database search programs. Nucleic Acids Res. 1997;25:3389–402.
16. Cheng J, Liu C, Koopman WJ, Mountz JD. Characterization of human Fas gene. Exon/intron organization and promoter region. J Immunol. 1995;154:1239–45.
17. Altschul SF, Gish W, Miller W, Myers EW, Lipman DJ. Basic local alignment search tool. J Mol Biol. 1990;215:403–10.
18. Tamura K, Stecher G, Peterson D, Filipski A, Kumar S. MEGA6: molecular evolutionary genetics analysis version 6.0. Mol Biol Evol. 2013;30:2725–9.
19. Stamatakis A. RAxML version 8: a tool for phylogenetic analysis and post-analysis of large phylogenies. Bioinformatics. 2014;30:1312.
20. Ronquist F, Huelsenbeck JP. MrBrayes 3: Bayesian phylogentetic inference under mixed models. Bioinformatics. 2003;19:1572–4.
21. Martin DP, Murrell B, Golden M, Khoosal A, Muhire B. RDP4: detection and analysis of recombination patterns in virus genomes. Virus Evol. 2015;1:vev003.
22. Han MV, Thomas GW, Lugo-Martinez J, Hahn MW. Estimating gene gain and loss rates in the presence of error in genome assembly and annotation using CAFE 3. Mol Biol Evol. 2013;30:1987–97.
23. Drummond AJ, Suchard MA, Xie D, Rambaut A. Bayesian phylogenetics with BEAUti and the BEAST 1.7. Mol Biol Evol. 2012;29:1969–73.
24. Reep RL, Stoll ML, Marshall CD, Homer BL, Samuelson DA. Microanatomy of facial vibrissae in the Florida manatee: the basis for specialized sensory function and Oripulation. Brain Behav Evol. 2001;58:1–14.
25. Schweizer J, Langbein L, Rogers MA, Winter H. Hair follicle-specific keratins and their diseases. Exp Cell Res. 2007;313:2010–20.
26. Chapalain V, Winter H, Langbein L, Le Roy JM, Labrèze C, Nikolic M, Schweizer J, Taïeb A. Is the loose anagen hair syndrome a keratin disorder? A clinical and molecular study. Arch Dermatol. 2002;138:501–6.
27. Peters T, Sedlmeier R, Büssow H, Runkel F, Lüers GH, Korthaus D, Fuchs H, Hrabé de Angelis M, Stumm G, Russ AP, Porter RM, Augustin M, Franz T. Alopecia in a novel mouse model RCO3 is caused by mK6irs1 deficiency. J Invest Dermatol. 2003;121:674–80.
28. Wojcik SM, Longley MA, Roop DR. Discovery of a novel Murine keratin 6 (K6) Isoform explains the absence of hair and nail defects in mice deficient for K6a and K6b. J Cell Biol. 2001;154:619–30.
29. McLean WH, Eady RA, Dopping-Hepenstal PJ, McMillan JR, Leigh IM, Navsaria HA, Higgins C, Harper JI, Paige DG, Morley SM. Mutations in the rod 1A domain of keratins 1 and 10 in bullous congenital ichthyosiform erythroderma (BCIE). J Invest Dermatol. 1994;102:4–30.
30. Raykova D, Klar J, Azhar A, Khan TN, Malik K, Iqbal M, Tariq M, Baig SM, Dahl N. Autosomal recessive transmission of a rare KRT74 variant causes hair and nail ectodermal dysplasia: allelism with dominant woolly hair/hypotrichosis. PLoS One. 2014;9:e93607.
31. Torchard D, Blanchet-Bardon C, Serova O, Langbein L, Narod S, Janin N, Goguel AF, Bernheim A, Franke WW, Lenoir GM. Epidermolytic palmoplantar keratoderma cosegrates with a keratin 9 mutation in a pedigree with breast and ovarian cancer. Nat Genet. 1994;6:106–10.
32. Langbein L, Rogers MA, Praetzel-Wunder S, Bockler D, Schirmacher P, Schweizer J. Novel type I hair keratins K39 and K40 are the last to be expressed in differentiation of the hair: completion of the human hair keratin catalog. J Invest Dermatol. 2007;127:1532–5.

6

Sea urchin growth dynamics at microstructural length scale revealed by Mn-labeling and cathodoluminescence imaging

Przemysław Gorzelak[1]*(iD), Aurélie Dery[2], Philippe Dubois[2] and Jarosław Stolarski[1]

Abstract

Background: Fluorochrome staining is among the most widely used techniques to study growth dynamics of echinoderms. However, it fails to detect fine-scale increments because produced marks are commonly diffusely distributed within the skeleton. In this paper we investigated the potential of trace element (manganese) labeling and subsequent cathodoluminescence (CL) imaging in fine-scale growth studies of echinoderms.

Results: Three species of sea urchins (*Paracentrotus lividus*, *Echinometra* sp. and *Prionocidaris baculosa*) were incubated for different periods of time in seawater enriched in different Mn^{2+} concentrations (1 mg/L; 3 mg/L; 61.6 mg/L). Labeling with low Mn^{2+} concentrations (at 1 mg/L and 3 mg/L) had no effect on behavior, growth and survival of sea urchins in contrast to the high Mn^{2+} dosage (at 61.6 mg/L) that resulted in lack of skeleton growth. Under CL, manganese produced clearly visible luminescent growth fronts in these specimens (observed in sectioned skeletal parts), which allowed for a determination of the average extension rates and provided direct insights into the morphogenesis of different types of ossicles. The three species tend to follow the same patterns of growth. Spine growth starts with the formation of microspines which are simultaneously becoming reinforced by addition of thickening layers. Spine septa develop via deposition of porous stereom that is rapidly (within less than 2 days) filled by secondary calcite. Development of the inner cortex in cidaroids begins with the formation of microspines which grow at ~3.5 μm/day. Later on, deposition of the outer polycrystalline cortex with spinules and protuberances proceeds at ~12 μm/day. The growth of tooth can be rapid (up to ~1.8 mm/day) and starts with the formation of primary plates (pp) in plumula. Later on, during the further growth of pp in aboral and lateral directions, secondary extensions develop inside (in chronological order: lamellae, needles, secondary plate, prisms and carinar processes), which are increasingly being solidified towards the incisal end. Interradial growth in the ambital interambulacral test plates exceeds meridional growth and inner thickening.

Conclusions: Mn^{2+} labeling coupled with CL imaging is a promising, low-cost and easily applicable method to study growth dynamics of echinoderms at the micro-length scale. The method allowed us to evaluate and refine models of echinoid skeleton morphogenesis.

Keywords: Biomineralization, Calcite, Labeling, CL, Manganese

* Correspondence: pgorzelak@twarda.pan.pl
[1]Institute of Paleobiology, Polish Academy of Sciences, Twarda 51/55, 00-818 Warsaw, Poland
Full list of author information is available at the end of the article

Background

A number of different methodological approaches have been used to study the growth of echinoderms at different length scales. These include indirect examination of natural growth lines and more direct tagging techniques (via plastic tube slipped over the skeleton, tag inserted into a drilled skeleton, passive integrated transponder tags, coded wire tags, fluorochrome chemical markers and stable isotopes) [1–3]. Among these methods, fluorescent dyes, such as tetracycline and calcein, had attracted considerable attention [4–11]. Both of these chemicals bind to calcium ions and are incorporated into the newly formed calcium carbonate skeleton during biomineralization. Under fluorescence microscopy, tetracycline and calcein fluoresce yellow and green, respectively. However, one of the major disadvantages of using fluorochrome markers in fine-scale biomineralization studies of echinoderms is that produced marks are often diffuse, making it very difficult to measure the growth

(cf. Fig. 1 in [9]). Notably, using fluorescent dyes, only millimeter-scale growth fractions can be detected. Furthermore, such chemical markers may stress echinoderms and perturb biomineralization [10]. Recently, Gorzelak et al. [2, 3] introduced a new method of labeling the growing echinoderm skeletons with stable isotope ^{26}Mg and NanoSIMS imaging at sub-micrometer spatial resolution. However, although NanoSIMS provides a level of sensitivity unmatched by any other techniques (spatial resolution up to about ~50 nm), it is not a widely accessible analytical tool and the cost of stable isotope is high. Thus, a low-cost, highly efficient and more easily applicable method is still desired for biomineralization studies of echinoderms.

Cathodoluminescence is the emission of a photon of characteristic wavelengths during excitation by an incident electron bombardment. CL-luminescence properties of calcite are usually attributed to different proportions of incorporated Mn^{2+} replacing Ca^{2+} in the crystal lattice. Mn^{2+}

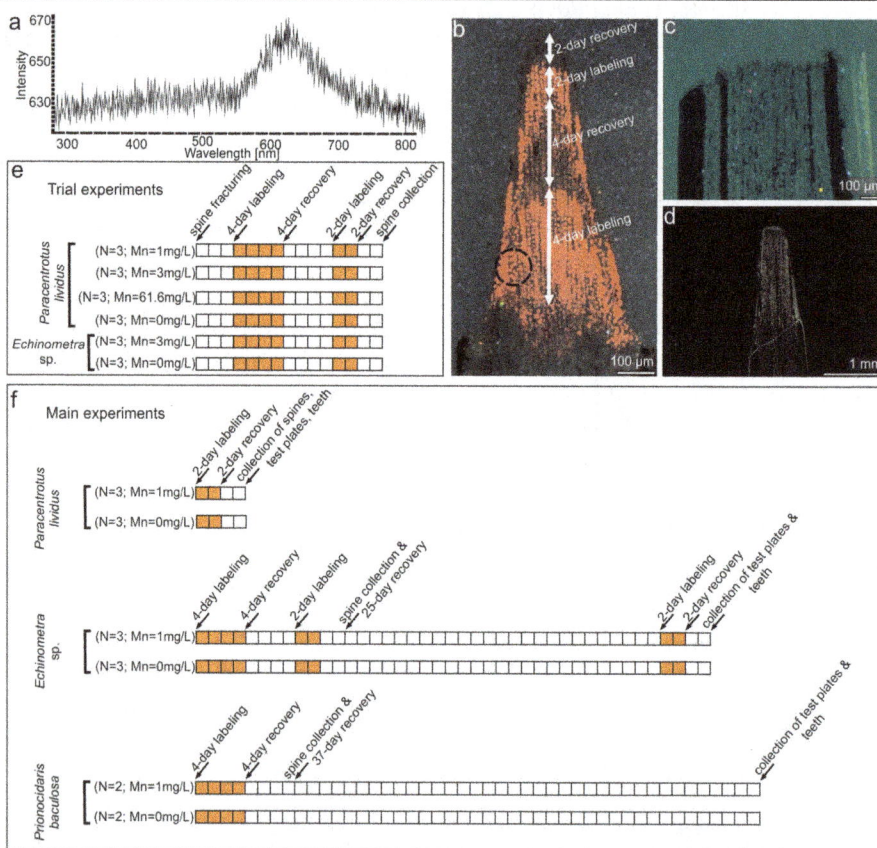

Fig. 1 Research methods and design of experiments. **a** example of CL-activated UV–VIS spectrum of a fragment of luminescent spine (marked with a black circle in (**b**) of *Echinometra* sp. labeled with Mn^{2+} at 3 mg/L showing Mn^{2+} emission maximum at 632 nm (Mn^{2+} activation in Ca^{2+} and Mg^{2+} positions in calcite), **b** lateral view of spine showing no growth differences during and between labeling events (orange-red skeletal regions indicate enhanced Mn^{2+} concentrations due to twice Mn^{2+} labeling; dark ragions indicate growth in normal (without Mn^{2+}) seawater), **c** lateral view of spine of *Paracentrotus lividus* labeled with Mn^{2+} at 61.4 mg/L without any signs of regeneration, **d**, lateral view of spine of control specimen of *Paracentrotus lividus* without any signs of luminescence, cutting fracture is delineated with dotted line, experimental design of trial (**e**) and main (**f**) experiments; each box represents a 24 h period

is considered the most important activator element whereas Fe^{2+} is the most important quencher element in carbonates [12]. The activation of Mn^{2+} in calcite can easily be recognized by a broad emission band in the yellow-orange range (wavelength range: 605–620 nm). However, in the case of echinoderm Mg-calcite, Mn^{2+} typically substitutes two different lattice positions (i.e., Ca^{2+} position in Mg-free calcite „main-structure" and Mg^{2+} position in magnesite-like „sub-structure"; [13, 14]) leading to the shift of Mn^{2+} peak to higher wavelenghts (orange-red emission band centered at ~632 nm, see Fig. 1a, b).

Concentration of mangenese in seawater is typically very low (0.1–0.2 µg/L; see [15, 16]). Therefore, recent biogenic carbonates, although not without exceptions, are typically non-luminescent or very weakly luminescent [17–26]. Owing to this fact, labeling with enhanced Mn^{2+} concentration in seawater proved to be a useful method in studying growth dynamics of some invertebrates, in particular some species of molluscs [19–21]. With respect to echinoderms, it has been argued that most of them are non-luminescent or reveal light blue zones of intrinsic luminescence [12, 18, 23, 27] although it has been shown that test plates of some field collected echinoids from offshore environments may also display orange luminescence zoning [28].

In this paper, we test the feasibility of Mn-labeling method for studying growth dynamics of echinoderms for the first time. Our results provide new insights into the morphogenesis of various types of echinoderm ossicles.

Methods

Mn-labeling experiments were performed in the Laboratory of Marine Biology in Bruxelles (Université Libre de Bruxelles, Belgium) using three species of sea urchins (*Paracentrotus lividus*, *Echinometra* sp. and *Prionocidaris baculosa*). Specimens of *Paracentrotus lividus*, 2.5–3.5 cm in ambital diameter, were obtained from the aquaculture facility in Luc-sur-Mer (English Channel, Normandy, France). Before the beginning of experiments the specimens were transported to the Laboratory of Marine Biology in Bruxelles and maintained for about two weeks in an aerated, closed circuit aquarium, containing ~1000 l of natural seawater under controlled and constant temperature (~16 °C), salinity (~33.5 psu), and pH conditions (~8.0), similar to the conditions in the aquaculture facility. Commercially obtained specimens of *Echinometra* sp. (3–3.5 cm in ambital diameter) and *Prionocidaris baculosa* (~1.5 cm in ambital diameter) collected from the coastal regions of Indonesia (presice locality unknown) and Cebu in the Philippines, respectively, were maintained for at least 2 weeks in two aerated aquariums, each containing about 100 l of natural seawater under constant temperature (~25.5 °C), salinity (~35 psu), and pH conditions (~8).

To test feasibility of the Mn-labeling method for studying growth dynamics of sea urchins, trial experiments on regenerating spines of *Paracentrotus lividus* and *Echinometra* sp. were conducted first. These initial experiments involved the following steps: (1) spine fracturing; (2) 3 days (72 h) recovery; (3) 4 days (96 h) Mn^{2+} labeling experiment; (4) 4 days (96 h) recovery; (5) 2 days (48 h) Mn^{2+} labeling experiment; (4) 2 days (48 h) recovery after which the experiment ended and the spines were removed from the tests. These steps are explained in details below.

Twelve specimens of *Paracentrotus lividus* and 6 specimens of *Echinometra* sp. were prepared 3 days before the beginning of the labeling experiment by cutting distal tips of three long primary spines. Trial Mn-labeling experiments started 3 days after cutting the spine tips when the process of their regeneration initiated. Twelve specimens of *Paracentrotus lividus* were incubated separately for 4 days in small beakers, each containing 1 l of natural seawater mixed with different concentrations of dissolved $MnCl_2$, $4H_2O$ (Sigma-Aldrich) resulting in 4 different nominal Mn^{2+} concentrations (control seawater without added Mn^{2+}, 1 mg/L, 3 mg/L, 61.6 mg/L), with 3 individuals per concentration. Low concentrations (1 mg/L and 3 mg/L) were used because they seemed to be sufficiently effective to enable mark detection under CL [17]. Importantly, it was previously demonstrated that higher Mn concentrations may induce mortality, stress, developmental defects or even growth inhibition in echinoderms [29–32]. For comparison purpose, we also used a very high manganese concentration (61.6 mg/L) which in previous reports was shown to prevent skeleton growth in echinoid embryos ([29–31]; see also Fig. 1c).

The beakers were covered by hard plastic covers and equipped with an air bubbling system. The specimens of *Echinometra* sp. were only incubated at Mn^{2+} = 3 mg/L and control seawater without added Mn^{2+} (3 individuals per concentration). After the first incubation period, all sea urchins were transferred to new beakers containing 1 l of natural seawater without Mn^{2+}. After 4 days of recovery, a second 2-day labeling procedure was repeated using the same Mn^{2+} concentrations. Finally, the specimens were again returned to normal conditions for 2 days, after which the experiment was terminated and selected spines with regenerated tips were removed from the urchins (Fig. 1e).

Dissected spines were first soaked in NaOH 1 M for 2 h, and then in 2.5% NaClO for 15 min in order to remove soft tissues. They were then rinsed in MilliQ water for 90 s and dried at 50 °C for ~48 h. Variously oriented thin sections were prepared by polishing the spines down to about ~25 µm. Thin sections were finally coated with carbon and examined with a cathodoluminescence (CL) microscope at the Institute of Paleobiology of the Polish Academy of Sciences in Warsaw. This microscope, linked to a Kappa video camera for recording images, is equipped with a hot cathode and is integrated with UV-VIS specto-graph. An electron energy of 14 keV and, depending on a sample, beam currents between 0.7–0.15 mA were used for both CL microscopy and spectroscopy. The exposure time for recording images was about 3 s. Integration time for CL-emission spectra of luminescent ossicles was 100 s.

During the trial experiments all sea urchins labeled with low concentration of manganese (1 mg/L and 3 mg/L) were actively relocating within beakers, did not loose any spine and their podia were actively moving, which suggested that the animals were probably not significantly stressed by increased level of Mn^{2+} as sup-ported by very similar growth rates during and between labeling events (Fig. 1b, see Results section). However, sea urchins subjected to high Mn^{2+} concentration (61.6 mg/L) were clearly stressed, in that growth was ceased (Fig. 1c), some spines were lost and no moving with podia was observed.

Having shown that low concentration of Mn^{2+} appears to be optimal for mark detection in that the two labeling events in the pilot samples were clearly visible under CL imaging (Fig. 2), we performed additional labeling exper-iments to further explore growth dynamics and morpho-genesis of different ossicle types in three different sea urchin species. For this purpose, we used 9 new speci-mens of *Paracentrotus lividus*, 6 specimens of *Echinome-tra* sp., which were previously used in the above trial experiments, and 4 new specimens of *Prionocidaris baculosa* (Fig. 1f). The specimens were incubated for ei-ther 2 (*Paracentrotus lividus* and *Echinometra* sp.) or 4 (*Prionocidaris baculosa*) days in seawater with the low Mn^{2+} concentrations (1 mg/L). After this immersion period, the specimens of *Paracentrotus lividus* and *Echino-metra* sp. were transferred for 2 days to beakers containing normal seawater without Mn^{2+}. They were then sacrificed and 3 types of ossicles (non-regenerative spine, tooth and interambulacral ambital test plate from each individual) were prepared for CL analyses in a manner similar to that described above. After 4-days labeling of the specimens of *Prionocidaris baculosa*, they returned to normal conditions for 4 days after which a few primary spines were removed from the urchins and prepared for CL analyses. These indi-viduals were sacrificed after 41 days from the date of the end of Mn-labeling after which interambularal test plates and teeth were also collected for additional CL analyses.

During all labeling experiments, the specimens of *Para-centrotus lividus* were kept at 17 °C, whereas the speci-mens of *Echinometra* sp. and *Prionocidaris baculosa* were maintained at 25 °C (Table 1). All specimens were kept under 12 h–light/12 h–dark photoperiod. Three param-eters (temperature, salinity, and electromotive force-emf) were monitored three times per day using respect-ively a WTW Multi 340i multimeter equipped with a

conductivity cell and integrated temperature sensor and a Metrohm pH-meter (826 pH mobile) equipped with a microelectrode (reference 6.0224.100, Metrohm), and calibrated with CertiPUR® buffer solutions pH 4.00 and 7.00 (Merck, Darmstadt, Germany). All emf measure-ments were converted to pH in total scale according to DelValls and Dickson's [33] method with TRIS/AMP buffers (kindly provided by the Biogeochemistry and Earth System Modeling Laboratory of the Université Libre de Bruxelles, Belgium). Differences between these parameters were assessed using one-way ANOVA with Tukey's multiple comparison tests. These parameters were very comparable and constant throughout all experi-ments (Table 1). Seawater in each beaker was replaced every day in all the above experiments. Alkalinity values (ranging from 2.4 to 2.6 mmol/kg) were determined by a potentiometric titration with HCl 0.1 M with 0.7 M NaCl using a Titrino 718 STAT Metrohm (Switzerland), and were calculated using the Gran function [34].

Before the beginning of all labeling experiments, the sea urchins were fed ad libitum with Sea Urchin Diets (Zeigler Bros) but during trial and further growth experi-ments all specimens were fed by only one Zeigler pellet every two days.

Microstructural details of selected samples, etched in 0.1% formic acid solution, dried and coated with carbon [35], were observed in a Phillips XL30 scanning electron microscope (SEM) at the Institute of Paleobiology, Polish Academy of Sciences in Warsaw.

Although our research was focused on qualitative as-pects of morphogenesis of various ossicle types, some quantitative data, e.g., the average extension rates during and between labelings (expressed in µm/day) were also collected. These data were analyzed with the Wilcoxon two-sample paired signed rank tests in PAST [36] due to not normally distributed data. We set the significance level to $\alpha = 0.05$. Thin sections are housed at the Institute of Paleobiology, Polish Academy of Sciences, Warsaw (ZPALV.42Mn-CL).

Results

Trial experiments

The lengths of regenerated tips in *Paracentrotus lividus* and *Echinometra* sp. measured from SEM microphoto-graphs were highly variable between spines of the same individual, between individuals within the same treat-ments, and between individuals from different treatments (Tables 2, 3). The average longitudinal extension rates of regenerating spines in controls and specimens labeled with low Mn^{2+} concentration (1 and/or 3 mg/L) were rather high (Tables 2, 3). On the other hand, no regen-eration was observed in all specimens labeled in sea-water with the highest Mn^{2+} concentration (61.6 mg/L).

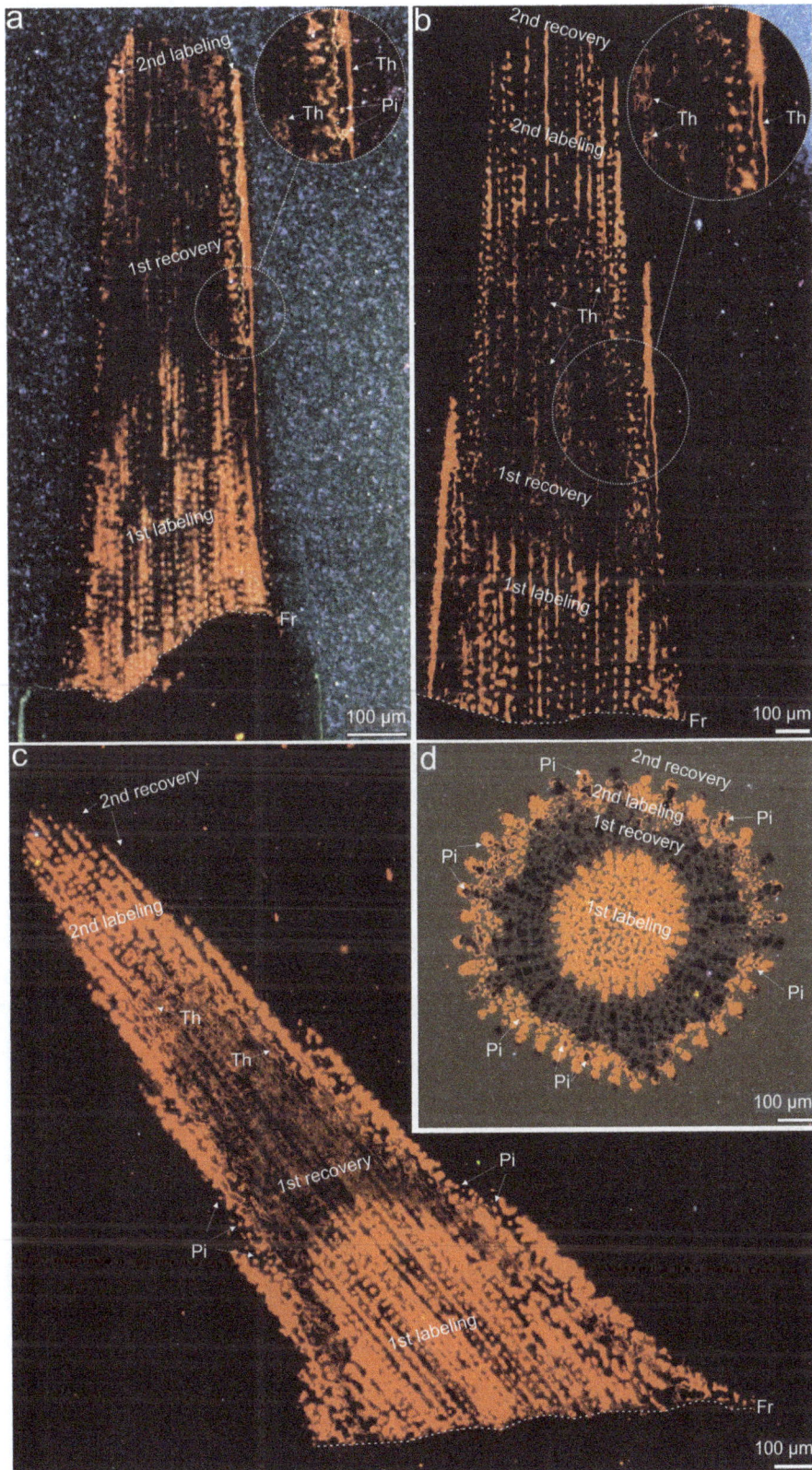

Fig. 2 (See legend on next page.)

(See figure on previous page.)

Fig. 2 CL images of polished sea urchin spines labeled during trial marking experiments. **a** lateral view with enlarged microregion of a spine showing regenerated tip of *Paracentrotus lividus* labeled with Mn^{2+} at 1 mg/L, **b** lateral view with enlarged microregion of a spine showing regenerated tip of *Paracentrotus lividus* labeled with Mn^{2+} at 3 mg/L, **c** lateral view of spine showing a regenerated tip of *Echinometra* sp. labeled with Mn^{2+} at 3 mg/L, **d** cross-section of a regenerated tip of spine of *Echinometra* sp. labeled with Mn^{2+} at 3 mg/L. Abbreviations: Th - thickening increments, Pi - pore infilling deposit Fr - cutting fracture (delineated with dotted line). Orange-red skeletal regions indicate enhanced Mn^{2+} concentrations due to twice Mn^{2+} labeling; dark regions indicate growth in normal (without Mn^{2+}) seawater

Table 1 Conditions (mean values) in beakers during trial and main experiments

Trial experiment	Species	Temperature	SD	Salinity	SD	pH_T	SD
	Paracentrotus lividus	15,88	1,07	33,4 [a]	0,13	7,95	0,09
Controls	*Paracentrotus lividus*	15,83	1,12	33,51	0,13	7,96	0,09
	Paracentrotus lividus	15,79	1,1	33,53	0,16	7,95	0,11
	Paracentrotus lividus	15,77	1,11	33,52	0,11	7,94	0,9
Mn 1 mg/L	*Paracentrotus lividus*	15,73	1,12	33,51	0,12	7,95	0,1
	Paracentrotus lividus	15,68	1,13	33,56	0,13	7,95	0,11
	Paracentrotus lividus	15,78	1,12	33,56	0,1	7,96	0,11
Mn 3 mg/L	*Paracentrotus lividus*	15,67	1,16	33,59	0,14	7,97	0,12
	Paracentrotus lividus	15,81	1,14	33,56	0,12	7,96	0,12
	Paracentrotus lividus	15,84	1,15	33,58	0,1	7,97	0,11
Mn 61.6 mg/L	*Paracentrotus lividus*	15,86	1,15	33,59	0,1	7,95	0,12
	Paracentrotus lividus	15,84	1,15	33,59	0,1	7,96	0,11
Main experiment							
	Paracentrotus lividus	17,75	0,39	33,48	0,21	8,11	0,06
Controls	*Paracentrotus lividus*	17,69	0,43	33,53	0,24	8,1	0,06
	Paracentrotus lividus	17,73	0,35	33,54	0,26	8,11	0,06
	Paracentrotus lividus	17,58	0,41	33,52	0,2	8,11	0,06
Mn 1 mg/L	*Paracentrotus lividus*	17,67	0,38	33,51	0,16	8,1	0,07
	Paracentrotus lividus	17,57	0,42	33,58	0,36	8,11	0,07
	Paracentrotus lividus	17,56	0,34	33,62	0,28	8,09	0,08
Mn 3 mg/L	*Paracentrotus lividus*	17,43	0,46	33,63	0,32	8,11	0,06
	Paracentrotus lividus	17,45	0,42	33,63	0,3	8,11	0,06
	Echinometra sp.	25,05	0,54	34,46	0,17	8,1	0,06
Controls	*Echinometra* sp.	25,04	0,52	34,48	0,2	8,1	0,05
	Echinometra sp.	25,06	0,51	34,43	0,2	8,11	0,06
	Echinometra sp.	25,06	0,51	34,43	0,19	8,11	0,06
Mn 1 mg/L	*Echinometra* sp.	25,07	0,52	34,47	0,24	8,1	0,05
	Echinometra sp.	25,07	0,52	34,45	0,23	8,09	0,05
Controls	*Prionocidaris baculosa*	25,16	0,37	35,29	0,22	7,96	0,05
	Prionocidaris baculosa	25,1	0,37	35,29	0,22	7,96	0,05
Mn 1 mg/L	*Prionocidaris baculosa*	25,16	0,36	35,3	0,21	7,95	0,06
	Prionocidaris baculosa	25,09	0,36	35,3	0,21	7,95	0,06

SD = standard deviations ($n = 36$). Conditions within individual beakers for a given experiment were not significantly different from each other, except the salinity in a beaker no 1. (marked with [a]) which is significantly different from each other

Table 2 Lengths of spine tips of *Paracentrotus lividus* regenerated during trial experiments (3 spines per individual/ 3 individuals per treatment) and calculated average longitudinal extension rates (ALER) per treatment

	Mn = 1 mg/L	Mn = 3 mg/L	Mn = 61.4 mg/L	Mn = 0 mg/L
Spine1	2581 mm	1819 mm	0 mm	2491 mm
Spine2	2435 mm	2,52 mm	0 mm	2774 mm
Spine3	2087 mm	2687 mm	0 mm	2756 mm
Spine1	1,65 mm	1,45 mm	0 mm	2457 mm
Spine2	1,69 mm	2,67 mm	0 mm	2867 mm
Spine3	1,83 mm	1,61 mm	0 mm	2894 mm
Spine1	1118 mm	0,922 mm	0 mm	1911 mm
Spine2	3414 mm	3462 mm	0 mm	2313 mm
Spine3	2923 mm	1386 mm	0 mm	2648 mm
ALER	183 µm/day	172 µm/day	0 µm/day	214 µm/day

Figure 2 summarizes results of CL analyses of trial experiments showing examples of spines sectioned either along (Fig. 2 a-c) or perpendicular (Fig. 2d) to the long axis. Under CL, all regenerated spines of both species labeled with low Mn^{2+} concentrations (1 mg/L; Fig. 2a or 3 mg/L; Fig. 2 b-d) reveal two bright orange-red luminescent stereom increments that are separated by one area with dark stereom trabeculae which only occasionally display luminescent outermost thickening trabecular layers. In the peripheral part of the spine, where septa are being formed, the skeleton shows irregular patchy luminescence with some brighter or darker spots. This is especially well visible in transverse sections (Fig. 2d).

The emission spectra of the luminescent regions peaked at ~632 nm (Fig. 1a), consistent with manganese-activated luminescence in echinoderm Mg-calcite [13]. By contrast, the spines with regenerated tips from control

Table 3 Lengths of the spine tips of *Echinometra* sp. regenerated during trial experiments (3 spines per individual/3 individuals per treatment) and calculated average longitudinal extension rates (ALER) per treatment

	Mn = 3 mg/L	Mn = 0 mg/L
Spine1	0,481 mm	4742 mm
Spine2	0,053 mm	1806 mm
Spine3	1929 mm	1835 mm
Spine1	1367 mm	0,083 mm
Spine2	1894 mm	0,12 mm
Spine3	1948 mm	0,05 mm
Spine1	2521 mm	2387 mm
Spine2	4656 mm	2305 mm
Spine3	5,32 mm	2262 mm
ALER	187 µm/day	144 µm/day

specimens are entirely non-luminescent (Fig. 1d). Thus, the two observed luminescent increments alternated with two non-luminescent growth fronts correspond to the two successive Mn-labeling events (4-day and 2-day) and two recovery periods (4-day and 2-day), respectively. As stressed above, no growth increments above the cutting fracture were observed in the spines of specimens incubated in seawater with the highest Mn^{2+} concentration (61.6 mg/L) (Fig. 1c). Under CL, these spines are entirely non-luminescent.

Main experiments

Figures 3, 4 and 5 show distribution of Mn-induced luminescent increments in various ossicle types of three sea urchin species (*Paracentrotus lividus*, *Echinometra* sp. and *Prionocidaris baculosa*) that were grown during the further dynamic labeling experiments with low Mn^{2+} concentrations (1 mg/L).

Paracentrotus Lividus

A single (2-day) labeling event is clearly distinguishable in all types of examined ossicles. The average longitudinal extension rate for a normal (non-regenerative) spine growth, measured in the central axis, is 106 µm/day (Fig. 3a). In the periphery, continuous luminescent layers corresponding to septa, with patchy appearance in some places, extend down to ca. 13 mm below the level of microspine deposition in the distalmost part of the spine (Fig. 3a). In transverse sections, patchy luminescent parts of the septa and some stereom bridges are visible (Fig. 3b-e).

Distribution and intensity of Mn-induced luminescence in the teeth of *Paracentrotus lividus* is not uniform. In particular, high intensity and rather uniform luminescence is observed near plumula/shaft boundary, where new primary plates with lamellae were being formed (Fig. 3 f). Intense luminescence extends down and encompasses also prisms, needles and carinal processes in the keel (Fig. 3g-i). Luminescence gradually disappears towards the adoral part of the tooth shaft. Herein, luminescence is restricted only to some isolated spaces between needle-prisms, primary, secondary and carinar plates. Most adorally, near the incisal end, an outermost layer corresponding to the abaxial crust is only luminescent (Fig. 3i). The estimated longitudinal extension rate of teeth in this species is ~1.8 mm/day. However, the growth rate of teeth in this and other species examined in this study need to be treated with caution because the least solidified part of the tooth, i.e., plumula, is extremely fragile and tends to be lost during preparation of thin sections. Furthermore, non-uniform luminescence due to simultaneous thickening process in adoral direction makes it difficult to measure the growth.

Three luminescent growth fronts are observed in interambulacral ambital plates (Fig. 3j, k). Sharp Mn-label is

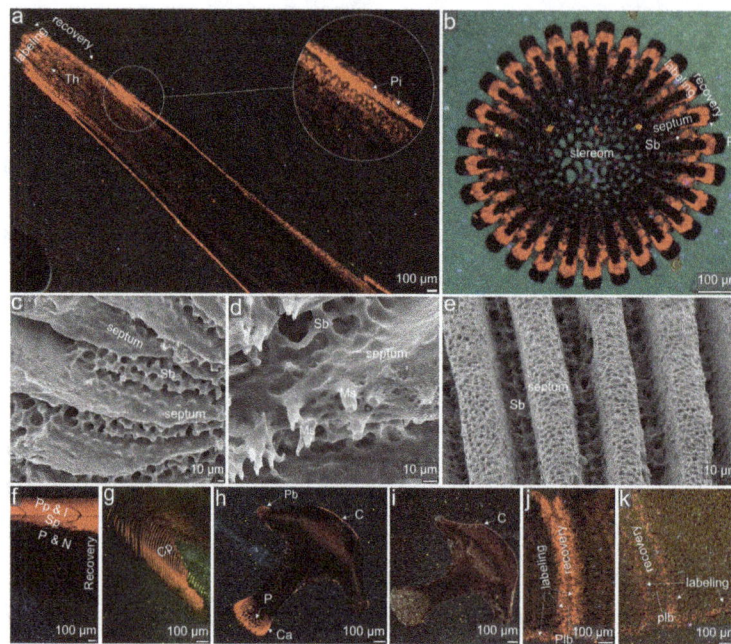

Fig. 3 CL and SEM images of *Paracentrotus lividus* labeled with Mn^{2+} at 1 mg/L during the second labeling experiment. **a** lateral view of a polished spine with enlarged microregion showing septa formation, **b** cross-section of a polished spine, **c-e** micromorphology of septa revealed after formic acid etching, **f**, **g** cross-sections showing lateral view of the polished tooth, **h**, **i** transverse cross-sections through the polished tooth (at the level of proximal shaft (~6 mm from the aboral end and ~8 mm from the adoral tip) and near the incisal end (~11 mm from the aboral end and ~3 mm from the adoral tip), respectively), **j**, **k** cross-sections near the polished ambital plates showing a contact between two test plates (adapical and interadial sutures, respectively). Abbreviations: Th - thickening increments, Pi - pore infilling deposit, Ms - microspines, Sb - stereom bridges, C - outer crust, Pb - pillar bridges, P - pisms, Cp - carinar process plates, Sp - secondary plates, Plb - plate boundary, L - lamellae, N - needles. Orange-red skeletal regions indicate enchanced Mn^{2+} concentrations due to Mn^{2+} labeling; dark regions indicate growth in normal (without Mn^{2+}) seawater

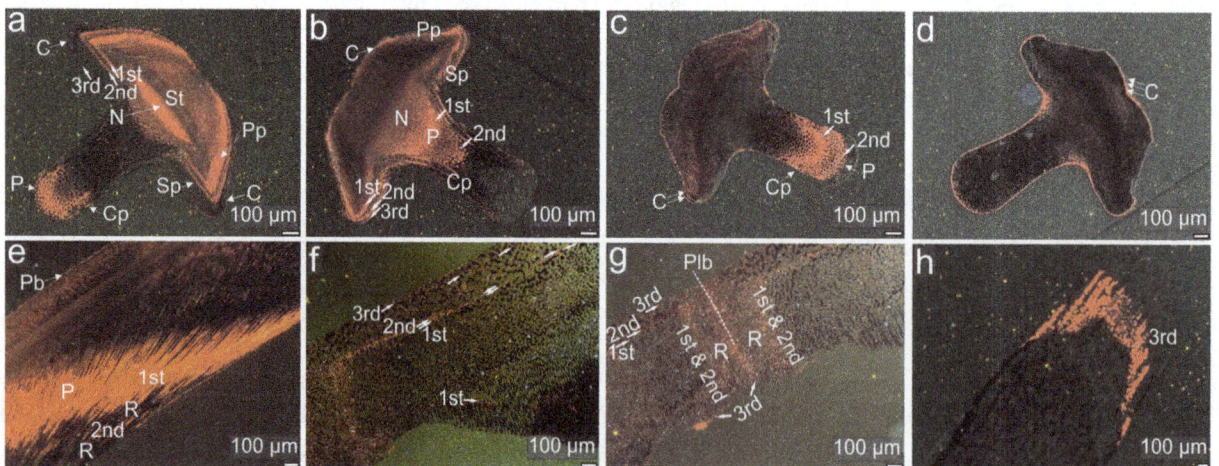

Fig. 4 CL images of polished ossicles of *Echinometra* sp. labeled with Mn^{2+} at 1 mg/L during the second labeling experiments. **a-d** transverse cross-sections through a tooth (from the level near shaft/plumula boundary (~8 mm from the adoral tip) towards the incisal end (~6, 4 and 2 mm from the adoral tip, respectively)), **e** lateral view of tooth near the proximal part of shaft, **f** cross-section of an ambital interambulacral plate in lateral view, **g** cross-section near an ambitus showing a contact (adapical suture) between two test plates, **h** lateral view of a growing spine. Abbreviations: 1st, 2nd, 3rd - successive labeling events (indicated by solid arrows); R - interlabeling recovery, C - outer crust, Pb - pillar bridges, P - pisms, Cp - carinar process plates, Sp - secondary plates, St - stone part, Pp - primary plates, Plb - plate boundary, L - lamellae, N - needles. Orange-red skeletal regions indicate enhanced Mn^{2+} concentrations due to Mn^{2+} labeling; dark regions indicate growth in normal (without Mn^{2+}) seawater

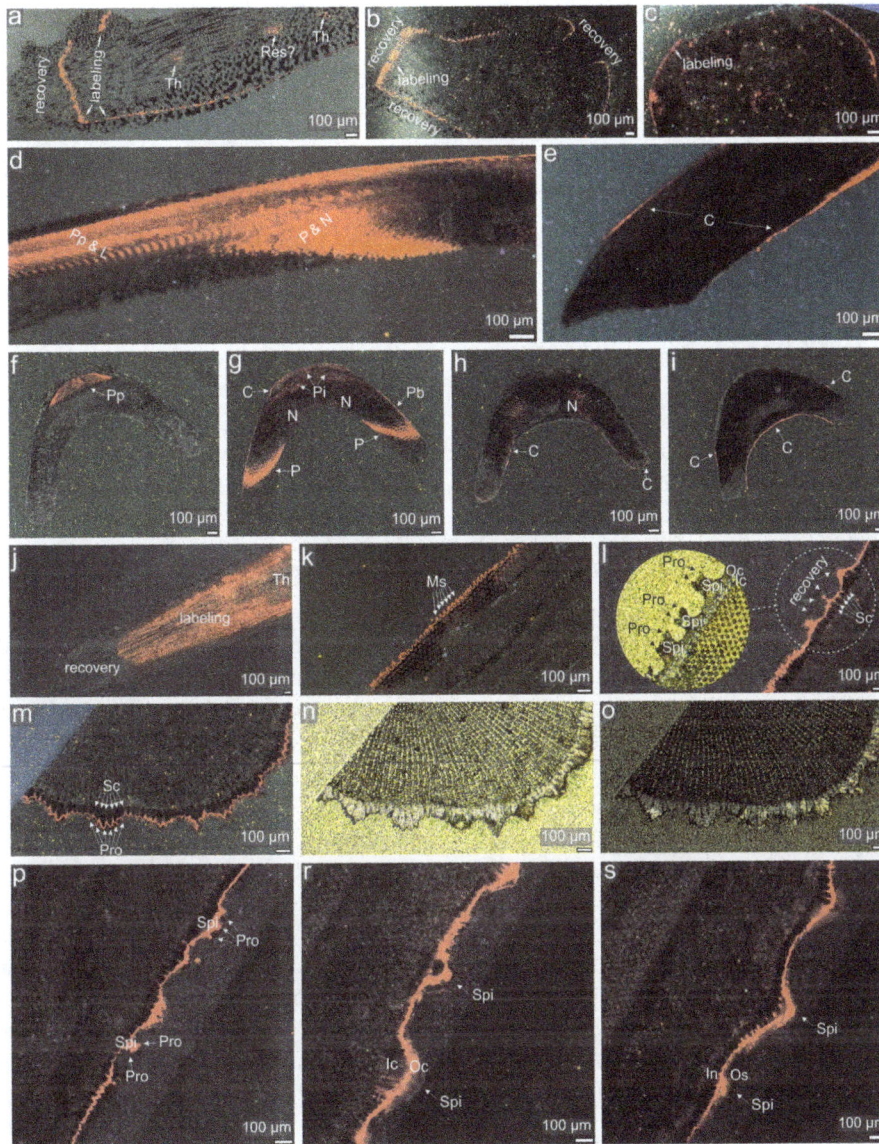

Fig. 5 CL and optical images of polished ossicles of *Prionocidaris baculosa* labeled with Mn²⁺ at 1 mg/L during the second marking experiments. **a** cross-section of an ambitus interambulacral plate in lateral view, **b** cross-section of a young interambulacral plate near the apical disc in meridional view, **c** cross-section of an ambitus interambulacral plates with enlarged central tubercle in lateral view, **d** lateral view of tooth near the proximal part of the shaft, **e** lateral view of tooth near the incisal end, **f-i** transverse cross-sections through a tooth (from the level near shaft/plumula boundary (~7 mm from the adoral tip) towards the incisal end (~5, 3 and 1 mm from the adoral tip, respectively)), **j** lateral view of a distal end of growing spine, **k** lateral view of a growing spines showing an initial phase of cortex formation at the middle height of the spine, **l** lateral view (at the middle height of the spine) of a growing spines showing later phase of cortex formation with enlargement of selected microregion, **m-o** cross-sections of a growing spine showing later phase of cortex formation (at the middle height of spine) under CL, optical and polarizing microscope views, respectively, **p-s** lateral views (at the middle height of spine) of a growing spine showing late phase of cortex formation. Abbreviations: Th - thickening increments, Res? - possible deposition of calcite on previously resorbed stereom, Ms - microspines, C - outer crust, Pb - pillar bridges, P - prisms, Pp - primary plates, L - lamellae, N - needles, Spi - spinule, Pro - protuberances, Sc - stromatic channels, Ic - inner cortex, Oc - outer cortex. Orange-red skeletal regions indicate enhanced Mn²⁺ concentrations due to Mn²⁺ labeling; dark regions indicate growth in normal (without Mn²⁺) seawater

continuously distributed along the inner plate margin forming a distinct thin layer. Luminescent growth fronts are somewhat more expanded in interradial and meridional directions. Approximated equatorial growth in adambulacral direction (~35 µm/day) exceeds meridional growth (~14 µm/day) and inner plate thickening (~4 µm/day).

Luminescence is not observed in external part of the plate, where primary tubercles are present.

Echinometra sp.
Three labeling events and 3 successive recovery periods (4-day labeling, 4-day recovery, 2-day labeling, 27-day

recovery, 2-day labeling and 2-day recovery) are visible in teeth and ambital interambulacral plates under CL (Fig. 4a-g). Distribution and intensity of Mn-induced luminescence in the tooth of *Echinometra* sp. are similar to those observed in *Paracentrotus lividus* (Fig. 4a-e). In the most adoral part of the shaft, however, not only abaxial crustic layer is luminescent but also adaxial outermost layer (Fig. 4d). The longitudinal growth of teeth can be approximated as ~1 mm/day.

Similar to *Paracentrotus lividus* interradial growth of ambital plates in *Echinometra* sp. (~23 μm/day) exceeds meridional growth (~7 μm/day) and inner plate thickening (2 μm/day) (Fig. 4f, g). Likewise, Mn-labels, in a form of continuously distributed layers are present in the inner side of the plate. Luminesent growth fronts are less distinct in interradial and meridional directions, where luminescent newly formed strereom trabeculae merge with the older ones that reveal luminescence only in the outermost thickening stereom layers. Luminescence is also observed externally on the plate surface. However, it commonly encompasses some stereom bars near the plate boundaries, not central tubercles, like mamelons.

The last 2-day labeling event is visible in one of a few examined spines (Fig. 4h). The average longitudinal extension rate for a normal (non-regenerative) spine growth in this species, measured in the central axis, is ~175 μm/day. In contrast to *Paracentrotus lividus* spine, Mn-label does not extend down below deposition of the distalmost stereom microspines.

Prionocidaris baculosa
A singe (4-day) labeling event is detectable in all types of ossicles under CL (Fig. 5). Calculated interradial extension rate of ambital plate is ~9 μm/day, whereas meridional growth rate (~3 μm/day) and inner plate thickening (~1 μm/day) are much slower (Fig. 5a). Continuous luminescent growth fronts, encompassing both stereom bars near the plate boundaries (Fig. 5a, b) and central tubercles (Fig. 5c), are also observed on the plate surfaces. Surprisingly, some irregular patches of luminescent stereom or luminescent outer thickening trabecular layers are also present in the central part of the sectioned plate (Fig. 5a). In the case of teeth, the most luminescent areas can be observed near shaft/plumula boundary (Fig. 5d). Herein, a ca. 2 mm in length luminescent growth front, composed of primary plates and lamellae, extends down in adaxial direction to prisms and needles, and to the continuous abaxial and adaxial crustic layers (Fig. 5e-i).

Among 4 examined primary spines, two revealed luminescent part of the cortex at different stage of its development (Fig. 5j-s). The average longitudinal extension rate of the „open stereom" for one of these spine, measured in the central axis, is ~175 μm/day. In this spine, continuous luminescent growth fronts extend from the

level of stereom meshwork formation down to the milled ring. Average extension rates for the inner cortical layer and spinule development are ~3.5 μm/day and 12 μm/day, respectively.

Comparison of the average extension rates during and between the labeling events
In order to test the influence of Mn exposure on growth rates in some ossicles, statistical comparisons between the average extension rates during and between the labeling events in four selected microstructural areas in four ossicles of three examined species were performed (at least 20 measurements per each ossicle in a given direction collected using imageJ). Results clearly show no statistically significant differences (Wilcoxon tests: $p > 0.05$; Fig. 6). Admittedly, however, such estimations of extension rates need to be treated with caution because different orientation of the cut surface may bias interpretations. Furthermore, small portions of the plumula and even potentially distalmost tips of some spines, that were grown during the second recovery period, appeared to be lost during sample preparation, thus at least in the case of spines, comparisons were only made between the skeleton grown during the first labeling event and successive first recovery period.

Discussion
In the case of labeling mollusc shells most authors recommended to use high-doses of manganese (≥ 25 mg/L) [19–21, 24–26]. However, care has to be taken because such a high concentration of Mn^{2+} appears to be stressful for sea urchins perturbing their biomineralization [29–31].

Little is known about lower activation limit of Mn^{2+} in calcite. Available data suggest a range 20–1000 ppm [37], much higher than typical skeletal Mn^{2+} content in most echinoderms [38]. As stressed by Richter et al. [12] Mn^{2+} detection limit can be lowered by technical improvements, i.e., by using the so-called hot cathode, instead of the so-called cold cathode. In our labeling experiments we incubated sea urchins with seawater enriched with low concentrations of manganese (1 and/or 3 mg/L) that were subsequently analyzed with the aid of hot cathode. The use of such a low Mn^{2+} concentration and hot cathode in subsequent CL analyses was necessary to avoid stress and growth inhibition [29–31], and to ensure Mn-label detection. Indeed, our data show that the sea urchins labeled at high Mn^{2+} concentrations (61.6 mg/L) were not regenerating their spines. This observation is consistent with previous studies on sea urchin embryo suggesting that that high Mn^{2+} concentrations lead to Ca^{2+} pump dysfunction resulting in a strong depletion of Ca^{2+} in the Golgi regions causing morphological abnormalities (at ≥ 7.7 mg/L), or inhibition of skeleton growth (at ≥ 61.6 mg/L) [29–31]. On the other hand, our sea urchins labeled at low Mn^{2+}

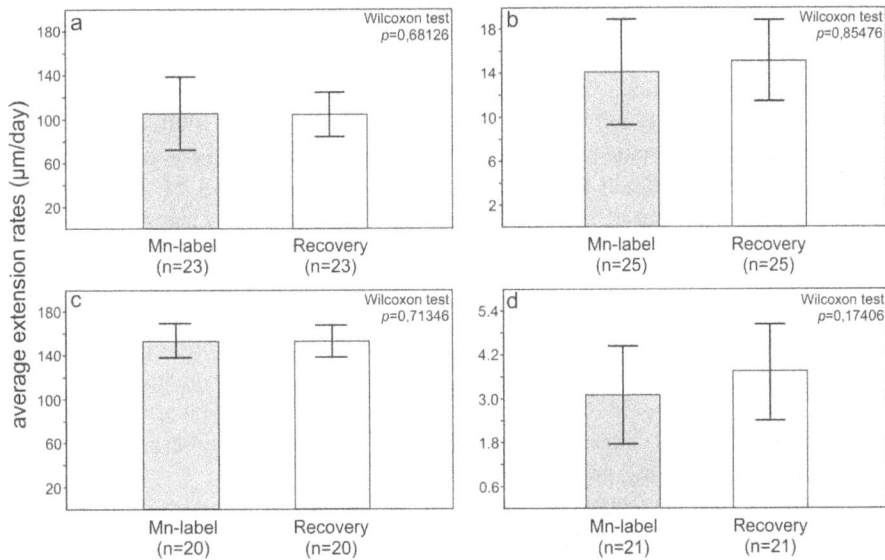

Fig. 6 Barcharts showing the average extension rates (±SD) during and between the labeling events for four selected microstructural areas in three examined species. Results of the Wilcoxon tests indicate no statistically significant differences. **a** longitudinal extension rates measured during the first labeling event and subsequent recovery period in *Paracentrotus lividus* (spine illustrated on Fig. 2a), **b** extension rates of test plate in meridional direction during the first labeling event and subsequent recovery period in *Paracentrotus lividus* (plate illustrated on Fig. 3j), **c** longitudinal extension rates of regenerating spine measured during the first labeling event and subsequent recovery period in *Echinometra* sp. (spine illustrated on Fig. 2c), **d** lateral extension rates of inner cortex measured during the first labeling event and subsequent recovery period in *Prionocidaris baculosa* (spine illustrated on Fig. 5m)

concentrations (1 and/or 3 mg/L) did not seem to be visibly stressed and went on growing. Notably, the absence of growth-rate differences during and between labeling events suggests that the Mn-labeling did not affect growth rate at least in some of the studied ossicles that were subjected to statistical comparisons (Fig. 6). In order to futher test the effect of Mn-labeling procedure on the sea urchin physiology, future analyses on respiration rates and immunotoxicological studies are planned. However, it seems that larger volumes of seawater, shorter labeling-time scale and lower concentrations of manganese may all minimize potential stress. Notably, our sea urchins labeled at 1 mg/L [Mn^{2+}] revealed bright luminescence using a beam current 0.15 mA, an electron energy 14 keV and the exposure time for recording images 3 s. Thus, it seems that labeling sea urchins at much lower concentration can still be detectable by increasing the exposure time and a beam current to the upper limit 0.2 mA. In fact, it has been recently shown that oysters labeled in seawater with low Mn^{2+} concentration (0.017 mg/L), formed a new shell that was barely visible but still detectable by cathodoluminescence [19].

Implications for morphogenesis of echinoid ossicles

Our study highlights the potential of using a combination of low Mn^{2+} doses in labeling experiments and hot cathode in subsequent CL imaging in biomineralization studies of echinoderms. Admittedly, caution is required in applying this method to all echinoderms because it

has been shown that some test plates of Recent echinoids, having increased levels of skeletal Mn, may show CL luminescence [28]. Indeed, although most echinoderms are non-luminescent [18, 27], skeletal concentration of Mn and Fe may differ between species and/or between populations of the same species from different environments [39]. Thus, the use of control specimens is strongly advised to assure validity of CL results.

For the specimens investigated in this study, we demonstrated that this method proved to be successful. This allowed for a much more detailed view of the growth dynamics and morphogenesis of echinoid skeleton at the micro-length scale. In the following we briefly review published data on morphogenesis of echinoderm skeleton and discuss some of the major implications of our results for understanding echinoderm biomineralization.

Spine growth

Our data demonstrate that average longitudinal extension rates of regenerating spines in *Paracentrotus lividus* (172 or 183 μm/day depending on a treatment) and *Echinometra* sp. (187 μm /day) are very similar to "normal" non-regenerative spine growth during experiments (106 and 175 μm/day, respectively; cf. Figs. 3a, 4h). Likewise, longitudinal extension rate of primary spine of *Prionocidaris baculosa* (175 μm/day; Fig. 5j) is very comparable. Consistently, previous study also reported that *Paracentrotus lividus* regenerates their spines at similar extension rate (130 μm/day; [2]). Notably, extension rates of spines in other sea urchin species (*Strongylocentrotus purpuratus* (160 μm/day) and *Arbacia punctulata* (260 μm/day)) reported by previous authors [40] are within the same order of magnitude. In three examined species, morphogenesis of stereom meshwork in spine is also basically the same, and follows a growth model of spine based on *Paracentrotus lividus* recently introduced by Gorzelak et al. [2]. Growth of the spine stereom starts with the formation of thin microspines which fuse together at regular intervals by lateral bridges. Simultaneously, thickening of the previously formed stereom trabeculae, in the form of deposition of continuous layers, proceeds on a time scale of about 1 μm/day. Thickening process, involving both longitudinal strereom trabeculae and lateral bridges, may extend down to about 1 mm below deposition of massive microspines. Overall, similar extension rates and stereom morphogenesis in various unrelated echinoid species clearly suggest some common regulatory factors and pathways involved in spine development.

Our data also provided direct insights into the timing and mechanisms of septa development. CL images revealed that septa in *Paracentrotus lividus* and *Echinometra* sp. can be formed via two, but not clearly separated from each other, phases of biomineralization, i.e., deposition of porous stereom that is secondarily filled by calcite as first suggested by [40]. The secondary process of peripheral pore occlusion in radial sectors may proceed rapidly (within less than 2 days) and can be simultaneous with the formation of "open" stereom in the central and peripheral areas, as well as transverse bridges joining the septa. Later on, most proximally the outer part of septa can be thickened by deposition of more or less continuous sheets and layers.

In *Prionocidaris baculosa*, external parts of mature spines do not display septa but are covered by the so-called cortex, characteristic for all cidaroids [41, 42]. So far, little was known about morphogenesis and growth dynamics of this structure. Dery et al. [43] based on SEM comparisons of different spines removed from cidaroid tests hypothesized the following steps in cortex development: (i) microspines formation, (ii) successive thickening of microspines by addition of lamellar layers, (iii) spinule development by further thickening of microspines in selected regions, (iv) formation of lateral protuberances eventually fusing between adjacent spinules. Our data proved that cidaroid cortex indeed starts with formation of microspines which are gradually becoming thickened externally (Fig. 7). Later on, the outer cortex, revealing conchoidal fracture in the bottom and the overlaying fine irregular crystals forming spinules with lateral protuberances, develop. This outermost cortical layer has a polycrystalline nature as revealed by the cross-polarized light (Fig. 5o) [41, 42].

The above data appear to be mostly consistent with the hypothetical developmental model by Dery et al. [43]; nevertheless they also provide first direct insights into spatiotemporal development of cortex. For instance, the cortex develops simultaneously with the formation of new open stereom meshwork in the apical part of the spine. From the level of distalmost stereom meshwork formation, cidaroid cortex, in the form of continuous growth front with regularly spaced pores of stromatic channels, extends down at ~3.5 μm/day along the entire spine surface to the level of the milled ring. Thickening rate can be more pronounced (~12 μm/day) at more or less regular intervals giving rise to spinule development of the so-called outer cortex (Fig. 7) which may extend laterally to form protuberances.

Teeth

Morphology and development of echinoid teeth were described extensively in the literature [44–51] but the accurate growth dynamics of individual tooth components are still not well understood. In general, it has been argued that the timing of tooth growth depends on both intrinsic and extrinsic factors. Depending on the species the growth rate may range from 1 to 1.5 mm per week [44, 52]. Furthermore, when food-limited, the growth of echinoid jaws can be much faster [53–55]. In echinoid teeth three zones can be distinguished, i.e., plumula, shaft and incisal end (Fig. 8a). The plumula is the growing zone of the tooth. Herein, triangular parallel arrays of slightly curved primary plates (pp), initially separated from each other, are being formed (Fig. 8b, c). Later on, these plates are thought to grow in aboral and lateral direction to form the "tongue-like" stage from which secondary plates (sp) extend (Fig. 8d-f). Subsequently, sp gradually curve towards the adaxial area and merge to form carinal processes in the keel which is not pronounced in cidaroids. Both pp and sp are stacked within each other resembling ice-cream cones (Fig. 8c). Simultaneously, lamella-needle-prism complexes are being formed from the adhesive points. Lamellae attach to the primary plates near the midline of the flange, whereas needles and prisms protrude axially and proximally at about right angles forming the so-called stone part, and a part of the keel. A tooth zone, where the keel reaches its fullest size, is termed a shaft. Herein, pillar-bridges are mostly present. They are located on the abaxial surfaces to which collagenous tissues attach and connect the tooth with the pyramids (jaws). Towards the distal part, tooth is becoming increasingly compact and solid due to conspicuous progressive thickening of the previously formed plates and infilling of cavity systems by calcite during the second stage of biomineralization. Within the least porous distalmost part of the tooth, in the so-called incisal or oral end, which

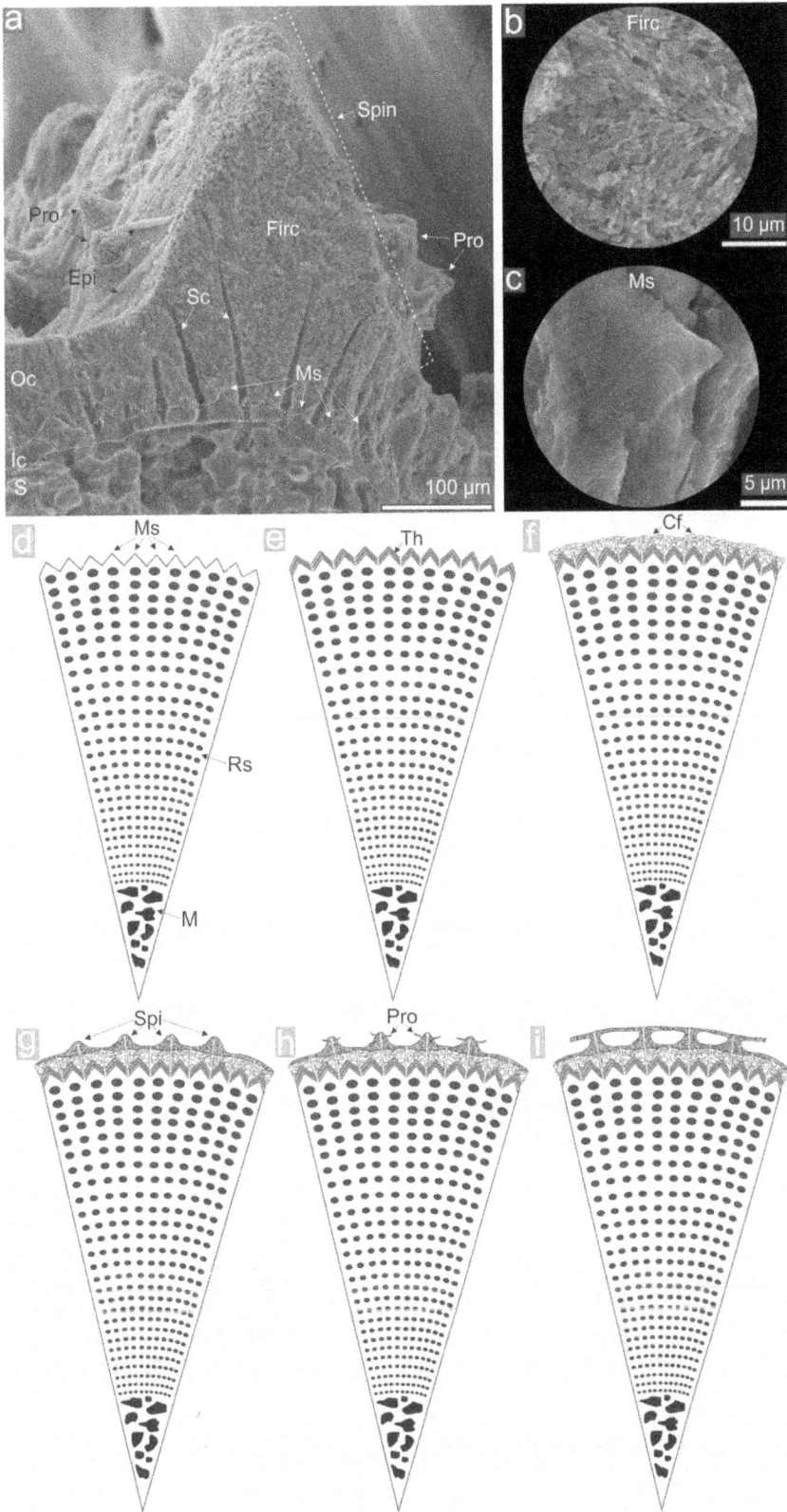

Fig. 7 (See legend on next page.)

(See figure on previous page.)

Fig. 7 Model of cortex development. **a** SEM image showing basic morphological features of cortex, **b** enlargement of fine irregular polycrystalline structure from the outermost cortex, **c** enlargement of the layered microspine from the inner cortex. **d-i** successive phases of cortex formation: microspine development (d), thickening of microspines (e), deposition of more massive stereom with conchoidal fracture on layered microspines between stromatic channels (f), development of spinules (g) formation of protuberances from spinules (h), eventual fusion of neighboring protuberances (i). Abbreviations: Spin - spinule, Pro - proturberances, Epi - epibionts, S - stereom, Ic - inner cortex, Oc - outer cortex, Ms - microspines, Firc - fine irregular crystal structure, Sc - stromatic channels, M - medulla, Rs - rectilinear stereom

continuously wears through abrasion during grazing, no mineral deposition is thought to occur.

The above literature data are mostly based on indirect structural observations of teeth removed from different sea urchins. Results presented here thus provide a more direct data on morphogenesis and growth dynamics of echinoid teeth. We demonstrate that the growth of tooth may vary considerably according to the species (~1.8 mm/day in *Paracentrotus lividus*, ~1 mm/day in *Echinometra* sp. and ~0.5 mm/day in *Prionocidaris baculosa*). In general, reported growth rates are higher than those previously reported in the literature (~0.24 mm/day) but this

can be attributed to rather food-limited conditions implemented during experiments (one Zeigler pellet per every two days). From the CL images a highly complex growth process becomes apparent. Different tooth elements are beginning to form in a sequential order (primary plates, lamellae, needles, secondary plate, prisms and carinar processes). Later on, however, these elements extend and thicken contemporaneously in different directions and at different rates. CL images revealed that while new plates are being formed in plumula, older structural elements, deeper in the shaft extend and thicken simultaneously. During simultaneous addition of new prisms and carinal processes externally extending the flanges and keel, the older, inner ones at the same height level are simultaneously becoming thickened. The calculated rate of infilling of cavity systems between prisms by calcite in the keel and flanges in available transverse sections intersected at comparable tooth height is within a range of ~1.5–3 µm/day in three examined species. Longitudinal extension rates of individual prisms are highly heterogenous and depends on the level from which they grow. In general, prisms may extend up to about 90 times more than they widen, consistent with theoretical predictions by Robach et al. [56]. Intriguingly, contemporaneous deposition of continuous outermost layers extending down to the distal part of the shaft may also take place. This demonstrates that the odontoblasts can be still active even in the distal part of the tooth, just above the incisal end.

Test plate

Growth of echinoid test has been investigated with the aid of radioisotopes and fluorescent markers by many authors [4, 57, 58, 59, 60, 61, 62] whose works inspired other researchers to develop several theoretical growth models of echinoid test [63–73]. According to extensive literature data, new coronar plates, which are added at the apical end of the corona, are subsequently shifted towards the peristome increasing their size in three dimensions. Extension rates of test plates depend on several factors, including the type of plate, their position within the test, and direction of the growth front. In general, circumferential (in interradial direction) growth is known to exceed meridional extension. The maximum interradial growth is usually found in ambital plates whereas maximum meridional growth is found in young plates near the apical disc. The growth of echinoid plates

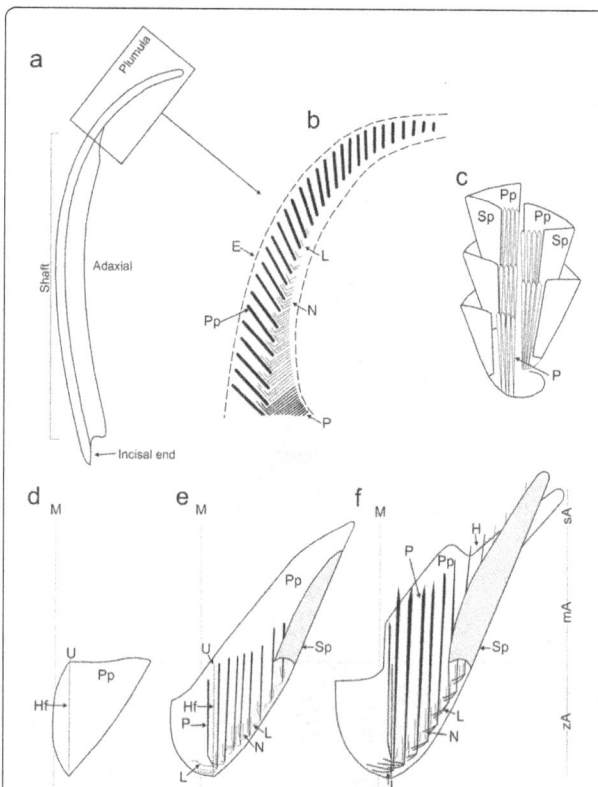

Fig. 8 Model of tooth development (compiled and modified after [45–49]). **a** lateral view of a tooth of *Paracentrotus lividus*, **b** enlargement of plumula region, **c** cone-like tooth elements, sticking alternately one within the other, **d-f** growth of individual tooth element from initial growth phase (**d**) to late stage (**f**). Abbreviations: E - epidermis, Pp - primary plate, L - lamellae, N - needles, P - prism, Sp - secondary plate, H - pillar bridges, M - median plane, Hf - main fold of pp., U - umbo, zA - central part of pp., mA - middle part of pp., sA - side part of pp.

at the macroscale and theoretical framework for understanding various shapes of echinoid tests are well characterized, and thus this was not an ultimate goal of our study. In this study we focused on stereom development in individual test plate (mostly interambulacral ambitals) at the microscale because this aspect is not fully understood.

We showed that ambital plates thicken very slowly from the inner side. Thickening of test plate proceeds via deposition of continuously distributed layer on previously formed trabeculae. The thickening rate is the slowest in the central part of the plate, and progressively increases towards the plate margins. The growth fronts, in the form of galleried stereom increments, expand more rapidly in interradial and meridional directions (Fig. 9). The three species are characterized by similar growth ratios: interradial/meridional (*Paracentrotus lividus*: 2.4, *Echinometra* sp.: 3.3; *Prionocidaris baculosa*: 3); meridional/thickening (*Paracentrotus lividus*: 3.5, *Echinometra* sp.: 3.5; *Prionocidaris baculosa*: 3); interradial/thickening (*Paracentrotus lividus*: 8.8, *Echinometra* sp.: 11.5; *Prionocidaris baculosa*: 9). In general, the average extension rates of three species examined in this study are comparable to those previously reported in the literature [7, 59]. In some species (*Echinometra* sp.) thickening of the previously formed stereom trabeculae, in the form of deposition of outermost trabecular layers, may also occur. New stereom is rarely formed on external part of the ambital plates where central tubercles, such as mamelons are fully developed. However, in young individuals, such as in the case of *Prionocidaris baculosa* examined in this study, deposition of continuous growth front involving mamelons may also

occur on the surface of ambital plates and is obviously more pronounced in the newly formed plates near the apical disc (Fig. 5a, b). It is noteworthy that in *Prionocidaris baculosa* some newly grown isolated stereom increments in small voids or thickening of the previously formed trabeculae can proceed in the central parts of the plate.

Conclusions

In this paper we introduced a promising method for labeling the growing echinoderm skeleton with manganese coupled with cathodoluminescence imaging to obtain information about biomineralization processes at the micro-length scale. Three sea urchin species were incubated in seawater enriched with Mn^{2+} which is the most important activator element of cathodoluminescence in calcite. Mn^{2+}-induced luminescent marks do not diffuse, allowing growth measurements to be made at the microscale resolution. This approach allowed for a refinement of the previous growth models of echinoid spine, tooth and test plate. Spine skeletogenesis begins with the formation of microspines which extend at least 100 times faster than simultaneous thickening process of individual stereom trabeculae. Development of septa proceeds via secondary peripheral pore occlusion in radial sectors. The timescale of formation of septa can be on the order of less than few days. Morphogenesis of cortex in cidaroids starts with the formation of microspines which thicken at ~3.5 µm/day along the entire spine surface to the level of milled ring. Later on, the thickening rate can be more pronounced (~12 µm/day) at more or less regular intervals giving rise to spinule

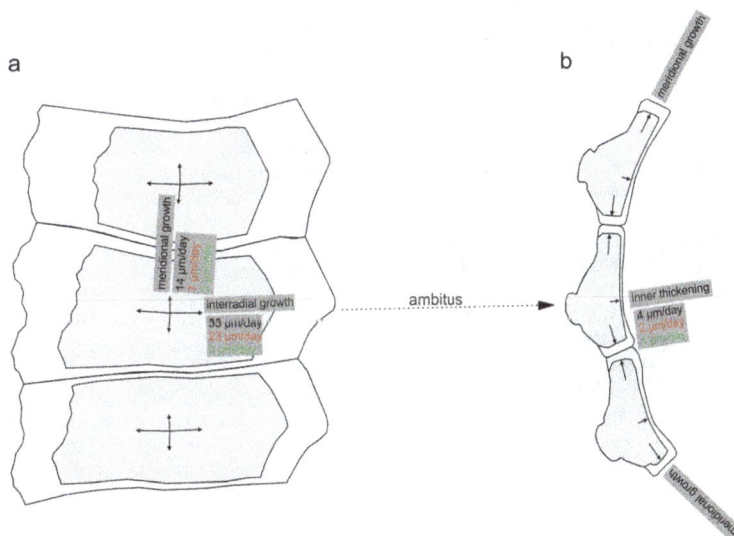

Fig. 9 Model of interambulacral ambital plate development (modified after [58]). **a** tangential view, **b** meridional view. Arrows indicate directions of the growth fronts. Mean growth rates per each species in a given direction are highlighted in color (black = *Paracentrotus lividus*, red = *Echinometra* sp., green = *Prionocidaris baculosa*)

development which may extend laterally to form protuberances. Formation of tooth is a highly complex process and involves formation of new plates in plumula and simultaneous addition of new plates in the shaft extending the keel, and progressive infilling of internal porous structure towards the incisal end. Formation of echinoid ambital plate falls into a readily identifiable pattern: interradial growth exceeds meridional growth and inner thickening process. Mn-labeling coupled with subsequent CL imaging is the first highly efficient method which enables a clear mark detection of the growing echinoid tooth.

Acknowledgements
We thank: Saloua M'Zoudi, Mathieu Bauwens and Philippe Pernet (all from Université Libre de Bruxelles, Faculté des Sciences, Laboratoire de Biologie Marine, Bruxelles, Belgium) for technical assistance during the experiments. We also thank three anonymous reviewers for very useful remarks.

Funding
This work was funded by the National Science Centre (NCN) grant no DEC-2011/03/N/ST10/04798 and was performed in the NanoFun laboratory co-financed by the European Regional Development Fund within the Innovation Economy Operational Programme POIG.02.02.00–00-025/09. The study was supported by FNRS Grant J.0219.16 SOFTECHI. A. Dery is a Research Assistant of the Université Libre de Bruxelles and Ph. Dubois is a Research Director of the National Fund for Scientific Research (FRS-FNRS; Belgium).

Authors' contributions
PG designed study, performed experiments, analysed data, and was a major contributor in writing the manuscript. AD and PD helped in performing experiments, and together with JS, provided intellectual contributions to interpretation and writing the manuscript. All authors read and approved the final manuscript.

Competing interests
The authors declare that they have no competing interests.

Author details
[1]Institute of Paleobiology, Polish Academy of Sciences, Twarda 51/55, 00-818 Warsaw, Poland. [2]Laboratoire de Biologie marine, Faculté des Sciences, Université Libre de Bruxelles, CP 160/15, av., F.D.Roosevelt, 50, B-1050 Bruxelles, Belgium.

References
1. Ebert TA. Growth and survival of postsettelement sea urchins. In: Lawrence JM, editor. Edible sea urchins: biology and ecology. Amsterdam: Elsevier; 2007. p. 95–134.
2. Gorzelak P, Stolarski J, Dubois P, Kopp C, Meibom A. [26]Mg labeling of the sea urchin regenerating spine: insights into echinoderm biomineralization process. J Struct Biol. 2011;176:119–26.
3. Gorzelak P, Stolarski J, Dery A, Dubois P, Escrig S, Meibom A. Ultra- and microscale growth dynamics of the cidaroid spine of Phyllacanthus imperialis revealed by [26]Mg labeling and NanoSIMS isotopic imaging. J Morphol. 2014;275:788–96.
4. Kobayashi S, Taki J. Calcification in sea urchins I. A tetracycline investigation of growth of the mature test in Strongylocentrotus intermedius. Calcif Tissue Res. 1969;4:210–23.
5. Ebert TA. Relative growth of sea urchin jaws: an example of plastic resource allocation. Bull Mar Sci. 1980;30:467–74.
6. Ellers O, Johnson AS. Polyfluorochrome marking slows growth only during the marking month in the green sea urchin Strongylocentrotus droebachiensis. Invertebr Biol. 2009;128:126–44.
7. Johnson AS, Salyers JM, Alcorn NJ, Ellers O, Allen DJ. Externally visible fluorochrome marks and allometries of growing sea urchins. Invertebr Biol. 2013;132:251–69.
8. Russell MP. Life history traits and resource allocation in the purple sea urchin Strongylocentrotus purpuratus (Stimpson). J Exp Mar Biol Ecol. 1987;108:199–216.
9. Russell MP, Meredith RW. Natural growth lines in echinoid ossicles are not reliable indicators of age: a test using Strongylocentrotus droebachiensis. Invertebr Biol. 2000;119:410–20.
10. Russell MP, Urbaniak LM. Does calcein affect estimates of growth rates in sea urchins? In: Heinzeller T, Nebelsick JH, editors. Echinoderms. München. London: Taylor & Francis; 2004. p. 53–7.
11. Russell MP, Ebert TA, Petraitis PS. Field estimates of growth and mortality of the green sea urchin, Strongylocentrotus droebachiensis. Ophelia. 1998;48:137–53.
12. Richter DK, Goette T, Goetze J, Neuser RD, Neuser RD. Progress inapplication of cathodoluminescence (CL) in sedimentary petrology. Miner Petrol. 2003;79:127–66.
13. Habermann D, Neuser RD, Richter DK. Quantitative high resolution spectra analysis in sedimentary calcite. In: Pagel M, Barbin V, Blanc P, Ohnenstetter D, editors. Cathodoluminescence in geosciences. New York, Tokyo: Springer, Berlin Heidelberg; 2000. p. 331–58.
14. El Ali A, Barbin G, Cervelle B, Ramseyer K, Bouroulec J. Mn^{2+}-activated luminescence in dolomite, calcite and magnesite: quantitative determination of manganese and site distribution by EPR and CL spectroscopy. Chem Geol. 1993;104:189–202.
15. Bender ML, Klinkhammer GP, Spencer DW. Manganese in seawater and the marine manganese balance. Deep-Sea Res. 1977;24:799–812.
16. Nordstrom DK, Plummer LN, Wigley TML, Ball JW, Jenne EA, Bassett RL, Crerar DA, Florence TM, Fritz B, Hoffman M, Holdren GR Jr, Lafon GM, Mattigod SV, McDuff RE, Morel F, Reddy MM, Sposito G, Thrailkill J. Comparison of computerized chemical models for equilibrium calculations in aqueous systems. In: Jenne EA, editor. Chemical modelling in aqueous systems, ACS symposium series 93. American Chemical Society: Washington; 1979. p. 857–92.
17. Barbin V, Ramseyer K, Debenay JP, Schein E, Roux M, Decrouez D. Cathodoluminecence of recent biogenic carbonates: an environmental and ontogenic fingerprint. Geol Mag. 1991;128:19–26.
18. Barbin V Cathodoluminescence of carbonates shells: biochemical vs diagenetic process. In: Pagel M, BV Blanc, Ohnenstetter D, editors. Cathodoluminescence in geosciences. Springer, Berlin; 2000. p. 303–329.
19. Barbin V, Ramseyer K, Elfman M. Biological record of added manganese in seawater: a new efficient tool to mark in vivo growth lines in the oyster species Crassostrea gigas. Int J Earth Sci. 2008;97:193–9.
20. Mahé K, Bellamy E, Lartaud F, de Rafélis M. Calcein and manganese experiments for marking the shell of the common cockle (Cerastoderma edule): tidal rhythm validation of increments formation. Aquat Living Resour. 2010;23:239–45.
21. Lartaud F, de Rafelis M, Ropert M, Emmanuel L, Geairon P, Renard M. Mn labelling of living oysters: artificial and natural cathodoluminescence analyses as a tool for age and growth rate determination of C. gigas (Thunberg, 1793) shells. Aquaculture. 2010;300:206–17. doi:10.1016/j.aquaculture.2009.12.018.
22. Lartaud F, Pareige S, de Rafelis M, Feuillassier L, Bideau M, Peru E, Romans P, Alcala F, Le Bris N. A new approach for assessing cold-water coral growth in situ using fluorescent calcein staining. Aquat Living Resour. 2013;26:187–96.
23. Barbin V. Application of cathodoluminescence microscopy to recent and past biological materials: a decade of progress. Miner Petrol. 2013;107:353–62.
24. Hawkes GP, Day RW, Wallace MW, Nugent KW, Bettiol AA, Jamieson DN, Williams MC. Analyzing the growth and form of mollusc shell layers, in situ, by cathodoluminescence microscopy and Raman spectroscopy. J Shellfish Res. 1996;15:659–66.
25. Langlet D, Alunno-Bruscia M, De Rafelis M, Renard M, Roux M, Schein E, Buestel D. Experimental and natural cathodoluminescence in the shell of Crassostrea gigas from Thau lagoon (France): ecological and environmental implications. Mar Ecol Prog Ser. 2006;317:143–56.
26. Auzoux-Bordenave S, Brahmi C, Badou A, de Rafelis M, Huchette S. Shell growth, microstructure and composition over the development cycle of the European abalone Haliotis tuberculata. Mar Biol. 2015;162:687–97.
27. Gorzelak P, Krzykawski T, Stolarski J. Diagenesis of echinoderm skeletons: constraints on paleoseawater Mg/Ca reconstructions. Glob Planet Change. 2016;144:142–57.
28. Richter DK, Zinkernagel U. Zur Anwendung der Kathodolumineszenz in der Karbonatpetrographie. Geol Rundsch. 1981;70:1276–302.

29. Pinsino A, Matranga V, Trinchella F, Roccheri MC. Sea urchin embryos as an in vivo model for the assessment of manganese toxicity: developmental and stress response effects. Ecotoxicology. 2010;19:555–62.

30. Pinsino A, Roccheri MC, Costa C, Matranga V. Manganese interferes with calcium, perturbs ERK signalling and produces embryos with no skeleton. Toxicol Sci. 2011;123:217–30.

31. Pinsino A, Matranga V, Roccheri MC. Manganese: a new emerging contaminant in the environment. In: Srivastava J, editor. Environmental Contamination. Rijeka: In Tech Open Access Publisher; 2012. p. 17–36.

32. Nilsson Sköld H, Baden SP, Looström J, Eriksson SP, Hernroth BE. Motoric impairment following manganese exposure in asteroid echinoderms. Aquat Toxicol. 2015;167:31–7.

33. DelValls TA, Dickson AG. The pH of buffers based on 2-amino-2-hydroxymethyl-1,3-propanediol ('tris') in synthetic sea water. Deep-Sea Res Pt I. 1998;45:1541–54.

34. Gran G. Determination of the equivalence point in potentiometric titrages-part II. Analyst. 1952;77:661–71.

35. Stolarski J. 3-dimensional micro- and nanostructural characteristics of the scleractinian corals skeleton: a biocalcification proxy. Acta Palaeontol Pol. 2003;48:497–530.

36. Hammer Ø, Harper DAT, Ryan PD. PAST: paleontological statistics software package for education and data analysis. Palaeontol Electron. 2001;4(1):9pp. http://palaeo-electronica.org/2001_1/past/issue1_01.htm

37. Füchtbauer H, Richter DK. Karbonatgesteine. In Füchtbauer H, editor. Sedimente und Sedimentgesteine, Stuttgart: Schweizerbart; 1988. p. 233–434.

38. Lebrato M, McClintock JB, Amsler MO, Ries JB, Egilsdottir H, Lamare M, Amsler CD, Challener RC, Schram JB, Mah CL, Cuce J, Baker BJ. From the Arctic to the Antarctic: the major, minor, and trace elemental composition of echinoderm skeletons. Ecology. 2013;94:1434.

39. Bray L, Pancucci-Papadopulou MA, Hall-Spencer JM. Sea urchin response to rising pCO2 shows ocean acidification may fundamentally alter the chemistry of marine skeletons. Mediterr Mar Sci. 2014;15:510–9.

40. Heatfield BM. Growth of the calcareous skeleton during regeneration of spines of the sea urchin Strongylocentrotus purpuratus (Stimpson); a light and scanning electron microscope study. J Morphol. 1971;134:57–90.

41. Märkel K, Kubanek F, Willgallis A. Polykristalliner Calcit bei Seeigeln (Echinodermata, Echinoidea). Z Zellforsch. 1971;119:355–77.

42. Märkel K, Roser U. The spine tissues in the echinoid Eucidaris tribuloides. Zoomorphology. 1983;103:25–41.

43. Dery A, Guibourt V, Catarino AI, Compère P, Dubois P. Properties, morphogenesis, and effect of acidification on spines of the cidaroid sea urchin Phyllacanthus imperialis. Invertebr Biol. 2014;133:188–99.

44. Märkel K. Morphologie der Seeigelzahne. II. Die gekielten Zahne der Echinacea (Echinoder- mata, Echinoidea). Z Morph Tiere. 1969;66:1–50.

45. Märkel K. The tooth skeleton of Echinometra mathaei (Blainville) (Echinodermata, Echinoidea). Annot Zool Jap. 1970;43:188–99.

46. Märkel K, Gorny P, Abraham K. Microarchitecture of sea urchin teeth. Fortschr Zool. 1977;24:103–14.

47. Märkel K. On the teeth of the recent cassiduloid Echinolampas depressa gray, and on some liassic fossil teeth nearly identical in structure (Echinodermata, Echinoidea). Zoomorphology. 1978;89:125–44.

48. Kniprath E. Ultrastructure and growth of the sea urchin tooth. Calcified Tissue Res. 1974;14:211–28.

49. Ziegler A, Stock SR, Menze BH, Smith AB. Macro- and microstructural diversity of sea urchin teeth revealed by large-scale micro-computed tomography survey. Proc SPIE. 2012;8506:85061G.

50. Stock SR, Ignatiev KI, Dahl T, Veis A, De Carlo F. Three-dimensional microarchitecture of the plates (primary, secondary, and carinar process) in the developing tooth of Lytechinus variegatus revealed by synchrotron X-ray absorption microtomography (microCT). J Struct Biol. 2003;144:282–300.

51. Stock SR. Sea urchins have teeth? A review of their microstructure, biomineralization, development and mechanical properties. Connect Tissue Res. 2014;55:41–51.

52. Holland ND. An autoradiographic investigation of tooth renewal in the Purple sea urchin (Strongylocentrotus purpuratus). J Exp Zool. 1965;158:275–82.

53. Levitan DR. Skeletal changes in the test and jaws of the sea urchin Diadema antillarum in response to food limitation. Mar Biol. 1991;111:431–5.

54. Lewis CA, Ebert TA, Boren ME. Allocation of 45calcium to body components of starved and fed purple sea urchins (Strongylocentrorus purpuratus). Mar Biol. 1990;105:213 22.

55. Ebert TA, Hernández JC, Clemente S. Annual reversible plasticity of feeding structures: cyclical changes of jaw allometry in a sea urchin. Proc R Soc B. 2014;281:20132284. http://dx.doi.org/10.1098/rspb.2013.2284

56. Robach JS, Stock SR, Veis A. Structure of first- and second-stage mineralized elements in teeth of the sea urchin Lytechinus variegatus. J Struct Biol. 2009;168:452–66.

57. Pearse JS, Pearse VB. Growth zones in the echinoid skeleton. Am Zool. 1975;15:731–53.

58. Märkel K. Wachstum des Coronarskeletes von Paracentrotus lividus Lmk. (Echinodemmata, Echinoidea). Zoomorphology. 1975;82:259–80.

59. Märkel K. Experimental morphology of coronar growth in regular echinoids. Zoomorphology. 1981;97:31–52.

60. Dafni J. A biomechanical model for the morphogenesis of regular echinoid tests. Paleobiol. 1986;12:143–60.

61. Ebert TA. Allometry, design and constraint of body components and of shape in sea urchins. J Nat Hist. 1988;22:1407–25.

62. Gage JD. Skeletal growth zones as age-markers in the sea urchin Psammechinus miliaris. Mar Biol. 1991;110:217–28.

63. Thompson DAW. On growth and form. Cambridge: Cambridge University Press; 1917.

64. Moss ML, Meehan M. Growth of the echinoid test. Acta Anat. 1968;69:409–44.

65. Raup DM. Theoretical morphology of echinoid growth. J Paleontol. 1968;42:50–63.

66. Seilacher A. Constructional morphology of sand dollars. Paleobiol. 1979;5:191–221.

67. Telford M. Domes, arches and urchins: the skeletal architechture of echinoids (Echinodermata). Zoomorphology. 1985;105:114–24.

68. Telford M. Structural models and graphical simulation of echinoids. Rotterdam: Balkema; 1994.

69. Baron CJ. What functional morphology cannot explain: a model of sea urchin growth and a discussion of the role of morphogenetic explanations in evolutionary biology. In: Dudley EC, editor. The unity of evolutionary biology. Proceedings of the Fourth International Congress of Systematic and Evolutionary Biology. Dioscorides. Portland; 1990. p. 471–488.

70. Ellers O. A mechanical model of growth in regular sea urchins: predictions of shape and a developmental morphospace. Proc R Soc Lond B Biol Sci. London. 1993;254:123–9.

71. Zachos LG. An equilibrium theory of echinoid plate geometry. GSA Abstracts with Programs 2007;39:501.

72. Zachos LG. A new computational growth model for sea urchin skeletons. J Theor Biol. 2009;259:646–57.

73. Abou Chakra M, Stone JR. Holotestoid: a computational model for testing hypotheses about echinoid skeleton form and growth. J Theor Biol. 2011;285:13–125.

7

Evaluation of the physiological activity of venom from the Eurasian water shrew *Neomys fodiens*

Krzysztof Kowalski[1*], Paweł Marciniak[2], Grzegorz Rosiński[2] and Leszek Rychlik[1]

Abstract

Background: Animal toxins can have medical and therapeutic applications. Principally, toxins produced by insects, arachnids, snakes and frogs have been characterized. Venomous mammals are rare, and their venoms have not been comprehensively investigated. Among shrews, only the venom of *Blarina brevicauda* has been analysed so far, and blarina toxin has been proven to be its main toxic component. It is assumed that *Neomys fodiens* employs its venom to hunt larger prey. However, the toxic profile, properties and mode of action of its venom are largely unknown. Therefore, we analysed the cardio-, myo- and neurotropic properties of *N. fodiens* venom and saliva of non-venomous *Sorex araneus* (control tests) in vitro in physiological bioassays carried out on two model organisms: beetles and frogs. For the first time, we fractionated *N. fodiens* venom and *S. araneus* saliva by performing chromatographic separation. Next, the properties of selected compounds were analysed in cardiotropic bioassays in the *Tenebrio molitor* heart.

Results: The venom of *N. fodiens* caused a high decrease in the conduction velocity of the frog sciatic nerve, as well as a significant decrease in the force of frog calf muscle contraction. We also recorded a significant decrease in the frog heart contractile activity. Most of the selected compounds from *N. fodiens* venom displayed a positive chronotropic effect on the beetle heart. However, one fraction caused a strong decrease in the *T. molitor* heart contractile activity coupled with a reversible cardiac arrest. We did not observe any responses of the insect heart and frog organs to the saliva of *S. araneus*. Preliminary mass spectrometry analysis revealed that calmodulin-like protein, thymosin β-10, hyaluronidase, lysozyme C and phospholipase A2 are present in the venom of *N. fodiens*, whereas thymosin β4, lysozyme C and β-defensin are present in *S. araneus* saliva.

Conclusion: Our results showed that *N. fodiens* venom has stronger paralytic properties and lower cardioinhibitory activity. Therefore, it is highly probable that *N. fodiens* might use its venom as a prey immobilizing agent. We also confirmed that *S. araneus* is not a venomous mammal because its saliva did not exhibit any toxic effects.

Keywords: Mammalian venom, Natural toxins, *Neomys fodiens*, Toxicity in vitro, Salivary glands, Shrews, *Sorex araneus*

Background

Many animal toxins have been discovered in the last century [1]. Even several species of toxic birds, such as *Pitohui kirhocephalus* or *Ifrita kowaldi*, have been discovered [2]. However, despite the availability of new proteomic, genomic and chromatographic separation techniques, most venomous and poisonous animals remain unstudied [3–5]. Principally, toxins produced by insects, arachnids, snakes and frogs have been

investigated thus far [3, 6]. Venoms produced by mammals remain mostly uncharacterized [3, 5, 6].

Only a few mammalian species, such as the short-tailed shrew, two species of water shrews and two species of solenodons (Soricomorpha), as well as the platypus (Monotremata), are proven to produce venoms [7–9]. Additionally, according to recent research, three species of vampire bats (Chiroptera) and three species of lorises (Primates) are considered venomous as well [5, 9–11]. Only the platypus venom has been comprehensively studied with a focus on its composition, function and evolution [12]. Some studies on toxic components and function of the vampire bat and loris (especially the slow loris) venom

* Correspondence: kowalski.biol@gmail.com
[1]Department of Systematic Zoology, Institute of Environmental Biology, Adam Mickiewicz University, Umultowska 89, 61-614 Poznań, Poland
Full list of author information is available at the end of the article

have been performed [10, 11]. Likewise, the biological activities of the venoms produced by few insectivorous mammals (from the order Soricomorpha) have been reported [7–9]. In general, studies on the structure of venomous glands and the toxic activity of solenodon venom are scarce [13, 14]. Among shrews, only the venom from the short-tailed shrew *Blarina brevicauda* has been analysed thus far [8, 15]. The main toxic component of this venom is blarina toxin (BLTX) — a serine protease with tissue kallikrein-like activity [8]. Moreover, blarinasin was purified and characterized in *B. brevicauda* venom [15]. However, despite the chemical similarity to BLTX, this protease is completely devoid of toxic activity. It is particularly surprising that the knowledge on venoms from two species of water shrews, *Neomys fodiens* and *N. anomalus*, quite common in Europe, is extremely poor [7, 9]. The toxic properties of *N. fodiens* and *N. anomalus* venoms from the submandibular salivary glands have been reported a few times in the past [16, 17]. Therefore, it is highly probable that, similar to *B. brevicauda*, the venoms of these two species contain compounds with toxic activity.

In the present paper, we aimed to analyse the profile and toxic activity of the venom from the Eurasian water shrew *N. fodiens*. The paralytic (neuro- and myotoxic) and cardiotoxic properties of the water shrew venom were identified by performing physiological bioassays on two model experimental animals: the mealworm beetle *Tenebrio molitor* and frogs (*Rana temporaria* and *Pelophylax* sp.). Additionally, control tests with the saliva of a non-venomous species, the common shrew *Sorex araneus*, were carried out.

Results

In vitro effects of the water shrew venom and its fractions on the heart contractile activity

The application of venom extract of the water shrew on the semi-isolated insect heart caused a small decrease in the heart contractile activity (−3.95% ± 1.51; Figs. 1a and 2a). After the application of *N. fodiens* extract on the frog heart, a small but significant decrease in the heart contractile activity was observed. During the 1st minute after venom application, the frequency of the heartbeat decreased by 1.73% ± 0.30 (Figs. 1b and 3b), whereas during 2nd minute by 1.36% ± 0.50 (Fig. 1b). There was no change in the insect heart activity after the application of saliva extract from the common shrew (−0.62% ± 0.40; Figs. 1a and 2c). Additionally, after the application of the common shrew saliva on frog heart we did not record any changes in the heart contractile activity (1st minute: −1.53% ± 0.87; 2nd minute: 0.28% ± 0.55; Figs. 1b and 3d). Moreover, the venom of the water shrew displayed a stronger negative chronotropic effect on the frog heart than *S. araneus* saliva.

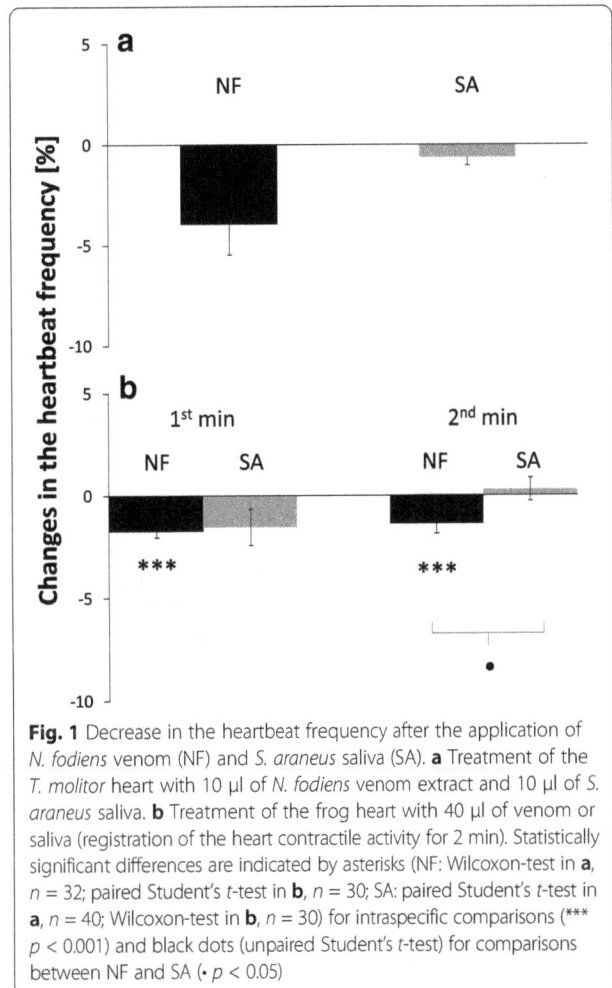

Fig. 1 Decrease in the heartbeat frequency after the application of *N. fodiens* venom (NF) and *S. araneus* saliva (SA). **a** Treatment of the *T. molitor* heart with 10 µl of *N. fodiens* venom extract and 10 µl of *S. araneus* saliva. **b** Treatment of the frog heart with 40 µl of venom or saliva (registration of the heart contractile activity for 2 min). Statistically significant differences are indicated by asterisks (NF: Wilcoxon-test in **a**, $n = 32$; paired Student's *t*-test in **b**, $n = 30$; SA: paired Student's *t*-test in **a**, $n = 40$; Wilcoxon-test in **b**, $n = 30$) for intraspecific comparisons (*** $p < 0.001$) and black dots (unpaired Student's *t*-test) for comparisons between NF and SA (• $p < 0.05$)

However, this result was significant only in the 2nd minute after sample application (Fig. 1b).

From 55 fractions obtained after the RP-HPLC (reverse phase high-performance liquid chromatography) separation of *N. fodiens* venom, 25 fractions were selected to analyse the cardioactivity in bioassays with the *T. molitor* heart (Fig. 4a). In the case of *S. araneus*, 16 fractions were assayed (Fig. 4b).

Among compounds from *N. fodiens* venom, fraction no. 5 caused a strong and highly significant decrease in the insect heart contractile activity (−43.7% ± 9.40; Fig. 5a). Moreover, in most cases, we observed a short and reversible cardiac arrest after the application of fraction no. 5 on the insect heart (Fig. 2b). Most of the other components (especially fractions no. 31, 34, 38 and 40) caused a small but significant increase in the heartbeat activity (Fig. 5a).

Among the components from the saliva of *S. araneus*, no fractions displayed cardioactivity (Fig. 5b). We did not observe any changes in the heartbeat frequency of the adult *T. molitor*. Only the application of fraction no. 29 caused an almost significant change. However, an

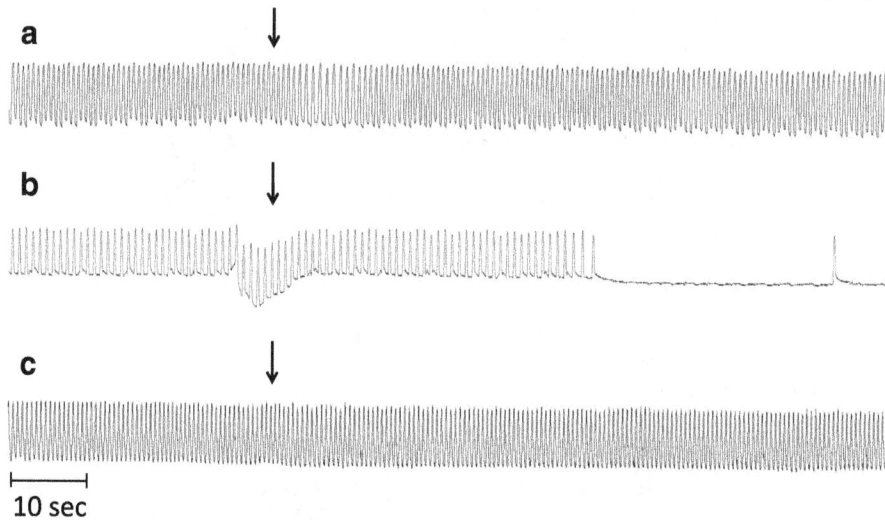

Fig. 2 Cardiomyograms displaying changes in the *T. molitor* heartbeat frequency after treatment with *N. fodiens* venom and *S. araneus* saliva. **a** Application of 10 μl of venom extract of *N. fodiens*. **b** Application of fraction no. 5 from *N. fodiens* venom. **c** Treatment with 10 μl of saliva extract of *S. araneus*. The sample application is indicated by an arrow

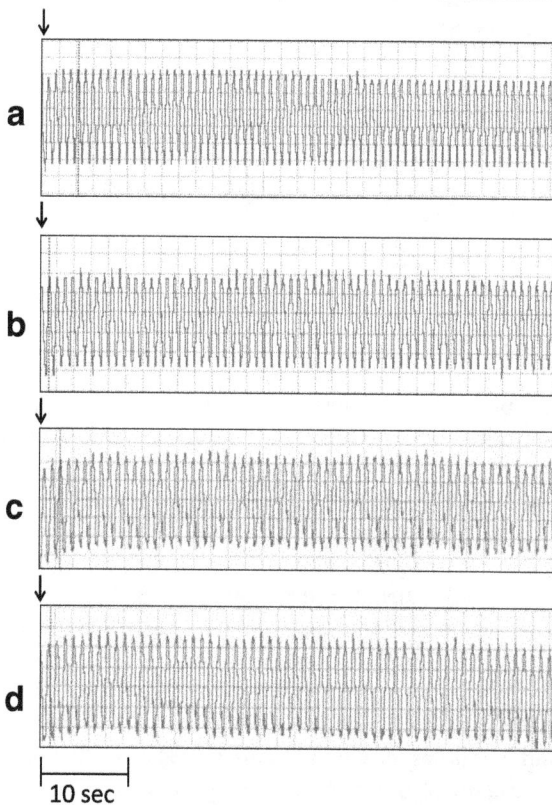

Fig. 3 Cardiomyograms displaying changes in the frog heartbeat frequency after treatment with *N. fodiens* venom and *S. araneus* saliva. The heart activity after application of 40 μl of Ringer's solution (control) (**a, c**), *N. fodiens* venom (in the 1st minute) (**b**), and *S. araneus* saliva (in the 1st minute) (**d**). The sample application is indicated by an arrow

increase in the insect heart contractile activity after the application of this fraction was very low (2.58% ± 1.14; Fig. 5b). The results of statistical analysis of this part are available in the Additional files section (see Additional file 1).

In vitro effects of venom from the water shrew on the frog calf muscle activity

The venom of the water shrew displayed a significant influence on the contractions of frog calf muscle. Immediately after the application of the venom on the muscle, the force of the muscle contraction was reduced by 9.86% ± 1.98. One minute after venom application, the muscle activity was still significantly decreased (−10.66% ±1.78; Fig. 6a). We did not observe such an influence of the saliva of the common shrew on the frog calf muscle contraction activity (Fig. 6a). Immediately after saliva application, the force of the muscle contraction was reduced by 3.50% ± 1.66. One minute later, it was reduced by 1.10% ± 1.19. Furthermore, *N. fodiens* venom displayed a stronger negative influence on muscle activity than saliva of the common shrew. This difference was almost significant immediately after sample application, whereas it was highly significant 1 min later (Fig. 6a).

In vitro effects of venom from the water shrew on the frog sciatic nerve activity

The venom of *N. fodiens* displayed a very strong negative effect on the frog sciatic nerve activity. Immediately after venom application, we observed a highly significant decrease in the nerve conduction velocity (−20.45% ± 2.98; Fig. 6b). Similarly, 1 min after venom application, the activity of the frog nerve was still significantly decreased

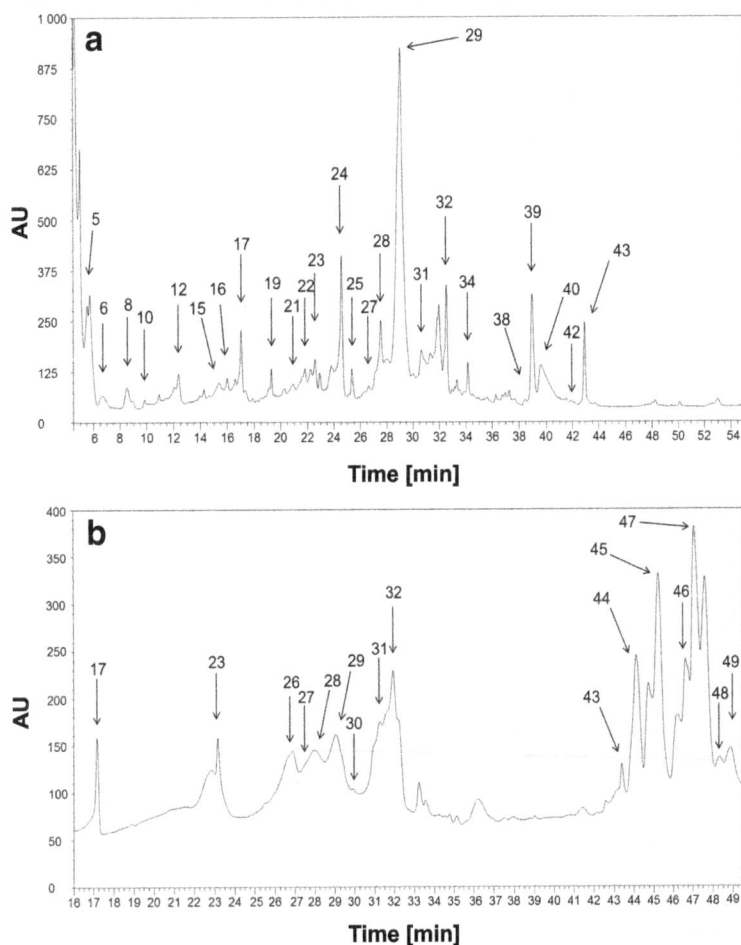

Fig. 4 Chromatograms displaying separation of the methanolic extract of the saliva of *N. fodiens* (**a**) and *S. araneus* (**b**). The analysed fractions are indicated by arrows

(−12.67% ± 2.28; Fig. 6b). We did not record such a change immediately after the application of *S. araneus* saliva on the frog sciatic nerve (−5.33% ± 1.08; Fig. 6b). By contrast, 1 min after treatment with the saliva of the common shrew, there was a small but significant increase in the nerve conduction velocity (3.43% ± 1.48; Fig. 6b). Moreover, the venom of the water shrew displayed a stronger negative neurotropic effect on the frog sciatic nerve than the saliva of the common shrew. This difference was highly significant both immediately after venom application and 1 min later (Fig. 6b).

Proteomic pre-identification of proteins from *N. fodiens* venom and *S. araneus* saliva

Preliminary mass spectrometry (MS) analysis enabled us to predict some protein/peptide molecules that are present in *N. fodiens* venom and might be involved in its toxic activity (Table 1). Calmodulin-like protein, thymosin β-10, hyaluronidase and β-nerve growth factor were found in the methanolic extract from *N. fodiens* venom.

Among the compounds determined in fractions, obtained by chromatographic separation, were hyaluronidase (fraction no. 39), cystatin C (fraction no. 5), coagulation factor VIII (fractions no. 5 and 39), lactyloglutathione lyase (fraction no. 39) and inhibitor of leech-derived tryptase (fraction no. 34). Additionally, lysozyme C was found in fraction no. 5, whereas phospholipase A2 and calmodulin-like protein were present in fraction no. 40. Phospholipase A2 (protein sequence coverage: 4%) was also detected in the methanolic extract of *N. fodiens*.

Thymosin β-10 and coagulation factor XI were present in the saliva of *S. araneus*, whereas cystatin-C was found in fractions no. 23, 28 and 29 (Table 1). Additionally, we found thymosin β-4 in fraction no. 23, lysozyme C in fraction no. 28 and β-defensin in fraction 29, as well as kallikrein 1-related peptidase in fraction no. 28. Nonetheless, these results should be treated with caution as further analyses (including obtaining a larger quantity of tissue material or more precise saliva purification) are required.

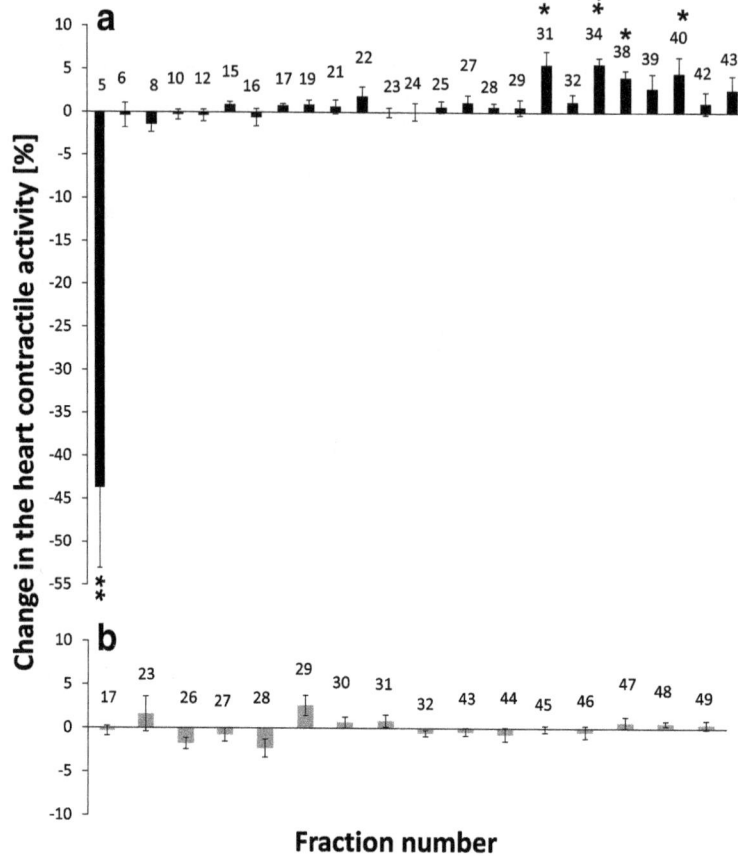

Fig. 5 Effects of fractions from *N. fodiens* venom (**a**) and *S. araneus* saliva (**b**) on the *T. molitor* heartbeat frequency. Statistically significant changes in the heart contractile activity are indicated by asterisks (Wilcoxon-test after Bonferroni correction: * $p' < 0.05$, ** $p' < 0.01$)

Discussion

At the end of the 50s and 60s of the last century, Pucek [16, 17] discovered that *N. fodiens* and *N. anomalus* produce in their salivary glands a potent venom that is toxic to mammals such as mice, voles and rabbits. The strongest toxic effects of saliva were revealed after intracerebral (mice and voles) and intravenous injections (rabbits). Following injection, experimental animals were overcome by a general depression and showed symptoms that included excessive urination, irregular respiration, paralysis of the hind limbs, convulsions and, finally, death [16, 17]. Similar symptoms were observed when the saliva of *B. brevicauda* was administered to mice, voles, rabbits and cats [7, 18, 19]. As noted by Lawrence [20], the venom of *B. brevicauda* displays predominantly neurotoxic activity. Our results also prove that the venom from the water shrew has stronger paralytic (neuro- and myotoxic) activity, whereas its cardiotropic effect on the frog and beetle heartbeat frequency is relatively small. At first, it might seem surprising, but it becomes more comprehensible when the food and metabolic requirements of shrews are taken into consideration.

It is well known that shrews must consume a large amount of food to meet their particularly high metabolic demands [21, 22]. Although shrews are mainly insectivorous, it has been repeatedly reported that they can also consume vertebrates, even larger than themselves, such as murid rodents, fish, frogs or lizards [23–26]. Therefore, it might be expected that venom could be a useful tool in catching and paralysing larger prey [7, 8]. Hunting larger prey might provide the food supply for a longer time, especially if the prey remains paralysed and decomposes considerably slowly. According to the optimal foraging theory, the hoarding of a food supply can save energy and time spending on prey foraging and catching. Furthermore, it might also minimize the risk of predation [27–29]. The food-storing habits of shrews have also been reported [21, 30–33]. According to Rychlik and Jancewicz [33], the water shrew usually hoards larger prey than other shrew species. Therefore, it is highly probable that *N. fodiens* can employ its venom to seize and immobilize such prey. Our results also support the hypothesis that *N. fodiens* venom may be a useful tool in the handling and paralysing of larger prey.

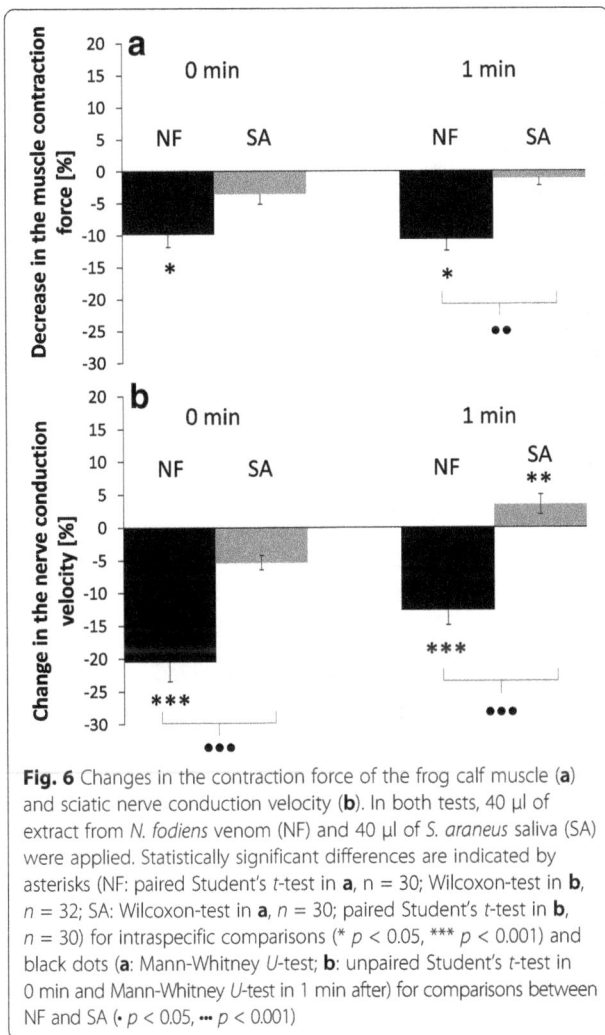

Fig. 6 Changes in the contraction force of the frog calf muscle (**a**) and sciatic nerve conduction velocity (**b**). In both tests, 40 µl of extract from *N. fodiens* venom (NF) and 40 µl of *S. araneus* saliva (SA) were applied. Statistically significant differences are indicated by asterisks (NF: paired Student's *t*-test in **a**, n = 30; Wilcoxon-test in **b**, *n* = 32; SA: Wilcoxon-test in **a**, *n* = 30; paired Student's *t*-test in **b**, *n* = 30) for intraspecific comparisons (* $p < 0.05$, *** $p < 0.001$) and black dots (**a**: Mann-Whitney *U*-test; **b**: unpaired Student's *t*-test in 0 min and Mann-Whitney *U*-test in 1 min after) for comparisons between NF and SA (• $p < 0.05$, ••• $p < 0.001$)

However, it should be noticed that Pucek [16, 17] observed the conspicuous effects of the water shrew venom after intracerebral and intravenous injections. We do not deny these results, but it seems that the recorded symptoms could partly result from mechanical damage caused by the insertion of a needle into the brain rather than from venom injection. Therefore, in our study, we analysed the responses of specific organs to venom application in in vitro bioassays instead of an examination of the whole organism response. Moreover, Pucek [16, 17] and Pearson [18] characterized only the properties of crude extracts from the salivary glands of shrews. In the present work, besides the investigation of the activity of salivary extract from the venom, for the first time, we examined the toxic activity of selected compounds from *N. fodiens* venom and *S. araneus* saliva fractionated by chromatographic separation.

In general, our results indicate that many components from *N. fodiens* venom have a weak but positive chronotropic influence on *T. molitor* heart contractile activity.

However, fraction no. 5 displayed a strong negative chronotropic effect on the heartbeat frequency coupled with a short and reversible cardiac arrest. It has been reported that *B. brevicauda* can use the venom as an insect-immobilizing agent [20, 34]. Thus, due to the cardioinhibitory activity of at least one molecule contained in the *N. fodiens* venom, we suppose that the water shrew can also employ its venom to hunt certain invertebrate prey, such as larger beetles, diplopods or crickets.

Most animal venoms comprise a mixture of bioactive compounds, such as proteins and peptides, salts or amino acids [35–37]. Therefore, it is highly probable that certain neurotoxins or proteins, similar to BLTX, may be present in the venom of *N. fodiens*. We are aware that obtaining a larger quantity of tissue material (by pooling extracts from salivary glands collected from at least 25 specimens) or more precise purification of fractionated samples are required to identify specific toxic molecules from the water shrew venom. Nevertheless, preliminary MS analysis did enable us to predict specific toxic compounds that presumably are present in *N. fodiens* venom and might be involved in its toxic activity. We suggest that calmodulin-like protein or cystatin C might contribute to the decrease in the frequency of the heartbeat. Zhang et al. [38] found that calmodulin may have acute and chronic effects on cardiac function. Cystatin-C has been proven to inhibit cysteine proteases and lead to neuro-degeneration and cardiovascular diseases [39]. In addition, we found lysozyme C and phospholipase A2 (PLA2) in fractions no. 5 and 40 from *N. fodiens* venom. Dufton [7] also reported lysozyme from the saliva of the water shrew, whereas PLA2 is widely distributed among elapid and viperid snake venoms [40, 41]. It has been proven that PLA2 molecules exhibit various pharmacological effects, such as cardio-, myo- and neurotoxicity, as well as pro- and anticoagulant effects [40, 41]. Therefore, it is possible that phospholipase A2 might also be responsible for the paralytic and cardiotropic symptoms recorded by us, but these conclusions are, by far, more speculative. Hyaluronidase, which is commonly present in animal venoms [42], might promote the spreading of these and the other components from *N. fodiens* venom [36, 42]. Additionally, it is highly possible to reveal kallikrein-like proteins (similar to BLTX) in the venom of the water shrew, especially because, in the present work, we found kallikrein 1-related peptidase in the saliva of the common shrew.

It is noteworthy that, in our study, the saliva of *S. araneus* has been analysed for the first time. We did not record any significant effects of the salivary extract from the common shrew on the heart, nerve or muscle activity. Similarly, none of the fractionated compounds from *S. araneus* saliva displayed cardiotropic effects on the *T. molitor* heart. These results prove that *S. araneus* is not

Table 1 Pre-identification of proteins from *N. fodiens* venom and *S. araneus* saliva based on tandem mass spectrometry analysis

Sample	Protein name	Species	Accession	Matched peptides	Protein sequence coverage [%]	Ion score	m/z	Identified peptides	Possible toxic activity
Neomys fodiens									
extract	calmodulin-like protein	*Mus musculus*	Q9D6P8	25	37	64	955	KEAFSLFDKD	Acute and chronic effects on cardiac function by regulation of the intracellular Ca^{2+} concentration [38]
						37	4086	R.SLGQNPTEAELQGMVNEIDKDGNGTVDFPEFLTMMSR.K + Oxidation (M)	
						57	4102	R.SLGQNPTEAELQGMVNEIDKDGNGTVDFPEFLTMMSR.K + 2 Oxidation (M)	
						97	1351	K.MKDTDSEEIR.E	
						81	1367	K.MKDTDSEEIR.E + Oxidation (M)	
						65	1092	K.DTDSEEIR.E	
	hyaluronidase-2	*Bos taurus*	Q8SQG8	14	8	24	980	HKMPLDPK	Facilitates spread of other venom proteins [36, 42]
	thymosin β-10	*Bos taurus*	P21752	11	11	22	862	K.TETQEK	Improves cardiac function, promotes vascularization and contractility in heart tissue [52, 53]
	β-nerve growth factor	*Mus musculus*	P01139	8	6	68	1153	KLQHSLDTALR.R	Unknown [36, 54]
						37	764	R.RLHSPR.V	
fraction no. 5	cystatin-C	*Rattus norvegicus*	P14841	14	8	28	1207	GTHTLTKSSCK	Inhibits cysteine proteases, failures in biological mechanisms controlling protease activities may led to many diseases such as neuro-degeneration or cardiovascular diseases [39]
	coagulation factor VIII	*Sus scrofa*	P12263	18	5	22	1427	ISALGKSAAGPLASGK	Acts as an anti-hemophilic factor [55]
	lysozyme C-1	*Sus scrofa*	P12067	6	8	24	930	YWCNDGK	Involved in an antimicrobial defence [10, 56]
fraction no. 31	hyaluronidase PH-20	*Myotis brandtii*	gi|521028001	14	11	47	1152	KDIEFYIPK	See above
fraction no. 34	chain E, leech-derived tryptase inhibitor Trypsin Complex	*Sus scrofa*	gi|3318722	48	21	177	2210	R.LGEHNIDVLEGNEQFINAAK.I	Prolongs the blood clotting time by thrombin and trypsin inhibition [57]
						115	2282	KIITHPNFNGNTLDNDIMLIKL	
						93	1045	KLSSPATLNSR.V	
						77	841	R.VATVSLPR.S	
						108	1515	KSSGSSYPSLLQCLKA	
						98	1051	KAPVLSDSSCKS	
fraction no. 39	coagulation factor VIII	*Sus scrofa*	P12263	19	7	23	1427	ISALGKSAAGPLASGK	See above
	lactylglutathione lyase	*Rattus norvegicus*	Q6P7Q4	17	25	78	1264	KDFLLQQTMLR.I	Involved in inflammation [58]
						54	1028	KKSLDFYTR.V	
						44	900	KSLDFYTR.V	
						57	976	KRFEELGVKF	
						65	2288	KGLAFVQDPDGYWIEILNPNKM	

Table 1 Pre-identification of proteins from *N. fodiens* venom and *S. araneus* saliva based on tandem mass spectrometry analysis *(Continued)*

Sample	Protein name	Species	Accession	Matched peptides	Protein sequence coverage [%]	Ion score	m/z	Identified peptides	Possible toxic activity
fraction no. 40	phospholipase A2	Oryctolagus cuniculus	P14422	6	6	27	1002	FAKFLSYK	Exhibits cardio-, myo- and neurotoxicity, as well as pro- and anticoagulant effects [40, 41]
	calmodulin-like protein	Mus musculus	Q9D6P8	14	5	25	1367	MKDTDSEEEIR	See above
					Sorex araneus				
extract	thymosin β-10	Rattus norvegicus	P63312	7	34	56	862	K.KTETQEKN	See above
						37	734	K.TETQEKN	
						48	875	K.ETIEQEK.R	
						77	1031	K.ETIEQEKR.S	
	coagulation factor XI	Mus musculus	Q91Y47	10	11	21	846	MICAGYK	Involved in blood clotting [59]
fraction no. 23	thymosin β-4	Oryctolagus cuniculus	P34032	69	88	46	1245	M.ADKPDMAEIEK.F	See thymosin β-10
						47	1652	M.ADKPDMAEIEKFDK.S + Oxidation (M)	
						34	862	K.KTETQEKN	
						38	734	K.TETQEKN	
						83	1371	K.TETQEKNPLPSKE	
						89	1512	K.NPLPSKETIEQEKQ	
						43	875	K.ETIEQEKQ	
						72	1348	K.ETIEQEKQAGES.-	
fraction no. 28	cystatin-C	Rattus norvegicus	P14841	14	8	27	1207	GTHTLTKSSCK	See above
	cystatin-C	Rattus norvegicus	P14841	11	6	43	1207	KGTHTLTKSSCKN	See above
	lysozyme C-1	Sus scrofa	P12067	10	19	22	787	AWVAWR	See above
	kallikrein 1-related peptidase b24	Mus musculus	Q61754	10	6	33	1204	KDKSNDLMLLRL	Might act as an inflammatory agent (increasing vascular permeability and lowering blood pressure) [8, 9]
fraction no.29	cystatin-C	Rattus norvegicus	P14841	14	8	27	1207	GTHTLTKSSCK	See above
	α-amylase 1	Mus musculus	P00687	8	11	25	780	DYVRTK	Unknown
	β-defensin 7	Mus musculus	Q91V70	8	11	24	760	FQIPEK	Exhibits a significant myo- and neurotoxic activity, modifying voltage-sensitive Na^+ channels, resulting in a potent analgesic effect [36]

a venomous mammal. It seems there are no venomous species among the genus of *Sorex*. Suspicion remains concerning the cinereus shrew (*S. cinereus*), but its saliva has not been characterized thus far [9]. Lopez-Jurado and Mateo [26] suggested that the Canarian shrew (*Crocidura canariensis*) produces toxic saliva. However, as in the case of *S. cinereus*, the saliva of *C. canariensis* has not been analysed. More studies are required to prove the toxicity of saliva from these two species. It seems that smaller shrews, such as the common shrew, do not need to produce molecules with toxic activity in their saliva because they are unlikely to subdue large vertebrate prey, such as murid rodents or frogs. Our behavioural data provide experimental confirmation of this prediction as well (Kowalski et al., unpublished observations).

Conclusions

Our results show that the venom of the water shrew displays stronger paralytic effects and lower cardioinhibitory activity. Therefore, we conclude that *N. fodiens* venom might be a useful tool in hunting and immobilizing larger prey. Additionally, for the first time, we separated and examined compounds from the water shrew venom and *S. araneus* saliva. Certain components from *N. fodiens* venom showed a weak positive impact on *T. molitor* heart contractile activity. However, we found that component(s) of sample no. 5 showed a strong negative chronotropic effect on the insect heart coupled with a reversible cardiac arrest. It enabled us to conclude that the water shrew can employ its venom to seize certain invertebrates, such as beetles or diplopods. Because none of the components from *S. araneus* saliva displayed toxic effects, we confirmed that the common shrew is not a venomous mammal. We suggest carrying out further analyses with a larger quantity of tissue material and more precise sample purification to identify proteins from the venom of *N. fodiens*. We are convinced that the present findings and further studies will enable us to better understand the role of venom in the animal world and mammal evolution in relation to toxicity. Moreover, because animal toxins can have applications in medicine and pharmacy [5, 43, 44], it is highly probable that our results also provide the opportunity for the design and development of new drugs, such as cardiac or neuromuscular blocking agents.

Methods
Animals
Shrews

Trapping sessions were performed in the suburbs of Poznań (western Poland) from April to October 2014–2016, excluding the coldest days. In total, we captured 25 water shrews and 25 common shrews. The captured

shrews were transported to the laboratory. Next, the animals were placed separately into large (39 × 21 × 28 cm; 23 L) terraria equipped with bedding (a mixture of peat, moss and sand). Each terrarium contained a shelter (flowerpot) and a bowl with water. Food (mealworms, minced beef, earthworms and snails) and water were provided *ad libitum*. The shrews were kept in the breeding room under standard laboratory conditions (temperature: 21 °C; humidity: 65–70%; artificial photoperiod: 12 L:12D). After about a week, they were killed using approved methods to obtain their submandibular salivary glands.

Frogs

Frogs [15 common frogs (*Rana temporaria*) and 37 frogs from the genus of *Pelophylax* sp.] were captured by a net near ponds and small water tanks located in the Morasko district of Poznań (western Poland).

Insects

Tenebrio molitor L. adults (4 weeks old) were obtained from a culture maintained at the Department of Animal Physiology and Development, Adam Mickiewicz University, Poznań, Poland, as described previously [45].

Venom/saliva collection and sample preparation

Shrews were killed by cervical dislocation, and their submandibular salivary glands were carefully isolated to obtain venom or saliva. Glands ($n = 15$; from each shrew species) designed for heart bioassays were transferred into 600 µl of insect Ringer's solution (RS: 274 mM NaCl; 19 mM KCl; 9 mM CaCl; 5 mM glucose, and 5 mM HEPES, pH 7.0; bioassays with extract on the semi-isolated beetle heart) or 600 µl of frog Ringer's solution (bioassays on frog organs). Glands ($n = 10$, from each shrew species) designed for chromatographic separation were transferred into 600 µl of methanol. Tissues were next homogenized, and samples were centrifuged at 10,000 × g and 4 °C for 30 min. The supernatants were collected, and the protein content was determined using a Direct Detect spectrometer (MERCK Millipore, Warsaw, Poland). Extracts with a final protein concentration of 1 µg/µl were used for bioassays.

Chromatographic separation

Supernatants suspended in methanol were used for peptide analysis by reverse phase high-performance liquid chromatography (RP-HPLC). Separation was performed using a Dionex Ultimate 3000 chromatographic system comprising a dual pump programmable solvent module. Supernatants were analysed using a BioBasic-18 analytical column (5 µm, 150 × 4.6 mm; Thermo Scientific). The samples were eluted with a gradient of 5–60% acetonitrile (ACN)/0.1% TFA with a flow rate of 0.5 ml/

min for 55 min. The eluent was monitored at 214 nm, and fractions were collected into 1.5-ml tubes. Next, ACN was evaporated, and samples were suspended in 70 µl of RS to determine the toxic activity of the separated components by performing bioassays on the semi-isolated *T. molitor* hearts.

In vitro insect heart bioassay

Cardioactivity of the water shrew venom was assayed on the hearts of adult *Tenebrio molitor* and frogs. The microdensitometric method was used to measure the chronotropic (change in frequency) effect of samples on the semi-isolated heart of the adult beetle [46]. Briefly, *T. molitor* adults were decapitated, and the abdomen was removed. The ventral body wall of the abdomen was excised. The fat body, digestive system, and Malpighian tubules were removed from the abdomen. The final preparation consisted of the dorsal vessel (the heart), wing muscles, body wall muscles, and the dorsal cuticle. The heart preparations with a regular heartbeat were selected and superfused in RS. The superfusion chamber with the heart preparation was installed into the microdensitometer MD-100 (Carl Zeiss, Jena, Germany). An open perfusion system with an injection port 70 mm above the superfusion chamber was used. The flow rate of the fresh RS was 140 µl/min, and excess solution was removed from the superfusion chamber using chromatographic paper (Whatman No. 3, UK). All tested samples were applied at the injection port with a Hamilton syringe (10 µl). Many applications of samples could be sequentially assayed in a single preparation. The open system was designed to enable the addition of the samples avoiding changes in pressure. After 10 min of initial stabilization, the activity of the isolated heart was recorded for 30 s. Next, the sample was applied, and the heart activity was recorded for a further 1.5 min. This procedure was repeated with 5-min intervals for each tested sample.

Computer software (Larwa) developed at the Department of Animal Physiology and Development was used to record and analyse the cardiomyograms [47]. The activities of the analysed samples were presented as the percentage change in the frequency of the beetle heart contractions after sample application.

In vitro frog heart bioassay

Frogs were decapitated, and the skin from the thorax was removed. The thorax wall was excised, and the pericardium was carefully removed (to not damage the heart). The mechanical responses of the spontaneously contracting, semi-isolated frog heart were measured by attaching one end of a thread to the apex of the heart using a clip and the other end to a force transducer (MLT 0420; ADInstruments, Australia). The contractile activity of the frog heart (the number of heart contractions/min) was recorded using a Power Lab 26T System (ADInstruments, Australia) attached to a computer equipped with Chart 5.5.4 software [48]. A given heart bioassay consisted of 5 steps:

(1) measurement of the heart contractile activity (HCA) for 1 min immediately after the isolation of the frog heart;
(2) measurement of HCA for 1 min after 40 µl of RS application on the semi-isolated frog heart (control test);
(3) measurement of HCA for 1 min after 40 µl of venom application on the semi-isolated frog heart;
(4) measurement of HCA for 1 min after step 3 (sometimes the effect of the venom appears in the 2nd minute after venom application);
(5) measurement of HCA for 1 min after 40 µl of RS application on the semi-isolated frog heart (to check the heart ability to recover).

This procedure was repeated with 5-min intervals for each tested sample. A maximum of 3 samples were applied on a single frog heart. The activities of the samples were presented as the percentage change in the frequency of the frog heartbeat after sample application.

In vitro frog muscle bioassay

After frog decapitation, the skin from the hind part of the body and hind legs was removed. Next, the calf muscle was carefully isolated and transferred into a tub connected to the electrical stimulator incorporated into the Power Lab 26T unit. To stimulate the contraction of frog muscle, a 500-mV voltage was applied. The mechanical responses of the isolated muscle were measured by attaching one end of a thread to the tendon of muscle and the other end to a force transducer (MLT 0420; ADInstruments, Australia). The force of the muscle contraction was recorded using a Power Lab 26T System (ADInstruments, Australia) attached to a computer equipped with Chart 5.5.4 software [49]. A given muscle bioassay consisted of 5 steps (occurring with 1-min intervals):

(1) measurement of the muscle contraction force (MCF) immediately after the isolation of the frog muscle;
(2) measurement of MCF immediately after 40 µl of RS application on the isolated frog muscle (control test);
(3) measurement of MCF immediately after 40 µl of venom application on the isolated frog muscle;
(4) measurement of MCF 1 min after venom application;
(5) measurement of MCF after 40 µl of RS application (to check the muscle ability to recover).

This procedure was repeated with 5-min intervals for each tested sample. A maximum of 3 samples were applied on a single frog calf muscle. The activities of samples were presented as the percentage change in the force of the muscle contraction after sample application.

In vitro frog nerve bioassay

After frog decapitation, the skin from the hind part of the body and hind legs was removed. The ventral wall of the body was excised, and the viscera were removed. Next, the nerve trunk was looped, and the sciatic nerve was carefully isolated (to not damage it) and transferred into a tub connected to the electrical Power Lab 26T as described previously. To stimulate the conduction velocity of the frog nerve, a 500-mV voltage was applied. The conduction velocity was recorded using the Power Lab 26T System (ADInstruments, Australia) attached to a computer [50]. The protocol of a given nerve bioassay was the same as that of a muscle assay.

This procedure was repeated with 5-min intervals for each tested sample. A maximum of 3 samples were applied on a single frog nerve. The activities of the samples were presented as the percentage change in the conduction velocity of the frog nerve after sample application.

Protein identification

Proteomic analysis was carried out at the Mass Spectrometry Laboratory, Institute of Biochemistry and Biophysics, Polish Academy of Sciences, Warsaw. Peptides from the extract of *N. fodiens* venom and *S. araneus* saliva were analysed by liquid chromatography coupled to tandem mass spectrometry LC-(MS-MS/MS) using a Nano-Acquity LC system (Waters, Milford, Massachusetts, USA) and an OrbitrapVelos mass spectrometer (Thermo Electron Corp., San Jose, CA). Before performing the analysis, the proteins were subjected to an ion-solution digestion procedure. Proteins were (1) reduced with 50 mM TCEP for 30 min at 60 °C, (2) alkylated with 200 mM MMTA for 30 min at room temperature and (3) digested overnight with trypsin (sequencing Grade Modified Trypsin - Promega V5111). Next, the samples were applied to an RP-18 precolumn (nanoACQUITY Symmetry® C18 - Waters 186,003,514) using water containing 0.1% TFA as a mobile phase and were transferred to a nano-HPLC RP-18 column (nanoACQUITY BEH C18 - Waters 186,003,545). The samples were eluted with a gradient of 0–35% acetonitrile in the presence of 0.05% formic acid with a flow rate of 250 nl/min for 180 min. The column was directly coupled to the ion source of the spectrometer working within data dependent on the MS to MS/MS switch. To ensure a lack of cross contamination from previous samples, each analysis was preceded by a blank run. The proteins were identified by a Mascot Search (Matrix Science, London, UK) against the SwissProt and NCBInr

databases. The search parameters were as follows: type of search: MS/MS Ion Search; enzyme specificity: trypsin; fixed methylthio modification of cysteine; variable modifications: methionine oxidation; mass values: monoisotopic; protein mass: unrestricted; peptide mass tolerance: 30 ppm; fragment mass tolerance: 0.1 D; number of missed cleavage sites allowed: 1; instrument type: HCD. Peptides with Mascot scores exceeding the threshold value of $p < 0.05$ were considered positively identified.

Data analysis

All data were presented as the mean values ± SEM (standard error of the mean) of the indicated number of replicates (n). The mean differences between the treatment and control group were determined using paired Student's *t*-test or Wilcoxon signed-rank test. To indicate statistically significant differences between *N. fodiens* venom and *S. araneus* saliva activity, unpaired Student's *t*-test or Mann-Whitney *U*-test was performed. Non-parametric tests (Wilcoxon signed-rank and Mann-Whitney) were performed when the datasets of non-normal distributions were compared. To counteract the problem of multiple comparisons, Bonferroni correction was performed. All statistical analyses were carried out using R software [51]. Differences were considered as statistically significant for *p*-values less than 0.05.

Acknowledgements

We are very grateful to M. Szymczak and J. Ziomek for help in laboratory work, to U. Eichert, G. Kowalski, S. von Merten, P. Kardynia, S. Dziemian-Zwolak, A. Kret, K. Bauer, A. Grozelier, P. Klimant, J. Borska, N-E. Karantanis and M. Pejka for help in field works.

Funding

The research was financially supported by grant no. 2015/17/N/NZ8/01567 from National Science Centre, Poland and, in part, by the budget of the Department of Systematic Zoology (Faculty of Biology, AMU, Poznań). The equipment used in proteomic analysis was sponsored in part by the Centre for Preclinical Research and Technology (CePT), a project co-sponsored by European Regional Development Fund and Innovative Economy, The National Cohesion Strategy of Poland.

Authors' contributions

All authors conceived and designed the experiments; KK and LR captured most experimental animals; KK and PM performed the experiments; KK and PM analysed the data; KK wrote the paper; PM, GR and LR revised and edited the manuscript. All authors read and approved the final manuscript.

Competing interest

The authors declare that they have no competing interests.

Author details

[1]Department of Systematic Zoology, Institute of Environmental Biology, Adam Mickiewicz University, Umultowska 89, 61-614 Poznań, Poland. [2]Department of Animal Physiology and Development, Institute of Experimental Biology, Faculty of Biology, Adam Mickiewicz University, Umultowska 89, 61-614 Poznań, Poland.

References

1. Ménez A, Bontems F, Roumestand C, Gilquin B, Toma F. Structural basis for functional diversity of animal toxins. P Roy Soc Edinb B. 1992;99:83–103.
2. Dumbacher JP, Spande TF, Daly JW. Batrachotoxin alkaloids from passerine birds: A second toxic bird genus (Ifrita kowaldi) from New Guinea. PNAS. 2000;97:12970–5.
3. Kita M. Bioorganic studies on the key natural products from venomous mammals and marine invertebrates. Bull Chem Soc Jpn. 2012;85:1175–85.
4. Von Reumont BM, Campbell LI, Jenner RA. Quo vadis venomics? A road map to neglected venomous invertebrates. Toxins. 2014;6:3488–551.
5. Rode-Margono JE, Nekaris KAI. Cabinet of curiosities: venom systems and their ecological function in mammals, with a focus on primates. Toxins. 2015;7:2639–58.
6. Uemura D, Kita M, Arimoto H, Kitamura M. Recent aspects of chemical ecology: natural toxins, coral communities, and symbiotic relationships. Pure Appl Chem. 2009;81:1093–111.
7. Dufton MJ. Venomous mammals. Pharmacol Ther. 1992;53:199–215.
8. Kita M, Nakamura Y, Okumura Y, Ohdachi SD, Oba Y, Yoshikuni M, Kido H, Uemura D. Blarina toxin, a mammalian lethal venom from the short-tailed shrew Blarina brevicauda: isolation and characterization. PNAS. 2004;101: 7542–7.
9. Ligabue-Braun R, Verli H, Carlini CR. Venomous mammals: a review. Toxicon. 2012;59:680–95.
10. Low DHW, Sunagar K, Undheim EAB, Ali SA, Alagon AC, Ruder T, Jackson TNW, Pineda Gonzales S, King GF, Jones A, Antunes A, Fry BG. Dracula's children: molecular evolution of vampire bat venom. J Proteome. 2013;89: 95–111.
11. Nekaris KAI, Moore RS, Rode EJ, Fry BG. Mad, bad and dangerous to know: the biochemistry, ecology and evolution of slow loris venom. J Venom Anim Toxins incl Trop Dis. 2013;19:21.
12. Whittington CM, Belov K. Tracing monotreme venom evolution in the genomics era. Toxins. 2014;6:1260–73.
13. Rabb GB. Toxic salivary glands in the primitive insectivore Solenodon. Nat Hist Misc. 1959;190:1–3.
14. Pournelle GH. Classification, biology and description of the venom apparatus of insectivores of the genera Solenodon, Neomys, and Blarina. In: Bücherl W, Buckley EE, Deulofeu V, editors. Venomous animals and their venoms. New York: Academic Press; 1968. p. 31–42.
15. Kita M, Okumura Y, Ohdachi SD, Oba Y, Yoshikuni M, Nakamura Y, Kido H, Uemura D. Purification and characterisation of blarinasin, a new tissue kallikrein-like protease from the short-tailed shrew Blarina brevicauda: comparative studies with blarina toxin. Biol Chem. 2005;386:177–82.
16. Pucek M. The effect of the venom of the European water shrew (Neomys fodiens fodiens Pennant) on certain experimental animals. Acta Theriol. 1959; 3:93–108.
17. Pucek M. Neomys anomalus Cabrera, 1907 - a venomous mammal. Bull Acad Pol Sci. 1969;17:569–73.
18. Pearson OP. On the cause and nature of a poisonous action produced by the bite of a shrew (Blarina brevicauda). J Mammal. 1942;23:159–66.
19. Pearson OP. A toxic substances from the salivary glands of a mammal (short-tailed shrew). Venoms. 1956;44:55–8.
20. Lawrence B. Brief comparison of short-tailed shrew and reptile poison. J Mammal. 1945;26:393–6.
21. Churchfield S. The natural history of shrews. 1st ed. UK: Comstock Publishing/ Cornell University Press; 1990.
22. Taylor JRE. Evolution of energetic strategies in shrews. In: Wójcik JM, Wolsan M, editors. Evolution of shrews. Białowieża: Mammal Research Institute of the Polish Academy of Sciences; 1998. p. 309–46.
23. Hamilton WI Ir The food of the Soricidae. J Mammal. 1930;11:26–39.
24. Buchalczyk T, Pucek Z. Food storage of the European water shrew, Neomys fodiens (Pennant, 1771). Acta Theriol. 1963;19:376–9.
25. Tomasi TE. Function of venom in the short-tailed shrew Blarina brevicauda. J Mammal. 1978;59:852–4.
26. Lopez-Jurado LF, Mateo JA. Evidence of venom in the Canarian shrew (Crocidura canariensis): immobilizing effects on the Atlantic lizard (Gallotia atlantica). J Zool. 1996;239:394–5.
27. MacArthur RH, Pianka ER. On optimal use of a patchy environment. Am Nat. 1966;100:603–9.
28. Schoener TW. Theory of feeding strategies. Ann Rev Ecol Syst. 1971;2:369–404.
29. Marten GG. An optimization equation for predation. Ecology. 1973;54:92–101.
30. Robinson DE, Brodie ED Jr. Food hoarding behavior in the short-tailed shrew Blarina brevicauda. Am Midl Nat. 1982;108:369–75.
31. Martin IG. Factors affecting food hoarding in the short-tailed shrew Blarina brevicauda. Mammalia. 1984;48:65–71.
32. Rychlik L. Changes in prey size preferences during successive stages of foraging in the Mediterranean water shrew Neomys anomalus. Behaviour. 1999;136:345–65.
33. Rychlik L, Jancewicz E. Prey size, prey nutrition, and food handling by shrews of different body size. Behav Ecol. 2002;13:216–23.
34. Martin IG. Venom of the short-tailed shrew (Blarina brevicauda) as an insect immobilizing agent. J Mammal. 1981;62:189–92.
35. Fry BG, Roelants K, Champagne DE, Scheib H, Tyndall JDA, King GF, Nevalainen TJ, Norman JA, Lewis RJ, Norton RS, Renjifo C, de la Vega RCR. The toxicogenomic multiverse: convergent recruitment of proteins into animal venoms. Annu Rev Genomics Hum Genet. 2009;10:483–511.
36. Fry BG, Casewell NR, Wüster W, Vidal N, Young B, Jackson TNW. The structural and functional diversification of the Toxicofera reptile venom system. Toxicon. 2012;60:434–48.
37. Casewell NR, Wüster W, Vonk FJ, Harrison RA, Fry BG. Complex cocktails: the evolutionary novelty of venoms. Trends Ecol Evol. 2013;28:219–29.
38. Zhang T, Brown JH. Role of Ca^{2+}/calmodulin-dependent protein kinase II in cardiac hypertrophy and heart failure. Cardiovasc Res. 2004;63:476–86.
39. Turk V, Stoka V, Turk D. Cystatins: biochemical and structural properties, and medical relevance. Front Biosci. 2008;1:5406–20.
40. Doley R, Zhou X, Kini RM. Snake venom phospholipase A_2 enzymes. In: Mackessy SP, editor. Handbook of venoms and toxins of reptiles. New York: CRC Press; 2010. p. 173–98.
41. Harris JB, Scott-Davey T. Secreted phospholipases A_2 of snake venoms: effects on the peripheral neuromuscular system with comments on the role of phospholipases A_2 in disorders of the CNS and their uses in industry. Toxins. 2013;5:2533–71.
42. Sannaningaiah D, Subbaiah GK, Kempaiah K. Pharmacology of spider venom toxins. Toxin Rev. 2014;33:206–20.
43. Koh DCI, Armugam A, Jeyaseelan K. Snake venom components and their applications in biomedicine. Cell Mol Life Sci. 2006;63:3030–41.
44. Marsh N, Williams V. Practical applications of snake venom toxins in haemostasis. Toxicon. 2005;45:1171–81.
45. Rosiński G, Wrzeszcz A, Obuchowicz L. Differences in trehalase activity in the intestine of fed and starved larvae of Tenebrio molitor L. Insect Biochem. 1979;9:485–8.
46. Rosiński G, Gäde G. Hyperglycaemic and myoactive factors in the corpora cardiaca of the mealworm Tenebrio molitor. J Insect Physiol. 1988;34: 1035–42.
47. Marciniak P, Grodecki S, Konopińska D, Rosiński G. Structure-activity relationship for the cardiotropic action of the Led-NPF-I peptide in the beetles Tenebrio molitor and Zophobas atratus. J Pept Sci. 2008;14:329–34.
48. Shatoor AS. Cardio-tonic effect of the aqueous extract of whole plant of Crataegus aronia syn: azarolus (L) on isolated rabbit's heart. Afr J Pharm Pharmacol. 2012;6:1901–9.
49. O'Rourke M, Chen T, Hirst DG, Rao P, Shaw C. The smooth muscle pharmacology of maximakinin, a receptor-selective, bradykinin-related nonadecapeptide from the venom of the Chinese toad, Bombina maxima. Regul Pept. 2004;121:65–72.
50. Hellyer SD, Selwood AI, Rhodes L, Kerr DS. Marine algal pinnatoxins E and F cause neuromuscular block in an in vitro hemidiaphragm preparation. Toxicon. 2011;58:693–9.
51. R Development Core Team. R. A language and environment for statistical computing. Vienna: R Foundation for Statistical Computing; 2015.
52. Bock-Marquette I, Saxena A, White MD, DiMaio JM, Srivastava D. Thymosin β4 activates integrin-linked kinase and promotes cardiac cell migration, survival and cardiac repair. Nature. 2004;432:466–72.
53. Ziegler T, Hinkel R, Stöhr A, Eschenhagen T, Laugwitz KL, le Noble F, David R, Hansen A, Kupatt C. Thymosin β4 improves differentiation and vascularization of EHTs. Stem Cells Int. 2017; Article ID 6848271, 10 pages, https://doi.org/10.1155/2017/6848271.
54. Fry BG. From genome to "venome": Molecular origin and evolution of the snake venom proteome inferred from phylogenetic analysis of toxin sequences and related body proteins. Genome Res. 2005;15:403–20.
55. Lenting PJ, van Mourik JA, Mertens K. The life cycle of coagulation factor VIII in view of its structure and function. Blood. 1998;92:3983–96.
56. van Hoek ML. Antimicrobial Peptides in Reptiles. Pharmaceuticals. 2014;7: 723–53.

8

Development of the nervous system in *Platynereis dumerilii* (Nereididae, Annelida)

Viktor V. Starunov[1,2*], Elena E. Voronezhskaya[3] and Leonid P. Nezlin[3*]

Abstract

Background: The structure and development of the nervous system in Lophotrochozoa has long been recognized as one of the most important subjects for phylogenetic and evolutionary discussion. Many recent papers have presented comprehensive data on the structure and development of catecholaminergic, serotonergic and FMRFamidergic parts of the nervous system. However, relatively few papers contain detailed descriptions of the nervous system in Annelida, one of the largest taxa of Lophotrochozoa. The polychaete species *Platynereis dumerilii* has recently become one of the more popular model animals in evolutionary and developmental biology. The goal of the present study was to provide a detailed description of its neuronal development. The data obtained will contribute to a better understanding of the basic features of neuronal development in polychaetes.

Results: We have studied the development of the nervous system in *P. dumerilii* utilizing histo- and immunochemical labelling of catecholamines, serotonin, FMRFamide related peptides, and acetylated tubulin. The first neuron differentiates at the posterior extremity of the protrochophore, reacts to the antibodies against both serotonin and FMRFamide. Then its fibres run forwards along the ventral side. Soon, more neurons appear at the apical extreme, and their basal neurites form the basel structure of the developing brain (cerebral neuropil and circumesophageal connectives). Initial development of the nervous system starts in two rudiments: anterior and posterior. At the nectochaete stage, segmental ganglia start to differentiate in the anterior-to-posterior direction, and the first structures of the stomatogastric and peripheral nervous system appear. All connectives including the unpaired ventral cord develop from initially paired nerves.

Conclusions: We present a detailed description of *Platynereis dumerilii* neuronal development based on anti-acetylated tubulin, serotonin, and FMRFamide-like immunostaining as well as catecholamine histofluorescence. The development of the nervous system starts from peripheral pioneer neurons at both the posterior and anterior poles of the larva, and their neurites form a scaffold upon which the adult central nervous system develops. The anterior-to-posterior mode of the ventral ganglia development challenges the primary heteronomy concept. Comparison with the development of Mollusca reveals substantial similarities with early neuronal development in larval Solenogastres.

Keywords: *Platynereis dumerilii*, Annelida, Trochophora, Neuronal development, Serotonin, FMRFamide, Tubulin, Confocal microscopy

Background

During the past decades, our understanding of bilaterian ontogeny, history and taxonomy has changed considerably. Based on a set of comparative morphological and genetic traits, three major groups of bilaterally symmetrical animals are currently identified: Deuterostomia (chordates, hemichordates and echinoderms), Ecdysozoa (arthropods, nematodes and several smaller phyla) and Lophotrochozoa (molluscs, annelids, nemerteans, sipunculids, phoronids and bryozoans) [1–6]. Popular animal models mainly belong to ecdysozoans and deuterostomes (e.g. *Caenorhabditis*, *Drosophila*, mouse), whereas lophotrochozoans are still largely ignored. Detailed morphological and developmental studies of model lophotrochozoans are required to address questions about the evolution in the three main bilaterian clades.

Recently, the annelid *Platynereis dumerilii* (Audouin & Milne Edwards, 1834) has been established as a promising animal model for developmental, evolutionary and

* Correspondence: starunov@gmail.com; nezlinl@mail.ru
[1]Department of Invertebrate Zoology, St-Petersburg State University, St-Petersburg 199034, Russia
[3]Institute of Developmental Biology, Rus. Acad. Sci, Moscow 119991, Russia
Full list of author information is available at the end of the article

ecological research [6–11]. This marine animal is well suited as a model system for several reasons.

1. *Platynereis* exhibits a canonical biphasic pelagobenthic life cycle involving a pelagic free swimming primary larva, and benthic adults. Egg cleavage follows the typical spiral model, and the canonical course of cleavages produces a typical trochophore larva [10]. Besides, *Platynereis* exhibits modes of development and body plans that are considered ancestral in many respects. This is also reflected in the level of genes, making this animal ideally suited for developmental comparative studies [6].
2. This animal will breed in a laboratory culture without access to the sea, and produce offspring continuously throughout the year. With 2000–3000 synchronously developing eggs, a single clutch can be split up into numerous samples for serial or parallel observations [10].
3. Both larval and juvenile stages are transparent and well suited for light microscopy including confocal laser scanning microscopy (CLSM) in combination with whole mount immunohistochemistry [11, 12].
4. *Platynereis* has a long story of classical embryological investigations, and many aspects of its development have been described in great detail [9, 12–15]. Thus, the study of this animal will provide comparative data that may contribute to better understanding of the basic plan of annelid development and evolution.

Our understanding of neuronal development in Lophotrochozoa has increased greatly over the last decades due to numerous comparative morphological studies, which used fluorescent histo- and immunocytochemical labelling in combination with confocal microscopy. Identification of catecholaminergic, serotonergic and FMRFamide-like immunoreactive nerve elements have been shown to be reliable neuron-specific markers that permit the detection of neurons, fibre tracts and isolated fibres in whole mount preparations [16, 17], while co-staining with the antibody against acetylated α-tubulin reveals gross morphology of the nervous system and ciliated structures [18]. Based on these techniques, detailed descriptions of neuronal development and larval nervous systems in representative species of molluscs [19–24], phoronids [25–28], bryozoans [29–31], brachiopods [32], echiurids [33, 34], and sipunculids [35] have been published.

Modern evolutionary studies have revealed substantial similarities in the molecular architecture of the nervous system in annelids, arthropods and vertebrates, suggesting that this architecture was present in the last common bilaterian ancestor and supporting a common origin of nervous system centralization in Bilateria [36–40]. Thus, the study of neuronal development in polychaetes may shed light on CNS evolution in Bilateria.

However, only few papers describe the development of the nervous system in polychaetes and the data is sometimes contradictory. Hay-Schmidt [41] describes in *Polygordius lacteus* (Polygordiidae) the development of a highly centralized serotonin- and FMRFamide-like immunoreactive (lir) nervous system, with the first neurons appearing within the developing central ganglia and nerve cords. Later, the study of neuronal development in planctotrophic larvae of errant *Phyllodoce maculata* (Phyllodocidae) [42] and sedentary *Pomatoceros lamarckii* (Serpulidae) [43] revealed that the first neurons to appearing in early trochophores was at the periphery. In both species, the first serotonergic cell appeared at the posterior end and the first FMRFa-lir cell at the anterior end of the larva. In larval *Sabellaria alveolata* (Sabellaridae) [44] the first serotonergic fibres were detected at later stages at the tip of each chaetal sac. Recently, Fischer and co-authors [11] provided an overview of neuronal development in lecitotrophic larvae of *Platynereis dumerilii* (Nereididae) until the late nectochaete stage based on anti-serotonin and anti-tubulin immunostaining. As to the catecholaminergic nervous system, its development remains completely unknown throughout the whole phylum Annelida, whereas several papers have reported distribution of catecholamine-synthesizing neurons in adult polychaetes [45–48] and medicinal leeches [49].

The goal of our research was to study in detail the neuronal development in *P. dumerilii* from a comparative and developmental perspective using immunocytochemical labelling with commercial antibodies against acetylated α-tubulin, FMRFamide and serotonin, as well as the histochemical staining of the catecholamines. Below, we have provided detailed descriptions of the development of anti-5-HT and anti-FMRFamide-like immunostainings in the nervous system of *P. dumerilii* from hatching until the formation of the juvenile worm with 10–11 segments. Catecholamine-containing structures were studied until the late nectochaete stage. Where possible, we have used the neuroanatomical terms suggested by Richter and co-authors [50]. Since the development of serotonergic elements of the nervous system until the mid-nectochaete stage was described earlier [11], we report it only in brief, focusing on newly observed details. Taking into account the low specificity of the FMRFamide antibody, which cross-reacts with many neuropeptides that have a RFamide terminus, we compared our results only with similar data from other trochozoan species. Detailed descriptions of the development of serotonergic, FMRFamidergic and catecholaminergic nervous elements in *Platynereis* will contribute to better understanding of the basic features of neuronal development and shed light on phylogenetic relationships within Trochozoa.

Results

Staging of development

Normal development of *P.dumerilii* has been described in great detail [11]. Since the pace of development is highly temperature dependent, we determined developmental stages according to the set of morphological and behavioural characters as described in [11]. Precise time points of fixation (in hours or days post-fertilisation) for each preparation are indicated in the figure legends.

FMRFa-like immunoreactivity (Figs. 1, 2 and 3)

The first FMRFa-like immunoreactive (FMRFa-lir) cells were detected in early trochophores shortly after hatching. Three neurons were located almost ventrally to the apical extreme (FMRFa-lir apical, *fa1–3*, Fig. 1a,b) and projected their basal neurites into a compact neuropil. In addition, one bipolar cell was located dorsally below the prototroch (FMRFa-lir dorsal, *fd*, Fig. 1a,c). Two neurites extended from the soma and ran in both directions beneath the prototrochal cells forming part of the trochal neurite bundle (prototroch nerve).

At the beginning of the mid-trochophore stage, two more cells that sent neurites into the trochal neurite bundle appeared in the lateral portions of the episphere (left and right FMRFa-lir lateral, *fl1–2*, Fig. 1d). Immunopositive neurites were detected in the apical neuropil, prototroch nerve and in the developing ventral cerebral commissure (*vc*, Fig. 1d). By 32 hpf, the set of FMRFa-lir cells remained unchanged (Fig. 1e): three asymmetrically located cells (*fa1–3*) in the apical region, the dorsal cell (*fd*) and the pair of lateral cells (*fl1* and *fl2*). No immunoreactivity was yet detected in the hyposphere.

At the late trochophore stage, the number of immunopositive cells in the apical organ increased to six and the first immunopositive cell bodies appeared in the region of the developing brain (ventral cerebral commissure) (*arrowheads*, Fig. 1f). At this time, faint FMRFa-like immunoreactivity was detected in the soma of the solitary cell located at the posterior extreme of the larva inside the telotroch ring (FMRFa-lir posterior, *fp*, Fig. 1g). The cell had an apical dendrite with surface cilia.

At the early metatrochophore stage, the same six cells in the apical organ, one pair of cells in the developing brain (cerebral ganglion), and the cell *fp* were detected (Fig. 1h,i). A pair of ciliated FMRFa-lir cells appeared at the ventral surface of the episphere (*fv1–2*, Fig. 1h), and their fibres projected into the prototrochal neurite bundle. Based on their position and the presence of the dendrite and projections to the prototrochal nerve, these cells represent the photoreceptor cells of the eyespot. In the hyposphere, the *fp* cell was still the only FMRFa-lir structure (Fig. 1h,j). Relative to other immunopositive structures, the intensity of staining in this cell was always very weak

in the soma and almost undetectable in the basal fibres. At the mid-metatrochophore stage, more cells appeared in the apical region, and a pair of cells appeared near each of the cells *fl1* and *fl2* (Fig. 1k).

At the late metatrochophore stage, more FMRFa-lir neurons appeared in the apical region, and the neuropil expanded (Fig. 1l). Immunopositive cells appeared ventrally to the posterior part of the stomatogastric nerve ring (Fig. 1l, *arrowhead*), and two groups of neurons appeared in association with the ventral nerve cords dorsal to the trochal neurite bundle (Fig. 1l, *arrows*). Immunoreactive fibres appeared in the posttrochal parts of the paired ventral cord. Two symmetrical FMRFa-lir neurons appeared in the first chaetigerous segment (Fig. 1l, *open arrows*; M, *arrows*). They were located ventrolateral to the cords and sent neurites into them. At this time, immunoreactivity in the posterior *fp* cell became very weak and soon disappeared (Fig. 1n).

At the early nectochaete stage, the adult CNS further developed, and more immunopositive elements were added to the paired ventral cord and the brain (Fig. 1o). The brain neuropil increased in size and became denser (Fig. 1p, asterisk). The number of immunopositive neurites in the ventral cord also increased, though these fibres did not reach the rear of the body. Fibres of FMRFa-lir neurons in the first chaetigerous segment projected into the ventral cord and segmental nerves (Fig. 1q, *arrows*). At this time, immunoreactivity in the dorsal *fd* cell started to fade (Fig. 1r, *arrow*). No immunoreactivity was detected in the posterior *fp* cell though the cell body could be visualized after anti-tubulin immunostaining being the only ciliated cell inside the telotroch ring (Fig. 1s, *arrow*).

At the mid-nectochaete stage, more FMRFa-lir neurons appeared in the head region. More neurites were added to the paired ventral cord and projected into its posterior end. Two pairs of immunopositive cells appeared in the second and third chaetigerous segments (Fig. 2a, *arrows*). In the dorsal part of the brain, cell bodies were detected anterior and posterior to the neuropil (Fig. 2b). In the ventral part, although no neurons were added the central neuropil grew in size (Fig. 2c). Immunopositive fibres were detected in the dorsal and ventral roots of the circumoesophageal connectives (Fig. 2b,c, *dr*, *vr*). The cells in the first and third segment were located near the roots of parapodial nerves (Fig. 2d). No FMRFa-lir innervation was as yet detected in the tentacular cirri, anal cirri and antennal stubs that were forming at this stage.

Starting from the late nectochaete stage, the number of immunoreactive neurons and neurites in the adult CNS, including the brain, dorsal neuropil and ventral cord started to increase (Fig. 2e-g), and FMRFa-lir innervation of the pharynx (*ph*) and gut (*g*) was first detected (Fig. 2f). In three-segmented errant

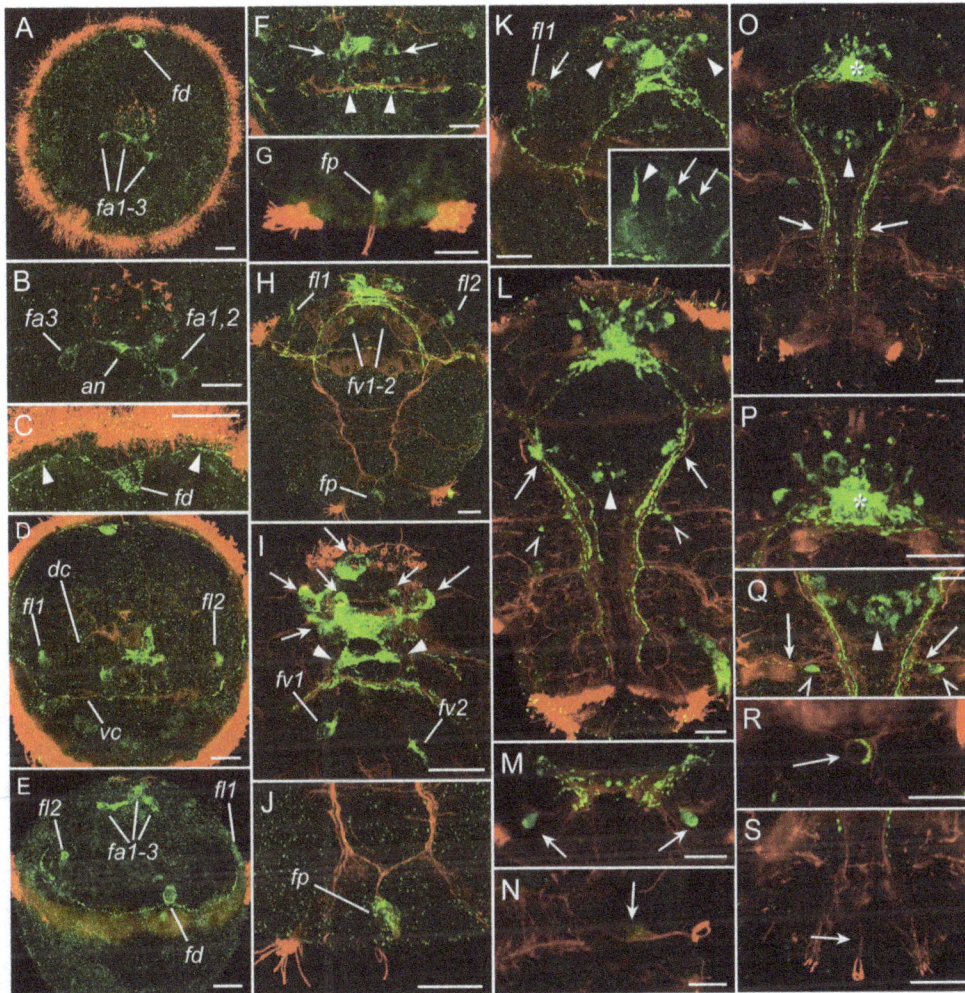

Fig. 1 Development of FMRFamide-like (*green*) and acetylated tubulin-like (*red*) immunoreactivity in *P. dumerilii*. Early trochophore (**a-c**), mid-trochophore (**d-e**), late trochophore (**f-g**), early metatrochophore (**h-j**), mid-metatrochophore (**k**), late metatrochophore (**l-n**), early nectochaete (**o-s**). **a:** 24 hpf, apical view. The first immunopositive cells are three apical neurons (*fa1–3*) and a solitary dorsal neuron (*fd*). **b:** 24 hpf, apical neurons at high magnification. The neurons *fa1,2* are located to the left and *fa3* to the right of apical neuropil (*an*). **c:** 24 hpf, high magnification of the dorsal neuron *fd*. Two neurites (arrowheads) extend from it in opposite directions beneath the prototroch. **d:** 28 hpf, apical view showing the cells *fl1* and *fl2*. *dc* and *vc* – dorsal and ventral cerebral commissures. **e:** 32 hpf, dorsal view showing the cells *fa1–3*, *fd*, and *fl2*. **f:** 36 hpf, apical view showing neurons in the apical organ (*arrows*) and ventral cerebral commissure (*arrowheads*). **g:** 36 hpf, ventral view of the solitary posterior cell (*fp*) with two apical cilia. **h:** 45 hpf, ventral view showing two ventral pretrochal cells (*fv1,2*) and lateral cells (*fl1,2*). **i:** 45 hpf, ventral view on the apical region showing apical neurons (*arrows*), cerebral ganglion (*arrowheads*) and ventral pretrochal cells (*fv1, fv2*). **j:** 45 hpf, ventral view on the posterior cell (*fp*). **k:** 50 hpf, ventral view of the pretrochal region showing FMRFa immunoreactivity in the apical region (*arrowheads*) and two cells (*arrow*) close to the cell *fl1*. *Inset*: High magnification of the two cells (*arrows*) close to the cell *fl1* (*arrowhead*). **l:** 56 hpf, ventral view. Two groups of neurons associated with the ventral nerve cord (*arrows*), ventral neurons behind the stomatogastric nerve ring (*arrowhead*), and two neurons in the first chaetigerous segment (*open arrowheads*) are seen. **m:** 56 hpf, the neurons of the first chaetigerous segment (*arrows*). **n:** 56 hpf, posterior cell *fp* (*arrow*). **o:** 72 hpf, ventral view showing the brain (*asterisk*), neurons behind the stomatogastric nerve ring (*arrowhead*), and paired ventral nerve cord (*arrows*). **p:** 72 hpf, brain region with dense neuropil (*asterisk*). **q:** 72 hpf, anterior part of the ventral nerve cord showing neurons behind the stomatogastric nerve ring (*arrowhead*), and two neurons in the first chaetigerous segment (*open arrowheads*) sending fibers into the ventral cord and segmental nerves (*arrows*). **r:** 72 hpf, dorsal cell *fd* (*arrow*). **s:** 72 hpf, posterior cell *fp* (*arrow*) with no FMRFa immunoreactivity. Scale bars = 20 μm

juvenile worms (Fig. 2h,i), a dense network of immunopositive fibres was detected in the ventral cord (*vc*) and subesophageal ganglion (*ceg*), and a plexus of FMRFa-lir neurones and neurites appeared in the dorsal portion of the body (Fig. 2i, *arrows*). At this time, FMRFa-lir neurites were first detected to innervate the parapodia (Fig. 3a, *arrows*) and the pygidium (Fig. 3b, *arrows*). At the periphery, the number of neurons and neurites continued to increase.

At the stage of cephalic metamorphosis, immunoreactive cells and fibres were detected in the central and peripheral nervous system (Fig. 3c-e, Additional file 1).

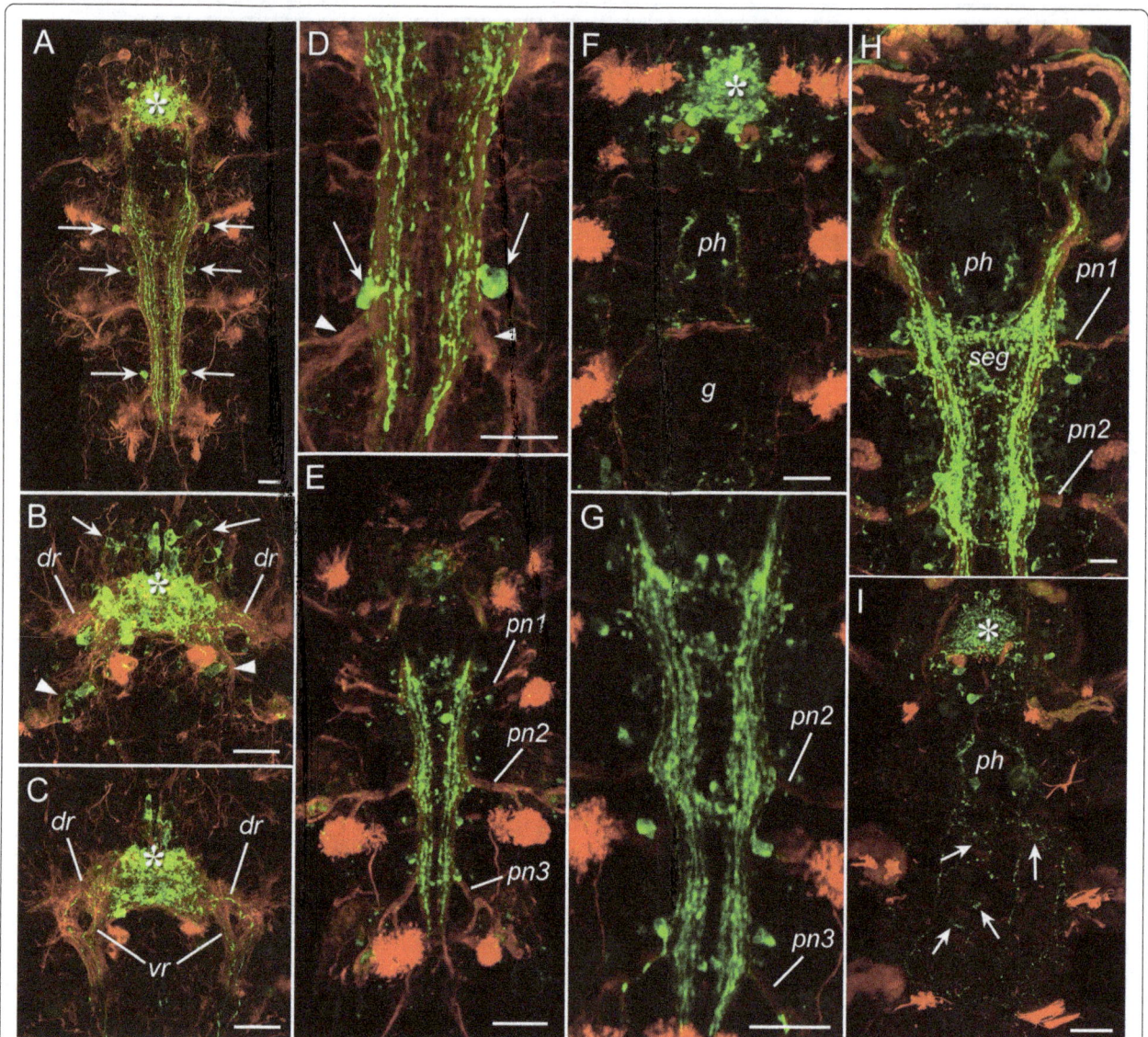

Fig. 2 Development of FMRFamide-like (green) and acetylated tubulin-like (red) immunoreactivity in *P. dumerilii*. Mid-nectochaete (**a-d**), late nectochaete (**e-g**), three-segmented errant juvenile worm (**h-i**). **a:** 3,5 dpf, ventral view showing the brain (*asterisk*) and the paired ventral cord. *Arrows* indicate the pairs of neurons in the larval segments 1–3. **b:** dorsal neuropil with FMRFa-lir neurons anterior (*arrows*) and posterior (*arrowheads*) to it; *dr* – dorsal roots of circumoesophageal connectives. **c:** Ventral neuropil; *dr* – dorsal roots of circumoesophageal connectives, *vr* – ventral roots of circumoesophageal connectives. **d:** Posterior part of the ventral cord with two neurons (*arrows*) in the third chaetigerous segment near the roots of parapodial nerves (*arrowheads*). **e:** 7 dpf, ventral view showing FMRF-lir structures in the body. *Pn1–3*– parapodial nerves of corresponding chaetigerous segments. **f:** dorsal view on the anterior body part showing brain neuropil (*asterisk*), innervation of the pharynx (*ph*) and gut (*g*). **g:** ventral nerve cords with adjacent FMRF-lir neurons. **h:** 16 dpf, FMRF-lir innervation of the ventral body part showing the subesophageal ganglion (*seg*) and pharynx (*ph*). **i:** FMRF-lir innervation of the dorsal body part showing brain neuropil (*asterisk*) with adjacent neurons, pharynx innervation (*ph*), and dorsal plexus with peripheral neurons (*arrows*). Scale bars = 20 μm

In the anterior region, FMRFa-lir neurons and neurites were present in almost all parts of the CNS including the brain (*br*), ventral nerve cord (*vnc*), circumesophageal connectives (*cc*), subesophageal ganglion (*ceg*), antennal (*an*), stomatogastric (*sn*) and segmental nerves (*arrows*). FMRFa-lir innervation was also found in the digestive system (pharynx and gut) and body wall (Fig. 3d,e).

At the stage of the small atokous worm, FMRFa-lir neurons and neurites were present in the central and peripheral nervous system, similar to the previous stage although the number of cells and neurites had increased (Fig. 3f-h, Additional file 2). A peripheral plexus of immunoreactive neurites and scattered neurons were detected along the length of the body.

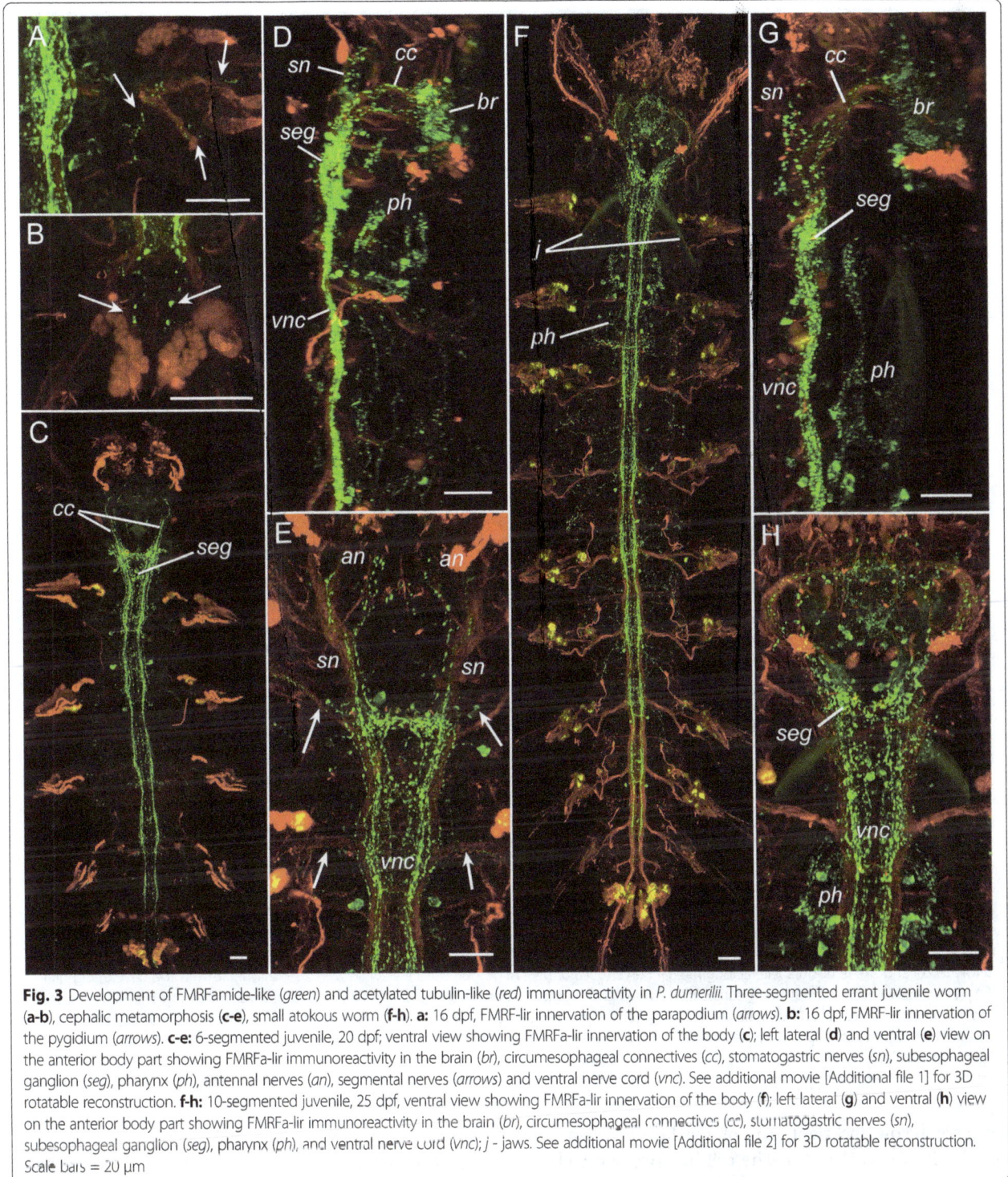

Fig. 3 Development of FMRFamide-like (*green*) and acetylated tubulin-like (*red*) immunoreactivity in *P. dumerilii*. Three-segmented errant juvenile worm (**a-b**), cephalic metamorphosis (**c-e**), small atokous worm (**f-h**). **a:** 16 dpf, FMRF-lir innervation of the parapodium (*arrows*). **b:** 16 dpf, FMRF-lir innervation of the pygidium (*arrows*). **c-e:** 6-segmented juvenile, 20 dpf; ventral view showing FMRFa-lir innervation of the body (**c**); left lateral (**d**) and ventral (**e**) view on the anterior body part showing FMRFa-lir immunoreactivity in the brain (*br*), circumesophageal connectives (*cc*), stomatogastric nerves (*sn*), subesophageal ganglion (*seg*), pharynx (*ph*), antennal nerves (*an*), segmental nerves (*arrows*) and ventral nerve cord (*vnc*). See additional movie [Additional file 1] for 3D rotatable reconstruction. **f-h:** 10-segmented juvenile, 25 dpf, ventral view showing FMRFa-lir innervation of the body (**f**); left lateral (**g**) and ventral (**h**) view on the anterior body part showing FMRFa-lir immunoreactivity in the brain (*br*), circumesophageal connectives (*cc*), stomatogastric nerves (*sn*), subesophageal ganglion (*seg*), pharynx (*ph*), and ventral nerve cord (*vnc*); *j* - jaws. See additional movie [Additional file 2] for 3D rotatable reconstruction. Scale bars = 20 μm

Serotonin-like immunoreactivity (Figs. 4 and 5)

The first 5-HT-lir neuron to appear was the posterior 5-HT-lir cell (serotonergic posterior, *sp*), detected at the protrochophore stage before hatching (17 hpf) (Fig. 4a). Its body was triangular or claviform, a short apical dendrite extended to the surface, and two basal fibres ran towards the prototroch along the ventral side of the larva. At the time of hatching (20 hpf), two cilia appeared on the apical dendrite of the *sp* cell (Fig. 4b, *arrowhead*), and its basal fibres reached the prototroch where each fibre bifurcated and ran under the prototroch in both directions. At this time, the first apical 5-HT-lir cell (serotonergic apical, *sa1*)

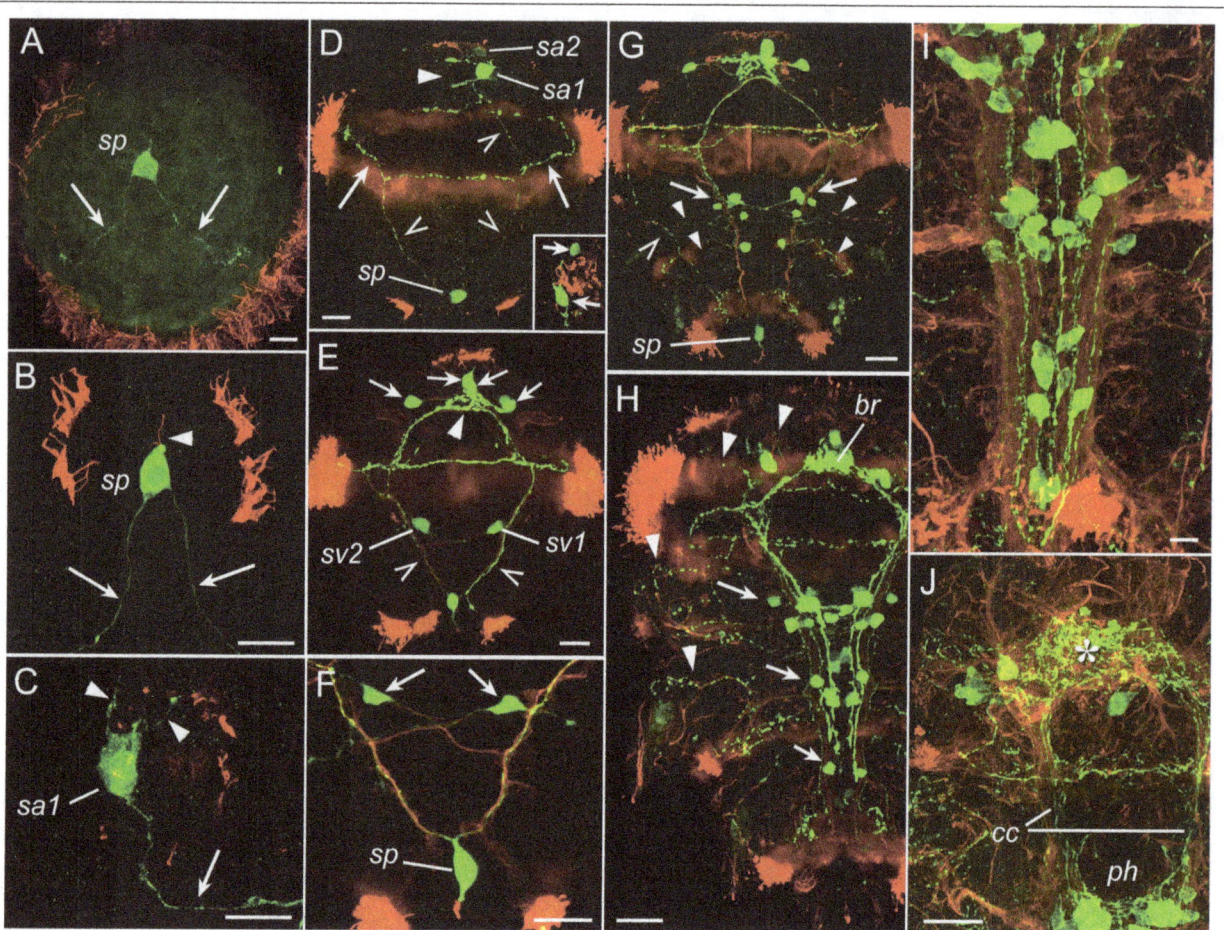

Fig. 4 Development of 5-HT-like (*green*) and acetylated tubulin-like (*red*) immunoreactivity in *P. dumerilii*. Protrochophore (**a**), early trochophore (**b,c**), mid-trochophore (**d**), early metatrochophore (**e,f**), mid-metatrochophore (**g**), early nectochaete (**h**), late nectochaete (**i,j**). **a:** 17 hpf, caudal view on the first posterior 5-HT-lir cell (*sp*) with two basal fibers (*arrows*). **b:** 20 hpf, the first posterior 5-HT-lir cell with two basal fibers (*arrows*) and two apical cilia (*arrowhead*) in the centre of the open-circuited telotroch ring. **c:** 20 hpf, apical view on the first apical 5-HT-lir cell (*sa1*) with two short fibers (*arrowheads*) running at the base of the horseshoe-shaped apical ciliary tuft, and the long fibre running towards the prototroch (*arrow*). **d:** 24 hpf, ventral view showing the cells *sp*., *sa1* and *sa2*, and 5-HT-lir fibres labelling the apical neuropil (*arrowhead*), prototroch nerve (*arrows*) and paired ventral cords (*open arrows*). Inset: 28 hpf, apical view showing the cells *sa1* and *sa2* (*arrows*). **e:** 44 hpf, ventral view showing four 5-HT-lir cells *sa1–4* (*arrows*) around the apical neuropil (*arrowhead*) and two ventral cells (*sv1* and *sv2*) near the ventral cords (*open arrows*). **f:** 44 hpf, high magnification of the posttrochal body part showing the posterior cell *sp*. and two ventral cells *sv1* and *sv2* (*arrows*) with their fibres running into the ventral cords crosswise. **g:** 46 hpf, ventral view showing 5-HT-lir cells associated with the ventral cords (*arrows*), transversal nerves (*arrowheads*) and the right dorsolateral longitudinal nerve (*open arrow*). **h:** 68 hpf, ventral view showing 5-HT-lir neurons in the developing brain (*br*), ventral nerve cord (*arrows*), and neurites in the head and segmental nerves (*arrowheads*). **i:** 3.5 dpf, ventral view showing 5-HT-lir neurons and neurites in the central and peripheral nervous system; *asterisk* - cerebral ganglion; *arrows* - ganglia of larval segments. **j:** 3.5 dpf, ventral view on the head region showing 5-HT-lir neuropil in the brain (*asterisk*) and innervation of the pharynx (*ph*). *cc* – circumoesophageal connectives. Scale bars = 20 μm

appeared below the right part of the horseshoe-shaped apical tuft of cilia (Fig. 4c). The cell had two short fibres running underneath the apical cilia (*arrowheads*), and a fibre directing towards the prototroch along the ventral side of the body (*arrow*). Soon after hatching (24 hpf), basal fibres of the two 5-HT-lir cells labelled the future paired ventral nerve cord and the prototroch nerve (Fig. 4d, *arrows*). At the mid-trochophore stage, another 5-HT-lir cell (*sa2*) appeared at the apical extreme, dorsally to the *sa1* cell (Fig. 4d). By the

late trochophore stage, two more neurons appeared in the apical hemisphere (Fig. 4e, *arrows*) and the basal neurites of all the cells projected into the developing apical neuropil.

At the early metatrochophore stage, the first pair of 5-HT-lir neurons appeared medially to the developing paired ventral cord at the level of the first chaetigerous segment (Fig. 4e). The cells were bipolar and connected to the cord crosswise. Each cell sent a neurite into the ipsi- and contralateral part of the cord (Fig. 4f). Several hours

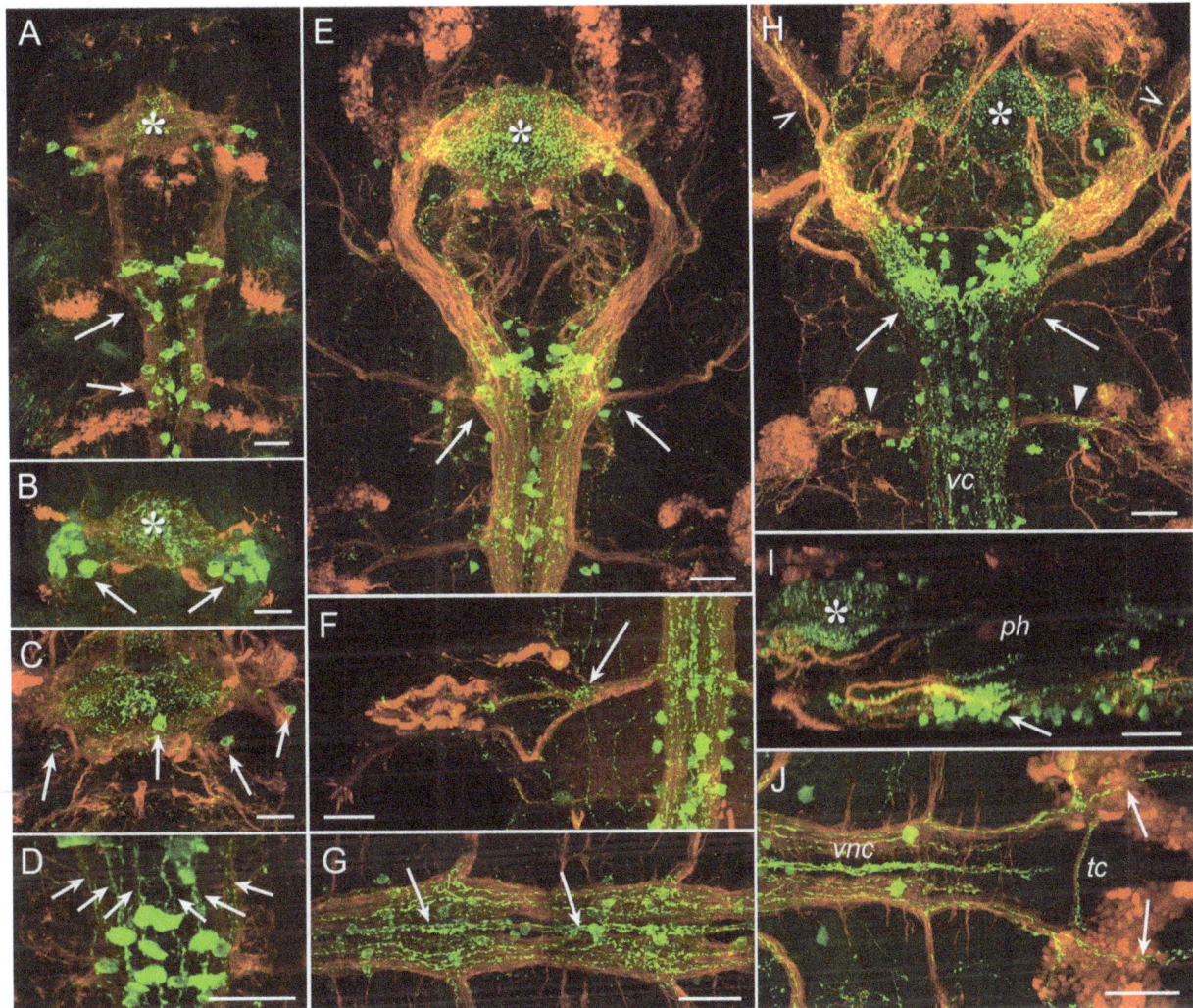

Fig. 5 Development of 5-HT-like (*green*) and acetylated tubulin-like (*red*) immunoreactivity in *P. dumerilii*. 3-segmented larva (**a-d**), and juvenile worm (**e-j**). **a:** 7 dpf, showing 5-HT-lir neurons in the cerebral ganglion (*asterisk*) and ventral cord (*arrows*). **b:** 7 dpf, dorsal part of the cerebral ganglion showing the neuropil (*asterisk*) and dorsolateral neurons (*arrows*). **c:** 7 dpf, dorsal part of the head showing apical neurons *sa1-sa4* (*arrows*). **d:** 7 dpf, 5-HT-lir neurites in the ventral cord group into six longitudinal bundles (*arrows*). **e:** 5-segmented juvenile,19 dpf, ventral view on the head region showing the brain (*asterisk*) and forming subesophageal ganglion (*arrows*). **f:** 7-segmented juvenile, 20 dpf, ventral view on the parapodium showing the parapodial ganglion (*arrow*). **g:** 7-segmented juvenile, 20 dpf, ventral nerve cord with a medial unpaired bundle of 5-HT-lir neurites (*arrows*). **h:** 10-segmented juvenile, 25 dpf, ventral view on the head region showing 5-HT-lir structures in the brain (*asterisk*), subesophageal ganglion (*arrows*), ventral cord (*vc*), tentacular nerves (*open arrows*), and parapodial nerves (*arrowheads*). **i:** 10-segmented juvenile, 25 dpf, lateral view on the head region showing 5-HT-lir structures in the brain (*asterisk*), subesophageal ganglion (*arrow*), and pharynx (*ph*). See additional movie [Additional file 3] for 3D rotatable reconstruction. **j:** 10-segmented juvenile, 25 dpf, ventral view on the caudal body end showing 5-HT-lir structures in the ventral cord (*vc*) and anal cirri (*arrows*); *tc* – terminal commissure. Scale bars = 20 μm

later, more 5-HT-lir cells appeared close to the ventral cord and two pairs of transverse nerves were detected at the level of the second and third chaetigerous segments (Fig. 4g). A pair of 5-HT-lir dorsolateral longitudinal nerves extended from the prototroch backward along both dorsolateral sides of the larva (Fig. 4g, *open arrow*).

Later, the number of cells and neurites in the ventral cord started to increase and, by the early nectochaete stage, up to 24 pairs of 5-HT-lir neurons were detected along the cord (Fig. 4h). Many 5-HT-lir neurites were detected in the nerves innervating the developing palps, anal cirri and frontal head region (Fig. 4h, arrowheads). By the late nectochaete stage, the number of immunopositive neurons in the brain and the ventral cord (ganglia of chaetigerous segments) had again increased (Fig. 4i). Immunopositive neuropil's developed in the brain and in the pharyngeal region, and numerous peripheral neurites could be seen all over the body (Fig. 4j).

In three-segmented juvenile worms, 5-HT-lir cells were detected in the brain and ventral cord (Fig. 5a).

Cerebral neurons were mostly located in the dorso-lateral parts of the brain (Fig. 5b), and weakly fluorescent perikarya of the apical cells *sa1-sa4* were still visible dorsally to the cerebral ganglion (Fig. 5c). Immunoreactive fibres in the ventral cord were arranged into six paired longitudinal bundles (Fig. 5d).

In 5- to 7-segmented juveniles, the ganglia of the first and second chaetigerous segment started to merge into the subesophageal ganglion (Fig. 5e, *arrows*). The parapodia of the first chaetigerous segment including the parapodial nerves transformed into the the third pair of tentacular cirri that develops from this segment, and other segmental nerves of this segment reduced. At this time, 5-HT-lir neurites appeared in the developing parapodial ganglia (Fig. 5f, *arrow*). The number of 5-HT-lir neurites in the ventral cord gradually increased, and a medial unpaired bundle appeared (Fig. 5g, *arrows*).

In 10-segmented worms (Fig. 5h-j, Additional file 3), 5-HT-lir structures in the anterior region were detected in the brain (*asterisk*), tentacular nerves (*open arrows*), subesophageal ganglion (*arrows*), ventral cord (*vc*) and parapodial nerves (*arrowheads*). At this time, 5-HT-lir innervation of the pharynx was detected (Fig. 5i). At the posterior body end, 5-HT-lir neurites were detected in the terminal commissure (*tc*) and innervated the anal cirri (Fig. 5j, *arrows*). No visible changes in the arrangement of 5-HT-lir nerve elements could be seen during the subsequent growth of the worms.

Catecholamine histofluorescence (Figs. 6 and 7)

In this method, the primary catecholamines, dopamine and noradrenaline, were converted to fluorescent 2-carboxymethyl-dihydroisoquinoline derivatives in a reaction with glyoxylic acid with subsequent air drying and heating. Blue-green fluorescence a characteristic for catecholamines, was first observed in a neurite under the prototroch at the time of hatching (Fig. 6a, *arrows*). Soon after hatching, two unipolar neurons were observed in the hyposphere close to the prototroch (catecholaminergic prototroch, *cp1,2*) with their neurites passing under the prototroch (Fig. 6b, *arrows*). Fluorescence was also observed in six cell-like structures in the posterior region of the trochophore (Fig. 6a–c, *pc*). No neurites were detected extending from these structures so we considered them to be non-nervous. The site of bright fluorescence was observed at the surface of the posterior end of the larva and was co-localized with the cilia of the telotroch (Fig. 6c). By the late trochophore stage, this fluorescence faded and became undetectable.

At the mid-trochophore stage, fluorescent neurites gradually appeared in the first and second brain commissures (Fig. 6d,e, arrows). At the end of this stage, the first neurites extending towards the apical organ were detected (Fig. 6e, *open arrow*). In late trochophores (Fig. 6f), CA-containing neurites were detected in the prototroch nerve (*arrowhead*), two cerebral commissures (*arrows*), and the apical organ (*open arrow*).

At the early metatrochophore stage, two more cells (catecholaminergic ventral, *cv1,2*) were detected slightly posterior to the first pair in the region of the developing paired ventral cord (Fig. 6g). In the episphere, the number of CA-containing neurites at the base of the apical organ (cerebral neuropil), and in the two brain commissures gradually increased (Fig. 6g-h). At this stage, the first fluorescent neurites were detected in the hyposphere. Initially, they were scattered in the ventral region (Fig. 6g), and later they were seen to be the part of the paired ventral cord (Fig. 6h, *arrows*). In the mid-metatrochophores, CA-containing neurites in the hyposphere started to arrange into two dorsolateral nerves (*arrows*) and three transverse commissures (*open arrows*) corresponding to the rudiments of the chaetigerous segments (Fig. 6i-j).

At the late metatrochophore stage, no additional CA-containing cells were detected, though the number of fluorescent neurites gradually increased (Fig. 6k). The neurites formed a neuropil in the brain and at the base of the larval eyes (Fig. 6l, *arrows*). Two pairs of short fibres extended dorsally from the second brain commissure. Fluorescent neurites in each half of the ventral cord were combined into two bundles (Fig. 6m, *arrows*), and more fibres appeared in the transverse commissures (Fig. 6m, *open arrows*). At the dorsal side (Fig. 6n), fluorescent fibres were detected in the prototroch nerve (*arrowheads*), dorsolateral nerves (*arrows*) and transverse commissures (*open arrows*).

During the early nectochaete stage, the number of CA-containing cells and neurites in the adult CNS gradually increased (Fig. 7a). The brain neuropil grew in size, so the brain commissures could no longer be recognized. Fourteen CA-containing neurons appeared in the head (Fig. 7b): four neurons near each eye (*arrows*), two pairs on both sides of the brain neuropil (*arrowheads*), and a pair in front (*open arrow*). In the body (Fig. 7c), more CA-containing neurites appeared in the ventral cord (*arrows*) and parapodial nerves (*arrowheads*). Starting from the mid-nectochaete stage, more CA-containing structures appeared in the CNS and at the periphery (Fig. 7d). Besides the brain and ventral cord, fluorescent neurites were detected to innervate the anterior regions of the digestive tract (Fig. 7e, *arrows*). Fluorescent neurons began to appear in the ganglia of all three chaetigerous segments. First, one pair of neurons was detected in each (Fig. 7f), and later, their number increased, so at the late nectochaete stage (Fig. 7g), the first segment contained five pairs of neurons (*arrows*), the second had three pairs (*arrowheads*), and the third had two pairs.

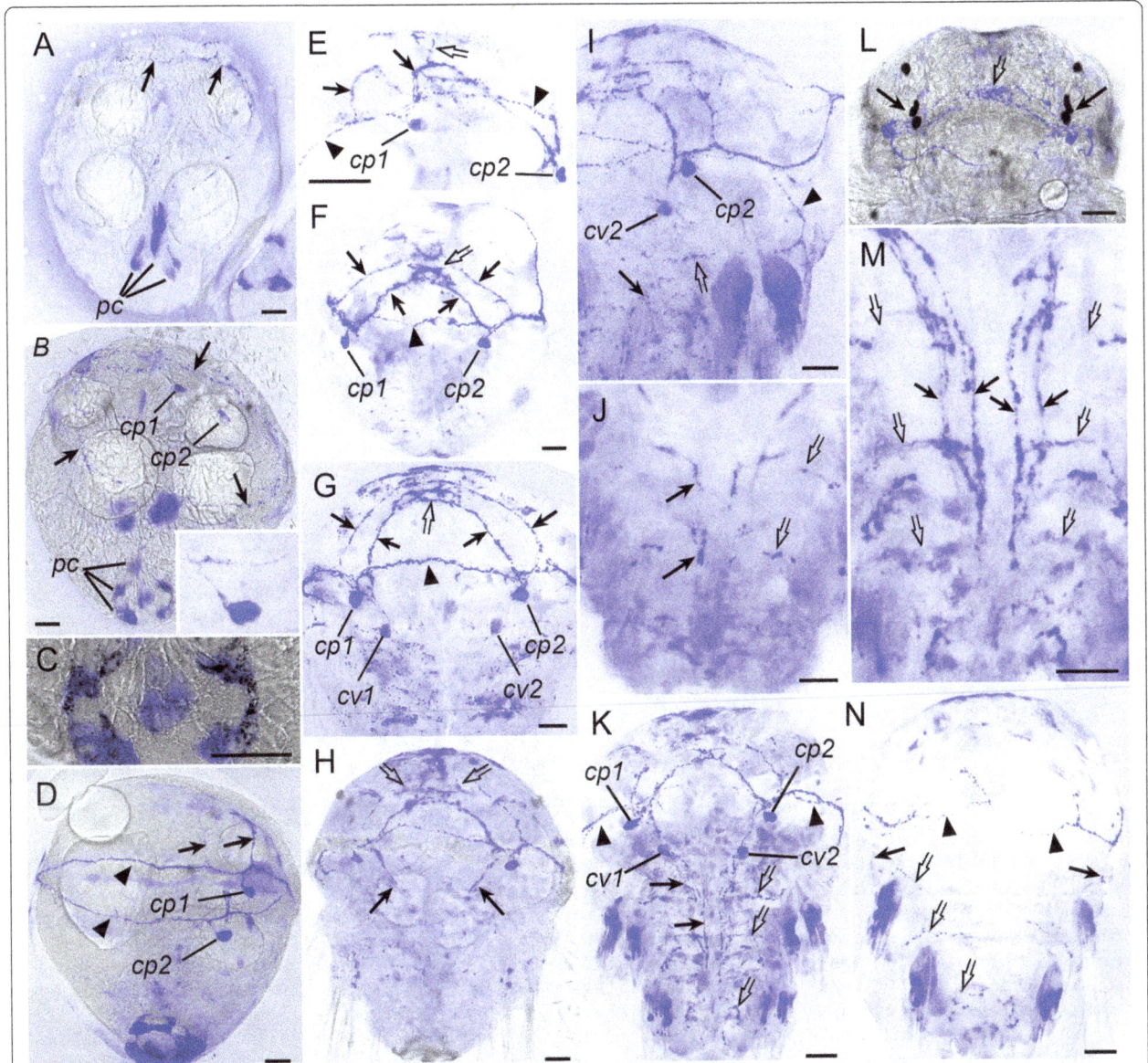

Fig. 6 Development of catecholamine containing elements (*blue*) in *P. dumerilii*. Early trochophore (**a-c**), mid-trochophore (**d,e**), late trochophore (**f**), early metatrochophore (**g,h**), mid-metatrochophore (**i,j**), late metatrochophore (**k-n**). **a-d,h** and **l** are combinations of fluorescence (*blue*) and transmission light image (*grey*). **a:** 20 hpf, ventrolateral view showing CA-containing neurite under the prototroch (*arrows*) and six presumably non-nervous structures at the posterior body end (posterior cells, *pc*). **b:** 20 hpf, two CF-containing cells (*cp1,2*) and their fibers underlying the prototroch (*arrows*). Inset: high magnification of the cell *cp1*. **c:** 20 hpf, fluorescent structures at the surface of the posterior body end. **d:** 28 hpf, lateral view showing the cells *cp1,2* and CA-containing neurites in the circular prototroch nerve (*arrowheads*) and two brain commissures (*arrows*). **e:** 32 hpf, lateral view on the anterior hemisphere showing the cells *cp1,2* and their neurites in the circular prototroch nerve (*arrowheads*), two brain commissures (*arrows*) and a solitary fiber going towards the apical organ (*open arrowhead*). **f:** 36 hpf, ventral view showing the cells *cp1,2* and CA-containing fibers in the prototroch nerve, (*arrowheads*), two brain commissures (*arrows*) and developing brain (*open arrow*). **g:** 40 hpf, ventral view showing the cells *cp1,2* and *cv1,2*, and CA-containing fibers in the prototroch nerve, (*arrowhead*), two brain commissures (*arrows*) and developing brain (*open arrow*). **h:** 44 hpf, ventral view showing CA-containing neurites in the apical region (*open arrows*) and in the region of developing ventral cord (*arrows*). **i:** 48 hpf, ventral view on the left size of anterior body part showing the cells *cp2* and *cv2*, and CA-containing neurites in the developing ventral cord (*arrow*), right dorsolateral nerve (*arrowhead*), and transverse commissure in the first chaetigerous segment (*open arrow*). **j:** 48 hpf, ventral view on the posterior body part showing CA-containing neurites in the developing ventral cord (*arrows*) and segmental nerves in the second and third chaetigerous segments (*open arrows*). **k:** 56 hpf, ventral view showing the cells *cp1,2* and *cv1,2*, as well as CA-containing neurites in the prototroch nerve (*arrowheads*), ventral cord (*arrows*) and three segmental nerves (*open arrows*). **l:** 56 hpf, dorsal view on the apical part showing brain neuropil (*open arrow*) and neurites at the base of larval eyes (*arrows*). **m:** 56 hpf, ventral view on the posterior body part showing CA-containing neurites in the ventral cord (*arrows*) and three segmental nerves (*open arrows*). **n:** 56 hpf, dorsal view showing CA-containing neurites in the prototroch nerve (*arrowheads*), dorsolateral nerves (*arrows*) and segmental nerves (*open arrows*). Scale bars = 20 μm

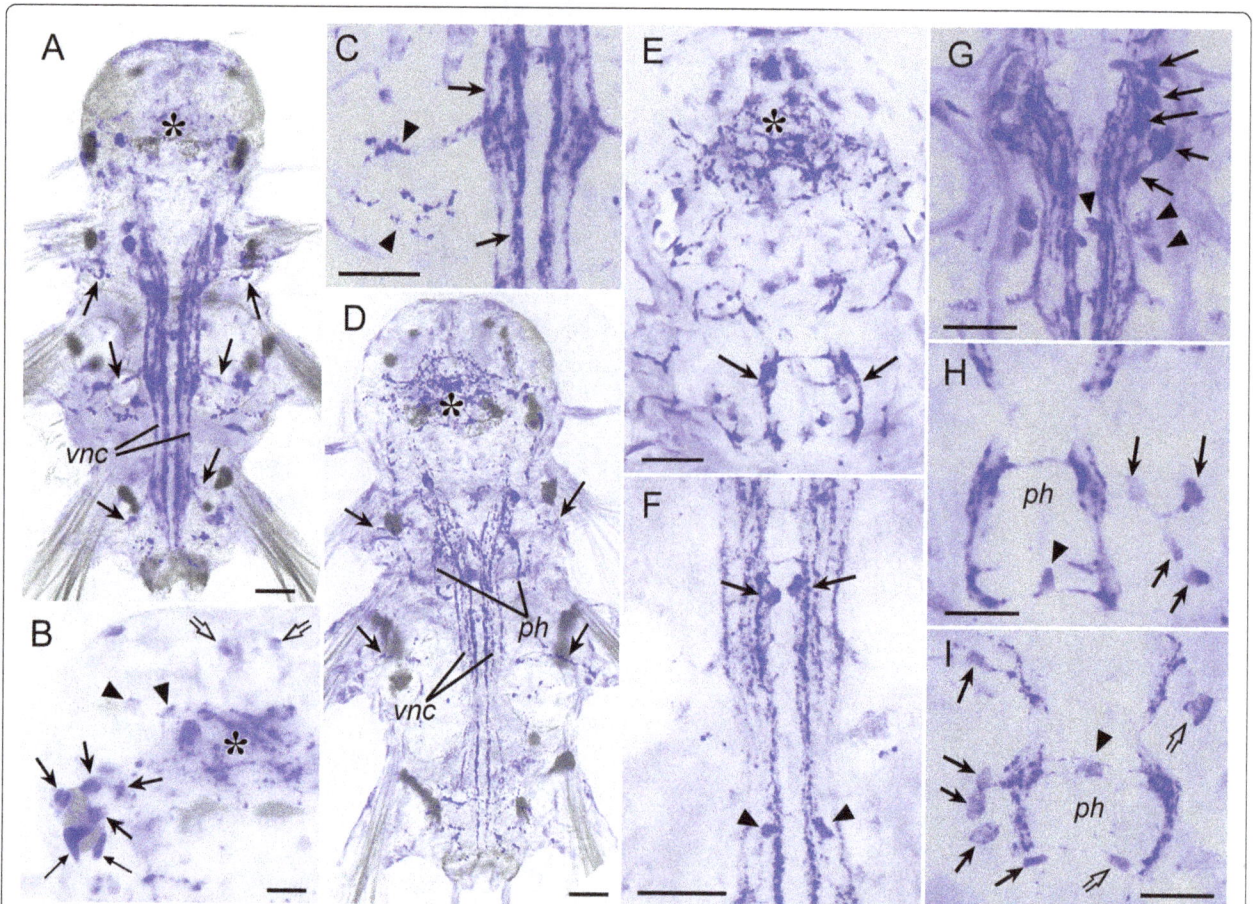

Fig. 7 Development of catecholamine containing elements (*blue*) in *P. dumerilii*. Early nectochaete (**a-c**), mid-nectochaete (**d-f**), and late nectochaete (**g-i**); ventral views. **a, b** and **d** are combinations of fluorescence (*blue*) and transmission light image (*grey*). **a:** 72 hpf, CA-containing structures in the brain (*asterisk*), ventral nerve cord (*vnc*) and larval segments; *arrows* indicate segmental nerves. **b:** 72 hpf, the head region showing the brain neuropil (*asterisk*), two neurons in front of the brain (*open arrows*), two neurons to the right of it (*arrowheads*), and four neurons near the right eye (*arrows*). The pigmented eye looks grey and has two autofluorescent structures (*small arrows*). **c:** 72 hpf, high magnification showing CA-containing structures in the paired ventral cord (*arrows*) and parapodial nerves (*arrowheads*). **d:** 3.5 dpf, CA-containing structures in the brain (*asterisk*), pharynx (*ph*), ventral nerve cord (*vnc*), and parapodia (*arrows*). **e:** 3.5 dpf, dorsal part of the head showing CA-containing neurites in the brain (*asterisk*) and pharynx (*arrows*). **f:** 3.5 dpf, high magnification of the ventral cord showing two pairs of CA-containing cells in the first (*arrows*) and second (*arrowheads*) chaetigerous segments. **g:** 5 dpf, high magnification of the ventral cord showing five pairs of CA-containing cells (*arrows*) in the first chaetigerous segment and three pairs (*arrowheads*) in the second segment (only neurons in the left half of the cord are indicated). **h:** 5 dpf, high magnification of the dorsal part of the pharynx (*ph*) showing four CA-containing cells to the left of it (*arrows*), and one medial cell (*arrowhead*). **i:** 5 dpf, high magnification of the ventral part of of the pharynx (*ph*) showing five CA-containing cells to the right of it (*arrows*), two controlateral cells (open arrows) and one medial cell (*arrowhead*). Scale bars = 20 μm

CA-containing cells were also detected in the anterior region of the digestive tract but significant deformation – which is inevitable during air drying – did not allow us to determine their number and precise location. Usually, four cells were visualised to the left or to the right of the dorsal part of the gut (Fig. 7h, *arrows*), and five cells to the left or to the right of the ventral part of the gut (Fig. 7i, *arrows*), although sometimes controlateral cells could be seen (Fig. 7i, *open arrows*). One medial cell was detected in the dorsal (Fig. 7h, *arrowhead*) and one in ventral part of the gut (Fig. 7i, *arrowhead*).

Dorsal sensory structure (Fig. 8)

Besides the apical sensory organ, another provisory sensory structure developed in the larval *P. dumerilii* (Fig. 8). At the mid-trochophore stage, anti-tubulin immunostaining revealed a bipolar cell of an unknown transmitter phenotype (Fig. 8a, *arrow*) close to the FMRFa-lir dorsal neuron *fd*. The soma of this cell was located slightly posterior to the prototroch nerve, a basal neurite ran into it, and a short apical fibre bore surface cilia. At the late trochophore stage, the bipolar sensory neuron sent a basal fibre into the apical neuropil (Fig. 8b, *double arrow*). This structure remained unchanged until the

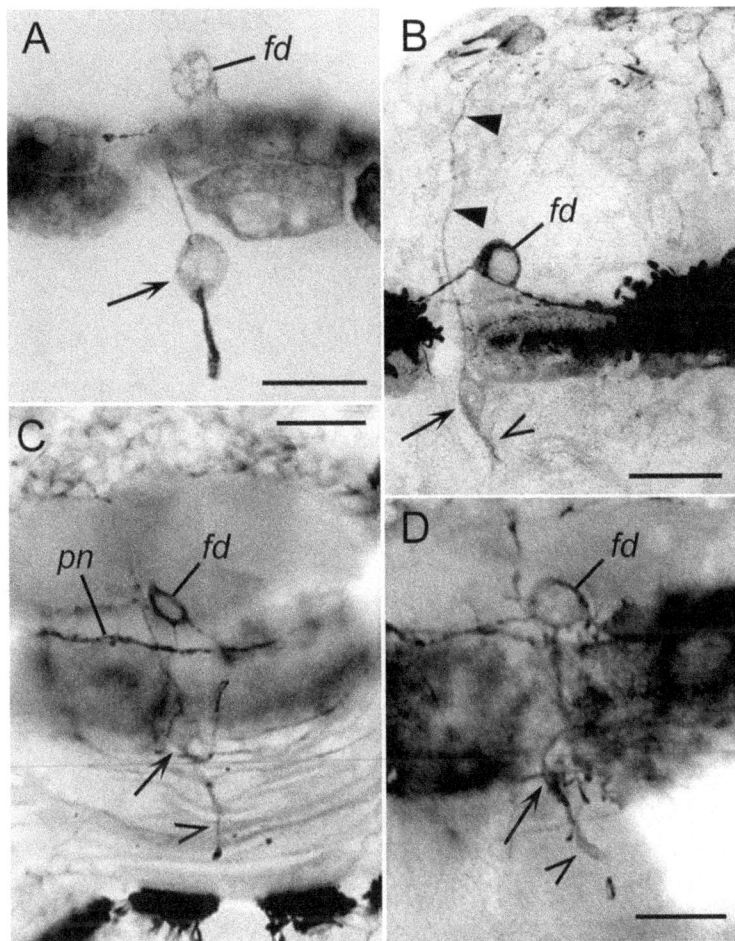

Fig. 8 Development of the dorsal sensory structure in *P. dumerilii*. Mid-trochophore (**a**), late trochophore (**b**), early nectochaete (**c**), and mid-nectochaete (**d**). Dorsal views, acetylated tubulin-like immunoreactivities are presented as negatives. **a:** 28 hpf, bipolar neuron (*arrow*) and the cell *fd* are located dorsally on both sides of the prototroch and send fibres into the prototroch nerve. **b:** 36 hpf, the cell *fd* and the bipolar sensory cell (*arrow*) with the short apical fibre (*open arrowhead*) and basal fibre (*arrowheads*) running into the apical neuropil. **c:** 68 hpf, the cell *fd* and the bipolar sensory cell (*arrow*) with the apical ciliated fibre (*open arrowhead*); *pn* – prototroch nerve. **d:** 72 hpf, the above listed structures show signs of degeneration. Scale bars = 20 μm

mid-nectochaete stage (Fig. 8c), and it later started to degenerate (Fig. 8d). Degeneration corresponded to the time when FMRFa-like immunoreactivity in the cell *fd* faded. At the late nectochaete stage, no trace of this structure was found.

Discussion

Our study provides a detailed description of neuronal development in larval and juvenile polychaete *Platynereis dumerilii* including the ontogeny of serotonergic and catecholamine containing neuronal elements as well as FMRFa-like immunoreactive structures. Many members of the RFamide neuropeptide family are known to be present in a variety of invertebrate animals including *P.dumerilii* [51, 52]. The commercial anti-FMRFamide antibody that we used is known to target several epitopes shared by different members of the family and is not specific for particular FMRFamide related peptides

[53]. Therefore it cannot be used to compare the distribution of specific peptides or track the evolution of the expression of certain neuropeptides genes. However, during the last few decades this antibody has been extensively used to study trochozoan neurogenesis (see refs above). We therefore consider these data useful for the comparative morphological analysis of neuronal development.

Our results on the development of serotonergic nervous elements are in accordance with the earlier study by Fischer and co-workers [11]. However, we present a more detailed description of 5-HT immunoreactive structures at the early stages of development. We also extended the study on the late developmental stages (juvenile worms). For all three chemically specific subsets of neurons we tried to identify the first neurons to appear as well as their projections and fate during metamorphosis (for the summary of neural development see Figs. 9, 10 and 11).

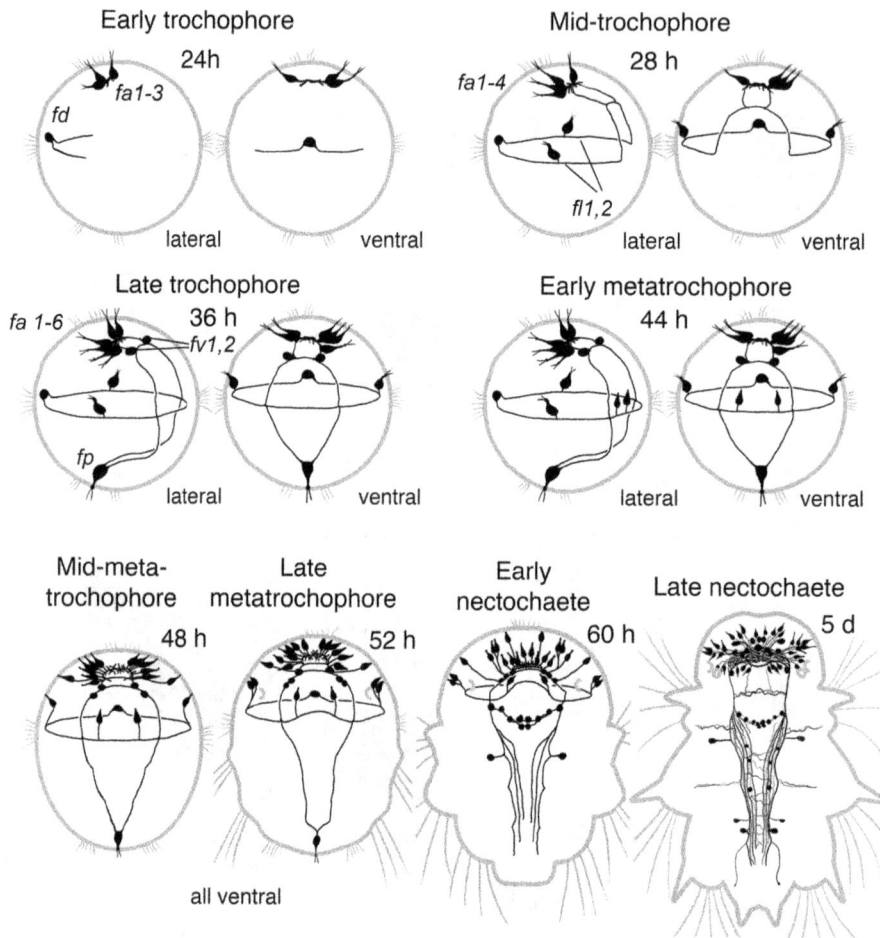

Fig. 9 Summary diagram of the ontogeny of FMRFamide-like structures in *P. dumerilii*. The earliest FMRFa-lir somata appear in early trochophores. Basal neurites of the solitary dorsal neuron *fd* outline the future prototroch nerve. Three apical sensory neurons *fa 1–3* send basal fibers into apical neuropil. By the late trochophore stage, the set of larval sensory neurons is supplemented with a solitary posterior sensory cell *p1*, two lateral cells *fl 1–2*, and three more apical cells *fa 4–6*. Besides, the first pair of central neurons *fv 1,2* appears in the developing brain. Basal fibers of these cells outline the prototroch nerve, ventral cord and cerebral commissures. At the metatrochophore stage, many more sensory cells and additional central neurons appear in the episphere. Starting from the nectochaete stage, the adult CNS develops rapidly. Apical end is always up. Relative dimensions are not maintained

The first cell we considered to be a neuron appears shortly before hatching at the posterior extreme of the larva and expresses serotonin (*sp1*). No other neuron-like structures were detected at this time by anti-tubulin immunostaining thus suggesting this cell to be the first neuronal structure. The cell has a short apical neurite and two long basal fibers, which run towards the prototroch. After hatching, two cilia appear at the end of its short apical neurite indicating the sensory nature of this cell. At the late trochophore stage, a similar FMRFa-lir cell (*fp1*) appears and we suggest that *sp1* and *fp1* are the same cell, which is both 5-HT- and FMRFa-lir.

1. Its position, size and morphology (a short apical and two basal fibres) are similar after both 5HT-tubulin and FMRFa-tubulin immunolabeling.

2. The cell always bears cilia, and no other cilia were detected inside the telotroch ring (compare Fig. 1j and Fig. 4f). Thus, the first neuron probably expresses 5HT and an FMRFa related peptide, the latter transiently at the stages from late trochophore to late metatrochophore.

Soon after hatching, the cell *sp1* is supplemented with the apical cell *sa1*. The neurites of these cells pioneer pathways upon which the adult nervous system subsequently develops: the brain (cerebral ganglion), paired ventral nerve cord, and trochal neurite bundle (prototroch nerve) (see Fig. 4d). It has been suggested recently that in the course of Lophotrochozan larval development, the earliest neurons appear at the periphery and their basal neurites form the scaffold upon which the future adult nervous system later develops [54]. In

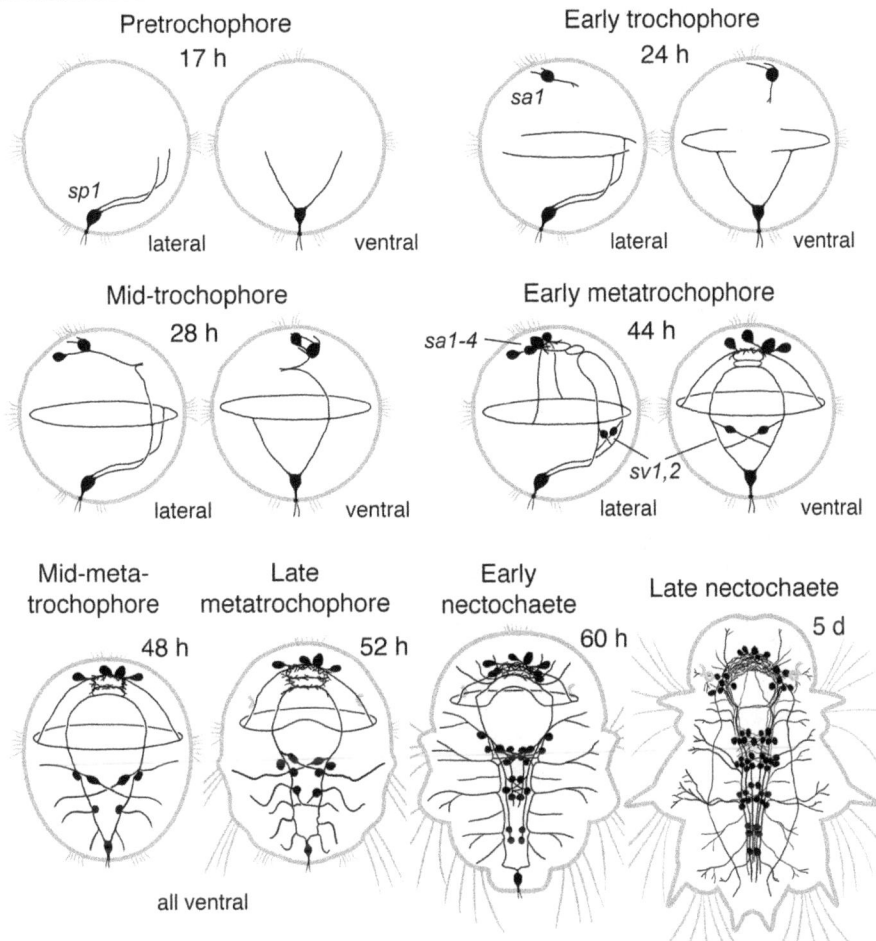

Fig. 10 Summary diagram of the ontogeny of serotonergic structures in *P. dumerilii*. The earliest 5HT-lir posterior sensory cell *sp1* appears at the pretrochophore stage before hatching. Soon after hatching, a solitary sensory cell *sa1* appears at the apical extreme and basal fibers of these two cells outline the future prototroch nerve, paired ventral cord and brain commissure. In metatrochophores, more sensory cells add in the apical region and along the developing ventral cord. Starting from the nectochaete stage, 5HT-lir cells and neurites rapidly develop in the adult CNS. Apical end is always up. Relative dimensions are not maintained

P.dumerilii, both pioneer neurons are also peripheral cells and their basal neurites label the pathways where the adult CNS will later develop thus supporting the concept of pioneer neurons.

The 5-HT-lir neuron *sp1* is strikingly similar to those, found in other polychaete species *Phyllodoce maculata* [42], and *Pomatoceros lamarckii* [43]. In *Sabellaria alveolata*, the soma of the pioneer neuron was not visualized, but the neuronal processes were detected as two bundles along the ventral side and underneath the prototroch [44]. Thus, a solitary posterior serotonergic pioneer neuron is probably a common feature among polychaetes. However, no co-localization of 5-HT- and FMRFa-like immunoreactivity was found in *P. maculata*, and *P. lamarckii*. At the apical extreme, early peripheral FMRFa- and 5-HT-lir neurons were also found in almost all polychaete species studied to date [42–44, 55–58]. The

only exception is *Owenia fusiformis*, which lacks 5 HT-lir cells in the apical organ [59, 60].

Two alternative hypotheses suggest that the annelid nervous system develops from either one anterior rudiment or two rudiments, anterior and posterior (for refs see [61]). Our data supports the idea that the scenario of larval neuronal development starting at two opposite poles, apical and caudal, could be a characteristic feature of polychaetes. An electron microscope study of early neuronal patterning in larval *Spirobranchus* has also revealed two initially separate parts of the nervous system: pretrochal and suboral [62]. The former precisely corresponds to early apical neurons in *P. dumerilii*, while the correspondence of the latter to the caudal system is not obvious. However, the description of *Spirobranchus* neuronal development by Lacalli started from later stages when caudal pioneering neurons could not be identified any more. Thus, we speculate that the suboral

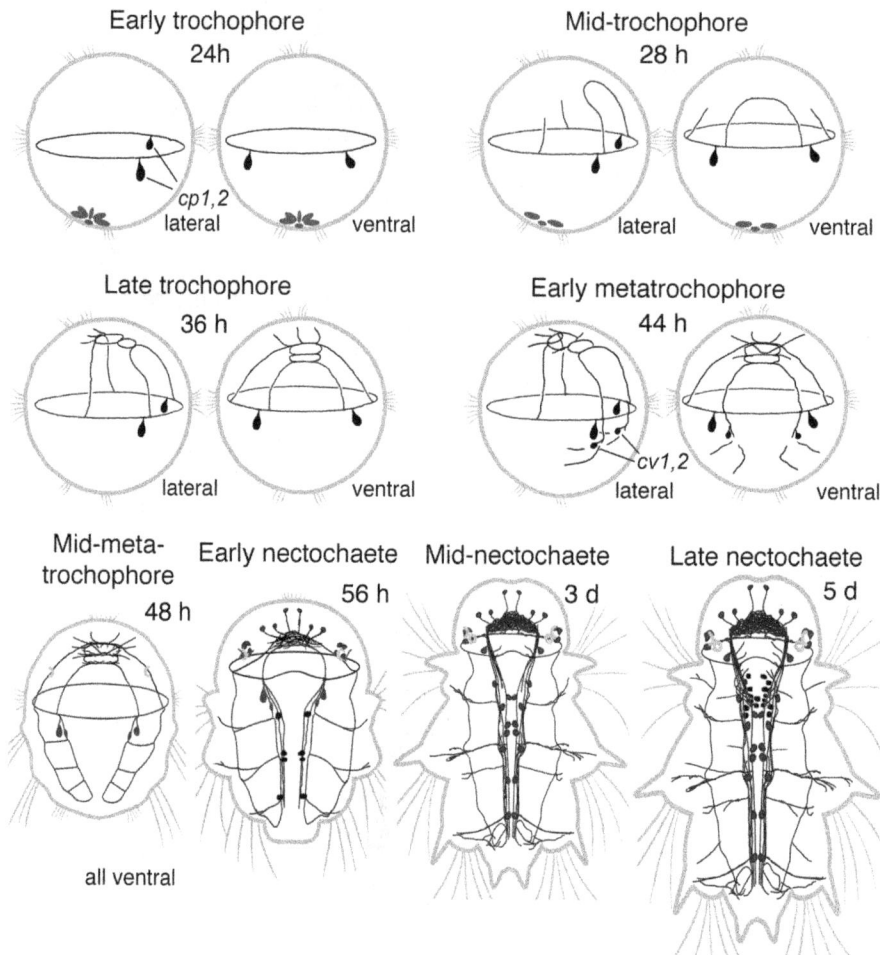

Fig. 11 Summary diagram of the ontogeny of catecholamine containing structures in *P. dumerilii*. The first CA cells *cp1,2* appear in early trochophores slightly behind the prototroch. Their basal fibers run into the prototroch nerve, and by the late trochophore stage, they outline the prototroch nerve, pretrochal part of the ventral nerve cord and brain commissures. In metatrochophores, two more cells *cv1,2* appear and by the late metatrochophore stage, the fibers of these cells outline the brain and brain commissures, prototroch nerve, ventral cord and three transverse commissures corresponding to larval segments. Starting from the nectochaete stage, more cells and neurites add in the developing central and peripheral adult nervous system. Apical end is always up. Relative dimensions are not maintained

part of the nervous system in *Spirobranchus* is derived from the caudal rudiment. Earlier, the "chimeric brain hypothesis" suggested that insect, annelid, and vertebrate nervous systems evolved from two opposite aggregations of neurons ("apical nervous system" versus "blastoporal nervous system") [39]. Our results are in agreement with this hypothesis.

The early dorsal FMRFa-lir cell *fd* has no analogues in other polychaete species studied to date. However, this cell is presumably the part of the dorsal sensory structure, which is very much like the posterior sensory organ found in another members of the order Phyllodocida, *P. maculata* [63]. Contrary to *P. dumerilii*, two 5-HT-lir and no FMRFa-lir neurons were found in the posterior sensory organ of *P. maculata*. The gross morphology, however, looks very similar when not counting the number of sensory neurons: one in *P. dumerilii* versus five in

P. maculata. One can speculate that, as compared to phyllodocid larvae with a long planktotrophic stage, the reduced free-swimming period of lecitotrophic *Platynereis* larvae resulted in partial reduction of the provisory sensory structures.

The ventral nerve cords initially appear as paired fibres of the posterior cell *sp1* (Fig. 4a-d), and the number of neurites in both cords gradually increases in the course of development. All histo- and immunostaining techniques exploited in this study revealed two symmetrical cords until the nectochaete stage (Figs. 1o,4h and 7f). Only in mid-nectochaetae, a central unpaired nerve bundle was detected after acetylated tubulin immunolabelling (Figs. 2a and 4i). However, paired symmetrical serotonergic nerve fibres were detected in the medial bundle even in 5- and 10-segmented juvenile worms (Fig. 5g,j). Thus, our data shows that the ventral nerve cord of *P. dumerilii* develops from initially paired nerves and the medial unpaired bundle appears later. It has been shown that during polychaete metamorphosis, the ventral neuroectoderm narrows and lengthens by mediolateral cell intercalation, and nerve bundles of the ventral cord cords approach each other [64]. Perhaps, for this reason, central bundles of the initially paired ventral cord come close and fuse together into the unpaired medial bundle. Thus, our results argue the suggestion that the unpaired median nerve belongs to the Annelida basic body plan [61, 65], and support the earlier idea of paired trunks within the ventral cord as the plesiomorphic condition [66].

CA-containing cells are sparse and, except the first pair cp1, 2, they appear at late developmental stages. No CA-containing cells were found in the apical organ at any stages. In general, the structure of all CA-containing neurons suggests they are not sensory: they are rounded and have no peripherally directed fibres. In other larval Trochozoa, CA-containing cells are usually sensory and located in the structures used for food detection: close to the mouth and on the internal surface of the ciliated lobes and tentacles [20, 23, 67–73]. We speculate that the lack of early sensory CA-containing cells in larval P. dumerilii resulted from transition to lecitotrophy.

The relatively small number of CA-containing cells in the ventral nerve cord is in accordance with the data for other polychaete species. Thus, only two and three pairs of such cells were found in Nephtys [45] and Ophryotrocha puerilis [46]. Similarly to P. dumerilii, four CA-containing nerve bundles were found in the ventral cord of O. puerilis, probably the functions of the catecholaminergic system in different polychaetes are similar.

In P. dumerilii, different parts of the future adult CNS are labelled by different pioneer neurons. Thus, according to our data, the paired ventral cord is outlined by the posterior serotonergic cell sp1, the prototroch nerve by the cells sp1, fd (FMRFa-lir dorsal), fl1,2(FMRFa-lir lateral) and cp1,2 (CA-containing ventrolateral). The adult brain including the cerebral ganglion and brain commissures are labelled by the cells fa1–4 (FMRFa-lir apical) and sa1–4 (serotonergic apical). Early stages of neuronal development in other polychaete species are similar in that the fibres of the early pioneer neurons outline the main parts of the future adult CNS [42–44, 58]. However, in planctotrophic larvae, numerous additional transitory larval sensory cells and additional meridianal nerves appear [42, 74]. Despite a relatively long free-swimming period, P. dumerilii larvae are lecitotrophic. One can speculate that additional neurons in planktotrophic larvae are involved in feeding behaviour.

For a long time, the pygidium of annelids was considered to be a simple structure composed of epidermal cells only [75]. However, a much more complex organization of the pygidium was described recently in adult P. dumerilii [76]. Our data however, expands on the differences in larval development. Thus, the commissure that connects the terminal parts of the ventral nerve cords (see Fig. 5j) was first detected at the nectochaete stage (See Fig. 1s and 4i). The presence of the terminal commissure was earlier described in Spirobranchus polycerus [62]. We speculate that the terminal commissure is a part of the basic annelid body plan. However, further investigations on the development of the pygidium in different annelid species are necessary.

A recent study in the development of the nervous system in Solenogastres (Mollusca, Neomeniomorpha) has shown that it is very similar to that of P. dumerilii: first neurons appear at both anterior and posterior poles, and the neurites of these neurons run towards the prototroch and form the scaffold for the future ventral nerve chords, so the CNS differentiates from both anterior and posterior poles [24]. In Polyplacophorans, which are sometimes united with Solenogastres as a basal clade Aculifera [77, 78], the scaffold is formed by two pairs of lateral cells being both 5-HT- and FMRFa-lir, and the CNS differentiates in the anterior-to-posterior direction. At the trochophore stage, the nervous system consists of a large apical organ, which later degenerates, cerebral ganglion, and paired ventral and ventrolateral cords with transverse commissures. The cerebral ganglion and the cords are the parts of the adult nervous system. The prototroch nerve is absent at all stages [79].

The contour of the mussel nervous system is formed by the pair of FMRFa-lir neurons located at the apical extreme of the trochophore. The nervous system differentiates in the anterior-to-posterior direction. The nervous system of trochophores and veligers consists of the apical organ, cerebral ganglion connected with it, and a pair of ventral cords. The prototroch nerve is also absent [23].

In the opistobranch gastropod Aplysia californica, the adult CNS is outlined by four posterior FMRFa neurons, and the nervous system of the trochophore consists of apical 5-HT-lir neurons and two ventral cords only [22, 80, 81]. In the freshwater pulmonate gastropod L. stagnalis, the trochophore nervous system includes two apical neurons, three posterior neurons, and a pair of longitudinal cords [82]. In both species, the nervous system differentiates in the posterior-to-anterior direction.

Comparison of general patterns of neuronal development in annelids and molluscs suggests that, despite of some differences, the sequence of events in the differentiation and development of the nervous system are always as follows: early peripheral neurons differentiate and their fibres form a scaffold upon which the future adult nervous system will later develop. The larval nervous system then develops, and later the adult nervous system forms along the pathways outlined by the early neurons. Finally, the larval nervous system and the early neurons disappear or partially incorporate into the adult nervous system [16, 20, 22, 23, 42, 43, 79, 82, 83]. This similarity suggests that the last common ancestor of Annelida and Mollusca had a biphasic life cycle with a planktonic larva and benthic adult [40, 84].

According to the primary heteronomy concept, all larval chaetigerous segments in polychaetes form simultaneously while postlarval segments form in the anterior-to-posterior direction [85]. This concept is also supported by *Hox* and *ParaHox* gene expression patterns, which are different in larval and postlarval segments in *Alitta virens* [86–88]. Conversely, expression patterns of the genes *engrailed* and *wingless* show the same anterior-to-posterior expression pattern in both larval and postlarval segments [89]. Our data demonstrates that contrary to the primary heteronomy concept, neurons differentiate in the anterior-to-posterior direction not only in postlarval but also in larval segments of *P. dumerilii*. Earlier, similar scenario of commissure appearance in ventral ganglia of *Sabellaria alveolata* larva was demonstrated [44]. Thus, simultaneous appearance of larval chaetigerous segments and the pattern of *Hox* genes expression support the primary heteronomy concept, though the sequence of neuronal development and expression of several other genes involved in segment formation contradict it. More data on the functions of *Hox* genes in postlarval development is necessary to resolve this contradiction.

Recently, *P. dumerilii* became a popular model animal for neurobiological studies and several novel methods were developed for its study. Characterization of neuropeptidome (repertoire of conserved proneuropeptides), identification of neuropeptides, reconstruction of individual neurons and neuronal circuits using serial-section transmission electron microscopy, and high resolution whole-body registration of gene expression allowed identification of many chemically specific neuronal subsets involved in regulation of ciliary locomotion, spatial orientation, circadian rhythms and settlement of the larva [7, 38, 52, 90–96]. Molecular topography of these subsets suggests they are evolutionary conserved [38–40, 96]. Immunostaining with anti-serotonin and FMRFamide antibodies as well as catecholamine histochemical staining visualizes only a small fraction of the neurons in the developing nervous system and results in an incomplete picture of the nervous system. However, most of the data for other Annelida and Lophotrochozoa in general was obtained using these immune- and histoshemical staining techniques, so we considered the presented results important for the comparative analysis. Comparative studies of neuronal clusters in basal branching annelid groups with different types of development may be useful for better understanding of the evolution of the bilaterian brain and its developmental plasticity.

Conclusions

We have presented a detailed description of *Platynereis dumerilii* neuronal development based on anti-acetylated tubulin, serotonin, and FMRFamide-like immunostaining as well as catecholamine histofluorescence. The results are summarized in Figs. 9, 10 and 11. Our data confirms and expands the existing results, and offers new information for comparative analysis of the nervous system ontogeny in Polychaeta and Trochozoa in general. According to our data, the development starts from the first pioneer neurons at the posterior and anterior poles of the larva, and the basal neurites of these neurons form a scaffold upon which the adult CNS later differentiates, starting from both anterior and posterior poles. Larval chaetigerous segments form from the anterior-to-posterior direction, contradicting the primary heteronomy concept. Comparison with Mollusca reveals substantial similarity with early neuronal development in larval Solenogastres. More comparative data on larval neuronal development in different groups of Trochozoa are necessary. Being a highly morphologically diverse taxon, Annelida are of particular interest, especially the groups that belong to the so-called basal branching annelids such as Chaetopteridae, Magelonidae and Amphinomidae [97, 98].

Methods

Platynereis dumerilii culture

P. dumerilii larvae were obtained from the laboratory breeding culture as described by Fischer and Dorresteijn [99]. Artificial sea water (ASW, Red Sea Coral Pro Salt) was used to culture adults and embryos. After spawning, the embryos were cultivated in glass Petri dishes 150 mm in diameter. Larvae and juvenile worms were fed on hard-boiled and homogenized quail eggs, and homogenized spinach. The speed of *P. dumerilii* development is known to be highly temperature dependent, thus minor temperature fluctuations cause significant changes in the pace of development, and complicate the comparison between studies [10, 11]. To avoid this problem, we cultivated all animals and developing embryos in a climate chamber at 18 °C, so the schedule of development corresponded to that described by Fisher and co-authors [11].

Sampling, fixation and staging

Samples were taken each hour slightly before and after hatching, every four hours until the nectochaete stage (ca. 3 days of development), and then every 12 h until the stage of small atokous worm (ca. 20–30 days). All larvae were collected on Nitex screens, rinsed in ASW, relaxed by gradually adding a 7.5% magnesium chloride aqueous solution, and fixed in fresh 4% paraformaldehyde in phosphate buffered saline (PBS) for 1 h at room temperature, washed in PBS, transferred into 70%

ethanol and stored at −20°. Besides the timing of development, each stage was additionally determined by the set of morphological characters described in [11].

Immunochemistry

After storage, the specimens were rinsed in PBS 3 × 15 min, blocked overnight in PBS with 10% normal goat serum, 0.25% bovine serum albumin, 1% Triton X-100 (TX), and 0.03% sodium azide, and incubated in a mixture of either anti-5-HT or anti-FMRFamide primary ABs (Immunostar, 20,080 and 20,091, respectively; both polyclonal and raised in rabbits and diluted 1:2000–3000) together with monoclonal anti-acetylated α-tubulin AB (Sigma, Cat. No. T-6793, developed in mouse, diluted 1:1000–1500) in a blocking solution for 1–3 days at 10 °C. The specimens were then washed in PBS and incubated in a mixture of goat anti-rabbit Alexa 488 conjugated IgG and goat anti-mouse Alexa 546 IgG (Molecular Probes) both diluted 1:800 in PBS-TX, overnight at 10 °C. The specimens were washed in PBS several more times and mounted on glass slides in glycerol or TDE [100]. If necessary, 2–5 µg/ml HOEHST 33258 (Invitrogen, H1398) was added to one of the last wash to label cell nuclei. This allowed unambiguous differentiation between neuron perikarya and aggregations of fibers. Replacement of the primary antibodies with non-immune serum did not result in any staining. Reversal of the colours of the secondary antibodies (anti-rabbit Alexa 546 IgG and anti-mouse Alexa 488 IgG) yielded identical staining patterns.

Catecholamine histochemistry

A glyoxylic acid fluorescent technique [101] was employed to visualize catecholamine-containing cells. Embryos were immersed in a freshly prepared, buffered glyoxylic acid-sucrose solution (500 mM sodium glyoxylate, 150 mM sucrose, 50 mM Tris buffer, pH 7.4) on glass slides at 4 °C. After 60 min of incubation, the solution was removed, and the embryos were air-dried at room temperature for 30 min. Preparations were then heated to 60 °C for 30 min, embedded in paraffin oil, and examined using Leica TCS SP5 laser scanning microscope (excitation at 405 nm, emission detection at 457–490 nm). For illustrations, the images were converted into negatives, so fluorescence of catecholamines looks blue. Controls in which larvae were once more examined after addition of distilled water to the preparation, showed no fluorescence characteristic of catecholamines.

Microscopy and image processing

All specimens were examined as whole-mounts on laser scanning microscopes Zeiss LSM 510, Leica TCS SP5 and Leica TCS SPE with high aperture oil immersion objectives using appropriate wavelength-filter configuration settings. No fewer than 100 embryos were examined at each stage for each of the antibodies. For each larva, 40–150 0.5 µm thick optical sections were taken and processed with Zeiss LSM IB, Leica LAS AF, Bitplane Imaris and ImageJ. Three-dimensional (3D), rotatable reconstructions were produced using Imaris and converted into AVI files. A series of optical sections were also projected into single images and exported as TIFF images. These images were then adjusted for contrast and brightness and assembled into plates using Adobe Photoshop CS.

Additional files

> **Additional file 1:** FMRFamide-like (green) and acetylated tubulin-like (magenta) immunoreactivity in the anterior body part of three-segmented juvenile of *P. dumerilii*, 3D rotatable reconstruction. (AVI 6162 kb)
>
> **Additional file 2:** FMRFamide-like (green) and acetylated tubulin-like (magenta) immunoreactivity in the anterior body part of ten-segmented juvenile of *P. dumerilii*, 3D rotatable reconstruction. (AVI 4623 kb)
>
> **Additional file 3:** 5-HT-like (green) and acetylated tubulin-like (magenta) immunoreactivity in the anterior body part of ten-segmented juvenile of *P. dumerilii*, 3D rotatable reconstruction. (AVI 3097 kb)

Acknowledgements
The study was partially performed at the core facilities "Chromas", "Centre for Molecular and Cell Technologies" and "Centre for Culturing Collection of Microorganisms" of Research Park of Saint-Petersburg State University.

Funding
This work was supported by the Russian Foundation for Basic Research (RFBR) Grants 16–34-60,134 to VVS, 15–29-02650 to EEV, and 14–04-00673 to LPN.

Authors' contributions
VVS conducted the fluorescent stainings, the CLSM analysis and contributed substantially to the interpretation of data and to the writing of the manuscript. EEV conducted the fluorescent stainings and contributed substantially to the interpretation of data and to the writing of the manuscript. LPN conducted the fluorescent stainings, the CLSM analysis and wrote the first draft of the manuscript. All authors read and approved the final manuscript.

Competing interests
The authors declare that they have no competing interests.

Author details
[1]Department of Invertebrate Zoology, St-Petersburg State University, St-Petersburg 199034, Russia. [2]Zoological Institute Rus, Acad. Sci, St-Petersburg 199034, Russia. [3]Institute of Developmental Biology, Rus. Acad. Sci, Moscow 119991, Russia.

References
1. Aguinaldo AMA, Turbeville JM, Linford LS, Rivera MC, Garey JR, Raff RA, et al. Evidence for a clade of nematodes, arthropods and other moulting animals. Nature. 1997;387:489–93.
2. Halanych K, Bacheller J, Aguinaldo A, Liva S, Hillis D, Lake J. Evidence from 18S ribosomal DNA that the lophophorates are protostome animals. Science. 1995;267:1641–3.
3. Field KG, Olsen GJ, Lane DJ, Giovannoni SJ, Ghiselin MT, Raff EC, et al. Molecular phylogeny of the animal kingdom. Science. 1988;239:748–53.
4. Kim CB, Moon SY, Gelder SR, Kim W. Phylogenetic relationships of annelids, molluscs, and arthropods evidenced from molecules and morphology. J Mol Evol. 1996;43:207 15.

5.   Peterson KJ, Cameron RA, Davidson EH. Bilaterian origins: significance of new experimental observations. Dev Biol. 2000;219:1–17.

6.   Tessmar-Raible K, Arendt D. Emerging systems: between vertebrates and arthropods, the Lophotrochozoa. Curr Opin Genet Dev. 2003;13:331–40.

7.   Jékely G, Colombelli J, Hausen H, Guy K, Stelzer E, Nédélec F, et al. Mechanism of phototaxis in marine zooplankton. Nature. 2008;456:395–9.

8.   Hardege JD. Nereidid polychaetes as model organisms for marine chemical ecology. Hydrobiologia. 1999;402:145–61.

9.   Dorresteijn AWC. Quantitative analysis of cellular differentiation during early embryogenesis Ofplatynereis dumerilii. Rouxs Arch Dev Biol. 1990;199:14–30.

10. Fischer A, Dorresteijn A. The polychaete Platynereis dumerilii (Annelida): a laboratory animal with spiralian cleavage, lifelong segment proliferation and a mixed benthic/pelagic life cycle. BioEssays. 2004;26:314–25.

11. Fischer AH, Henrich T, Arendt D. The normal development of Platynereis dumerilii (Nereididae, Annelida). Front Zool. 2010;7:31.

12. Dorresteijn AWC, O'Grady B, Fischer A, Porchet-Henneré E, Boilly-Marer Y. Molecular specification of cell lines in the embryo of Platynereis (Annelida). Roux's Arch Dev Biol. 1993;202:260–9.

13. Fisher A, Dorresteijn AW, Hoeger U. Metabolism of oocyte construction and the generation of histospecificity in the cleaving egg. Lessons from nereid annelids. Int J Dev Biol. 1996;40:421–30.

14. Arendt D, Tessmar K, de Campos-Baptista M-IM, Dorresteijn A, Wittbrodt J. Development of pigment-cup eyes in the polychaete Platynereis dumerilii and evolutionary conservation of larval eyes in Bilateria. Development. 2002; 129:1143–54.

15. Ackermann C, Dorresteijn A, Fischer A. Clonal domains in postlarval Platynereis dumerilii (Annelida: Polychaeta). J Morphol. 2005;266:258–80.

16. Croll R. Insights into early molluscan neuronal development through studies of transmitter phenotypes in embryonic pond snails. Microsc Res Tech 2000;578:570–8.

17. Wanninger A. Shaping the things to come: ontogeny of lophotrochozoan neuromuscular systems and the tetraneuralia concept. Biol Bull. 2009;216:293–306.

18. Jackson AR, MacRae TH, Croll RP. Unusual distribution of tubulin isoforms in the snail Lymnaea stagnalis. Cell Tissue Res. 1995;281:507–15.

19. Croll RP. Development of embryonic and larval cells containing serotonin, catecholamines, and FMRFamide-related peptides in the gastropod mollusc Phestilla sibogae. Biol Bull. 2006;211:232–47.

20. Dickinson AJG, Croll RP. Development of the larval nervous system of the gastropod Ilyanassa obsoleta. J Comp Neurol. 2003;466:197–218.

21. Nielsen C, Haszprunar G, Ruthensteiner B, Wanninger A. Early development of the aplacophoran mollusc Chaetoderma. Acta Zool. 2007;88:231–47.

22. Dickinson AJ, Croll RP, Voronezhskaya EE. Development of embryonic cells containing serotonin, catecholamines, and FMRFamide-related peptides in Aplysia californica. Biol Bull. 2000;199:305–15.

23. Voronezhskaya EE, Nezlin LP, Odintsova NA, Plummer JT, Croll RP. Neuronal development in larval mussel Mytilus trossulus (Mollusca: Bivalvia). Zoomorphology. 2008;127:97–110.

24. Redl E, Scherholz M, Todt C, Wollesen T, Wanninger A. 2014. Development of the nervous system in Solenogastres (Mollusca) reveals putative ancestral spiralian features. EvoDevo. 2014;5:48.

25. Temereva E, Wanninger A. Development of the nervous system in Phoronopsis harmeri (Lophotrochozoa, Phoronida) reveals both deuterostome- and trochozoan-like features. BMC Evol Biol. 2012;12:121.

26. Temereva EN, Tsitrin EB. Development, organization, and remodeling of phoronid muscles from embryo to metamorphosis (Lophotrochozoa: Phoronida). BMC Dev Biol. 2013;13:14.

27. Temereva EN, Tsitrin EB. Organization and metamorphic remodeling of the nervous system in juveniles of Phoronopsis harmeri (Phoronida): insights into evolution of the bilaterian nervous system. Front Zool. 2014;11:35.

28. Temereva EN, Tsitrin EB. Development and organization of the larval nervous system in Phoronopsis harmeri: new insights into phoronid phylogeny. Front Zool. 2014;11:3.

29. Gruhl A. Serotonergic and FMRFamidergic nervous systems in gymnolaemate bryozoan larvae. Zoomorphology. 2009;128:135–56.

30. Gruhl A. Neuromuscular system of the larva of Fredericella sultana (Bryozoa: Phylactolaemata). Zool Anzeiger A J Comp Zool. 2010;249:139–49.

31. Santagata S. Evolutionary and structural diversification of the larval nervous system among marine bryozoans. Biol Bull. 2008;215:3–23.

32. Altenburger A, Wanninger A. Neuromuscular development in Novocrania anomala: evidence for the presence of serotonin and a spiralian-like apical organ in lecithotrophic brachiopod larvae. Evol Dev. 2010;12:16–24.

33. Hessling R, Westheide W. Are Echiura derived from a segmented ancestor? Immunohistochemical analysis of the nervous system in developmental stages of Bonellia viridis. J Morphol. 2002;252:100–13.

34. Hessling R. Metameric organisation of the nervous system in developmental stages of Urechis caupo (Echiura) and its phylogenetic implications. Zoomorphology. 2002;121:221–34.

35. Wanninger A, Koop D, Bromham L, Noonan E, Degnan BM. Nervous and muscle system development in Phascolion strombus (Sipuncula). Dev Genes Evol. 2005;215:509–18.

36. Denes AS, Jékely G, Steinmetz PRH, Raible F, Snyman H, Prud'homme B, et al. Molecular architecture of annelid nerve cord supports common origin of nervous system centralization in bilateria. Cell. 2007;129:277–88.

37. Arendt D, Denes AS, Jékely G, Tessmar-Raible K. The evolution of nervous system centralization Philos. Trans R Soc Lond B Biol Sci 2008; 363:1523–1528.

38. Tomer R, Denes AS, Tessmar-Raible K, Arendt D. Profiling by image registration reveals common origin of annelid mushroom bodies and vertebrate pallium. Cell. 2010;142:800–9.

39. Tosches MA, Arendt D. The bilaterian forebrain: an evolutionary chimaera. Curr Opin Neurobiol. 2013;23:1080–9.

40. Marlow H, Tosches MA, Tomer R, Steinmetz PR, Lauri A, Larsson T, et al. Larval body patterning and apical organs are conserved in animal evolution. BMC Biol. 2014;12:7.

41. Hay-Schmidt A. The Larval Nervous System of Polygordius lacteus Scheinder, 1868 (Polygordiidae, Polychaeta): Immunocytochemical Data. Acta Zool. 1995;76:121–40.

42. Voronezhskaya EE, Tsitrin EB, Nezlin LP. Neuronal development in larval polychaete Phyllodoce maculata (Phyllodocidae). J Comp Neurol. 2003;455: 299–309.

43. McDougall C, Chen W-C, Shimeld SM, Ferrier DEK. The development of the larval nervous system, musculature and ciliary bands of Pomatoceros lamarckii (Annelida): heterochrony in polychaetes. Front Zool. 2006;3:16.

44. Brinkmann N, Wanninger A. Larval neurogenesis in Sabellaria alveolata reveals plasticity in polychaete neural patterning. Evol Dev. 2008;10:606–18.

45. Clark ME. Histochemical Localization of Monoamines in the Nervous System of the Polychaete Nephtys. Proc R Soc B Biol Sci. 1966;165:308–25.

46. Schlawny A, Hamann T, Müller MA, Pfannenstiel H-D. The catecholaminergic system of an annelid (Ophryotrocha puerilis, Polychaeta). Cell Tissue Res. 1991;265:175–84.

47. Schlawny A, Grünig C, Pfannenstiel H-D. Sensory and secretory cells of Ophryotrocha puerilis (Polychaeta). Zoomorphology. 1991;110:209–15.

48. Díaz-Miranda L, de Motta GE, García-Arrarás JE. Monoamines and neuropeptides as transmitters in the sedentary polychaete Sabellastarte magnifica: actions on the longitudinal muscle of the body wall. J Exp Zool. 1992;263:54–67.

49. Crisp KM, Klukas KA, Gilchrist LS, Nartey AJ, Mesce KA. Distribution and development of dopamine- and octopamine-synthesizing neurons in the medicinal leech. J Comp Neurol. 2002;442:115–29.

50. Richter S, Loesel R, Purschke G, Schmidt-Rhaesa A, Scholtz G, Stach T, et al. Invertebrate neurophylogeny: suggested terms and definitions for a neuroanatomical glossary. Front Zool. 2010;7:29.

51. Santama N, Benjamin PR, Burke JF. Alternative RNA splicing generates diversity of neuropeptide expression in thebrain of the snail Lymnaea: in situ analysis of mutually exclusive transcripts of the FMRFamide gene. Eur J Neurosci. 1995;7:65–76.

52. Conzelmann M, Williams EA, Krug K, Franz-Wachtel M, Macek B, Jékely G. The neuropeptide complement of the marine annelid Platynereis dumerilii. BMC Genomics. 2013;14:906.

53. Voronezhskaya EE, Elekes K. Expression of FMRFamide gene encoded peptides by identified neurons in embryos and juveniles of the pulmonate snail Lymnaea stagnalis. Cell Tissue Res. 2003;314:297–313.

54. Voronezhskaya EE, Ivashkin EG. Pioneer neurons: A basis or limiting factor of lophotrochozoa nervous system diversity? Russ J Dev Biol. 2010;41:337–46.

55. Brinkmann N, Wanninger A. Neurogenesis suggests independent evolution of opercula in serpulid polychaetes. BMC Evol Biol. 2009;9:270.

56. Meyer NP, Carrillo-Baltodano A, Moore RE, Seaver EC. Nervous system development in lecithotrophic larval and juvenile stages of the annelid Capitella teleta. Front Zool. 2015;12:15.

57. Helm C, Krause A, Bleidorn C. Immunohistochemical investigations of the development of *Scoloplos armiger* ("intertidalis clade") indicate a paedomorphic origin of *Proscoloplos cygnochaetus* (Annelida, Orbiniidae). Invertebr Biol. 2015;134:214–30.

58. Helm C, Schemel S, Bleidorn C. Temporal plasticity in annelid development–ontogeny of *Phyllodoce groenlandica* (Phyllodocidae, Annelida) reveals heterochronous patterns. J Exp Zool B Mol Dev Evol. 2013;320:166–78.

59. Helm C, Vöcking O, Kourtesis I, Hausen H. *Owenia fusiformis* – a basally branching annelid suitable for studying ancestral features of annelid neural development. BMC Evol Biol. 2016;16:129.

60. Hay-Schmidt A. The evolution of the serotonergic nervous system. Proc R Soc London Ser B Biol Sci. 2000;267:1071–9.

61. Orrhage L, Müller MCM. Morphology of the nervous system of Polychaeta (Annelida). Hydrobiologia. 2005;535-536:79–111.

62. Lacalli TC. Structure and organization of the nervous system in the trochophore larva of Spirobranchus. Philos Trans R Soc Lond Ser B Biol Sci. 1984;306:79–135.

63. Nezlin LP, Voronezhskaya EE. Novel, posterior sensory organ in the trochophore larva of *Phyllodoce maculata* (Polychaeta). Proc Biol Sci. 2003;270:159–62.

64. Steinmetz PRH, Zelada-Gonzáles F, Burgtorf C, Wittbrodt J, Arendt D. Polychaete trunk neuroectoderm converges and extends by mediolateral cell intercalation. Proc Natl Acad Sci U S A. 2007;104:2727–32.

65. Müller MCM. Polychaete nervous systems: Ground pattern and variations – cLS microscopy and the importance of novel characteristics in phylogenetic analysis. Integr Comp Biol. 2006;46:125–33.

66. Bullock TH, Horridge GA. Structure and function in the nervous systems of invertebrates. Vol. 2. San Francisco: WH Freeman; 1965.

67. Shunkina KV, Zaytseva OV, Starunov VV, Ostrovsky AN. Comparative morphology of the nervous system in three phylactolaemate bryozoans. Front Zool. 2015;12:28.

68. Santagata S. Structure and metamorphic remodeling of the larval nervous system and musculature of *Phoronis pallida* (Phoronida). Evol Dev. 2002;4:28–42.

69. Croll RP, Jackson DL, Voronezhskaya EE. Catecholamine-Containing Cells in Larval and Postlarval Bivalve Molluscs. Biol Bull. 1997;193:116–24.

70. Croll RP, Voronezhskaya EE, Hiripi L, Elekes K. Development of catecholaminergic neurons in the pond snail, *Lymnaea stagnalis*: II. Postembryonic development of central and peripheral cells. J Comp Neurol. 1999;404:297–309.

71. Voronezhskaya EE, Hiripi L, Elekes K, Croll RP. Development of catecholaminergic neurons in the pond snail, *Lymnaea stagnalis*: I. Embryonic development of dopamine-containing neurons and dopamine-dependent behaviors. J Comp Neurol. 1999;404:285–96.

72. Hay-schmidt A. Distribution of catecholamine-containing, serotonin-like and neuropeptide FMRFamide-like immunoreactive neurons and processes in the nervous system of the actinotroch larva of *Phoronis muelleri* (Phoronida). Cell Tissue Res. 1990;259:105–18.

73. Hay-Schmidt A. Catecholamine-containing, serotonin-like and neuropeptide FMRFamide-like immunoreactive cells and processes in the nervous system of the pilidium larva (Nemertini). Zoomorphology. 1990;109:231–44.

74. Nezlin LP. The golden age of comparative morphology: Laser scanning microscopy and neurogenesis in trochophore animals. Russ J Dev Biol. 2010;41:381–90.

75. Anderson DT. The comparative embryology of the Polychaeta. Acta Zool. 1966;47:1–42.

76. Starunov VV, Dray N, Belikova EV, Kerner P, Vervoort M, Balavoine G. A metameric origin for the annelid pygidium? BMC Evol Biol. 2015;15:1–17.

77. Smith SA, Wilson NG, Goetz FE, Feehery C, Andrade SCS, Rouse GW, et al. Resolving the evolutionary relationships of molluscs with phylogenomic tools. Nature. 2011;480:364–7.

78. Kocot KM, Cannon JT, Todt C, Citarella MR, Kohn AB, Meyer A, et al. Phylogenomics reveals deep molluscan relationships. Nature. 2011;477:452–6.

79. Voronezhskaya EE, Tyurin SA, Nezlin LP. Neuronal development in larval chiton *Ischnochiton hakodadensis* (Mollusca: Polyplacophora). J Comp Neurol Wiley Online Library. 2002;444:25–38.

80. Marois R, Carew TJ. Ontogeny of serotonergic neurons in *Aplysia californica*. J Comp Neurol. 1997;386:477–90.

81. Marois R, Carew TJ. Fine structure of the apical ganglion and its serotonergic cells in the larva of *Aplysia californica*. Biol Bull. 1997;192:388–98.

82. Croll RP, Voronezhskaya EE. Early elements in gastropod neurogenesis. Dev Biol. 1996;173:344–7.

83. AJG D, Nason J, Croll RP. Histochemical localization of FMRFamide, serotonin and catecholamines in embryonic *Crepidula fornicata* (Gastropoda, Prosobranchia). Zoomorphology. 1999;119:49–62.

84. Jägersten G. Evolution of the Metazoan Life Cycle. London: Academic Press; 1972.

85. Ivanoff PP. Die Entwicklung der larvalsegmente bei den Anneliden. Z Morphol Okol Tiere. 1928;10:62–161.

86. Kulakova M, Bakalenko N, Novikova E, Cook CE, Eliseeva E, Steinmetz PRH, et al. Hox gene expression in larval development of the polychaetes *Nereis virens* and *Platynereis dumerilii* (Annelida, Lophotrochozoa). Dev Genes Evol. 2007;217:39–54.

87. Bakalenko NI, Novikova EL, Nesterenko AY, Kulakova MA. Hox gene expression during postlarval development of the polychaete *Alitta virens*. Evodevo 2013;4:13.

88. Kulakova MA, Cook CE, Andreeva TF. ParaHox gene expression in larval and postlarval development of the polychaete *Nereis virens* (Annelida, Lophotrochozoa). BMC Dev Biol. 2008;8:61.

89. Prud'homme B, de Rosa R, Arendt D, Julien J-F, Pajaziti R, Dorresteijn AWC, et al. Arthropod-like expression patterns of engrailed and wingless in the annelid *Platynereis dumerilii* suggest a role in segment formation. Curr Biol. 2003;13:1876–81.

90. Conzelmann M, Jékely G. Antibodies against conserved amidated neuropeptide epitopes enrich the comparative neurobiology toolbox. EvoDevo. 2012;3:23.

91. Shahidi R, Williams EA, Conzelmann M, Asadulina A, Verasztó C, Jasek S, et al. A serial multiplex immunogold labeling method for identifying peptidergic neurons in connectomes. elife. 2015;4:e11147.

92. Williams EA, Conzelmann M, Jékely G. Myoinhibitory peptide regulates feeding in the marine annelid Platynereis. Front Zool. 2015;12:1.

93. Conzelmann M, Offenburger SL, Asadulina A, Keller T, Munch TA, Jekely G. Neuropeptides regulate swimming depth of Platynereis larvae. Proc Natl Acad Sci U S A. 2011;108:E1174–83.

94. Randel N, Asadulina A, Bezares-Calderón LA, Verasztó C, Williams EA, Conzelmann M, et al. Neuronal connectome of a sensory-motor circuit for visual navigation. elife. 2014;3:e02730.

95. Tosches MA, Bucher D, Vopalensky P, Arendt D. Melatonin signaling controls circadian swimming behavior in marine zooplankton. Cell. 2014;159:46–57.

96. Tessmar-Raible K, Raible F, Christodoulou F, Guy K, Rembold M, Hausen H, et al. Conserved sensory-neurosecretory cell types in annelid and fish forebrain: insights into hypothalamus evolution. Cell. 2007;129:1389–400.

97. Weigert A, Helm C, Meyer M, Nickel B, Arendt D, Hausdorf B, et al. Illuminating the base of the annelid tree using transcriptomics. Mol Biol Evol. 2014;31:1391–401.

98. Struck TH, Paul C, Hill N, Hartmann S, Hösel C, Kube M, et al. Phylogenomic analyses unravel annelid evolution. Nature. 2011;471:95–8.

99. Fischer A, Dorresteijn A. Culturing *Platynereis dumerilii*. 2016. http://www.staff.uni-giessen.de/~gf1307/breeding.htm. Accessed 18 May 2017.

100. Staudt T, Lang MC, Medda R, Engelhardt J, Hell SW. 2,2'-Thiodiethanol: A new water soluble mounting medium for high resolution optical microscopy. Microsc Res Tech. 2007;70:1–9.

101. Lindvall O, Björklund A. The glyoxylic acid fluorescence histochemical method: a detailed account of the methodology for the visualization of central catecholamine neurons. Histochemistry. 1974;39:97–127.

9

Mitochondrial acclimation potential to ocean acidification and warming of Polar cod (*Boreogadus saida*) and Atlantic cod (*Gadus morhua*)

Elettra Leo[1,2], Kristina L. Kunz[1,2,3], Matthias Schmidt[1,2], Daniela Storch[1], Hans-O. Pörtner[1,2] and Felix C. Mark[1*]

Abstract

Background: Ocean acidification and warming are happening fast in the Arctic but little is known about the effects of ocean acidification and warming on the physiological performance and survival of Arctic fish.

Results: In this study we investigated the metabolic background of performance through analyses of cardiac mitochondrial function in response to control and elevated water temperatures and PCO_2 of two gadoid fish species, Polar cod (*Boreogadus saida*), an endemic Arctic species, and Atlantic cod (*Gadus morhua*), which is a temperate to cold eurytherm and currently expanding into Arctic waters in the wake of ocean warming. We studied their responses to the above-mentioned drivers and their acclimation potential through analysing the cardiac mitochondrial function in permeabilised cardiac muscle fibres after 4 months of incubation at different temperatures (Polar cod: 0, 3, 6, 8 °C and Atlantic cod: 3, 8, 12, 16 °C), combined with exposure to present (400μatm) and year 2100 (1170μatm) levels of CO_2.
OXPHOS, proton leak and ATP production efficiency in Polar cod were similar in the groups acclimated at 400μatm and 1170μatm of CO_2, while incubation at 8 °C evoked increased proton leak resulting in decreased ATP production efficiency and decreased Complex IV capacity. In contrast, OXPHOS of Atlantic cod increased with temperature without compromising the ATP production efficiency, whereas the combination of high temperature and high PCO_2 depressed OXPHOS and ATP production efficiency.

Conclusions: Polar cod mitochondrial efficiency decreased at 8 °C while Atlantic cod mitochondria were more resilient to elevated temperature; however, this resilience was constrained by high PCO_2. In line with its lower habitat temperature and higher degree of stenothermy, Polar cod has a lower acclimation potential to warming than Atlantic cod.

Keywords: Arctic fish, RCP 8.5, Heart mitochondria, Mitochondrial capacity, Proton leak

Background

Ocean warming driven by anthropogenic CO_2 emissions influences the distribution of marine animals causing significant impacts on biodiversity and ecosystem structure [1, 2], such as local extinctions [3] and poleward migrations [4–6]. Fish (and other ectotherms) are particularly sensitive to fluctuations in temperature since their body temperature is in equilibrium with their environmental temperature [7]. Fish species distribution, in fact, is confined to a specific temperature window, due to the temperature dependency of physiological processes and to sustain maximal energy efficiency ([8] for review).

The increased CO_2 concentration in the atmosphere is one of the major causes for the global greenhouse effect and also causes a decrease in ocean pH, a phenomenon commonly known as ocean acidification [9]. High CO_2 partial pressure (PCO_2) is known to affect biological and physiological processes of marine organisms (e.g. [10–14]) and tolerances towards other stressors [15–17]. Moreover, high PCO_2 could provoke a narrowing of the thermal

* Correspondence: Felix.Christopher.Mark@awi.de
[1]Alfred Wegener Institute, Helmholtz Centre for Polar and Marine Research, Integrative Ecophysiology, Am Handelshafen 12, D-27570 Bremerhaven, Germany
Full list of author information is available at the end of the article

tolerance window of ectotherms, so that limits of its thermal acclimation capacity are met earlier [2, 18–21].

At the cellular level, exposure to high temperature can cause changes in the three dimensional structures of proteins, including the assembly states of multiprotein complexes and eventually protein denaturation and loss of activity [7]. Moreover, increasing temperatures can alter the cellular membranes packing order, which can cause changes in membrane-associated processes until a potential complete loss of function [22]. Furthermore, since cellular oxygen demand increases with increasing temperature, the production of mitochondrial reactive oxygen species (ROS) is likely to increase which can damage biological molecules, including lipids, proteins and DNA [23, 24]. Therefore, towards the upper limit of the thermal window, the cellular energetic costs for maintenance rise, increasing baseline energy turnover and allowing only for time-limited periods of passive tolerance. If high temperature persists over this period of passive tolerance, the costs of maintenance can only be covered at the expense of other functions such as growth and reproduction, decreasing the overall animal fitness [17]. Therefore, in light of ongoing ocean acidification and warming it is important to understand how fish respond to increasing habitat temperatures, their ability to adjust their thermal sensitivity and the role that high PCO_2 plays in thermal acclimation [2, 25].

The fish heart is highly aerobic and sensitive to temperature [26, 27]. Its capacity limits have been hypothesized to shape the warming-induced onset of sublethal thermal constraints in fishes [2, 28–31]. Recent studies have shown that high temperature leads to heart failure in various fish species like New Zealand triplefins and temperate and tropical wrasses [28, 29, 32, 33]. It was suggested that progressive impairment of several components of the mitochondrial function measured in permeabilised heart muscle fibres, such as oxidative phosphorylation (OXPHOS, respiratory state III), ATP production efficiency and the capacity of single complexes of the Electron Transport System (ETS) shape the temperature of heart failure (T_{HF}). High temperature changes the fluidity of mitochondrial membranes, which can entail increased proton leak through the inner membrane ([19] for review), resulting in decreased coupling ratios and causing decreased membrane potential [34, 35] and, as a consequence, inhibit the electrogenic transport of substrates, i. e. the transport of charged substrates like glutamate and malate that leads to the translocation of net charge across the membrane [36]. This indicates that mitochondrial metabolism is involved in functional constraints and thermal limitation of this tissue [28, 29, 32, 33]. Therefore, alterations in cardiac mitochondrial metabolism might lead to impaired cardiac energy turnover and, as a consequence, constraints in

cardiac performance and ultimately affect the fishes' thermal sensitivity.

Although an extensive literature has been produced on the effects of temperature on fish cellular metabolism and mitochondrial function (e.g. [8, 33, 37] and the literature therein), only few studies have addressed the effects of moderately elevated PCO_2 on them [30, 38–41]. Moreover, as ocean warming and ocean acidification caused by high PCO_2 are two sides of the same coin, they must be considered in combination in order to draw ecologically realistic conclusions [17, 42, 43].

Ocean acidification and warming trends are projected to exert particularly strong effects in the Arctic. As one of the consequences, temperate species may become established in Arctic habitats (by poleward migration), potentially displacing resident taxa [1, 4, 6]. For example, in the past decade the Northeast Arctic population of Atlantic cod (*Gadus morhua*, NEAC) has expanded its range into the Barents Sea [44, 45], on the North-east Greenland shelf [46] and in the coastal waters around Svalbard, which are inhabited by native Polar cod (*Boreogadus saida*), a key species in this region [1, 47].

Polar cod is a permanently cold adapted Arctic fish (thermal habitat around Svalbard ranging from –2 to +7 °C [48, 49]) while NEAC is a cold acclimated sub-Arctic population of temperate Atlantic cod expanding into the Arctic (habitat thermal range around Svalbard: 0–8 °C [1, 50]). Cold-acclimated and -adapted fish are known to have elevated mitochondrial densities. Among cold adapted species, extreme stenotherms such as high Antarctic fish, have high densities but low mitochondrial capacities and low proton leak in aerobic tissues [37, 51–53]. This may result in the low maintenance costs derived by proton leak and narrow thermal windows of these species and, as a consequence, cause high sensitivity to ocean warming [53, 54]. On the other hand, eurythermal cold adaptation ensures mitochondrial function over a wider range of temperatures at lower mitochondrial densities and maximized capacities [53, 55]. As a permanently cold adapted fish, Polar cod may therefore not be able to adjust mitochondrial capacities during warming to a similar extent as NEAC, which apparently has a higher capacity to adjust to higher temperatures by decreasing mitochondrial densities and capacities and thereby developing the metabolic plasticity necessary to acclimate to new conditions [56]. The differences in thermal response and, in particular, the ability to acclimate to higher temperatures will play a central role for their interaction in a changing ecosystem.

Hence, the aim of this study was to investigate the acclimation potential of Polar cod *Boreogadus saida* and Northeast Arctic cod (NEAC) *Gadus morhua* exposed to water temperatures and PCO_2 projected for the year 2100 in the Arctic i.e. 8 °C and 1170µatm PCO_2 (RCP 8.5 [57]). For a deeper understanding of the impact of ocean acidification and warming on the bioenergetics of the two species in relation to thermal tolerance, we further investigated mitochondrial function in the cardiac muscle of animals incubated for 4 months at four different temperatures (Polar cod: 0, 3, 6, 8 °C and Atlantic cod: 3, 8, 12, 16 °C), and two PCO_2 (400µatm and 1170µatm) in a cross factorial design. We used permeabilised cardiac muscle fibres to investigate a system resembling the living state as closely as possible [58–60], facilitating the extrapolation from measurements of cardiac mitochondrial capacities to their potential effects on the heart and eventually drawing conclusions on the effects of high temperature and high PCO_2 on the whole organism. Moreover, by analysing the mitochondrial function at the respective incubation temperature we could investigate the acclimation potential of the two species. We hypothesized that NEAC had higher thermal limits and a larger acclimation capacity than Polar cod and found accordingly that mitochondrial functions are constrained at lower temperatures in Polar cod than in NEAC. We discuss our results in light of the findings reported by Kunz et al. [61], who showed wider thermal windows for growth and standard metabolic rate (SMR) in NEAC than in Polar cod from the same acclimation experiment.

Methods
Animal collection
Juvenile Polar cod were collected by bottom trawl in combination with a fish lift [62] on January 17th, 2013 from the inner part of Kongsfjorden (Svalbard, 78° 97' N 12°51' E) at 120 m depth and a water temperature between 2 and 3 °C. They were kept at 3.3–3.8 °C in the facilities of the Tromsø Aquaculture Research Station, in Kårvik (Norway) until late April 2013 when they were transported to the aquarium facilities of the Alfred Wegener Institute (AWI) in Bremerhaven (Germany), where they were kept at 5 °C, 32 PSU and ambient PCO_2 until the start of the incubation.

Juvenile Northeast Arctic cod (NEAC) were caught in late August 2013 in several locations off Western Svalbard during RV Heincke cruise HE408 in Rijpfjorden (80° 15.42' N 22° 12.89' E), Hinlopenstretet (79° 30.19' N 18° 57.51' E), and Forlandsundet (78° 54.60' N 11° 3.66' E) at 0–40 m depth and water temperatures between 3.5 and 5.5 °C using a pelagic midwater trawl in combination with a fish lift [62]. The specimens were transported to the AWI facilities in Bremerhaven (Germany), where they were kept at 5 °C, 32 PSU and ambient PCO_2 until the start of the incubation.

Incubation
Polar cod incubation started in June 2013 and of NEAC in May 2014. After at least 4 weeks of acclimation to laboratory conditions (5 °C, 32 PSU and ambient PCO_2), individuals from both species were housed in single tanks and randomly allocated to the temperature and PCO_2 incubation set-up with a 12 h day/night rhythm. The respective PCO_2 conditions were pre-adjusted in a header tank containing ~200 l of seawater. Virtually CO_2-free pressurized air and pure CO_2 were mixed by means of mass flow controllers (4 and 6 channel MFC system, HTK, Hamburg, Germany) to achieve the desired PCO_2. Temperature was adjusted by 1 °C per day for each group starting from 5 °C. PCO_2 in the high PCO_2 group was adjusted within 1 day after the incubation temperature was reached. The animals were kept under incubation conditions for 4 months and fed *ad libitum* with commercial pellet feed (Amber Neptun, 5 mm, Skretting AS, Norway) every fourth day [61]. The sampling of Polar cod and NEAC took place after 4 days of fasting, due to sampling and experimental logistics three to six individuals of Polar cod and four to eight individuals of NEAC were sampled in one batch. Because of a failure in the power supply the group incubated at 3 °C and high PCO_2 died before the mitochondrial capacity could be investigated.

Average length and weight, as well as the number of the specimens per treatment at the time of sampling are given in Table 1.

CO_2 and carbonate chemistry
Temperature, salinity, DIC and pH (total scale) were measured once to twice a week in triplicates in order to monitor the seawater chemistry of the incubation. Temperature and salinity were measured with a WTW LF 197 multimeter (WTW, Weilheim, Germany). pH was measured with a pH meter (pH 3310, WTW, Weilheim, Germany) calibrated with thermally equilibrated NBS-buffers (2-point-calibration). The pH-values were then corrected to pH Total scale using pH-defined Tris-Buffer (Batch 4, Marine Physical Laboratory, University of California, San Diego, CA, USA).

DIC was measured by a Seal QuAAtro SFA Analyzer (800 TM, Seal Analytical, Mequon, United States of America). Calculations of the carbonate system were conducted using CO2sys [63], applying the K1, K2 constants after Mehrbach et al. [64], refitted after Dickson and Millero [65] and using KHSO4 dissociation constants after Dickson [66] assuming a pressure of 10 dbar.

Complete summaries of the seawater parameters and raw data for both species are available from the Open Access library PANGAEA [67, 68].

Table 1 Total length, body weight and number of fish (n) used for testing cardiac mitochondrial respiration in Polar cod (B. saida) and NEAC (G. morhua)

Acclimation	Species					
	B. saida			G. morhua		
	Total length (cm)	Body weight (g)	n	Total length (cm)	Body weight (g)	n
0 °C control	15.28 ± 0.37	22.88 ± 2.05	5	-	-	-
0 °C high	14.30 ± 0.64	19.22 ± 2.61	6	-	-	-
3 °C control	15.62 ± 0.98	27.16 ± 6.25	3	20.04 ± 0.92	60.84 ± 9.81	5
3 °C high	-	-	-	21.61 ± 0.46	78.19 ± 6.91	8
6 °C control	15.73 ± 0.21	25.21 ± 1.14	6	-	-	-
6 °C high	17.52 ± 0.61	32.17 ± 2.90	5	-	-	-
8 °C control	15.18 ± 0.72	20.52 ± 2.56	6	23.26 ± 1.75	99.04 ± 22.13	5
8 °C high	15.07 ± 0.47	18.76 ± 1.11	4	21.51 ± 0.82	80.51 ± 10.46	8
12 °C control	-	-	-	22.70 ± 0.80	98.70 ± 13.14	6
12 °C high	-	-	-	23.42 ± 0.72	100.75 ± 9.22	8
16 °C control	-	-	-	21.56 ± 0.69	81.48 ± 9.37	4
16 °C high	-	-	-	24.27 ± 1.91	133.13 ± 31.87	6

"control" and "high" indicate control (400µatm) and high (1170µatm) CO_2 concentrations. Values are given as means ± S.E.M

Preparation of permeabilised cardiac fibres

Fish were anaesthetized with 0.2 g l^{-1} tricaine methane sulphonate (MS222) and killed by a spinal cut behind the head plate. Hearts were rapidly excised and washed with ice-cold modified relaxing buffer BIOPS (2.77 mM CaK$_2$EGTA, 7.23 mM K$_2$EGTA, 5.77 mM Na$_2$ATP, 6.56 mM MgCl$_2$, 20 mM taurine, 15 mM Na$_2$-phospho-creatine, 20 mM imidazole, 0.5 mM dithiothreitol, 50 mM MES, 220 mM sucrose, pH 7.4, 380 mOsmol l^{-1}; modified after [69]). Hearts were then separated in fibres and placed in 2 ml ice-cold BIOPS containing 50 µg ml^{-1} saponin and gently shaken on ice for 20 min. Fibres were then washed three times for 10 min in 2 ml ice-cold modified mitochondrial respiration medium MIR05 (0.5 mM EGTA, 3 mM MgCl$_2$, 60 mM K-lactobionate, 20 mM taurine, 10 mM KH$_2$PO$_4$, 20 mM HEPES, 160 mM sucrose, 1 g l^{-1} bovine albumine serum, pH 7.4, 380 mOsmol l^{-1}) [29, 69].

Directly before experimentation, a subsample of about 10 mg fibres was blotted dry, weighed and introduced into the oxygraph sample chambers.

Mitochondrial respiration

Mitochondrial respiration was recorded using Oroboros Oxygraph-2 k™ respirometers (Oroboros Instruments, Innsbruck, Austria) and measured as weight-specific oxygen flux [pmol O$_2$ (mg fresh weight sec)$^{-1}$] calculated in real time using Oroboros DatLab Software 5.2.1.51 (Oroboros Instruments, Innsbruck, Austria).

All analyses were performed at the respective incubation temperatures, with cO_2 in a range from ~370 nmol ml^{-1} (100% air saturation) to 100 nmol ml^{-1} and PCO_2 at atmospheric levels.

A substrate-uncoupler-inhibitor titration (SUIT) protocol was used on the permeabilised cardiac fibres to investigate the partial contributions of the single components of the phosphorylation system [69]). NADH - Coenzyme Q oxidoreductase (Complex I, CI) and Succinate dehydrogenase (Complex II, CII) substrates (10 mM glutamate, 2 mM malate, 10 mM pyruvate and 10 mM succinate) were added. Saturating ADP (3 mM) was added to stimulate oxidative phosphorylation (OXPHOS). Cytochrome c (10 µM) was added to test the integrity of the outer membrane. Respiration state IV$^+$ was measured by addition of atractyloside (0.75 mM) or oligomycin (6 µM) (for Polar cod and NEAC respectively) and step-wise (1 µM each) titration of carbonyl cyanide p-(trifluoromethoxy) phenylhydrazone (FCCP) was used to uncouple mitochondria (ETS). Complex I, Complex II and Coenzyme Q – cytochrome c reductase (Complex III, CIII) were inhibited by the addition of rotenone (0.5 µM), malonate (5 mM) and antimycin a (2.5 µM), respectively. Lastly the activity of the Cytochrome c oxidase (Complex IV, CIV) was measured by the addition of the electron donor couple ascorbate (2 mM) and N,N,N^I,N^I-tetramethyl-p-phenylenediaminc (TMPD, 0.5 mM).

All chemicals were obtained from Sigma-Aldrich (Germany).

Data analysis

Mitochondrial respiration rates are expressed per mg fresh weight of cardiac fibres and the values are given as means ± S.E.M. OXPHOS coupling efficiency was calculated as [(OXPHOS-State IV$^+$) OXPHOS^{-1}] after Gnaiger [70].

Normal distribution of the data was assessed by Shapiro-Wilk test and homoscedasticity was evaluated by F-test or Bartlett test in case of two or more groups, respectively. Differences between PCO_2 treatments within the same temperature treatment were evaluated by Student's t-test (with Welch's correction in case of non-homoscedastic data). Differences across temperatures in the same PCO_2 treatment were evaluated with one-way ANOVA followed by Tukey's test for the comparison of means.

The level of statistical significance was set at $p < 0.05$ for all the statistical tests.

All statistical tests were performed using R 3.2.0 and the "stats" package [71].

Results

The maximal oxidative phosphorylation capacity (OXPHOS) of permeabilised heart fibres of both species is shown in Fig. 1. In Polar cod, the groups incubated under control PCO_2 showed significantly lower OXPHOS flux in the 0 °C acclimated fish than in all further incubation groups (3 °C, $p = 0.007$; 6 °C, $p = 0.007$; 8 °C, $p = 0.001$). Mitochondrial respiration was at a similar level in the groups incubated at 3, 6 and 8 °C ($p > 0.05$). High PCO_2 levels did not affect OXPHOS, with no differences between the OXPHOS of the groups incubated at the two PCO_2 levels within a temperature treatment ($p > 0.05$). The groups incubated under high PCO_2 displayed fluxes that were similar at 6 and 8 °C ($p > 0.05$) but significantly higher than in the 0 °C incubated group ($p = 0.04$, Fig. 1a).

Temperature had a significant effect on the OXPHOS of NEAC, with fluxes increasing with incubation temperature (control PCO_2: F = 4.74, $p = 0.02$; high

PCO_2: F = 3.78; $p = 0.02$, Fig. 1b). Moreover, the 16 °C/high PCO_2 incubated group showed a lower OXPHOS compared to the 16 °C/control PCO_2 group ($p = 0.03$). This resulted in a more evident plateauing of OXPHOS between 12 and 16 °C in the group incubated under high PCO_2. Comparing the two species, Polar cod had significantly higher OXPHOS capacities than NEAC at both 3 °C ($p = 0.01$, Fig. 1 blue box) and 8 °C (control PCO_2: $p = 0.04$; high PCO_2: $p = 0.04$, Fig. 1 red box).

In both species, state IV^+ was sensitive to temperature (Fig. 2): in Polar cod it remained unchanged in the groups incubated at 0, 3 and 6 °C ($p > 0.05$) but was significantly higher in animals incubated at 8 °C compared to the other incubation groups (6 to 8 °C/control PCO_2: $p = 0.01$; 6 to 8 °C/high PCO_2: $p = 0.04$) as shown in Fig. 2a. Quantifying State IV^+ as a percent fraction of OXPHOS, it was close to 20% and thus lowest in the 3 °C and 6 °C groups of $B. saida$, while at 0 and 8 °C the fraction of State IV^+ exceeded these values about two-fold as shown in Fig. 3.

In NEAC, State IV^+ increased along with incubation temperature (control PCO_2: F = 5.96; $p = 0.02$, high PCO_2: F = 12.43; $p < 0.001$) as depicted in Fig. 2b, however, State IV^+ increased under high PCO_2 at 8 °C compared to the group incubated under control PCO_2 at the same temperature ($p = 0.02$). Fractional values of State IV^+ in OXPHOS (Fig. 3) for the groups incubated under present levels of CO_2 revealed values close to 20% in the groups incubated to 3 and 8 °C and two-fold higher values after incubation to 12 and 16 °C. In the groups incubated under high PCO_2, State IV^+ of the group incubated at 8 °C showed values similar to the groups incubated to 12 and 16 °C (Fig. 3). In consequence, sensitivity to CO_2 varied with incubation temperature and

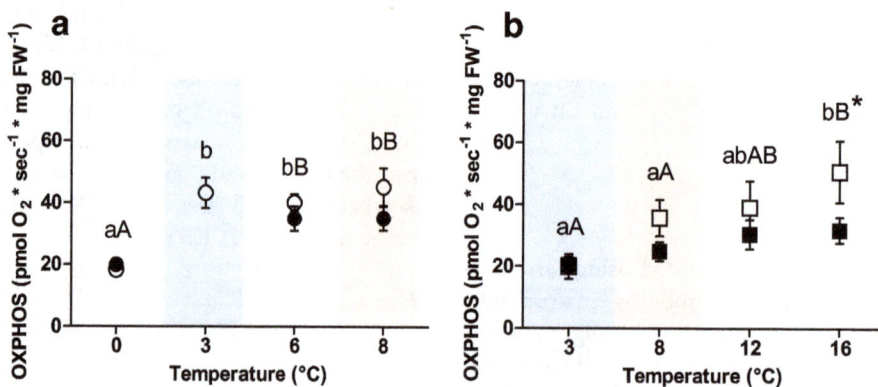

Fig. 1 Maximal oxidative phosphorylation capacity (OXPHOS) of permeabilised heart muscle fibres of (**a**) Polar cod (*B. saida*) and (**b**) NEAC (*G. morhua*). Different letters within panels indicate significant differences ($p < 0.05$) between temperature treatments; *lower case letters*: control PCO_2 (400µatm), *upper case letters*: high PCO_2 (1170µatm), * indicates significant differences ($p < 0.05$) between CO_2 groups at the same temperature. All values are reported as means ± S.E.M. (for *n* refer to Table 1). *Open symbols*: control PCO_2 (400µatm), *filled symbols*: high PCO_2 (1170µatm). *Circles*: Polar cod, *Squares*: NEAC. *Blue box*: cold shared incubation temperature (3 °C), *Red box*: warm shared incubation temperature (8 °C) between the two species

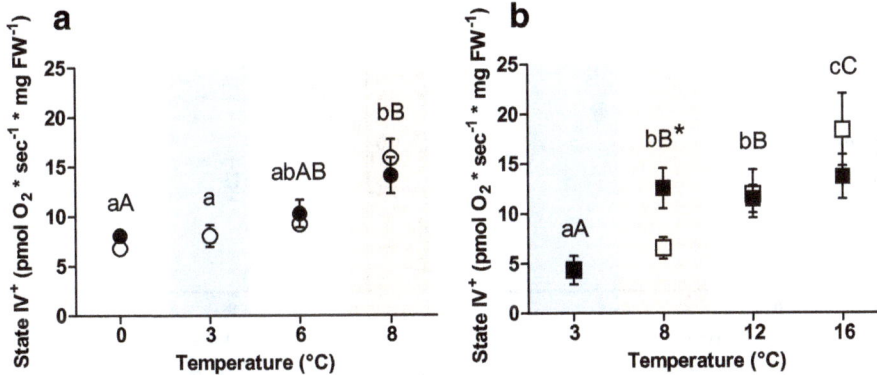

Fig. 2 State IV* of permeabilised heart muscle fibres of (**a**) Polar cod (*B. saida*) and (**b**) NEAC (*G. morhua*). Different letters within panels indicate significant differences ($p < 0.05$) between temperature treatments; *lower case letters*: control PCO_2 (400µatm), *upper case letters*: high PCO_2 (1170µatm), * indicates significant differences ($p < 0.05$) between CO_2 groups at the same temperature. All values are reported as means ± S.E.M. (for n refer to Table 1). *Open symbols*: control PCO_2 (400µatm), *filled symbols*: high PCO_2 (1170µatm). *Circles*: Polar cod, *Squares*: NEAC. *Blue box*: cold shared incubation temperature (3 °C), *Red box*: warm shared incubation temperature (8 °C) between the two species

was maximal but with opposite effects at 8 °C (stimulation of state IV+ above controls) and 16 °C (depression of OXPHOS below controls at 16 °C).

OXPHOS coupling efficiency in Polar cod under control PCO_2 was maximal in the group incubated to 3 °C (0.82 ± 0.02), and decreased at 8 °C to values comparable to the 0 °C group (control PCO_2: 0.61 ± 0.03, high PCO_2: 0.58 ± 0.05), mainly because of increased State IV+ at 8 °C (Fig. 2, 3 and 4). In NEAC (Fig. 4b), the OXPHOS coupling efficiency was maximal at 8 °C and control PCO_2 (0.81 ± 0.02) and minimal at 16 °C (0.64 ± 0.06). In the groups incubated under high PCO_2, the maximum of OXPHOS coupling efficiency fell to 3 °C (0.77 ± 0.03) and reached its minimum at 8 °C (0.46 ± 0.08) to rise again at 12 °C and 16 °C (0.58 ± 0.05 and 0.56 ± 0.05, respectively). However, these changes in OXPHOS coupling efficiency were not significant (control PCO_2: F = 5.27; $p = 0.82$, high

PCO_2: F = 9.7886, $p = 0.072$). At 8 °C, the OXPHOS coupling efficiency was significantly lower under high PCO_2 than in the control PCO_2 group ($p = 0.003$). Comparing the OXPHOS coupling efficiency between the two species, NEAC and Polar cod showed similar values in the 3 °C/control PCO_2 group (Fig. 4 blue box) and at 8 °C/high PCO_2 ($p > 0.05$, Fig. 4 red box), while the coupling efficiency was higher in NEAC incubated at 8 °C/control PCO_2 than in Polar cod incubated under the same conditions ($p < 0.001$, Fig. 4 red box).

The thermal sensitivity of Complex IV also differed between the two species (Fig. 5). In Polar cod, Complex IV capacity rose from 0 to 6 °C (control PCO_2: F = 67.29, $p < 0.001$) and decreased between 6 °C and 8 °C (control PCO_2: $p < 0.001$). This trajectory was only present as a non-significant trend in the groups incubated under high PCO_2 (F = 3.88, $p = 0.10$) because of the non-significant decrease of the mean

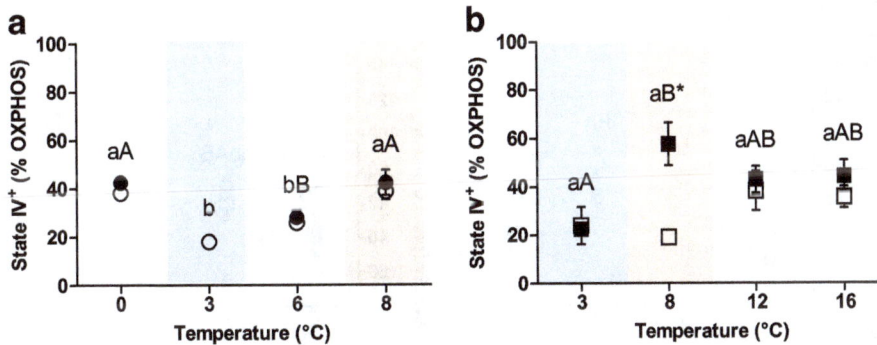

Fig. 3 Percentage of oxygen consumed by State IV+ in relation to OXPHOS in permeabilised heart muscle fibres of Polar cod (*B. saida*, panel **a**) and NEAC (*G. morhua*, panel **b**). Different letters within the panels indicate significant differences ($p < 0.05$) between temperature treatments; *lower case letters*: control PCO_2 (400µatm), *upper case letters*: high PCO_2 (1170µatm), * indicates significant differences ($p < 0.05$) between CO_2 groups at the same temperature. All values are reported as means ± S.E.M. (for n refer to Table 1). *Open symbols*: control PCO_2 (400µatm), *filled symbols*: high PCO_2 (1170µatm). *Circles*: Polar cod, *Squares*: NEAC. *Blue box*: cold shared incubation temperature (3 °C), *Red box*: warm shared incubation temperature (8 °C) between the two species

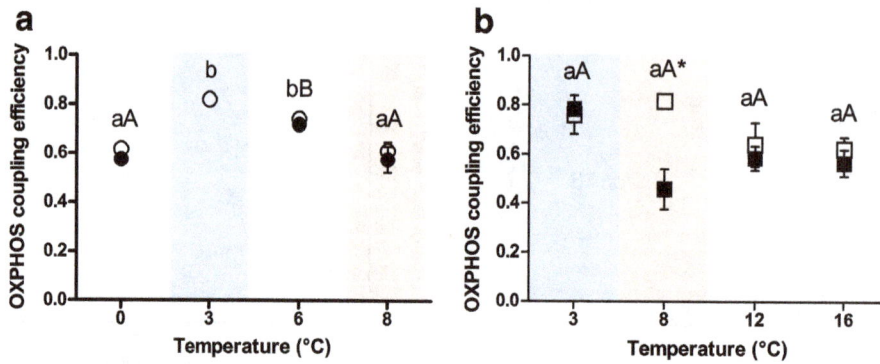

Fig. 4 OXPHOS coupling efficiency in permeabilised heart muscle fibres of (**a**) Polar cod (*B. saida*) and (**b**) NEAC (*G. morhua*). Different letters within panels indicate significant differences ($p <0.05$) between temperature treatments; *lower case letters*: control PCO_2 (400µatm), *upper case letters*: high PCO_2 (1170µatm), * indicates significant differences ($p <0.05$) between CO_2 groups at the same temperature. All values are reported as means ± S.E.M. (for *n* refer to Table 1). *Open symbols*: control PCO_2 (400µatm), *filled symbols*: high PCO_2 (1170µatm). *Circles*: Polar cod, *Squares*: NEAC. *Blue box*: cold shared incubation temperature (3 °C), *Red box*: warm shared incubation temperature (8 °C) between the two species

capacity of Complex IV at 6 °C/high PCO_2 compared to control PCO_2 at the same temperature ($p = 0.09$). In NEAC, Complex IV capacity increased with increasing temperatures in the groups incubated under control PCO_2 (F = 3.25, $p = 0.05$), but not in the groups incubated under high PCO_2 (F = 2.18, $p = 0.12$). At 16 °C, the capacity of NEAC Complex IV was lower under high PCO_2 ($p =0.099$) than under control PCO_2. Comparing the two species, the capacity of Complex IV was similar (non-significant differences) in all shared treatments (3 °C/control CO_2, 8 °C/control CO_2 and 8 °C/high CO_2: $p >0.05$, Fig. 5 blue and red boxes).

Discussion

Our study shows differences in mitochondrial metabolism between a cold-adapted Arctic and a cold-acclimated sub-Arctic fish from the same area, potentially leading to

differences in acclimation capacities to ocean acidification and warming.

Mitochondria from permeabilised heart fibres appeared to be affected mainly by the incubation temperature while high levels of CO_2 significantly affected mitochondrial respiration only in NEAC (*Gadus morhua*) and mainly at the highest investigated temperature (16 °C). NEAC OXPHOS and Complex IV capacities decreased under elevated CO_2 at high temperature, although the latter only as non-significant trend. This suggests that the noxious effects of high PCO_2 are stronger at the upper end of the thermal window and might affect the heat tolerance of NEAC [2, 17]. Furthermore, proton leak at 8 °C was higher in the group incubated under high PCO_2 than in the control PCO_2 group, indicating that overall mitochondrial efficiency might be affected through alterations of membrane characteristics. Elevated PCO_2 is reported to inhibit Citrate Synthase and

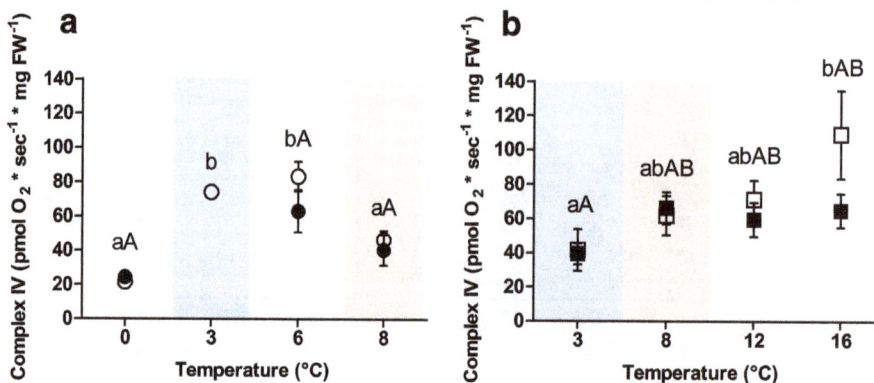

Fig. 5 Complex IV (Cytochrome c Oxidase) capacity. Panel **a**: permeabilised heart muscle fibres of Polar cod (*B. saida*). Panel **b**: permeabilised heart muscle fibres of NEAC (*G. morhua*). Different letters within the panels indicate significant differences ($p <0.05$) between temperature treatments; *lower case letters*: control PCO_2 (400µatm), *upper case letters*: high PCO_2 (1170µatm). All values are reported as means ± S.E.M. (for *n* refer to Table 1). *Open symbols*: control PCO_2 (400µatm), *filled symbols*: high PCO_2 (1170µatm). *Circles*: Polar cod, *Squares*: NEAC. *Blue box*: cold shared incubation temperature (3 °C), *Red box*: warm shared incubation temperature (8 °C) between the two species

Complex II in mammals and fish [40, 72, 73] with subsequent stimulation of the mitochondrial anaplerotic pathways to overcome this inhibition [40, 74]. The difference in sensitivity of the two species to elevated levels of CO_2 could be related to differences in preferential metabolic pathways, with Polar cod (*Boreogadus saida*) relying more than NEAC on anaplerotic pathways that feed directly into Complex I such as the oxidation of glutamate, pyruvate or palmitoyl carnitine [40, 73]. Further investigation, especially at the genetic level is needed. Furthermore, it is still unknown whether and to what extent elevated PCO_2 might alter the membrane characteristics and contribute to proton leak.

In Polar cod, OXPHOS of the groups incubated at 3-6-8 °C was higher at the respective incubation temperature than OXPHOS of the 0 °C treatments while the OXPHOS coupling efficiency was highest in the 3 °C group and lowest in the 0 and 8 °C groups. This indicates an optimum temperature for ATP production efficiency between 3 and 6 °C. At lower and higher temperatures, the increased proton leak in relation to OXPHOS created a less favourable ratio between ATP produced and oxygen consumed. These findings match those by Drost et al. [75], where heart rate of acutely warmed Polar cod increased until a first Arrhenius breakpoint at 3 °C. Heart rate still increased further but at a lower rate until 8 °C, passing a second break temperature. In our study, 8 °C corresponds to the highest rate of proton leak, and lowest Complex IV capacity, implying a direct participation of mitochondria in the thermal responses of the heart. The close similarity between the data from the acute study of Drost et al. [75], our 4-months incubation study and a study on behavioural thermal preference from Schurmann and Christiansen [76] indicates preferred temperatures of 3–6 °C within a thermal gradient from 0 to 8 °C for Polar cod, suggesting that Polar cod have only limited abilities to acclimate to higher temperatures.

In contrast, NEAC OXPHOS continued to increase with long-term incubation temperatures to even above those experienced within the natural habitat. This appears to occur without compromising OXPHOS coupling efficiency and reveals a higher acclimation potential than Polar cod, in line with the overall distribution area of Atlantic cod from temperate to (sub-) Arctic waters. This apparent plasticity is in line with the findings by Zittier et al. [77] in which NEAC specimens acclimated to 15 °C displayed critical temperatures (Tc, defined as the onset of the anaerobic metabolism, cf. Frederich & Pörtner [78]) about 10 °C higher than specimens kept at ambient temperature (4 °C). In Polar cod, the high proton leak at 8 °C is the main cause of reduced mitochondrial efficiency (OXPHOS coupling efficiency). This

increase in proton leak can be caused by loss of membrane integrity in response to changes in membrane fluidity [7, 79]. In a previous study, Martinez et al. [80] found increased proton permeability of the inner mitochondrial membrane of the Antarctic silverfish *Pleuragramma antarcticum* after warming. In addition, Strobel et al. [40] found that this may be due to an unchanged saturation index of the mitochondrial membrane, observed in liver of the Antarctic *Notothenia rossii* after warm acclimation. These findings suggest a limited ability of Antarctic stenothermal fish to acclimate to temperature changes. Similar patterns may constrain acclimation of cold-adapted Arctic fish. The decreased capacity of Complex IV at 8 °C in Polar cod implies that the interactions between the inner membrane and embedded enzymes may also be affected by high temperatures [80, 81]. In NEAC, proton leak was lower than in Polar cod and reached 40% of OXPHOS at 12 °C, while in Polar cod the same relative values were found at 8 °C under control PCO_2. A strong thermal response of proton leak may reflect high temperature sensitivity of the organism [52, 82–84], and thus a higher baseline proton leak combined with its steeper increase upon warming may point towards a stronger degree of cold adaptation in Polar cod.

The findings in this study contrast earlier results obtained in isolated mitochondrial suspensions where mitochondria remained fully functional beyond whole organism heat limits [82, 85]. The present findings suggest that mitochondria may display wider thermal limits in suspensions than when embedded in permeabilised fibres. Mitochondria in permeabilised fibres may still interact with other cellular organelles and are thus integrated into a more complex system than are isolated mitochondria. These considerations suggest that thermal tolerance is more constrained in permeabilized fibres than in isolated mitochondria. Such findings may thus be in line with the assumed narrowing of thermal windows once molecular and mitochondrial functions are integrated into larger units up to whole organism [86]. While the experiment was carried out at non-limiting PO_2 in the media (>100 nmol ml^{-1}) [87], diffusion gradients of oxygen and/or other substances within the permeabilised cardiac fibres may cause this hierarchy in thermal constraints. In a study on growth, mortality and standard metabolic rates (SMR) of the same Polar cod and NEAC as examined in this study, Kunz et al. [61] found higher SMR in Polar cod than in NEAC at the same incubation temperatures. This is mirrored in the mitochondrial respiration presented in this study, where OXPHOS capacity in Polar cod was larger than in NEAC at both 3 and 8 °C. In Polar cod, the SMR of the 3 and 6 °C groups were lower than in the groups incubated at 8 °C, which is mirrored in the pattern of cardiac State IV$^+$ respiration. At 8 °C the OXPHOS coupling efficiency (i.e. ATP production efficiency) decreased as State IV$^+$ increased and the capacity of Complex IV decreased. Maybe these findings indicate decreased cardiac mitochondrial efficiency that may limit cardiac function and promote heart failure, which is consistent with a drop in cardiac function [75] and the onset of heart failure in Polar cod at 8 °C. At this temperature, oxygen demand and mortality increased, and growth decreased in this species [61]. In fact, the estimated decrease in ATP production efficiency at 8 °C was paralleled by a reduced feed conversion efficiency and concomitant increase in SMR. This likely indicates a shift in energy allocation due to an impaired balance between energy production and demand, e.g. due to increased mitochondrial proton leak (see [88] for review). According to these findings, 8 °C is close to the long-term upper thermal tolerance limit for the Svalbard population of Polar cod, which is again in line with the observed increased mortality [61].

In NEAC, the parallel rise of whole organism SMR and cardiac fibre OXPHOS and the parallel decrease of OXPHOS and SMR at high PCO_2 compared to controls at 16 °C indicates that cardiac mitochondrial function is adjusted to the level of whole animal energy demand at different incubation temperatures and that the effects of high PCO_2 are greatest close to the upper thermal limit. Thermal constraints setting in at whole animal level may again relate to the thermal sensitivity of cardiac mitochondrial function [28, 29, 32, 33]. The fact that first performance limitations are observed in the 16 °C/high PCO_2 incubation may not be of direct relevance for the Svalbard stock of NEAC over the next century, but marks a potential southern distribution limit for the Barents Sea and Norwegian Sea.

Polar cod is a cold adapted species and the constraint on cardiac mitochondrial metabolism at 8 °C, concomitant with increased mortality indicates that the animal's thermal window matches its current habitat temperature range. In contrast, adult NEAC show the ability to broaden their thermal window beyond the present sub-Arctic habitat temperatures (see above). Because of the habitat temperature range of the two species is similarly wide but shifted to lower temperatures in Polar cod, combined with the high metabolic baseline cost (SMR) of Polar cod the two species may be classified as cold-adapted (Polar cod) or cold-acclimated (NEAC) eurytherms. NEAC appear to be much more plastic than Polar cod, thus, Polar cod may be more vulnerable to future ocean conditions than NEAC.

Conclusions

Future ocean acidification and warming may impair cardiac mitochondrial function of Polar cod (*Boreogadus saida*) and Northeast Arctic cod (NEAC, *Gadus morhua*) in somewhat different ways. In Polar cod, high temperature (8 °C) increases proton leak and thereby decreases ATP production efficiency, while high CO_2 levels did not have a significant effect. In NEAC, mitochondrial respiration remained functional at higher temperatures, but capacity was depressed by the combination of high temperature and high PCO_2. Furthermore, in NEAC, incubation temperature leads to variable mitochondrial response patterns under elevated PCO_2. The causes of the different responses to elevated PCO_2 in the heart of these two species remain to be identified, for example, the role of anaplerotic pathways and their regulation should be further investigated.

As a result of the degree of cold adaptation, Polar cod display high metabolic maintenance costs (indicating that it is cold-eurythermal) and low acclimation capacity, while NEAC is cold acclimated and benefits from a lower rate of metabolism and a higher plasticity to acclimate to increasing temperature. As a consequence, mitochondrial function of NEAC hearts may be less constrained by rising temperatures than Polar cod, indicating that NEAC could outperform and possibly replace Polar cod in the waters around Svalbard if ocean warming and acidification further increase towards the conditions predicted for the end of the century (8 °C and 1170µatm PCO_2). Since Polar cod has a key role in Arctic ecosystems [48], temperature driven changes in the distribution of this species can be an important component in the impacts of climate change on Arctic ocean ecosystems.

Acknowledgements

We thank Silvia Hardenberg, Nils Koschnick, Timo Hirse, Isabel Ketelsen and Heidrun Windisch for their support during the incubation and sampling procedures. We acknowledge the project Polarisation (Norwegian Research Council, 214184/F20) for providing Polar cod specimens and the crews of RV Heincke (HE 408) and RV Helmer Hanssen for the animal collection.

Funding

This project was funded by the German Federal Ministry of Education and Research (BMBF, FKZ 03F0655B) within the research program BIOACID phase II and by the PACES program of AWI.

Authors' contributions

EL, KK, MS, DS, HOP and FCM designed the study. EL, KK, MS carried out the animal incubations. EL performed the experiment on cardiac mitochondria and all data analyses and interpreted the results together with FCM. EL and FCM drafted the manuscript, KK MS DS and HOP contributed to writing the manuscript. All authors read and approved the final manuscript.

Competing interests

The authors declare that they have no competing interests.

Author details

[1]Alfred Wegener Institute, Helmholtz Centre for Polar and Marine Research, Integrative Ecophysiology, Am Handelshafen 12, D-27570 Bremerhaven, Germany. [2]University of Bremen, Fachbereich 2, NW 2/Leobener Strasse, D-28359 Bremen, Germany. [3]Alfred Wegener Institute, Helmholtz Centre for Polar and Marine Research, Bentho-Pelagic Processes, Am Alten Hafen 26, D-27568 Bremerhaven, Germany.

References

1. Renaud PE, Berge J, Varpe Ø, Lønne OJ, Nahrgang J, Ottesen C, Hallanger I. Is the poleward expansion by Atlantic cod and haddock threatening native polar cod, Boreogadus saida? Polar Biol. 2012;35:401–12.

2. Pörtner HO, Farrell AP. Physiology and climate change. Science. 2008;322:690–2.

3. Pörtner HO, Knust R. Climate change affects marine fishes through the oxygen limitation of thermal tolerance. Science. 2007;315:95–7.

4. Parmesan C. Ecological and evolutionary responses to recent climate change. Annu Rev Ecol Evol S. 2006;37:637–69.

5. Poloczanska ES, Brown CJ, Sydeman WJ, Kiessling W, Schoeman DS, Moore PJ, Bander K, Bruno JF, Buckley LB, Burrows MT, Duarte CM, Halpern BS, Holding J, Kappel CV, O'Connor MI, Pandolfi JM, Parmesan C, Schwing F, Thompson SA, Richardson AJ. Global imprint of climate change on marine life. Nat Clim Change. 2013. doi:10.1038/nclimate1958.

6. Fossheim M, Primicerio R, Johannesen E, Ingvaldsen RB, Aschan MM, Dolgov A. Recent warming leads to a rapid borealization of fish communities in the Arctic. Nat Clim Change. 2015. doi:10.1038/NCLIMATE2647.

7. Hochachka PW, Somero GN. Biochemical adaptation: mechanism and process in physiological evolution. Oxford: Oxford University Press; 2002.

8. Guderley H, St-Pierre J. Going with the flow or life in the fast lane: contrasting mitochondrial responses to thermal change. J Exp Biol. 2002;205:2237–49.

9. Caldeira K, Wickett ME. Anthropogenic carbon and ocean pH. Nature. 2003;425:365.

10. Di Santo V. Ocean acidification exacerbates the impact of global warming on embryonic little skate, Leucoraja erinacea (Mitchill). J Exp Mar Biol Ecol. 2014. doi:10.1016/j.jembe.2014.11.006.

11. Dixson DL, Abrego D, Hay ME. Chemically mediated behavior of recruiting corals and fishes: a tipping point that may limit reef recovery. Science. 2014;345:892–7.

12. Heuer RM, Grosell M. Physiological impacts of elevated carbon dioxide and ocean acidification on fish. Am J Physiol Regul Integr Compa Physiol. 2014;307:R1061–84.

13. Wittmann AC, Pörtner HO. Sensitivities of extant animal taxa to ocean acidification. Nat Clim Change. 2013;3:995–1001.

14. Przeslawski R, Byrne M, Mellin C. A review and meta-analysis of the effects of multiple abiotic stressors on marine embryos and larvae. Glob Chang Biol. 2015;21:2122–40.

15. Hoegh-Guldberg, Bruno JF. The Impact of Climate Change on the World's Marine Ecosystems. Science. 2010. doi: 10.1126/science.1189930.

16. Hutchins DA, Mulholland MR, Fu F-X. Nutrient cycles and marine microbes in a CO2-enriched ocean. Oceanography. 2009;22:128–45.

17. Pörtner HO. Oxygen-and capacity-limitation of thermal tolerance: a matrix for integrating climate-related stressor effects in marine ecosystems. J Exp Biol. 2010;213:881–93.

18. Metzger R, Sartoris FJ, Langenbuch M, Pörtner HO. Influence of elevated CO2 concentrations on thermal tolerance of the edible crab Cancer pagurus. J Therm Biol. 2007;32:144–51.

19. Pörtner HO. Integrating climate-related stressor effects on marine organisms: unifying principles linking molecule to ecosystem-level changes. Mar Ecol Prog Ser. 2012;470:273–90.

20. Flynn EE, Bjelde BE, Miller NA, Todgham AE. Ocean acidification exerts negative effects during warming conditions in a developing Antarctic fish. Conserv Physiol. 2015. doi:https://doi.org/10.1093/conphys/cov033.

21. Pimentel MS, Faleiro F, Dionisio G, Repolho T, Pousao-Ferreira P, Machado J, Rosa R. Defective skeletogenesis and oversized otoliths in fish early stages in a changing ocean. J Exp Biol. 2014;217:2062–70.

22. Hofmann GE, Todgham AE. Living in the now: physiological mechanisms to tolerate a rapidly changing environment. Annu Rev Physiol. 2010;72:127–45.

23. Abele D, Puntarulo S. Formation of reactive species and induction of antioxidant defence systems in polar and temperate marine invertebrates and fish. Comp Biochem Physiol Part A. 2004;138:405–15.

24. Mueller IA, Grim JM, Beers JM, Crockett EL, O'Brien KM. Inter-relationship between mitochondrial function and susceptibility to oxidative stress in red- and white-blooded Antarctic notothenioid fishes. J Exp Biol. 2011;214:3732–41.

25. Stillmann JH. Acclimation Capacity Underlies Susceptibility to Climate Change. Science. 2003. doi:10.1126/science.1083073.

26. Farrell AP. Cardiorespiratory performance during prolonged swimming tests with salmonids: a perspective on temperature effects and potential analytical pitfalls. Phil Trans R Soc B. 2007. doi:10.1098/rstb.2007.2111.

27. Ekström A, Brijs J, Clark TD, Gräns A, Jutfelt F, Sandblom E. Cardiac oxygen limitation during an acute thermal challenge in the European perch: Effects of chronic environmental warming and experimental hyperoxia. Am J Physiol Reg Int Comp Physiol. 2016. doi:10.1152/ajpregu.00530.2015.

28. Hilton Z, Clements KD, Hickey AJ. Temperature sensitivity of cardiac mitochondria in intertidal and subtidal triplefin fishes. J Comp Physiol B. 2010;180:979–90.

29. Iftikar FI, Hickey AJR. Do Mitochondria Limit Hot Fish Hearts? Understanding the Role of Mitochondrial Function with Heat Stress in Notolabrus celidotus. PLoS ONE. 2013. doi:10.1371/journal.pone.0064120.

30. Strobel A, Bennecke S, Leo E, Mintenbeck K, Pörtner HO, Mark FC. Metabolic shifts in the Antarctic fish Notothenia rossii in response to rising temperature and PCO2. Front Zool. 2012;9:28.

31. Rodnick KJ, Gamperl AK, Nash GW, Syme DA. 2014. Temperature and sex dependent effects on cardiac mitochondrial metabolism in Atlantic cod (Gadus morhua L.). J Therm Biol. 2014;44:110–8.

32. Iftikar FI, MacDonald JR, Baker DW, Renshaw GMC, Hickey AJR. Could thermal sensitivity of mitochondria determine species distributions in a changing climate? J Exp Biol. 2014. doi:10.1242/jeb.098798.

33. Iftikar FI, Morash AJ, Cook DG, Herbert NA, Hickey AJR. Temperature acclimation of mitochondrial function from the hearts of a temperate wrasse (Notolabrus celidotus). Comp Biochem Phys A. 2015. doi: 10.1016/j.cbpa.2015.01.017.

34. Brand MD. The efficiency and plasticity of mitochondrial energy transduction. Biochem Soc Trans. 2005. doi:10.1042/BST20050897.

35. Brand MD, Nicholls DG. Assessing mitochondrial dysfunction in cells. Biochem J. 2011. doi:10.1042/BJ20110162.

36. Vinogradov AD, Grivenn VG. The mitochondrial complex I: progress in understanding of catalytic properties. IUBMB Life. 2001;52:129–34.

37. Clarke A, Johnston NM. Scaling of metabolic rate with body mass and temperature in teleost fish. J Anim Ecol. 1999;68:893–905.

38. Michaelidis B, Spring A, Pörtner HO. Effects of long-term acclimation to environmental hypercapnia on extracellular acid–base status and metabolic capacity in Mediterranean fish Sparus aurata. Mar Biol. 2007;150:1417–29.

39. Deigweiher K, Koschnick N, Pörtner HO, Lucassen M. Acclimation of ion regulatory capacities in gills of marine fish under environmental hypercapnia. Am J Physiol Reg Int Comp Physiol. 2008. 10.1152/ajpregu.90403.2008.

40. Strobel A, Graeve M, Pörtner HO, Mark FC. Mitochondrial acclimation capacities to ocean warming and acidification are limited in the Antarctic nototheniid fish, Notothenia rossii and Lepidonotothen squamifrons. PLoS ONE. 2013. doi:10.1371/journal.pone.0068865.

41. Stapp LS, Kreiss CM, Pörtner HO, Lannig G. Differential impacts of elevated CO2 and acidosis on the energy budget of gill and liver cells from Atlantic cod, Gadus morhua. Comp Biochem Phys A. 2015. doi:10.1016/j.cbpa.2015.05.009.

42. Harvey BP, Gwynn-Jones D, Moore PJ. Meta-analysis reveals complex marine biological responses to the interactive effects of ocean acidification and warming. Ecol Evol. 2013. doi: 10.1002/ece3.516.

43. Stillman JH, Paganini AW. Biochemical adaptation to ocean acidification. J Exp Biol. 2015. doi:10.1242/jeb.115584.

44. Drinkwater K. Comparison of the response of Atlantic cod (Gadus morhua) in the high-latitude regions of the North Atlantic during the warm periods of the 1920s–1960s and the 1990s–2000s. Deep-Sea Res. 2009;Pt II 56:2087–96.

45. Kortsch S, Primicerio R, Fossheim M, Dolgov AV, Aschan M. Climate change alters the structure of arctic marine food webs due to poleward shifts of boreal generalists. Proc R Soc B. 2015;282:20151546.

46. Christiansen JS, Bonsdorff E, Byrkjedal I, Fevolden S-E, Karamushko OV, Lynghammar A, Mecklenburg CW, Møller PDR, Nielsen J, Nordström MC, Præbel K, Wienerroither RM. Novel biodiversity baselines outpace models of fish distribution in Arctic waters. Sci Nat. 2016. doi:10.1007/s00114-016-1332-9.

47. Olsen E, Aanes S, Mehl S, Holst JC, Aglen A, Gjøsæter H. Cod, haddock, saithe, herring, and capelin in the Barents Sea and adjacent waters: a review of the biological value of the area. ICES J Mar Sci. 2010;67:87–101.

48. Laurel BJ, Spencer M, Iseri P, Copeman LA. Temperature-dependent growth and behavior of juvenile Arctic cod (Boreogadus saida) and co-occurring North Pacific gadids. Polar Biol. 2015. doi: 10.1007/s00300-015-1761-5.

49. Mark FC, Rohardt G. Continuous thermosalinograph oceanography along HEINCKE cruise track HE451-1. Alfred Wegener Institute, Helmholtz Center

for Polar and Marine Research, Bremerhaven. 2016. doi:10.1594/
PANGAEA.863418.

50. Michalsen K, Johansen T, Subbey S, Beck A. Linking tagging technology
and molecular genetics to gain insight in the spatial dynamics of two
stocks of cod in Northeast Atlantic waters. ICES J Mar Sci. 2014. doi:10.
1093/icesjms/fsu083.

51. Johnston IA, Calvo J, Guderley YH. Latitudinal variation in the abundance
and oxidative capacities of muscle mitochondria in perciform fishes. J Exp
Biol. 1998;201:1–12.

52. Pörtner HO, van Dijk PLM, Hardewig I, Sommer A. Levels of metabolic cold
adaptation: tradeoffs in eurythermal and stenothermal ectotherms. In:
Davison W, Williams HC, editors. Antarctic ecosystems: models for wider
ecological understanding. Christchurch: Caxton; 2000. p. 109–22.

53. Pörtner HO, Bock C, Knust R, Lannig G, Lucassen M, Mark FC, Sartoris FJ. Cod
and climate in a latitudinal cline: physiological analyses of climate effects in
marine fishes. Climate Res. 2008;37:253–70.

54. Pörtner HO. Climate dependent evolution of Antarctic ectotherms: an
integrative analysis (EASIZ, SCAR). Deep-Sea Res Pt II. 2006;53:1071–104.

55. Blier PU, Lemieux H, Pichaud N. Holding our breath in our modern world:
will mitochondria keep the pace with climate changes? Can J Zool. 2014;92:
591–601.

56. Lucassen M, Koschnick N, Eckerle LG, Pörtner HO. Mitochondrial mechanisms
of cold adaptation in cod (Gadus morhua L.) populations from different
climatic zones. J Exp Biol. 2006;209:2462–71.

57. Pörtner HO, Karl DM, Boyd PW, Cheung WWL, Lluch-Cota SE, Nojiri Y, Schmidt
DN, Zavialov PO. Ocean systems. In: Field CB, Barros VR, Dokken DJ, Mach KJ,
Mastrandrea MD, Bilir TE, Chatterjee M, Ebi KL, Estrada YO, Genova RC, Girma B,
Kissel ES, Levy AN, MacCracken S, Mastrandrea PR, White LL, editors. Climate
Change 2014: Impacts, Adaptation, and Vulnerability. Part A: Global and
Sectoral Aspects. Contribution of Working Group II to the Fifth Assessment
Report of the Intergovernmental Panel on Climate Change. Cambridge and
New York: Cambridge University Press; 2014. p. 411–84.

58. Natori H. The property and contraction process of isolated myofibrils. Jikei
Med J. 1954;1:119–26.

59. Saida K, Nonomura Y. Characteristics of Ca2 + − and Mg2 + −induced
tension development in chemically skinned smooth muscle fibers. J Gen
Physiol. 1978;72(1):1–14.

60. Pesta D, Gnaiger E. High-resolution respirometry. OXPHOS protocols for human
cells and permeabilized fibres from small biopsies of human muscle. Methods
Mol Biol. 2012;810:25–58.

61. Kunz KL, Frickenhaus S, Hardenberg S, Johansen T, Leo E, Pörtner HO, Schmidt M,
Windisch HS, Knust R, Mark FC. New encounters in Arctic waters: a comparison of
metabolism and performance of polar cod (Boreogadus saida) and Atlantic cod
(Gadus morhua) under ocean acidification and warming. Polar Biol. 2016. doi: 10.
1007/s00300-016-1932-z.

62. Holst JC, McDonald A. FISH-LIFT: a device for sampling live fish with trawls.
Fish Res. 2000;48:87–91.

63. Lewis E, Wallace DWR. Program developed for CO2 system calculations.
Carbon Dioxide Information Analysis Center, Oak Ridge National Laboratory,
Oak Ridge. 1998; TN. ORNL/CDIAC-105.

64. Mehrbach C, Culberson CH, Hawley JE, Pytkowicz RN. Measurement of the
apparent dissociation constants of carbonic acid in seawater at atmospheric
pressure. Limnol Oceanogr. 1973;18:897–907.

65. Dickson AG, Millero FJ. A comparison of the equilibrium constants for
the dissociation of carbonic acid in seawater media. Deep Sea Res.
1987;34:1733–43.

66. Dickson AG. Standard potential of the reaction: AgCl (s) + ½ H2 (g) = Ag (s)
+ HCl (aq), and the standard acidity constant of the ion HSO4 in synthetic
sea water from 273.15 to 318.15 K. J Chem Thermodyn. 1990;22:113–27.

67. Schmidt M, Leo E, Kunz KL, Lucassen M Windisch HS, Storch D, Bock C,
Pörtner HO, Mark FC. (Table 1 + Table 2) Time series of seawater carbonate
chemistry calculated throughout incubation periods of Boreogadus saida
and Gadus morhua during exposure to different CO2 and temperature
conditions. 2016. doi:10.1594/PANGAEA.866369.

68. Leo E, Kunz K, Schmidt M, Storch D, Pörtner HO, Mark FC. Individual
mitochondrial functioning parameters from cardiac permeabilised fibers of
Polar cod (Boreogadus saida) and Atlantic cod (Gadus morhua) acclimated
to ocean acidification and warming. 2017. https://doi.pangaea.de/10.1594/
PANGAEA.873536. Accessed 4 Apr 2017.

69. Gnaiger E, Kuznetsov AV, Schneeberger S, Seiler R, Brandacher G, Steurer W,
Margreiter R. Mitochondria in the cold. In: Heldmaier M, Klingenspor M,
editors. Life in the Cold. Heidelberg: Springer; 2000. p. 431–42.

70. Gnaiger E, Boushel R, Søndergaard H, Munch-Andersen T, Damsgaard R,
Hagen C, Diéz-Sánchez C, Ara I, Wright-Paradis C, Schrauwen P,
Hesselink M, Calbet JAL, Christiansen M, Helge JW, Saltin B.
Mitochondrial coupling and capacity of oxidative phosphorylation in
skeletal muscle of Inuit and Caucasians in the arctic winter. Scand J
Med Sci Spor. 2015;25:126–34.

71. R Core Team. R: A language and environment for statistical computing. R
Foundation for Statistical Computing, Vienna, Austria. 2015; URL: http://
www.R-project.org/. Accessed 4 Apr 2017.

72. Simpson DP. Regulation of renal citrate metabolism by bicarbonate ion and
pH: observations in tissue slices and mitochondria. J Clin Invest. 1967;46:225.

73. Wanders RJA, Meijer AJ, Groen AK, Tager JM. Bicarbonate and the Pathway
of Glutamate Oxidation in Isolated Rat-Liver Mitochondria. Eur J Biochem.
1983. doi: 10.1111/J.1432-1033.1983.Tb07455.X.

74. Langenbuch M, Pörtner HO. Energy budget of hepatocytes from Antarctic
fish (Pachycara brachycephalum and Lepidonotothen kempi) as a function of
ambient CO2: pH-dependent limitations of cellular protein biosynthesis? J
Exp Biol. 2003. doi:10.1242/jeb.00620.

75. Drost HE, Carmack EC, Farrell AP. Upper thermal limits of cardiac function
for Arctic cod Boreogadus saida, a key food web fish species in the Arctic
Ocean. J Fish Biol. 2014;84:1781–92.

76. Schurmann H, Christiansen JS. Behavioral thermoregulation and swimming
activity of two Arctic teleosts (subfamily Gadinae)-the polar cod (Boreogadus
saida) and the navaga (Eleginus navaga). J Therm Biol. 1994;19:207–12.

77. Zittier Z. Einfluss der Temperatur auf das Wachstum von Fischen
unterschiedlicher Entwicklungsstadien. Diploma thesis, Universität Bremen.
2006. hdl:10013/epic.28057.

78. Frederich M, Pörtner HO. Oxygen limitation of thermal tolerance defined by
cardiac and ventilatory performance in spider crab, Maja squinado. Am J
Physiol Reg Int Comp Physiol. 2000;279:R1531–8.

79. Hazel JR. Thermal adaptation in biological membranes: is homeoviscous
adaptation an explanation? Annu Rev Physiol. 1995;57:19–42.

80. Martinez E, Menze MA, Torres JJ. Mitochondrial energetics of benthic and
pelagic Antarctic teleosts. Mar Biol. 2013. doi: 10.1007/s00227-013-2273-x.

81. O'Brien J, Dalhoff E, Somero GN. Thermal resistance of mitochondrial
respiration: hydrophobic interactions of membrane proteins may limit
mitochondrial thermal resistance. Physiol Zool. 1991;64:1509–26.

82. Hardewig I, Peck LS, Pörtner HO. Thermal sensitivity of mitochondrial
function in the Antarctic Notothenioid Lepidonotothen nudifrons. J Comp
Physiol B. 1999;169:597–604.

83. Salin K, Luquet E, Rey B, Roussel D, Voituron Y. Alteration of mitochondrial
efficiency affects oxidative balance, development and growth in frog (Rana
temporaria) tadpoles. J Exp Biol. 2012. doi:10.1242/jeb.062745.

84. Salin K, Auer SK, Rey B, Selman C, Metcalfe NB. Variation in the link between
oxygen consumption and ATP production, and its relevance for animal
performance. Proc. R. Soc. B. 2015. doi: 10.1098/rspb.2015.1028.

85. Weinstein RB, Somero GN. Effects of temperature on mitochondrial
function in the Antarctic fish Trematomus bernachii. J Comp Physiol.
1998;168B:190–6.

86. Pörtner HO. Climate variations and the physiological basis of temperature
dependent biogeography: systemic to molecular hierarchy of thermal
tolerance in animals. Comp Biochem Phys A. 2002;132:739–61.

87. Gnaiger E. Bioenergetics at low oxygen: dependence of respiration and
phosphorylation on oxygen and adenosine diphosphate supply. Resp
Physiol. 2001;128:277–97.

88. Pörtner HO, Mark FC, Bock C. Oxygen limited thermal tolerance in fish?
Answers obtained by nuclear magnetic resonance techniques. Respir
Physiol Neurobiol. 2004;141:243–60.

10

Reproductive axis gene regulation during photostimulation and photorefractoriness in Yangzhou goose ganders

Huanxi Zhu[1], Zhe Chen[1], Xibin Shao[2], Jianning Yu[1], Chuankun Wei[1], Zichun Dai[1] and Zhendan Shi[1*]

Abstract

Background: The Yangzhou goose is a long-day breeding bird that has been increasingly produced in China. Artificial lighting programs are used for controlling its reproductive activities. This study investigated the regulations of photostimulation and photorefractoriness that govern the onset and cessation of the breeding period.

Results: Increasing the daily photoperiod from 8 to 12 h rapidly stimulated testis development and increased plasma testosterone concentrations, with peak levels being reached 2 months after the photoperiod increase. Subsequently, testicular activities, testicular weight, spermatogenesis, and plasma testosterone concentrations declined steadily and reached to the nadir at 5 months after the 12-hour photoperiod. Throughout the experiment, plasma concentrations of triiodothyronine and thyroxine changed in reciprocal fashions to that of testosterone. The stimulation of reproductive activities caused by the increasing photoperiod was associated with increases in gonadotropin-releasing hormone (GnRH), but decreases in gonadotropin-inhibitory hormone (GnIH) and vasoactive intestinal peptide (VIP) gene messenger RNA (mRNA) levels in the hypothalamus. In the pituitary gland, the levels of follicle-stimulating hormone (FSH) and luteinizing hormone (LH) mRNA abruptly increased during the longer 12-hour photoperiod. The occurrence of photorefractoriness was associated with increased GnIH gene transcription by over 250-fold, together with increased VIP mRNA levels in the hypothalamus, and then prolactin and thyroid-stimulating hormone in the pituitary gland. FSH receptor, LH receptor, and StAR mRNA levels in the testis changed in ways paralleling those of testicular weight and testosterone concentrations.

Conclusions: The seasonal reproductive activities in Yangzhou geese were directly stimulated by a long photoperiod via upregulation of GnRH gene transcription, downregulation of GnIH, VIP gene transcription, and stimulation of gonadotrophin. Development of photorefractoriness was characterized by hyper-regulation of GnIH gene transcription in the hypothalamus, in addition of upregulation of VIP and TRH gene transcription, and that of their receptors, in the pituitary gland.

Keywords: Yangzhou goose ganders, Reproductive activities, Photoperiod, Gene mRNA expressions

Background

Most birds exhibit well-defined seasonal changes in gonadal development, body mass, molting, metabolism, and other physiological parameters [1]. Such seasonal physiological changes are a means of coping with the seasonal fluctuations of environmental factors, such as temperature and food availability, in order to improve survival capacity [2–7]. Seasonality of reproductive activity and that of

other physiologic processes in the majority of birds and mammals is presumed to arise from an interaction between endogenous circannual rhythms and a variety of environmental changes, the most important of which is the daily photoperiod [8].

The classic theory of photoperiodic regulation of seasonal reproductive activities in birds proposes that light signals are perceived by photoreceptors in the deep brain, and induce the secretion of thyroid-stimulating hormone (TSH) from the pars tuberalis, which then acts on ependymal cells to induce the thyroid hormone-activating enzyme type 2 deiodinase. This enzyme

* Correspondence: zdshi@jaas.ac.cn
[1]Laboratory of Animal Improvement and Reproduction, Institute of Animal Science, Jiangsu Academy of Agricultural Sciences, Nanjing 210014, China
Full list of author information is available at the end of the article

catalyzes the conversion of thyroxine (T4) to triiodothyronine (T3) [9], which initiates the nervous impulses that lead to the synthesis and release of gonadotropin-releasing hormone (GnRH) [10, 11]. GnRH is then transported by portal blood circulation to the anterior pituitary gland, where it stimulates the synthesis and release of the gonadotropins, luteinizing hormone (LH) and follicle-stimulating hormone (FSH) [12–14]. LH and FSH are responsible for the stimulation of gonad growth and development, as well as for the production of sex steroid hormones [15–18]. The photoperiodic regulation of the annual reproductive cycle requires two kinds of physiologic responses, namely photostimulation and photorefractoriness. The former leads to the activation of the reproductive system and brings animals into the breeding season, and the latter inhibits reproductive activities, terminating the breeding season [3, 14]. In addition to positive regulation by GnRH, the reproductive system is negatively regulated by gonadotropin-inhibitory hormone (GnIH), whose secretion is also subject to photoperiodic regulation [13, 19, 20].

GnIH, which can inhibit LH secretion [20, 21], is produced from the hypothalamic paraventricular nucleus (birds) [22–24] or the dorsomedial nucleus of the hypothalamus (mammals) [25], and is contained in nerve fibers extending to various brain regions, including the pre-optic area and the median eminence, where GnRH perikarya are located. Prolactin (PRL) and its releasing hormone, vasoactive intestinal peptide (VIP), which is secreted by the hypothalamus, are also involved in the regulation of seasonal reproductive activities [26]. For example, the secretion of VIP and PRL, as well as their gene expression, are highly responsive to increases in photoperiod [27–29], and peak PRL concentrations in blood coincide with the onset of gonadal regression [14, 30, 31]. Furthermore, the long photoperiod regulated testicular regression, which was severely retarded, while molting of feathers was blocked in European starlings actively immunized against VIP, which inhibited pituitary PRL secretion [32].

Moreover, it is well established that thyroid hormones play an important role in the regulation of seasonal breeding and other physiological activities such as growth and molting [33, 34]. In many seasonal breeding animals and birds, there is a reciprocal relationship between the plasma concentrations of thyroid hormones and testosterone [34–38]. Based on previous findings showing that thyroidectomy could prevent the development of photorefractoriness in both avian and mammalian species [34, 39, 40], recent investigations in Japanese quail (*Coturnix japonica*) showed that light-induced thyroid hormone synthesis in the mediobasal hypothalamus (MBH) is responsible for the regulation of the neuroendocrine axis involved in seasonal reproduction [41, 42].

The thirty some Chinese geese (*Anser cygnoides*) breeds throughout the country all exhibit strong seasonality in breeding activities, despite the fact that they have been domesticated for more than six thousand years [43]. Although these breeds are considered to be of the same genetic origin, as suggested by mitochondrial DNA typing [44], they exhibit divergent breeding seasonality, depending on the geographical location of their habitat. For example, northern breeds are typically long-day breeding birds, whereas the southern breeds are short-day types (Fig. 1) [45]. Such diverse breeding seasonality makes the Chinese geese breeds good model fowls for studying the photoperiodic regulation of seasonal reproduction. Yangzhou goose, a synthetic breed that is

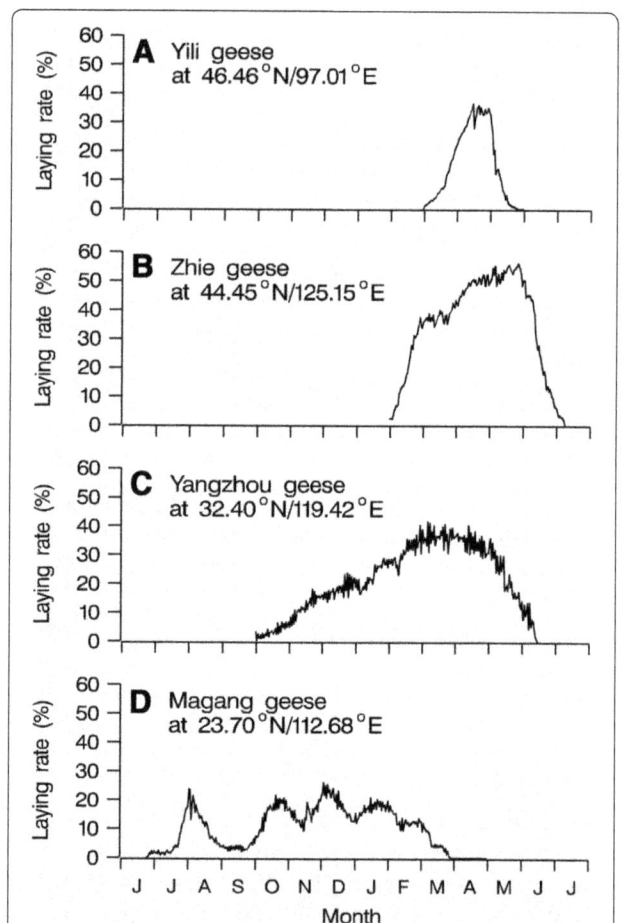

Fig. 1 Representative laying patterns of four domestic geese breeds of different reproductive seasonality. Type 1 (**a**) laying in the long-day seasons of March to May (from J.H. Li, unpublished laying record of a flock of 500 geese). Type 2 (**b**) lays entirely during the long-day seasons of spring and early summer (from W. Wu, unpublished laying record of a flock of 650 geese). Type 3 (**c**) starts laying during autumn, but peaks in spring and ends in early summer (from Z.D. Shi, unpublished laying record of a flock of 520 geese). Type 4 (**d**) starts laying in summer when the day length shortens, and ends in the following spring after spring equinox when daily photoperiod increases (from Z.D. Shi, unpublished laying record of a flock of 850 geese)

widely produced in China, is a long-day breeding fowl, whose egg laying activity starts in autumn when the daily photoperiod decreases, peaks between February and March when the photoperiod lengthens, and ends between May and June when the daily photoperiod further increases and becomes greater than 14 to 16 h [45]. The reproductive seasonality of the Yangzhou goose can be regulated by an artificial photoperiod, and this has been used to induce out-of-season breeding in order to improve economic efficiency of production. In spite of this, the mechanisms of development of photosensitivity and photorefractoriness, which coordinate the formation and length of the reproductive state and are therefore important for breeding and economic efficiency, remain unknown, as do the endocrine, neuroendocrine, and molecular mechanisms that underlie the transduction of photoperiodic signals during the profound fluctuations in reproductive activities. Previous studies and information are scarce, but a preliminary study [46] showed that a long photoperiod of 14 h (14 L:10D) rapidly induced photorefractoriness, and caused reproductive activity to cease not long after it was stimulated. Therefore, in our study, we used the equatorial 12L:12D photoperiod to induce and to maintain reproductive activity. However, to further test the effects of a short-to-long, and a-long-to-longer photoperiod, we designed different photoprotocols in two experimental groups.

This study was designed to unravel the mechanisms of photostimulation and photorefractoriness by investigating the changes in the concentration of plasma testosterone and thyroid hormones, the artificial photoprogram-induced waxing and waning of the testes, and the transcription patterns of relevant genes in the hypothalamus, pituitary gland, and testes.

Methods
Experimental design and animals
The experiments were carried out on Sunlake Swan Farm (119°58′E, 31°48′N), Henglin Township, Changzhou, Jiangsu Province, China. Two mechanically ventilated goose barns with an automatic lighting program were used to host the ganders used in the study. On March 1, 2015, a flock of Yangzhou goose ganders ($n = 210$) of 112 days of age and the same genetic origin were equally divided into two groups: A and B.

Group A ganders were initially exposed to a short photoperiod of 8 h (8L:16D) for 56 days, from March 1, 2015, to April 25, 2015 (Fig. 2a). Subsequently, the photoperiod was increased to 12 h (12L:12D) for 218 days, from April 26, 2015, to November 29, 2015 (Fig. 2a). In the final phase, the daily photoperiod was further increased to 16 h (16L:8D) for 38 days, from November 30, 2015, until the end of the experiments on January 6, 2016 (Fig. 2a). Group B ganders also

underwent a three-phase photo-treatment (Fig. 2b). The first phase was the same as in group A, namely an 8-hour short photoperiod (8L:16D), but lasted for 86 days instead, from March 1, 2015, to May 29, 2015 (Fig. 2a). The second phase consisted of a 12-hour daily photoperiod (12L:12D) and lasted for 184 days, from May 30, 2015, to November 29, 2015 (Fig. 2b). Finally, from November 30, 2015, to January 6, 2016, group B ganders were exposed to an 8-hour short photoperiod (8L:16D) for 38 days (Fig. 2b). The second phase was extended to 218 days in group A and 184 days in group B, thus far exceeding the time required for progressive development of photorefractoriness under a 12L:12D cycle, which is 130–150 days, and induced considerable reproductive regression in both groups to a similar extent. Provision of the 8-hour short photoperiod was achieved by confining the ganders in the mechanically ventilated barns from 4:00 to 8:00 am, and also from 16:00 to 20:00 pm. For provision of the 12-hour photoperiod, the ganders were confined in the barns from 4:00 to 6:00 am, and also from 18:00 to 20:00 pm. The 16-hour long photoperiod treatment consisted of the natural illumination during the daytime plus the supplementary illumination of 80 to 100 lux by fluorescent tubes at times after sunset and before sunrise.

The birds were fed ad libitum with mixed feed of 12.5% crude protein, supplemented with green grass whenever possible. Feed was given during daytime, but ganders always had free access to drinking water. During the course of the experiment, blood samples were collected via wing veins into heparinized syringes every 14 days. Plasma was separated from the blood within 3 h of sample collection by centrifugation at $2000 \times g$, and stored at -20 °C until the measurements of hormone concentrations were conducted.

Tissue collection, microscopy, and histological evaluation
On days 51, 67, 88, 131, 205, 250, and 313 of the experiment, eight ganders from each group were randomly selected and slaughtered for tissue sample collection. Immediately after collection, tissues from the hypothalamus, pituitary, and testes were snap frozen in liquid nitrogen, and stored at -80 °C until gene expression assays were performed.

A piece of testicular tissue ($\sim$0.125 cm^3) was sliced from the left testis of each gander and was immediately fixed in 10% buffered neutral formalin solution for 24 h, and subsequently used for histology studies using an automated tissue processor (Shandon Excelsior ES, P09046, ThermoScientific, Germany). The processing involved standard step-wise dehydration with alcohol of increasing concentrations (50, 60, 95%, and absolute alcohol, respectively), clearing in xylene (two changes), infiltration, and embedding in molten paraffin wax.

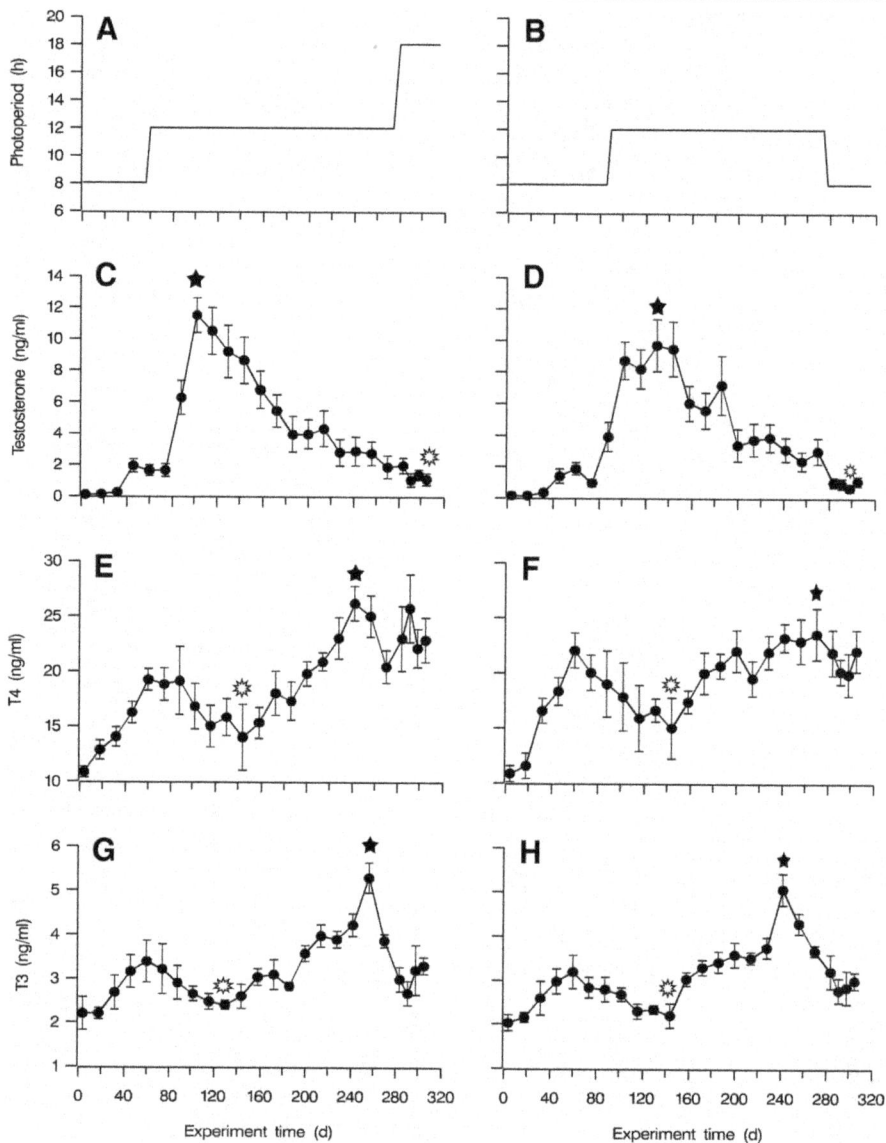

Fig. 2 Plasma concentrations of testosterone (**c** and **d**), triiodothyronine (**e** and **f**), and thyroxine (**g** and **h**) in ganders under two artificial photoperiods. *Vertical bars* indicate standard error of the mean. The artificial photoperiod treatment (**a**) consisted of a short photoperiod of 8 h (8L:16D) for 56 days from March 1, 2015 until April 25, 2015, followed by a long photoperiod of 12 h (12L:12D) for 218 days from April 26, 2015 until November 29, 2015, and finished with a long photoperiod of 16 h (16L:8D) from November 30, 2015 until the end of the experiment, on January 6, 2016. Another artificial photoperiod treatment (**b**) consisted of a short photoperiod of 8 h (8L:16D) that lasted for 86 days from March 1, 2015 until May 29, 2015. The second phase of a 12-hour daily photoperiod (12L:12D) lasted for 184 days from May 30, 2015 until November 29, 2015. The last photophase consisted of an 8-hour (8L:16D) short photoperiod for 38 days from November 30, 2015 until January 6, 2016. The long photoperiod treatment consisted of natural illumination during the daytime plus supplementary illumination (80–100 lux) by fluorescent tubes at times after sunset and before sunrise. *Star*: high hormone concentrations; *Lace box*: low hormone concentrations

Tissue sections (5 μm) were mounted on glass slides and stained with hematoxylin and eosin using an automated slide stainer (Shandon VaristainGermini ES, A78000013, ThermoScientific, Germany). Stained sections were individually examined under a bright field Olympus BX63 light microscope (OLYMPUSBX63, Olympus Corporation, Tokyo) at 10× and 40× magnification for changes in the diameter of the seminiferous tubule, and the numbers of spermatogonia, spermatocytes, and elongated spermatids.

Measurements of hormone concentrations

Plasma testosterone concentrations were determined by enzyme-linked immunosorbent assay using the Quantitative Diagnostic Kit for testosterone (North Institute of Biological Technology, Beijing, China). The assay sensitivity was 0.1 ng/mL, and the intra- and inter-assay variation coefficients were both below 15%. Serial dilutions of gander plasma samples resulted in an inhibition curve parallel to the standard curve. The r-values of the assay standard curves were greater than 0.99.

Blood concentrations of total T3 were also measured by enzyme-linked immunosorbent assay using the Quantitative Diagnostic Kit for 3,5,3'-triiodothyronine (North Institute of Biological Technology, Beijing, China). The sensitivity of the assay was 0.5 ng/mL, and the intra- and inter-assay variation coefficients were both below 10%. Similarly, blood concentrations of total T4 were measured by enzyme-linked immunosorbent assay using the Quantitative Diagnostic Kit for thyroxin (North Institute of Biological Technology, Beijing, China). The sensitivity of the assay was 0.5 ng/mL, and the intra- and inter-assay variation coefficients were both below 15%. The r-values of the assay standard curves were all greater than 0.99.

RNA isolation, complementary DNA synthesis, and quantitative real-time polymerase chain reaction

Total RNA from hypothalamus, pituitary, and testis tissues was extracted with Trizol using a commercial kit according to the manufacturer's instructions (RNAiso Plus, Code No. 9108, Takara, Japan). For RNA extraction, chloroform (0.2 mL) was added to the Trizol reagent (Code.no 9108, Takara, Japan), the mixture was vigorously shaken, and after 15 min, centrifuged at $12,000 \times g$ for 15 min at 4 °C. Isopropanol (0.5 mL) was then added to the separated upper aqueous phase and centrifuged at $12,000 \times g$ for 10 min at 4 °C. The pellet obtained was washed with 75% ethanol, air dried, and dissolved in 20 mL diethylpyrocarbonate-treated sterile water. RNA quality was assessed by electrophoresis on a 1.2% agarose gel, and the total RNA concentration was determined by measuring the absorbance at 260 nm using a spectrophotometer (NanoDrop 2000c Thermo-Scientific, Germany). A total of 5 μg RNA was reversely transcribed using the Takara PrimerScriptTM RT reagent kit (Perfect Real Time) (RR037, Takara, Japan) according to the manufacturer's instructions.

Quantitative real-time polymerase chain reaction analysis was used to measure the levels of messenger RNA (mRNA) of various genes in samples from the hypothalamus, pituitary, and testes from ganders in each photoperiod phase (Fig. 2a and b). The species source and function of the genes tested are shown in Table 1. Gene-specific primers were designed using the Primer 3.0 software (www.ncbi.nlm.nih.gov/tools/primer-blast/) based on the BLAST, Ensemble, and GenBank databases (Table 2). Expression of β-actin mRNA was used as a reference gene. All polymerase chain reactions (total volume of 20 μL) consisted of 10 mL SYBR Premix Ex

TaqII (Takara, Japan), 1 mL complementary DNA, 10 pmole of each forward and reverse primers (Table 2), and 7 mL ultrapure water. The thermal cycling profile used was 95 °C for 30 s, 40 cycles of 94 °C for 5 s, and 60 °C for 30 s. Fluorescence yields obtained from three replicate reactions of each complementary DNA sample were analyzed using the Mastercycler ep realplex (Eppendorf, Germany); furthermore, eight biological replicates were used to ensure the validity and accuracy of the experimental results. The relative expression levels of different genes in the tissues were calculated according to the $2^{-\Delta\Delta CT}$ method [47].

Statistical analysis

The results were mean values ± standard error of 12 (Fig. 2) and 8 (Figs. 3, 4, 5 and 6) replicate samples in each treatment groups. Differences of plasma concentrations of testosterone, T3, and T4 between photo-programs were analyzed with two-way analyses of variance (two-way ANOVA). Gene expression levels were analyzed with one-way analyses of variance (one-way ANOVA) with main treatments of the two photo-programs and the serial sampling of the experiment. Differences of each effect means were compared by the mean ± standard error of mean and considered significant at $P < 0.05$. Statistical analyses were performed using IBM SPSS software (ver. 11.0; IBM SPSS, Armonk, NY, USA).

Results

Concentrations of plasma testosterone

In the first half of the 56-day 8-hour short photoperiod phase, plasma testosterone concentrations were low, below 0.5 ng/mL, in ganders from both groups, A and B (Fig. 2c–d). With the continuation of the 8-hour photoperiod, the concentrations were slightly elevated to approximately 2 ng/mL, and at the end of the 8-hour photoperiod phase, on day 86 of the experiment, to 4 ng/mL in group B (Fig. 2d). At the same time, in group A ganders, which were already exposed to a 12-hour photoperiod since one month, plasma testosterone concentrations had already increased to significantly higher levels of approximately 6 ng/mL ($P = 0.024$, $F = 10.743$) (Fig. 2c). Subsequently, as both gander groups continued to be exposed to the second 12-hour photoperiod phase, plasma testosterone concentrations continued to increase rapidly, and they reached peak levels (8–11 ng/mL) from day 100 to day 140 of the experiment. Toward the end of the 12-hour photoperiod phase or on day 274 of the experiment, testosterone concentrations in both groups progressively decreased to lower levels, approximately 4 ng/mL. Thereafter, in the third photoperiod phase (days 275–313 of the experiment) plasma testosterone concentrations further decreased

Table 1 Species source and functions in literature of the genes tested in this study

Gene name	Species	Tissue distribution	Gene function
GnRH-I	*Anser cygnoides*	Hypothalamus	GnRH regulates the secretion of the gonadotropins LH and FSH [66].
GnIH	*Anser cygnoides*	Hypothalamus	GnIH inhibits LH secretion and reduces testis weigh [20].
VIP	*Gallus gallus*	Hypothalamus	VIP in the brain acts as a neuroendocrine factor and regulates PRL secretion [51].
TRH	*Gallus gallus*	Hypothalamus	TRH regulates thyroid stimulating hormone secretion [67].
GnRH Receptor	*Gallus gallus*	Pituitary	GnRH is a hypothalamic decapeptide that centrally controls reproduction by binding to GnRH receptors on pituitary gonadotropes and stimulating the secretion of LH and FSH [68].
GnIH Receptor	*Anser cygnoides*	Pituitary	GnIH acts directly on the pituitary via the GnIH receptor and inhibits gonadotropin release [69].
VIP Receptor-I	*Gallus gallus*	Pituitary	VIP is a hypothalamic polypeptide that controls reproduction by binding to VIP receptors on pituitary gonadotropes and stimulating PRL secretion [70].
TRH Receptor	*Gallus gallus*	Pituitary	TRH acts directly on the pituitary via the TRH receptor and controls thyroid-stimulating hormone (TSH) secretion e [71].
FSH beta	*Anser cygnoides*	Pituitary	FSH stimulates gonadal growth and estrogen secretion by Sertoli cells.
LH beta	*Anser cygnoides*	Pituitary	LH controls estrogen and androgen production by mature ovarian follicles, and regulates androgen production by Leydig cells [72].
PRL	*Anser cygnoides*	Pituitary	PRL inhibits gene expression of steroidogenic enzymes and reduces testis weight [73, 74].
TSH beta	*Anser cygnoides*	Pituitary	TSH is a glycoprotein released from the adenohypophysis that activates iodine uptake, thyroid hormone synthesis, and the release of thyroid hormones from the thyroid gland [75].
LH Receptor	*Anser cygnoides*	Testis	LH receptor is one of the three glycoprotein hormone receptors that is necessary for critical reproductive processes, including gonadal steroidogenesis, oocyte maturation and ovulation, and male sex differentiation [76].
FSH Receptor	*Anser cygnoides*	Testis	FSHR is a transmembrane receptor that interacts with FSH, and its activation is necessary for the hormonal functions of FSH [77].
3-beta HSD	*Anser cygnoides*	Testis	3-beta HSD catalyzes an obligatory step in the biosynthesis of all classes of hormonal steroids, namely, the oxidation/isomerization of 3-beta-hydroxy-5-ene steroids into the corresponding 3-keto-4-ene steroids in gonadal as well as in peripheral tissue [78].
StAR	*Anser cygnoides*	Testis	StAR plays a critical role in steroid hormone synthesis, and it is thought to increase the delivery of cholesterol to the inner mitochondrial membrane where P450scc resides [79].

Abbreviations: *GnRH* gonadotropin-releasing hormone, *GnIH* gonadotropin-inhibitory hormone, *VIP* vasoactive intestinal peptide, *TRH* thyrotropin releasing hormone, *FSH* follicle-stimulating hormone, *LH* luteinizing hormone, *PRL* prolactin, *TSH* thyroid-stimulating hormone, *3-beta HSD* three beta-hydroxysteroid dehydrogenase, *StAR* steroidogenic acute regulatory protein

to levels below 2 ng/mL in both groups (Fig. 2c–d), despite their divergent treatments.

Concentrations of plasma T4 and T3

The general pattern of the concentrations of plasma T4 and T3, and that of their relative ratio (T4/T3), was reciprocal to that of testosterone concentrations. For example, in the first short 8-hour photoperiod phase, both plasma T4 and T3 concentrations increased from their initial low levels and peaked on day 60 of the experiment. Similarly, the values of the T4/T3 ratio (5.86 and 7.07 for groups A and B respectively) were significantly higher than those at the beginning of the experiment ($P = 0.031$, $F = 20.121$ and $P = 0.047$, $F = 25.339$, for groups A and B respectively). The concentrations of both hormones started then to decrease, reaching a low plateau value at around day 140 of the experiment (2–3 months into the 12-hour photoperiod phase), but they subsequently increased, reaching peak levels around days 240 to 260 of the experiment, or almost the end of the 12-hour photoperiod phase, especially for T3. Until day 270, the values of the T4/T3 ratio were maintained at relatively high levels: 9.5611 and 7.27 for groups A and B, respectively. Then they started to decrease again, especially for T3, but displayed a small increase during the final phase of photoperiod change, around days 297 to 313 of the experiment (Fig. 2f). Throughout the experiment, the changing patterns of T4 and T3 concentrations were highly similar in both groups A and B.

Table 2 Primers used in the real-time quantitative PCR assay of gene transcription

Gene name	Accession number	Primer sequences (5'-3')	Annealing temperature (℃)	PCR product (bp)
β-actin	L08165.1	upstream: TGACGCAGATCATGTTTGAGA	60	159
		downstream: GCAGAGCGTAGCCCTCATAG		
GnRH-I	EF495207.1	upstream: CTGGGACCCTTGCTGTTTTG	60	232
		downstream: AGGGGACTTCCAACCATCAC		
GnIH	KC514473.1	upstream: ATCTACCTAGGCATGCTCCAA	58	115
		downstream: ACAGGCAGTGACTTCCCAAAT		
VIP	DQ023159	upstream: ACCAGTGTCTACAGCCATCTTTTG	58	204
		downstream: AGGTGGCTCAGCAGTTCATCTACA		
TRH	NM_001030383.2	upstream: GCAAGAGGGGCTGGAATGAT	58	133
		downstream: ATGGCAGACTGCTGAAGGTC		
GnRH Receptor	KJ659046.1	upstream: TCTGCTGGACCCCCTACTAC	60	127
		downstream: TCCAGGCAGGCATTGAAGAG		
GnIH Receptor	KF696709.1	upstream: GTCGTCATGTACACCCGCAT	56	103
		downstream: TCTTGCGAGACACCTTCCTC		
VIP Receptor-I	NM_001097523.1	upstream: TACTGCGTCATGGCCAACTT	58	153
		downstream: TGTCCAAGCGGTGATGAACA		
TRH Receptor	NM_204930.1	upstream: CTATGGCTATGTGGGGTGCC	60	189
		downstream: ACTGAGGCGAAAGACCAGAC		
FSH beta	EU252532	upstream: GTGGTGCTCAGGATACTGCTTCA	60	209
		downstream: GTGCAGTTCAGTGCTATCAGTGTCA		
LH beta	DQ023159	upstream: GACCCGGGAACCGGTGTA	58	90
		downstream: AGCAGCCACCGCTCGTAG		
PRL	DQ023160	upstream: TGCTCAGGGTCGGGGTTTCA	56	218
		downstream: GCTTGGAGTCCTCATCGGCAAGTT		
TSH beta	FJ797681.1	upstream: CTCTGTCCCAAAACGTGTGC	56	121
		downstream: CCACACTTGCAGCTTATGGC		
LH Receptor	XM_013192443.1	upstream: CGGATACACAACGATGCCCT	60	74
		downstream: GACTCCAGTGCCGTTGAAGA		
FSH Receptor	KC477215.1	upstream: CCTAGCCATTGCTGTGCATTT	60	106
		downstream: TGCCAGGTTGCTCATCAAGG		
3-beta HSD	KC310447.1	upstream: TGTGACGTTCCTGTACCGTG		153
		downstream: TTACAACGGGTACACGCCTC		
StAR	KF958133.1	upstream: GGAGCAGATGGGAGACTGGA	56	177
		downstream: CGCCTTCTCGTGGGTGAT		

Body weight, testis morphometry, and histology

In both gander groups, body weight exhibited a two-phase variation (Fig. 3a). Before day 131 of the experiment, the weight of live ganders was maintained around 3.95 kg, but toward the end of the experiment on day 313, it increased to significantly higher values, in the range of 4.73–4.93 kg ($P = 0.005$, $F = 16.215$), for both gander groups (Fig. 3a).

The weight of the left testis in both groups was below 2 g on day 51, during the 8-hour photoperiod (Fig. 3b). For the ganders of group B, it remained the same until day 67 of the experiment. In the ganders of group A however, the increase in the photoperiod from 8 to 12 h for just 12 days promptly stimulated testicular growth, and the weight of the left testis increased to approximately 6 g (Fig. 3b), and continued to increase until the day 88 of the experiment, when it reached a peak value significantly higher than that observed in the ganders of group B ($P = 0.000$, $F = 17.952$). In the latter group, the peak weight of 11 g was reached only on day 131 of the experiment (Fig. 3b). Thereafter, testicular weight decreased in both groups. However, at the end of the experiment (day 313) the testicular weight of ganders in

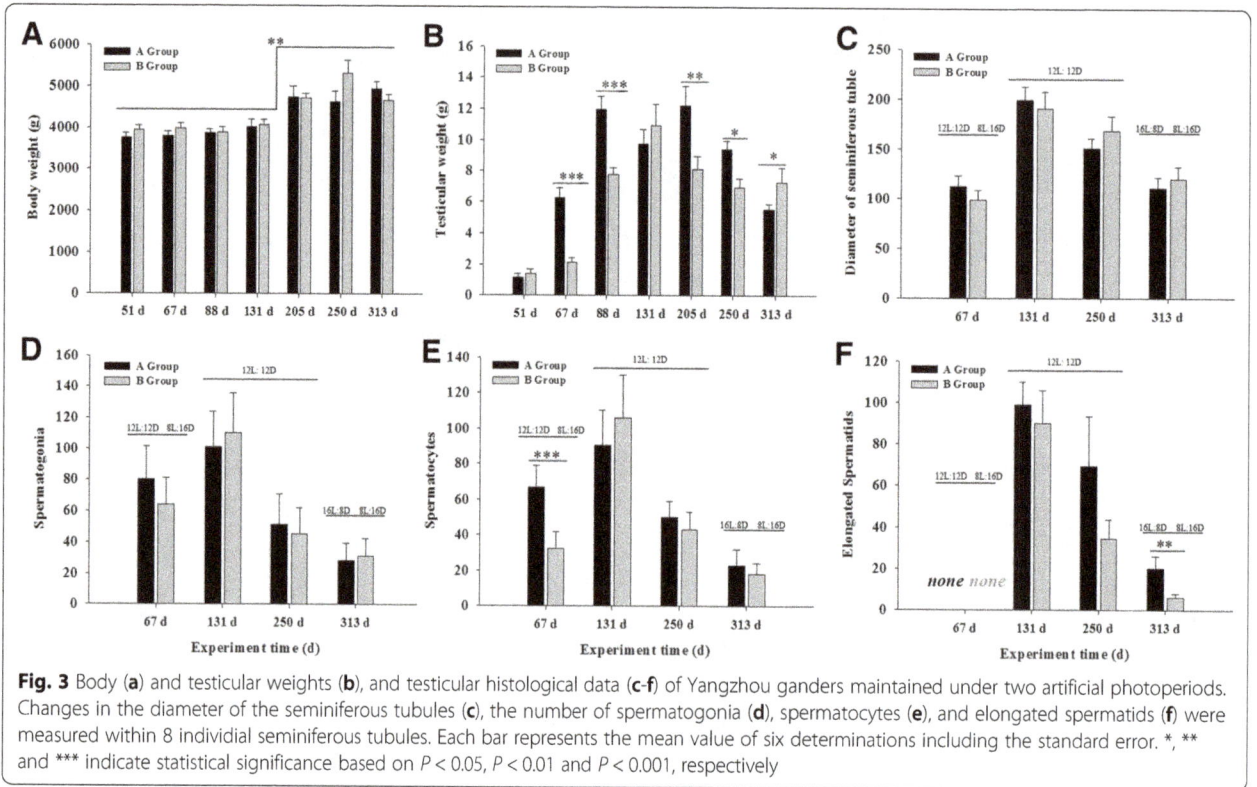

Fig. 3 Body (**a**) and testicular weights (**b**), and testicular histological data (**c-f**) of Yangzhou ganders maintained under two artificial photoperiods. Changes in the diameter of the seminiferous tubules (**c**), the number of spermatogonia (**d**), spermatocytes (**e**), and elongated spermatids (**f**) were measured within 8 individial seminiferous tubules. Each bar represents the mean value of six determinations including the standard error. *, ** and *** indicate statistical significance based on $P < 0.05$, $P < 0.01$ and $P < 0.001$, respectively

group A was significantly lower than that of ganders in group B, when the two groups were exposed to photoperiods of 16 and 8 h, respectively (Fig. 3b).

Microscopic observation of testis histological sections (Fig. 4) revealed that for both gander groups, the seminiferous tubule diameter (Fig. 3c) and the number of spermatogonia (Fig. 3d), spermatocytes (Fig. 3e), and elongated spermatids (Fig. 3f) changed in a similar fashion throughout the experiment: they increased from low values on day 67, to peak values on day 131, and decreased thereafter to low values again until the end of experiment (Fig. 4). It should be noted that on day 67, the spermatocyte number (Fig. 3e) was significantly higher in group A, which was already exposed to a

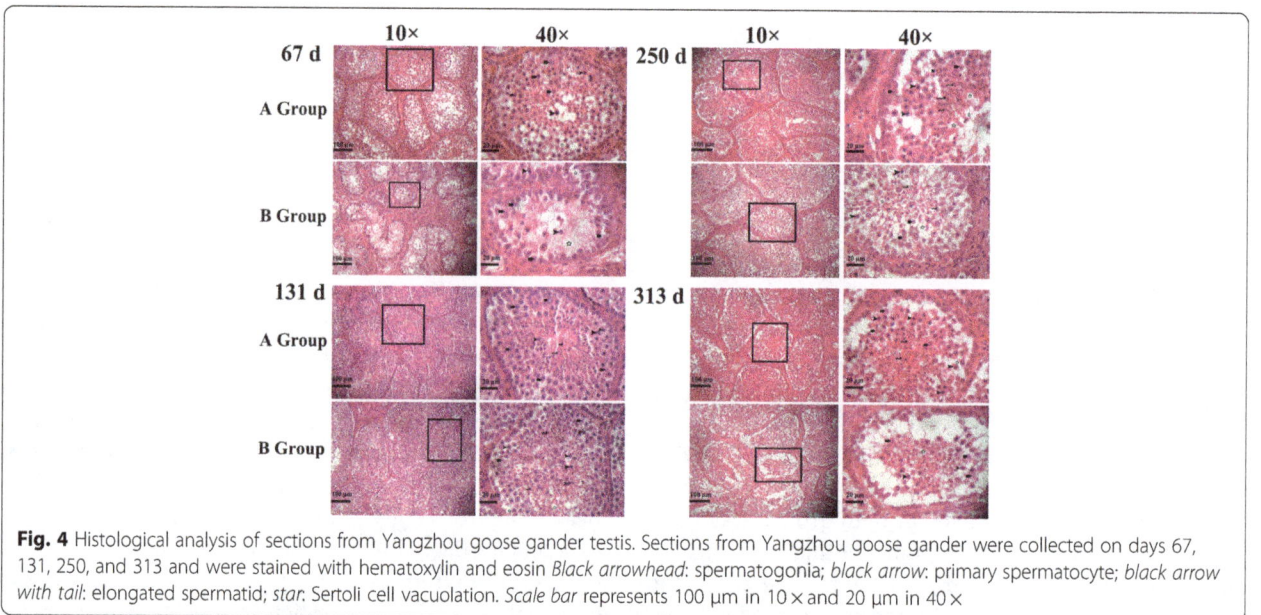

Fig. 4 Histological analysis of sections from Yangzhou goose gander testis. Sections from Yangzhou goose gander were collected on days 67, 131, 250, and 313 and were stained with hematoxylin and eosin *Black arrowhead*: spermatogonia; *black arrow*: primary spermatocyte; *black arrow with tail*: elongated spermatid; *star*: Sertoli cell vacuolation. *Scale bar* represents 100 µm in 10× and 20 µm in 40×

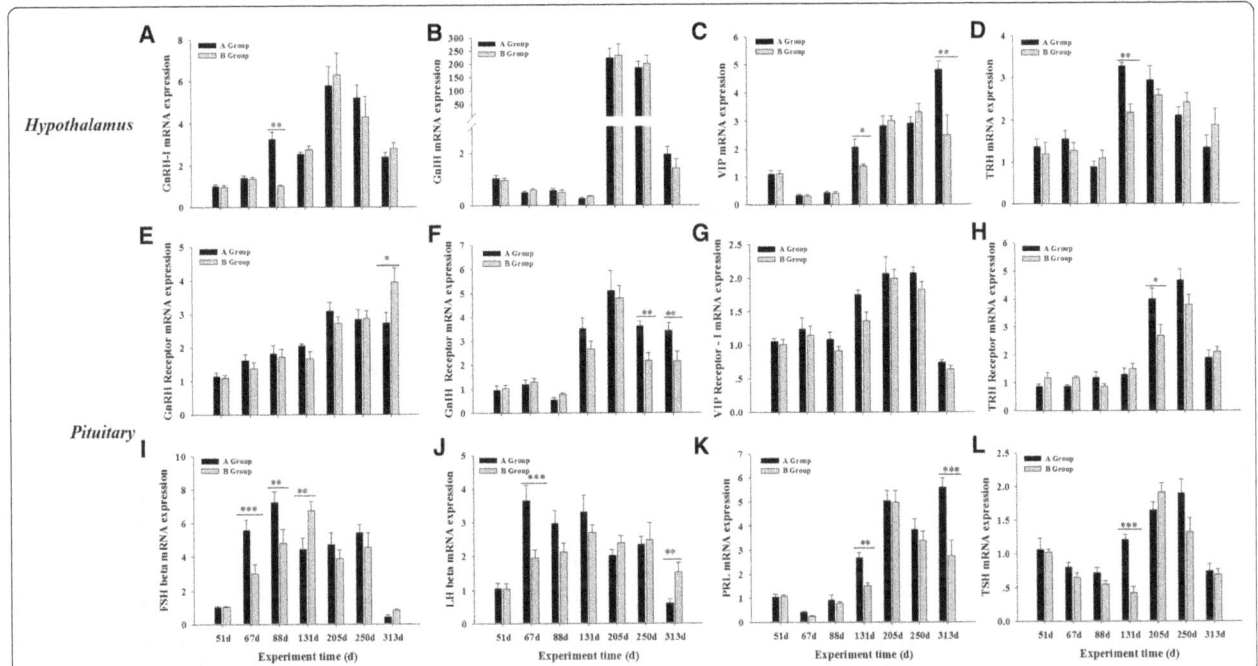

Fig. 5 Hypothalamic (**a-d**) and pituitary (**e-l**) mRNA levels in Yangzhou goose ganders under two artificial photoperiods. Graphs **a** to **d** represent mRNA expression levels of hypothalamic releasing hormones GnRH-I, GnIH, VIP and TRH, respectively, while **e** to **h** their corresponding receptors in the pituitary gland, and **i** to **l** the FSH, LH, PRL and TSH pituitary hormones or beta subunits wherever applicable. Each value represents the average of data from eight independent culture experiments. Data are shown as mean values ± standard error of the mean. *, ** and *** indicate statistical significance based on $P < 0.05$, $P < 0.01$ and $P < 0.001$, respectively, between the treatments

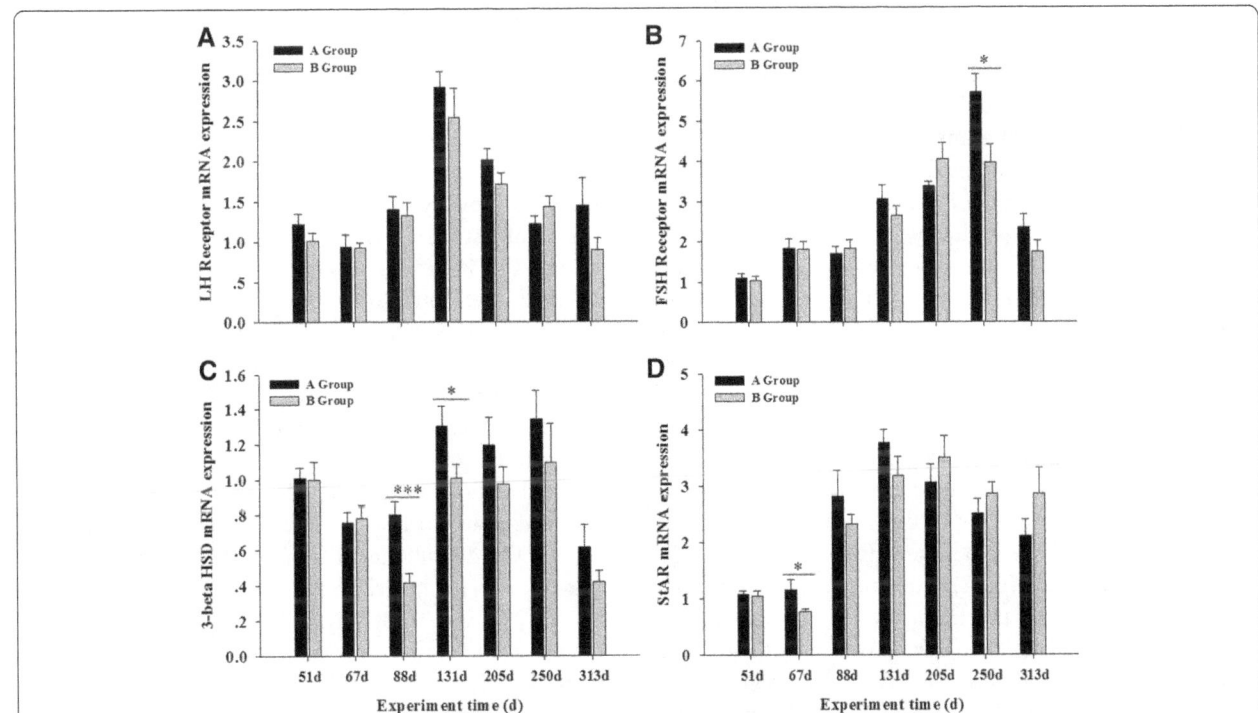

Fig. 6 Testicular mRNA levels of LH receptor (**a**), FSH receptor (**b**), 3-beta HSD (**c**) and StAR (**d**) in Yangzhou goose ganders under two artificial photoperiods. Each value represents the average of data from eight independent culture experiments. Data are shown as mean values ± standard error of the mean. *, ** and *** indicate statistical significance based on $P < 0.05$, $P < 0.01$ and $P < 0.001$, respectively. between the treatments

12-hour photoperiod, than in group B, which was still exposed to an 8-hour photoperiod ($P = 0.000$, $F = 24.965$), as was also the case with the seminiferous tubule diameter (Fig. 3c). On day 313, when the experiment ended, there were still more elongated spermatocytes in group A than in group B (Fig. 4 and Fig. 3f). In addition to the changes in diameter, the number of vacuoles in the seminiferous tubules was higher both at the beginning and end of the experiment than at the peak of the reproductive activity on day 131, where the seminiferous tubules were more compact and in a cell-filled state.

Gene expression levels
Hypothalamus and pituitary genes
All of the 16 genes tested exhibited photoperiod-dependent mRNA level regulation throughout the experiment (Fig. 5). The hypothalamic GnRH-I mRNA levels changed similarly in ganders from groups A and B (Fig. 5a): they increased from the beginning of the experiment to reach a peak level on day 205, during the 12-hour photoperiod, and then steadily decreased, with an exception on day 88, when they were significantly higher in group A than in group B. Similarly, GnIH gene transcription levels varied in a strikingly similar manner in both groups (Fig. 5b): they steadily decreased from their already low levels at the beginning of the experiment to a minimum value on day 131, during the 12-hour photoperiod, and then remarkably increased by 200-fold to a maximum from day 205 to day 250, before they finally decreased again. The changes in VIP and TRH gene transcription in the hypothalamus exhibited changes similar to that of GnIH throughout the experiment (Fig. 5c–d): their mRNA levels decreased to or remained at low levels until day 88 of the experiment, and then increased to higher levels on day 131, during peak reproductive activities. However, on two occasions, namely on days 131 and 313 of the experiment, the VIP mRNA levels were significantly higher in group A than in group B ($P = 0.013$, $F = 13.542$ and $P = 0.002$, $F = 20.189$), while the same was true for the TRH mRNA levels on day 131.

In the pituitary gland, the GnRH receptor mRNA levels increased steadily from day 51, during the short 8-hour photoperiod, to a plateau value at day 250 in group A, whereas in group B, they further increased to a peak value at the end of the experiment (Fig. 5e). The GnIH receptor mRNA levels (Fig. 5f) remained low before day 88, but they started to rise from day 131, and reached a peak level on day 205, then decreased in both groups. However, on days 250 and 313, the post-peak mRNA levels were higher in group A than in group B ($P = 0.001$, $F = 24.773$ and $P = 0.004$, $F = 7.200$ for days 250 and 313 respectively) (Fig. 5f). Throughout the experiment, the mRNA levels of the VIP receptor changed in a manner

similar to that of the GnIH receptor, as did the TRH receptor mRNA levels (Fig. 5g–h).

FSH beta mRNA (Fig. 5i) expression levels were initially low under the 8-hour photoperiod, and gradually increased to the peak on day 131, and then decreased to the nadir on day 313 in Group B ganders. The FSH mRNA levels on days 67 and 88 were significantly higher in ganders of group A than in those of group B ($P = 0.000$, $F = 11.710$ and $P = 0.005$, $F = 6.744$ for days 67 and 88 respectively). Similarly, in group A, the LH beta mRNA levels (Fig. 5j) increased from their low levels during the 8-hour photoperiod on day 51 to peak levels on day 67, just 12 days after the photoperiod was increased from 8 to 12 h. This peak was maintained until day 131 and then gradually decreased to a minimum on day 313 (end of the experiment). In addition, on day 67, they were significantly higher in group A than in group B ($P = 0.000$, $F = 25.221$), whose LH beta transcription patterns changed similarly to those of FSH beta. The PRL mRNA levels (Fig. 5k) remained low before day 88 in both groups, then started to increase from day 131 to a peak level on day 205, and thereafter decreased. In group A, on days 131 and 313, PRL transcription was significantly higher than in group B ($P = 0.002$, $F = 12.867$ and $P = 0.000$, $F = 21.107$ for days 131 and 313 respectively). The levels of TSH beta mRNA (Fig. 5l) changed in both groups in a manner similar to that of GnIH, except on day 131, where the expression was abruptly and significantly higher in group A than in group B ($P = 0.000$, $F = 20.946$).

Testicular genes
The FSH receptor mRNA levels (Fig. 6a) increased steadily from the beginning of the experiment; they reached peak levels during days 205–250, and then decreased to low levels at end of experiment. In both groups of ganders, the LH receptor mRNA levels (Fig. 6b) also increased, but they peaked on day 131, and then decreased until the end of experiment. This changing pattern was also true for the StAR mRNA levels (Fig. 6c), except that on day 67 they were significantly higher in group A than in group B ($P = 0.012$, $F = 5.480$). The mRNA levels of 3-beta hydroxysteroid dehydrogenase remained low or decreased before day 88, and then increased to peak levels from day 131 to 250, and finally decreased again to low levels at the end of experiment (Fig. 6d). On days 88 and 131, the levels were also higher in group A than in group B ($P = 0.000$, $F = 16.329$ and $P = 0.037$, $F = 3.879$ for days 88 and 131 respectively).

Discussion
This study systematically investigated the molecular mechanisms associated with testicular development and regression in response to artificial photoprograms in Yangzhou goose ganders. A wide spectrum of variables

was assessed, including testis histology, hormone plasma concentrations, and gene transcription patterns. In addition to the general responses to an increase in photoperiod (i.e. testicular development and subsequent regression), some minor responses in the timing of testicular development and gene transcription were also observed in relation to subtle changes in photoperiod. These results reinforce the proposition that the Yangzhou goose is a long-day breeding bird, and provide new insights into the molecular mechanisms that mediate the photoperiod-regulated seasonal changes in reproductive activities.

It was also noted that after 40 days of exposure to an 8-hour short photoperiod, testosterone concentrations slightly increased in both gander groups studied (A and B). This phenomenon may be linked to the weak expression of gonadal or reproductive activities during short days after dissipation of refractoriness to the long days [4, 45, 48]. In addition, in group B and on day 88 of the experiment, refractoriness to a short photoperiod also led to spontaneous testicular development immediately before the photoperiod was increased. Subsequently, an increase in daily photoperiod from 8 to 12 h resulted in rapid growth of the testes, accompanied by an increase in testicular weight from less than 2 g during the 8-hour short photoperiod to approximately 6 g in just 10 days, and to 12 g in 21 days. The delay of 28 days in photoperiod increase for group B with respect to that of group A postponed testicular growth at a similar pace. Following 40 days of exposure to a 12-hour photoperiod, the ganders reached peak reproductive activities as represented by the high levels of plasma testosterone and active spermatogenesis in the testis at days 100–150 of the experiment. After that time point, refractoriness to the long photoperiod gradually commenced, as reflected by a decrease in testicular weight, disappearance of spermatozoa in the seminiferous tubules of the testes, and especially by the steady decrease in the concentration of plasma testosterone. The 12-hour equatorial photoperiod was adopted in this study not only to induce the reproductive activities of ganders, but also to test how photorefractoriness develops in the Yangzhou goose. Clearly, ganders develop a kind of "absolute photorefractoriness" similar to that reported in starlings [49], in contrast to the "relative photorefractoriness" observed in quails [49] and sheep [50]. Therefore, at the end of the experiment, the testosterone concentration decreased to minimal levels in both groups A and B. Such changes should be the result of different mechanisms, i.e. the further refractoriness to long photoperiod in group A by an increase in photoperiod to 16 h (as was indicated by upregulation of VIP and PRL mRNA levels), and the inhibition of reproductive activity in group B by a decrease in photoperiod.

It has been proposed that the endocrine mechanism mediating the photoperiod-dependent regulation of seasonal reproduction is through the coordinated production and secretion of the gonadotrophins LH and PRL by the pituitary gland [14]. Increases in photoperiod first drive LH secretion, which stimulates gonadal development and initiates the breeding season, and then PRL secretion, which initiates the development of photorefractoriness, and after reaching peak levels, reduces LH secretion and leads to gonad regression. Since production and secretion of LH is regulated by the hypothalamic-releasing hormones GnRH and GnIH [13, 21, 25], and that of PRL by VIP [32, 51], these hormones must be involved in the photoperiod-dependent regulation of seasonal breeding [3, 13, 14]. Therefore, in the present study, the mRNA levels of GnRH, GnIH, and VIP were measured in the hypothalamus of the ganders, and those of their receptors in the pituitary gland. In both groups of ganders, initiation of reproductive activities after prolonged exposure to a short 8-hour photoperiod, or a state of refractoriness to the short photoperiod, concomitantly occurred with a steady decrease in the mRNA levels of VIP and GnIH. Switching from an 8-hour to a 12-hour photoperiod upregulated the transcription of hypothalamic GnRH, but further downregulated that of GnIH, which reached a minimum when ganders were in a fully active reproductive state. Changes in GnRH/GnIH gene transcription in response to an increase in photoperiod were accordingly reflected by changes in FSH and LH mRNA levels in the pituitary gland. It must be noted that the earlier exposure to a 12-hour long photoperiod in group A was associated with an earlier upregulation of both FSH and LH mRNA levels. This could explain the earlier testicular growth and rise in the concentrations of plasma testosterone observed in the ganders of group A.

Notwithstanding the above observations, the post-peak decrease in reproductive activities after a prolonged exposure to a 12-hour long photoperiod (day 205 of the experiment) was associated with a greater than 300-fold upregulation in the mRNA levels of GnIH. This increase in GnIH gene transcription and hence secretion of the hormone may be the driving force in inducing reproductive regression during the development of refractoriness, despite a 2- to 3-fold increase in GnRH mRNA levels after continued exposure to a long photoperiod. The temporal elevation of GnRH mRNA levels from day 205 to day 250 may not be photoperiod-driven, but could rather be caused by a decreased negative control feedback arising from the diminished plasma testosterone concentrations. Whereas the declined GnRH expression at end of experiment could be the true effect of photoperiodic, by the refractoriness under 16 h photoperiod or by the inhibition under 8 h photoperiod.

On the other hand, the inhibition by GnIH of pituitary gonadotrophin synthesis could be mediated via two pathways, one by the direct effect on pituitary gland and the other indirectly via reducing GnRH secretion by inhibiting GnRHneurons in the hypothalamus [13]. The inhibitory effect of GnIH on gonadotrophin expression and secretion could be at a maximum level starting from day 205 of the experiment, that is approximately 150 days after switching to the long 12-hour photoperiod, if pituitary GnIH receptor mRNA levels were also included. Another factor that is important for the development of refractoriness to a long photoperiod is PRL, whose mRNA levels reached their highest value on day 205 of the experiment, when the mRNA levels of the GnIH/ GnIH receptor also reached their highest value and could exert their maximal inhibitory impact on gonadotrophin secretion. The mRNA levels of VIP and the VIP receptor, which stimulate pituitary PRL secretion, were already upregulated from day 131 of the experiment. As ganders in group A were exposed to a 12-hour photoperiod one month earlier than ganders in group B, the VIP/VIP receptor mRNA levels were also observed to rise earlier in the former group. Furthermore, at the end of the experiment on day 313, both VIP and PRL mRNA levels were further upregulated in group A, in response to an increase in photoperiod from 12 h to 16 h 35 days earlier than in group B. Such an upregulation did not occur in the ganders of group B, which experienced a decrease of photoperiod from 12 h to 8 h.

The photostimulation and refractoriness of reproductive activities were also analyzed in terms of testicular steroidogenesis gene transcription patterns. The transcription of LHR, which mediates the gonadotrophic effects of LH, displayed a typical rise-and-fall pattern, following that of the changes in testicular weight and plasma testosterone concentration. A similar effect was observed for StAR and 3-beta hydroxysteroid dehydrogenase transcription. Thus, toward the end of the experiment, when both LH beta and LH receptor mRNA levels significantly subsided, so did the mRNA levels of the steroidogenic genes StAR and 3-beta hydroxysteroid dehydrogenase. Depletion of these key enzymes of steroidogenesis would result in diminished testosterone production, as shown by the steady decline of plasma testosterone concentration toward the end of the experiment. This, in turn, would impair spermatogenesis, resulting in testis atrophy and reduced testis weight. Of the testicular genes tested, the FSH receptor mRNA levels peaked during days 205–250 of the experiment; that is, during the testicular regression process. This illustrates that the biological role of FSH in the regulation of testicular functions may occur during the early stages of spermatogenesis (i.e., germ cell mitosis before the spermatocyte stage) [52]. Testosterone is more important during the later stages of spermatogenesis, when

morphologic changes lead to spermatid formation. As a result, on day 313 of the experiment, when testosterone and FSH levels were minimal, no spermatids were observed in the testes of ganders in either group.

The thyroid hormones function as molecular switches in the regulation of seasonal reproduction [11, 53], and are involved in both photoperiod-induced reproductive stimulation and refractoriness. In Japanese quail, removal of thyroid hormones by thyroidectomy reduces the maximal extent of testis development induced by a long photoperiod [49] In the quail, it was discovered that TSH is synthesized in the pars tuberalis of the pituitary gland in response to an increase in photoperiod, and through retrograde uptake, it stimulates the ependymal cells in MBH to express type 2 iodothyroninedeiodinase, which converts T4 to T3 [41, 54]. Such locally produced T3, stimulates the release of GnRH into the portal vascular system by inducing morphological or plasticity changes in GnRH-secreting neurons [41], resulting in elevated LH secretion and full reproductive activities [11, 13]. On the other hand, thyroid hormones have a permissive role in the development of photorefractoriness that leads to reproductive regression in both mammals and birds [34, 39, 55–57]. However, when there is a complete lack of thyroid hormones, reproductive activities still change in response to daily photoperiod in both thyroidectomized birds [49] and red deer [34]. Therefore, it appears that the photoperiod-dependent regulation of seasonal reproduction or a GnRH pulse generator should still be operating at brain centers higher than MBH, while local T3 production in the MBH might amplify or modify this central effect. It is possible that systemic effects, in contrast to the above-discussed local actions of thyroid hormones in the MBH, may mediate reproductive inhibiting actions. In several animal species living in temperate zones, such as ducks [36], Moscovy drakes [37] and geese [38], there was a seasonal rise of plasma thyroid hormones around the time of reproductive regression. In domestic ganders, it has been shown that the concentrations of T4 and T3 decreased rapidly with increasing testosterone concentration, and with the approach of the non-breeding season, testosterone concentration decreased rapidly, whereas T4 concentration showed the opposite trend [38]. In the present study, the changes in plasma concentrations of both T4 and T3 exhibited a pattern opposite to that of testosterone. The seasonal fluctuations of thyroid hormones might be an aspect of the general seasonal physiological changes; alternatively, high thyroid hormone levels during the non-breeding season might facilitate, if not altogether cause, the termination of the breeding season, and facilitate the manifestation of other seasonal events such as growth [34], as discussed further below. It should also be noted that the involvement of thyroid hormones in the refractoriness to a 12-hour long photoperiod, observed in this study, occurred with a concomitant upregulation of the TRH, TRH receptor, and TSH mRNA levels, as well as of the VIP, VIP receptor, and PRL mRNA levels. Indeed, in European starlings, abolishing the development of photorefractoriness by thyroidectomy also prevented the corresponding increase in PRL secretion [39].

Thyroid hormones, PRL, and GnIH not only stimulate the development of photorefractoriness that leads to reproductive regression, but can also regulate feed intake and growth. GnIH has been shown to stimulate feed intake in both animals [58] and fowls [59, 60]. Based on this, it was proposed that may be a molecular switch between feeding, and hence somatic growth, and reproduction [58]. Furthermore, both thyroid hormones and PRL are essential for long photoperiod-induced seasonal increases in feed intake and live weight gain simultaneously to reproductive regression [34, 61], whereas PRL has been found to be orexigenic in domestic cows [62, 63]. It appears that thyroid hormones and PRL also fulfill the roles of molecular switches between growth and reproduction. Consequently, at the late stages of the experiment, the reproductive system regression accompanied by a significant weight increase in ganders could well be the result of actions of these three molecular switches. It was proposed that the seasonal somatic growth in wild animals that occur at the same time of reproductive regression [34, 64, 65] promotes animal survival throughout the hardships of winter. It appears that this strategy remains operative in the domestic goose as well, and involves the same endocrine regulatory factors.

Conclusions

In summary, reproductive stimulation of long-day breeding Yangzhou goose ganders was characterized by GnRH upregulation and GnIH and VIP downregulation in the hypothalamus during long or increasing photoperiods. This, in turn, stimulated the transcription of the gonadotrophin gene, whereas inhibited that of the PRL gene in the pituitary gland, which is known to stimulate testicular development by upregulating genes important for steroidogenesis and spermatogenesis. On the other hand, development of photorefractoriness was characterized by hyper-regulation of GnIH transcription in the hypothalamus, in addition to transcription upregulation of VIP, TRH, and their receptors in the pituitary gland. Consequently, gonadotrophin transcription was downregulated, whereas that of PRL and TSH was upregulated. The combined effects of these changes resulted in reproductive system regression.

Acknowledgements
We thank all anonymous reviewers for their suggestions and criticisms which improved the quality of the manuscript

Funding
This study was supported by the National Science Foundation of China (grant no. 31372314), Fund for Independent Innovation of Agricultural Sciences in Jiangsu Province (CX(15)1008), and Research Fund of Jiangsu Academy of Agricultural Science (ZX(15)1007). HXZ was supported by Postdoctoral Research Funding Program of Jiangsu Province (1501079C).

Authors' contributions
HXZ, XBS and ZDS designed the study. HXZ, ZC, JNY, CKW and ZCD carried out the animal experiment, collected and analyzed the samples. HXZ and ZDS prepared the manuscript with correction input by ZC. All authors read and approved the final manuscript.

Competing interests
The authors declare that they have no competing interests.

Author details
[1]Laboratory of Animal Improvement and Reproduction, Institute of Animal Science, Jiangsu Academy of Agricultural Sciences, Nanjing 210014, China. [2]Sunlake Swan Farm, Changzhou 213101, China.

References
1. Wingfield JC, Farner DS. Endocrinology of reproduction in wild species. Avian Biol. 1993;9:327.
2. Dixit AS, Singh NS. Photoperiod as a proximate factor in control of seasonality in the subtropical male Tree Sparrow, Passer montanus. Front Zool. 2011;8:44–56.
3. Dawson A, Sharp PJ. Photorefractoriness in birds—photoperiodic and non-photoperiodic control. Gen Comp Endocrinol. 2007;153:378–84.
4. Hahn TP, Boswell T, Wingfield JC, Ball GF. Temporal flexibility in avian reproduction. In Current ornithology. Springer; 1997: 39–80.
5. Hahn T, Pereyra M, Katti M, Ward G, MacDougall-Shackleton S. Effects of food availability on the reproductive system. New Delhi: Functional Avian Endocrinology Narosa Publishing House; 2005. p. 167–80.
6. Dawson A. Seasonal differences in the secretion of luteinising hormone and prolactin in response to n-methyl-dl-aspartate in starlings (sturnus vulgaris). J Neuroendocrinol. 2005;17:105–10.
7. Dawson A. The effect of temperature on photoperiodically regulated gonadal maturation, regression and moult in starlings–potential consequences of climate change. Funct Ecol. 2005;19:995–1000.
8. Sharp PJ. Photoperiodic control of reproduction in the domestic hen. Poult Sci. 1993;72:897–905.
9. Ikegami K, Yoshimura T. Circadian clocks and the measurement of daylength in seasonal reproduction. Mol Cell Endocrinol. 2012;349:76–81.
10. Yoshimura T. Neuroendocrine mechanism of seasonal reproduction in birds and mammals. Anim Sci J. 2010;81:403–10.
11. Yoshimura T. Thyroid hormone and seasonal regulation of reproduction. Front Neuroendocrinol. 2013;34:157–66.
12. Chaiseha Y, Halawani MEE. Neuroendocrinology of the female turkey reproductive cycle. J Poult Sci. 2005;42:87–100.
13. Bédécarrats GY, Baxter M, Sparling B. An updated model to describe the neuroendocrine control of reproduction in chickens. Gen Comp Endocrinol. 2015;227:58–63.
14. Sharp PJ, Blache D. A neuroendocrine model for prolactin as the key mediator of seasonal breeding in birds under long- and short-day photoperiods. Can J Physiol Pharmacol. 2003;81:350–8.
15. Burns JM. The influence of chicken and mammalian luteinizing hormones on ovarian estrogen synthesis in the pullet (Gallus domesticus). Comp Biochem Physiol A Physiol. 1972;42:1011–7.
16. Brown N, Baylé J-D, Scanes C, Follett B. Chicken gonadotrophins: their effects on the testes of immature and hypophysectomized Japanese quail. Cell Tissue Res. 1975;156:499–520.
17. Maung SL, Follett BK. The endocrine control by luteinizing hormone of testosterone secretion from the testis of the Japanese quail. Gen Comp Endocrinol. 1978;36:79–89.
18. Phillips A, Scanes CG, Hahn DW. Effect of androgens and gonadotropins on progesterone secretion of chicken granulosa cells. Comp Biochem Physiol A Physiol. 1985;81:847–52.
19. Bédécarrats GY, McFarlane H, Maddineni SR, Ramachandran R. Gonadotropin-inhibitory hormone receptor signaling and its impact on reproduction in chickens. Gen Comp Endocrinol. 2009;163:7–11.

20. Ciccone NA, Dunn I, Boswell T, Tsutsui K, Ubuka T, Ukena K, Sharp P. Gonadotrophin inhibitory hormone depresses gonadotrophin α and follicle-stimulating hormone β subunit expression in the pituitary of the domestic chicken. J Neuroendocrinol. 2004;16:999–1006.

21. Tsutsui K, Saigoh E, Ukena K, Teranishi H, Fujisawa Y, Kikuchi M, Ishii S, Sharp PJ. A novel avian hypothalamic peptide inhibiting gonadotropin release. Biochem Biophys Res Commun. 2000;275:661–7.

22. Bentley G, Perfito N, Ukena K, Tsutsui K, Wingfield J. Gonadotropin-inhibitory peptide in song sparrows (melospiza melodia) in different reproductive conditions, and in house sparrows (passer domesticus) relative to chicken-gonadotropin-releasing hormone. J Neuroendocrinol. 2003;15:794–802.

23. Ukena K, Ubuka T, Tsutsui K. Distribution of a novel avian gonadotropin-inhibitory hormone in the quail brain. Cell Tissue Res. 2003;312:73–9.

24. Ubuka T, Bentley GE, Ukena K, Wingfield JC, Tsutsui K. Melatonin induces the expression of gonadotropin-inhibitory hormone in the avian brain. Proc Natl Acad Sci U S A. 2005;102:3052–7.

25. Kriegsfeld LJ, Mei DF, Bentley GE, Ubuka T, Mason AO, Inoue K, Ukena K, Tsutsui K, Silver R. Identification and characterization of a gonadotropin-inhibitory system in the brains of mammals. Proc Natl Acad Sci U S A. 2006;103:2410–5.

26. Sharp PJ. Photoperiodic regulation of seasonal breeding in birds. Ann N Y Acad Sci. 2005;1040:189–99.

27. Mauro LJ, Youngren OM, Proudman JA, Phillips RE, El Halawani ME. Effects of reproductive status, ovariectomy, and photoperiod on vasoactive intestinal peptide in the female turkey hypothalamus. Gen Comp Endocrinol. 1992;87:481–93.

28. Silverin B, Viebke P. Low temperatures affect the photoperiodically induced LH and testicular cycles differently in closely related species of tits (Parus spp.). Horm Behav. 1994;28:199–206.

29. Deviche P, Saldanha CJ, Silver R. Changes in brain gonadotropin-releasing hormone-and vasoactive intestinal polypeptide-like immunoreactivity accompanying reestablishment of photosensitivity in male dark-eyed juncos (Junco hyemalis). Gen Comp Endocrinol. 2000;117:8–19.

30. Sharp P, Sreekumar K. Photoperiodic control of prolactin secretion. In: A Dawson and CM Chaturvedi (Eds), Avian Endocrinology. New Delhi: Narosa; 2001. p. 245–255.

31. Shi ZD, Huang YM, Liu Z, Liu Y, Li XW, Proudman JA, Yu RC. Seasonal and photoperiodic regulation of secretion of hormones associated with reproduction in Magang goose ganders. Domest Anim Endocrinol. 2007;32:190–200.

32. Dawson A, Sharp PJ. The role of prolactin in the development of reproductive photorefractoriness and postnuptial molt in the European starling (Sturnus vulgaris). Endocrinology. 1998;139:485–90.

33. Nicholls T, Goldsmith A, Dawson A. Photorefractoriness in birds and comparison with mammals. Physiol Rev. 1988;68:133–76.

34. Shi ZD, Barrell GK. Requirement of thyroid function for the expression of seasonal reproductive and related changes in red deer (Cervus elaphus) stags. J Reprod Fertil. 1992;94:251–9.

35. Boissinagasse L, Maurel D, Boissin J. Seasonal variations in thyroxine and testosterone levels in relation to the moult in the adult male mink (Mustela vison Peale and Beauvois). Can J Zool. 1981;59:1062–6.

36. Jallageas M, Assenmacher I. Further evidence for reciprocal interactions between the annual sexual and thyroid cycles in male Peking ducks. Gen Comp Endocrinol. 1979;37:44–51.

37. Sharp PJ, Klandorf H, Mcneilly AS. Plasma prolactin, thyroxine, triiodothyronine, testosterone, and luteinizing hormone during a photoinduced reproductive cycle in mallard drakes. J Exp Zool. 1986;238:409–13.

38. Zeman M, Košutzký J, Miček Ľ, Lengyel A. Changes in plasma testosterone, thyroxine and triiodothyronine in relation to sperm production and remex moult in domestic ganders. Reprod Nutr Dev. 1990;30:549–57.

39. Goldsmith AR, Nicholls TJ. Thyroidectomy prevents the development of photorefractoriness and the associated rise in plasma prolactin in starlings. Gen Comp Endocrinol. 1984;54:256–6.

40. Follett B, Nicholls T. Acute effect of thyroid hormones in mimicking photoperiodically induced release of gonadotropins in Japanese quail. J Comp Physiol B. 1988;157:837–43.

41. Yoshimura T, Yasuo S, Watanabe M, Iigo M, Yamamura T, Hirunagi K, Ebihara S. Light-induced hormone conversion of T4 to T3 regulates photoperiodic response of gonads in birds. Nature. 2003;426:178–81.

42. Nakao N, Ono H, Yamamura T, Anraku T, Takagi T, Higashi K, Yasuo S, Katou Y, Kageyama S, Uno Y. Thyrotrophin in the pars tuberalis triggers photoperiodic response. Nature. 2008;452:317–22.

43. Bo WC. The origin of Chinese domestic geese. Agric Archaeol. 1996;3:268–72.

44. Wang JW, Qiu XP, Zeng FT, Shi XW, Zhang YP. Genetic differentiation of domestic goose breeds in China. Acta Genet Sin. 2005;32:1053–9.

45. Shi ZD, Tian YB, Wu W, Wang ZY. Controlling reproductive seasonality in the geese: a review. Worlds Poult Sci J. 2008;64:343–55.

46. Rousselot-Pailley D, Sellier N. Influence de quelques facteurs zootechniques sur la fertilité des oies. In: BRILLARD, J.P. (Ed.) Control of fertility in domestic birds. Les Colloques de L'INRA 54. 1990; p. 145–155.

47. Livak KJ, Schmittgen TD. Analysis of relative gene expression data using real-time quantitative PCR and the $2^{-\Delta\Delta CT}$ method. Methods. 2001;25:402–8.

48. Péczely P, Czifra G, Sepr"Odi A, Teplán I. Effect of low light intensity on testicular function in photorefractory domestic ganders. Gen Comp Endocrinol. 1985;57:293–300.

49. Follett BK, Nicholls TJ. Influences of thyroidectomy and thyroxine replacement on photoperiodically controlled reproduction in quail. J Endocrinol. 1985;107:211–21.

50. Jackson G, Jansen H, Kao C. Continuous exposure of Suffolk ewes to an equatorial photoperiod disrupts expression of the annual breeding season. Biol Reprod. 1990;42:63–73.

51. Sharp PJ, Sterling RJ, Talbot RT, Huskisson NS. The role of hypothalamic vasoactive intestinal polypeptide in the maintenance of prolactin secretion in incubating bantam hens: observations using passive immunization, radioimmunoassay and immunohistochemistry. J Endocrinol. 1989;122:5-NP.

52. Mclachlan RI, O'Donnell L, Meachem SJ, Stanton PG, Kretser DMD, Pratis K, Robertson DM. Hormonal regulation of spermatogenesis in primates and man: insights for development of the male hormonal contraceptive. J Androl. 2002;23:149–62.

53. Dardente H, Wyse CA, Birnie MJ, Dupré SM, Loudon ASI, Lincoln GA, Hazlerigg DG. A molecular switch for photoperiod responsiveness in mammals. Curr Biol. 2010;20:2193–8.

54. Nakao N, Ono H, Yoshimura T. Thyroid hormones and seasonal reproductive neuroendocrine interactions. Reproduction. 2008;136:1–8.

55. Wieselthier AS, Tienhoven AV. The effect of thyroidectomy on testicular size and on the photorefractory period in the starling (Sturnus vulgaris L.). J Exp Zool. 1972;179:331–8.

56. Pant K, Chandola-Saklani A. T3 fails to mimic certain effects of T4 in munia birds: physiological implications for seasonal timing. Comp Biochem Physiol C Pharmacol Toxicol Endocrinol. 1995;111:157–64.

57. Dahl G, Evans N, Thrun L, Karsch F. Thyroxine is permissive to seasonal transitions in reproductive neuroendocrine activity in the ewe. Biol Reprod. 1995;52:690–6.

58. Smith JT, Ross Young I, Veldhuis JD, Clarke IJ. Gonadotropin-inhibitory hormone (GnIH) secretion into the ovine hypophyseal portal system. Endocrinology. 2012;153:3368–75.

59. Tachibana T, Sato M, Takahashi H, Ukena K, Tsutsui K, Furuse M. Gonadotropin-inhibiting hormone stimulates feeding behavior in chicks. Brain Res. 2005;1050:94–100.

60. Fraley GS, Coombs E, Gerometta E, Colton S, Sharp P, Li Q, Clarke I. Distribution and sequence of gonadotropin-inhibitory hormone and its potential role as a molecular link between feeding and reproductive systems in the Pekin duck (Anas platyrhynchos domestica). Gen Comp Endocrinol. 2013;184:103–10.

61. Curlewis JD, Loudon AS, Milne JA, Mcneilly AS. Effects of chronic long-acting bromocriptine treatment on liveweight, voluntary food intake, coat growth and breeding season in non-pregnant red deer hinds. J Endocrinol. 1988;119:413–20.

62. Lacasse P, Ollier S. The dopamine antagonist domperidone increases prolactin concentration and enhances milk production in dairy cows. J Dairy Sci. 2015;98:7856–64.

63. Lacasse P, Ollier S, Lollivier V, Boutinaud M. New insights into the importance of prolactin in dairy ruminants. J Dairy Sci. 2016;99:864–74.

64. Hoffman R, Davidson K, Steinberg K. Influence of photoperiod and temperature on weight gain, food consumption, fat pads and thyroxine in male golden hamsters. Growth. 1981;46:150–62.

65. Vriend J, Borer K, Thliveris J. Melatonin: its antagonism of thyroxine's antisomatotrophic activity in male Syrian hamsters. Growth. 1986;51:35–43.

66. Dunn IC, Millam JR. Gonadotropin releasing hormone: forms and functions in birds. Avian Poult Biol Rev. 1998;9:61–85.

67. Wang JT, Xu SW. Effects of cold stress on the messenger ribonucleic acid levels of corticotrophin-releasing hormone and thyrotropin-releasing hormone in hypothalami of broilers. Poult Sci. 2008;87:973.

68. Sun YM, Flanagan CA, Illing N, Ott TR, Sellar R, Fromme BJ, Hapgood J, Sharp P, Sealfon SC, Millar RP. A chicken gonadotropin-releasing hormone receptor that confers agonist activity to mammalian antagonists. Identification of D-Lys(6) in the ligand and extracellular loop two of the receptor as determinants. J Biol Chem. 2001;276:7754–61.

69. Yin H, Ukena K, Ubuka T, Tsutsui K. A novel G protein-coupled receptor for gonadotropin-inhibitory hormone in the Japanese quail (Coturnix japonica): identification, expression and binding activity. J Endocrinol. 2005;184:257–66.

70. Kansaku N, Shimada K, Ohkubo T, Saito N, Suzuki T, Matsuda Y, Zadworny D. Molecular cloning of chicken vasoactive intestinal polypeptide receptor complementary DNA, tissue distribution and chromosomal localization. Biol Reprod. 2001;64:1575–81.

71. De GB, Geris KL, Manzano J, Bernal J, Millar RP, Abou-Samra AB, Porter TE, Iwasawa A, Kühn ER, Darras VM. Involvement of thyrotropin-releasing hormone receptor, somatostatin receptor subtype 2 and corticotropin-releasing hormone receptor type 1 in the control of chicken thyrotropin secretion. Mol Cell Endocrinol. 2003;203:33.

72. Leska A, Dusza L. Seasonal changes in the hypothalamo-pituitary-gonadal axis in birds. Reprod Biol. 2007;7:99–126.

73. Buntin JD, Advis JP, Ottinger MA, Lea RW, Sharp PJ. An analysis of physiological mechanisms underlying the antigonadotropic action of intracranial prolactin in ring doves. Gen Comp Endocrinol. 1999;114:97–107.

74. Tabibzadeh C, Rozenboim I, Silsby JL, Pitts GR, Foster DN, El HM. Modulation of ovarian cytochrome P450 17 alpha-hydroxylase and cytochrome aromatase messenger ribonucleic acid by prolactin in the domestic turkey. Biol Reprod. 1995;52:600.

75. Nakamura K, Iwasawa A, Kidokoro H, Komoda M, Zheng J, Maseki Y, Inoue K, Sakai T. Development of thyroid-stimulating hormone beta subunit-producing cells in the chicken embryonic pituitary gland. Cells Tissues Organs. 2004;177:21–8.

76. Puett D, Angelova K, Costa MRD, Warrenfeltz SW, Fanelli F. The luteinizing hormone receptor: Insights into structure-function relationships and hormone-receptor-mediated changes in gene expression in ovarian cancer cells. Mol Cell Endocrinol. 2011;329:47–55.

77. Simoni M, Gromoll J, Nieschlag E. The follicle-stimulating hormone receptor: biochemistry, molecular biology, physiology, and pathophysiology. Endocr Rev. 1997;18:739.

78. Zhang Y, Liu H, Yang M, Hu S, Li L, Wang J. Molecular cloning, expression analysis and developmental changes in ovarian follicles of goose 3β-hydroxysteroid dehydrogenase 1. Anim Prod Sci. 2014;54:992.

79. Lin D, Sugawara T, Clark BJ, Stocco DM, Saenger P, Rogol A, Miller WL. Role of steroidogenic acute regulatory protein in adrenal and gonadal steroidogenesis. Science. 1995;267:1828.

Optic-nerve-transmitted eyeshine, a new type of light emission from fish eyes

Roland Fritsch[1][*] (iD), Jeremy F. P. Ullmann[2,3], Pierre-Paul Bitton[1], Shaun P. Collin[4][†] and Nico K. Michiels[1][*][†]

Abstract

Background: Most animal eyes feature an opaque pigmented eyecup to assure that light can enter from one direction only. We challenge this dogma by describing a previously unknown form of eyeshine resulting from light that enters the eye through the top of the head and optic nerve, eventually emanating through the pupil as a narrow beam: the Optic-Nerve-Transmitted (ONT) eyeshine. We characterize ONT eyeshine in the triplefin blenny *Tripterygion delaisi* (Tripterygiidae) in comparison to three other teleost species, using behavioural and anatomical observations, spectrophotometry, histology, and magnetic resonance imaging. The study's aim is to identify the factors that determine ONT eyeshine occurrence and intensity, and whether these are specifically adapted for that purpose.

Results: ONT eyeshine intensity benefits from locally reduced head pigmentation, a thin skull, the gap between eyes and forebrain, the potential light-guiding properties of the optic nerve, and, most importantly, a short distance between the head surface and the optic nerves.

Conclusions: The generality of these factors and the lack of specifically adapted features implies that ONT eyeshine is widespread among small fish species. Nevertheless, its intensity varies considerably, depending on the specific combination and varying expression of common anatomical features. We discuss whether ONT eyeshine might affect visual performance, and speculate about possible functions such as predator detection, camouflage, and intraspecific communication.

Keywords: Marine visual ecology, Eye anatomy, Eyeshine, Optic nerve, Light guidance, Tripterygiidae, *Tripterygion delaisi*

Background

Vision implies the presence of photoreceptors that absorb and transform light energy into a neural signal that can be interpreted by the brain. Advanced visual abilities as in vertebrates, however, also depend on the presence of the melanin containing retinal pigment epithelium (RPE) behind the photoreceptors. It absorbs excess light to prevent scattering within the eye and shields the photoreceptors against light coming from behind the eye, improving image contrast and resolution. This explains why the pupils of camera-type eyes are typically black. Some eyes, however, do not function strictly unidirectionally and may show stunningly bright pupils, from which light appears to be emitted by the eye, a phenomenon called *eyeshine*. Figure 1 provides an overview of previously described types of eyeshine [1, 2] and the new type described in this study. Based on the general mechanism, we broadly categorize the different types as either reflection-based (Fig. 1a-d), or transmission-based (Fig. 1e and f).

Reflection-based eyeshine can either be caused by an iridescent cornea (Fig. 1a), or a reflective layer within the eye (Fig. 1b-d) [1]. Iridescent-cornea-reflected (ICR) eyeshine differs from the other types in that it is produced outside the eyecup. It occurs in many shallow-water fishes and is assumed to serve flare reduction, as filter, or to camouflage the pupil [3]. Reflective layers behind the photoreceptors cause the well-known eyeshine featured in many animals, cats being a prominent example [1, 2]. There is variation in the layer type involved (*tapetum lucidum*, *stratum argenteum*), the location of the reflectors (RPE, inner choroid, outer choroid), what reflective structures are responsible (extracellular fibres or intracellular platelets, needles, cuboids, spherules), and how these reflect the incident light (diffusely or specularly) [1, 2, 4]. In all cases the eye seems to glow from within because light, which is not absorbed by the retinal photoreceptors, is reflected back

* Correspondence: roland.fritsch@uni-tuebingen.de;
roland.fritsch@uqconnect.edu.au; nico.michiels@uni-tuebingen.de
†Equal contributors
[1]Institute of Evolution and Ecology, University of Tübingen, 72076 Tübingen, Baden-Württemberg, Germany
Full list of author information is available at the end of the article

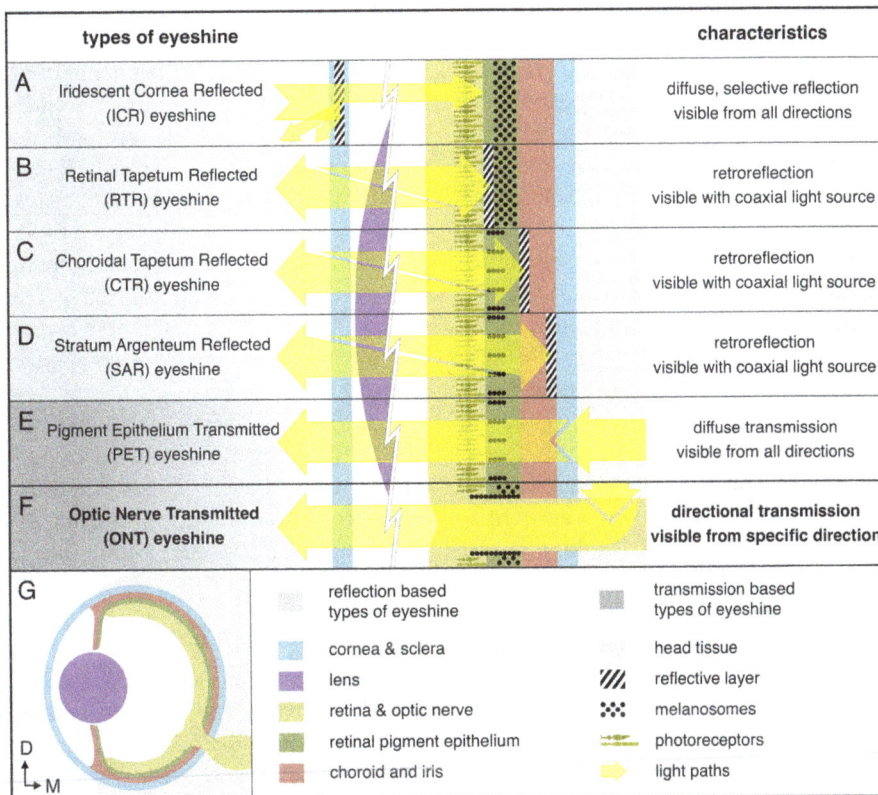

Fig. 1 Types of eyeshine in vertebrate eyes. **a** ICR eyeshine constitutes a case of iridescence produced by arrays of regularly arranged collagen fibres in the cornea. Light reflection is only partial, specular, and both wavelength- and angle-dependent. **b** Retinal-tapetum-reflected (RTR) eyeshine is caused by an intracellular *tapetum lucidum* in the inner part of the RPE. Tapetal reflection can be specular and/or diffuse. The eyes' optics project the reflected light back to its source, an effect known as retroreflection. **c** In CTR eyeshine, the *tapetum lucidum* lies within the inner portion of the choroid. For light to reach the tapetum, the RPE must transmit light at least partially (e.g., through melanosome motility). The choroidal tapetum can be cellular (specialised, reflective cells) or fibrous (regularly arranged, extracellular collagen fibres), both of which can generate strong retroreflection. **d** SAR eyeshine has similar properties and requirements as CTR, but is caused by the *stratum argenteum*, a distinct, cellular, reflective layer in the outer choroid. **e** PET eyeshine occurs when the RPE and choroid are (partially) unpigmented, and the eyes are in an exposed, dorsal position. Ambient light shines through the fundus and is seen as PET eyeshine from a wide range of directions. PET eyeshine regularly co-occurs with SAR eyeshine. **f** In optic-nerve-transmitted (ONT) eyeshine, light enters the eye through the ON. Pigmentation of the RPE and choroid are irrelevant for ONT eyeshine, but pigment sheaths around the intraocular ON protect the photoreceptors from transmitted light. ONT eyeshine can co-occur with any other type of eyeshine. Since the light is emitted from the optic disc, its projection and visibility are restricted to a narrow beam of the same shape as the disc. **g** Cross-section through an eye for a general overview of the involved structures. *D* dorsal, *M* medial

through the photoreceptors. The remaining light is then emitted out of the pupillary aperture along the incident optic pathway. In eyes with focusing optics, the reflection is directed towards the light's original source, an effect called retroreflection. Consequently, the reflection is visible only to an observer whose viewing axis is near-coaxial with the illuminating light [5]. These forms of reflective eyeshine (types B-D, Fig. 1) increase the retina's probability of photon capture by doubling the effective length of the photoreceptors' outer segments. This is particularly useful under dim light for crepuscular, nocturnal, and deep sea species [1]. Alternatively, reflection-based eyeshine might also allow species to maintain a certain photon capture efficiency while reducing investment in receptor length and photopigment density [6]. Other possible functions include improvement of polarization sensitivity [7] and camouflage in the pelagic environment [8].

Choroidal-tapetum-reflected (CTR; Fig. 1c) and stratum-argenteum-reflected eyeshine (SAR; Fig. 1d) require the RPE to be at least partially translucent. In scorpionfishes (Scorpaenidae) and toadfishes (Batrachoididae) this is achieved by the melanosomes congregating in RPE cell processes enveloping the cones, leaving the somata transparent, and thus allowing incoming light to reach the *argenteum* [9]. If, additionally, the eye bulges out of the skull dorsally and exposes its back, downwelling light can partially pass through, causing pigment-epithelium-transmitted (PET) eyeshine (Fig. 1e). To an observer, the resultant eyeshine is similar to reflective eyeshine, but without the retroreflective property. The effect is generally less bright due to light absorption while passing through several layers of tissue. Its occurrence in the toadfish's light-adapted eyes questions photon-catch enhancement and favours camouflage as probable function [9].

Here, we characterize a previously undescribed phenomenon of apparent light emission from the eyes as a new type of eyeshine. It is generated by ambient light that passes through the dorsal cranium and the extra- and intraocular sections of the optic nerve (ON) into the eye. There it is emitted from the optic disc, the area where all ganglion cell axons converge to form the ON, and leaves the eye through the pupil. Due to this light path, we call it optic-nerve-transmitted (ONT) eyeshine (Fig. 1f). Even though ONT eyeshine is also transmission-based, its properties differ crucially, and it can occur independently of the presence of the other types, even when RPE and choroid are densely pigmented, as the ON bypasses both. ONT eyeshine does not involve reflection and thus does not require illumination from the direction of the observer. Because the transmitted light is projected into the environment by the lens, the effect can only be seen from a specific, narrow angle. We describe ONT eyeshine qualitatively and quantitatively in the triplefin *Tripterygion delaisi*, a species with particularly strong ONT eyeshine. To determine whether *T. delaisi* evolved special features that facilitate its ONT eyeshine, we examine the anatomical structures along the light path using spectrophotometry, histology, and magnetic resonance imaging (MRI). Specifically, we investigated whether general skull structure increases transmittance, whether the ON's morphological characteristics could allow it to act as a light guide, and whether the optic disc shows features that prevent or control interference of the ONT eyeshine with vision. To ascertain whether these structures and characteristics are specific adaptations only found in *T. delaisi*, or just common features of small fishes, we compared its cranial and ocular anatomy with that of three similarly sized teleost species with varying combinations of traits that might

influence ONT eyeshine: *Tripterygion melanurus* (Tripterygiidae) is closely related and very similar to *T. delaisi* but has a more densely pigmented head; *Parablennius zvonimiri* (Blenniidae) is more distantly related but shares a similar, crypto-benthic ecology; the black-and-white morph of *Amphiprion ocellaris* (Pomacentridae) was included as an assumed negative control. Of the studied species, it is most distantly related to *T. delaisi*, belongs to a family of free-swimming, bentho-pelagic fishes, and was also chosen for having the darkest head pigmentation among the study species. These few species cannot and are not meant to provide a comprehensive analysis, but should suffice to reveal the most relevant factors that generate and modify the ONT eyeshine.

Finally, we discuss whether the phenomenon is simply a by-product of small size or if it could serve a function, like detection of targets featuring reflective eyeshine, camouflage, intraspecific communication, or enhancing visual performance.

Results and discussion

Due to the diversity of techniques used in this study and of factors involved in ONT eyeshine, we combine the presentation and discussion of the gathered data by the main relevant aspects. Data are presented as mean ± standard deviation. In-text we refer to the summarized data sets in Tables 1, 2 and 3. A comprehensive overview of all anatomical data can be found in the additional online material.

General mechanism and properties of ONT eyeshine

For ONT eyeshine to occur, a fish's head must be subjected to downwelling light. A portion of that light is transmitted through the head and is in part collected by

Table 1 Summarized absolute and relative anatomical data of specimens used for histology and MRI (see also Figs. 5 and 6)

Species	Value type	Total length [mm]	Head diameter [mm]	Body volume [mm³]	Skull (bone) thickness [μm] (% of HD)	Dermis thickness [μm] (% of HD)	ON depth [mm] (% of HD)	ON layers [N]	ON layer thickness [μm] (% of ONØ)	ON CSA [mm²] (% of head CSA)
T. delaisi	absolute	53.6 ± 4.6	6.9 ± 0.4	684 ± 133	54.0 ± 6.9	94.2 ± 8.0	1.57 ± 0.04	7.8 ± 0.8	54.9 ± 13.4	0.135 ± 0.040
	relative				(0.76 ± 0.09)	(1.33 ± 0.12)	(22.1 ± 2.0)		(13.2 ± 2.1)	(0.35 ± 0.07)
10	N	10	10	10	3	3	3	3	3	3
T. melanurus	absolute	40.5 ± 5.1	4.9 ± 06	258 ± 86	38.3 ± 7.5	35.3 ± 6.2	1.09 ± 0.05	6.5 ± 0.7	40.1 ± 11.1	0.065 ± 0.035
	relative				(0.78 ± 0.01)	(0.72 ± 0.004)	(22.3 ± 3.0)		(14.2 ± 0.03)	(0.36 ± 0.12)
4	N	4	4	4	2	2	2	2	2	2
P. zvonimiri	absolute	42	5.55	339	84.2 ± 54.7[a]	260.6 ± 160.4[a]	1.38 ± 0.21[a]	4.5[a]	63.5 ± 8.4[a]	0.124 ± 0.037[a]
	relative				(1.52 ± 0.99)[a]	(4.70 ± 2.89)[a]	(24.9 ± 3.8)[a]		(16.0 ± 2.1)[a]	(0.51 ± 0.15)[a]
1	N	1	1	1	1	1	1	1	1	1
A. ocellaris	absolute	52.9 ± 5.2	10.0 ± 1.2	1419 ± 503	255.5 ± 39.5	275.0 ± 80.9	2.16 ± 0.31	10.2 ± 0.8	42.7 ± 4.3	0.096 ± 0.007
	relative				(2.50 ± 0.21)	(2.67 ± 0.52)	(21.1 ± 1.6)		(12.2 ± 1.7)	(0.12 ± 0.03)
9	N	9	9	9	4	4	4	3	3	3

Data represent mean ± SD of N individuals or of repeated measurements within single individual ([a]). Reduced N caused by splitting individuals over different methods.
CSA cross-sectional area, head CSA calculated as $\frac{\pi}{4} HD^2$; *HD* head diameter, calculated as mean of head width and height; *ON* optic nerve; *ON Ø* optic nerve diameter

Table 2 Summarized spectrophotometric measurements data of ONT eyeshine transmission efficiency (see also Fig. 3)

Species (N)	Standard length [mm]	Head diameter [mm]	ON depth [mm]	DWS [photons/s/sr/mm²]	ONT eyeshine [photons/s/sr/mm²] (relative to DWS)	PET eyeshine [photons/s/sr/mm²] (relative to DWS)
T. delaisi (5)	46.4 ± 3.8	7.3 ± 0.9	1.61 ± 0.15	1.51 ± 0.34 ×10¹⁹	3.15 ± 1.81 ×10¹⁷ (0.0203 ± 0.0088)	7.01 ± 2.23 ×10¹⁶ (0.0047 ± 0.0010)
T. melanurus (2)	30.5 ± 6.4	4.4 ± 0.6	0.98 ± 0.13	1.04 ± 0.11 ×10¹⁹	2.09 ± 0.46 ×10¹⁷ (0.0205 ± 0.0065)	2.24 ± 0.55 ×10¹⁶ (0.0021 ± 0.0003)
P. zvonimiri (5)	38.6 ± 6.5	6.0 ± 1.2	1.49 ± 0.23	1.47 ± 0.17 ×10¹⁹	1.56 ± 0.12 ×10¹⁷ (0.0108 ± 0.0018)	3.04 ± 1.12 ×10¹⁶ (0.0022 ± 0.0011)
A. ocellaris (5)	25.5 ± 2.2	5.6 ± 0.1	1.18 ± 0.09	1.46 ± 0.16 ×10¹⁹	1.66 ± 0.63 ×10¹⁷ (0.0112 ± 0.0036)	1.13 ± 0.26 ×10¹⁶ (0.0008 ± 0.0001)

ON depth was estimated by applying relative ON depths from Table 1 to head diameters in this data set. Head diameter was calculated as the mean of head width and height measurements. *DWS* diffuse white standard, *sr* steradian

the ON, passes through the optic disc, and exits the eye through the lens and pupil (Fig. 2; Additional files 1 and 2). Inspecting the retina of *T. delaisi* and the other study species with an endoscope while illuminating the fish's head dorsally demonstrates that the light indeed emanates from the optic disc (Fig. 3a-d; Additional file 3). Entering the eye through the ON and optic disc means that ONT eyeshine can be produced independently of a tapetum, stratum argenteum, or partially light-transmissive pigment epithelium.

ONT eyeshine further differs in two crucial aspects from the other transmission-based (PET) eyeshine. First, it is emitted as a narrow beam in a specific direction, a consequence of the fact that it is focused by the lens, becoming a projection of the optic disc's image. The resulting ONT eyeshine beam extended over $3.5 \pm 0.3°$ ($n = 6$) horizontally and $19.6 \pm 3.3°$ ($n = 5$) vertically in *T. delaisi*, mirroring the optic disc's elongated shape (Figs. 2c and 3a). Second, ONT eyeshine is brighter than PET eyeshine. While PET light needs to pass the sclera, choroid, pigment epithelium, and retina, which in combination absorb or scatter most of the light, ONT light is transmitted through *T. delaisi*'s unpigmented intraocular ON and thus bypasses all above tissue layers. This was supported by the results of the spectrophotometric measurements of ONT and PET eyeshine (Table 2). *T. delaisi*'s optic disc transmitted on average $2.03 \pm 0.88\%$ ($n = 5$) of incident light (approximated as the radiance of a diffuse white standard in the same position), while the surrounding retina's transmission was $0.47 \pm 0.10\%$, i.e., PET eyeshine was only one fourth as bright as ONT eyeshine (Fig. 3e and f). *T. melanurus* had a very similar ONT eyeshine transmission

$(2.05 \pm 0.65\%, n = 2)$, but a much weaker PET eyeshine with only one tenth that transmission. Contrary to our expectations, both the individuals of the blenny *P. zvonimiri* and the clownfish *A. ocellaris* exhibited marked ONT eyeshine during the spectrophotometric measurements. *P. zvonimiri*'s ONT eyeshine was the dimmest, with $1.08 \pm 0.18\%$ ($n = 5$) transmission, and one fifth that much PET light passing through the surrounding retina. *A. ocellaris* featured a comparably bright ONT eyeshine ($1.12 \pm 0.36\%$, $n = 5$), but virtually no PET eyeshine as its retina transmitted only one fifteenth of that. The bright ONT eyeshine in *A. ocellaris*, our assumed negative control, can be attributed to the small size of the individuals of *A. ocellaris* available for measurement. The individuals used for anatomical investigations were twice the size and showed no visually discernible eyeshine (compare Tables 1 and 2). Specific transmission increased markedly with wavelength in all species (Fig. 3f), presumably due to the lower absorption and scattering of longer wavelength light in biological tissue [10].

Light transmissive features of the skull
Systematic illumination of ventral and dorsal point locations of the head of *T. delaisi* identified a small, triangular region directly between and behind the eyes that produced the strongest ONT eyeshine (Fig. 4a). Four features of this region facilitate light transmission. First, the skin and cranial bones are extremely thin (Fig. 5a1; Table 1; 94.2 ± 8.0 μm and 54.0 ± 6.9 μm, $n = 3$). Corrected for head size, these values correspond to $1.33 \pm 0.12\%$ and $0.76 \pm 0.09\%$ of the head diameter, which is much thinner than those of *Parablennius zvonimiri* and *Amphiprion ocellaris* (Table 1).

Table 3 Optic nerve torsion data summarised per genus, compare Fig. 6

Genus	Number	ON length [mm]	Starting angle [°]	Torsion [°/mm]	Total torsion [°]
Tripterygion	5	0.81 ± 0.16	19.3 ± 9.6	93.4 ± 8.2	75.5 ± 20.5
Parablennius	1	0.96	12.5	78.7	75.7
Amphiprion	3	1.40 ± 0.09	11.6 ± 20.0	82.5 ± 12.2	116.2 ± 24.8

Parameters were first averaged per individual (not shown), then per genus. *T. delaisi* and *T. melanurus* were grouped because of their virtually identical internal anatomy

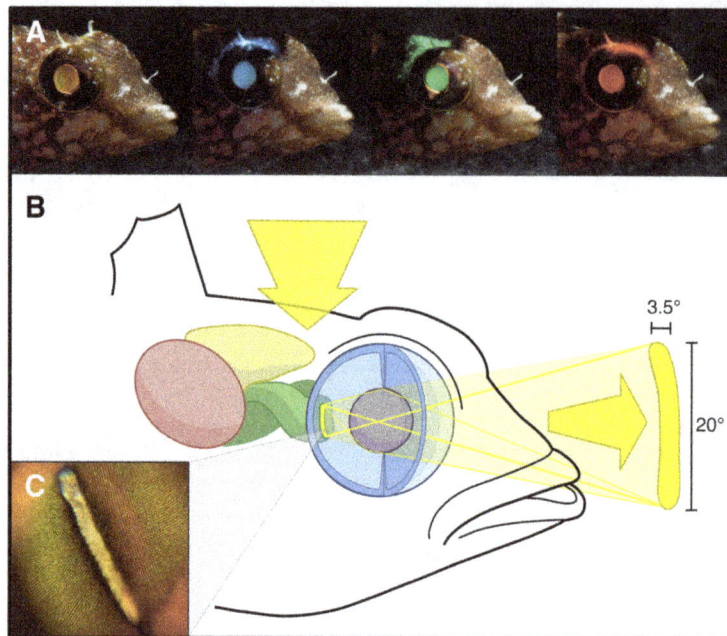

Fig. 2 Laboratory demonstration and origin of ONT eyeshine. **a** *Tripterygion delaisi* illuminated dorsally with different spectra, illustrating that the ONT eyeshine is generated by skull illumination and based on transmission of the incident light (Additional file 2). **b** Schematic light path of ONT eyeshine in *T. delaisi*. Light, which enters through the dorsal head surface, passes through the anterior brain and extraocular ONs, exits through the optic disc, and is projected into the environment as a narrow beam of light. *red*: optic tectum; *buff*: telencephalon; *green*: optic nerves; *blue*: eye; *purple*: lens; *yellow*: ONT eyeshine light path. **c** View into the eye of *T. delaisi* through an endoscope (without internal light source) under the same illumination as in A$_{white}$, showing the light-emitting, elongated optic disc (see also Additional file 3). A similar image can be seen directly with the human eye. Without small-aperture or micro-lens optics (e.g., endoscopes), the entire pupil seems to emit light since the optic disc's image is blurred as a consequence of focusing on the outside of the fish with a wide aperture lens (as in a, taken with DSLR camera)

Second, the skin is only partially pigmented (Fig. 4a). Third, fat tissue, which would scatter light more strongly than other tissue because of its higher refractive index, is absent from the skin and skull (Fig. 5a1). Fourth, the interstitial gap between skull, eyes, and telencephalon leaves the anterior ON exposed to light entering the head (Fig. 5a2).

T. melanurus is morphologically very similar to *T. delaisi*, except for differences in colouration and size (Fig. 5b-b3). The equally strong ONT eyeshine (Table 2) could result from the used individuals' smaller head size compensating the darker head pigmentation (Fig. 5b). The other species' weaker ONT eyeshine can be explained by increased pigmentation, thicker skin, and thicker skull bones (Fig. 5c, c1, d, and d1). In addition, *A. ocellaris*' bones are chambered and parts of its skull contain fat tissue (Fig. 5d1), both of which might increase light loss through scattering due to the heterogeneity of the tissue and its refractive index. The fact that *P. zvonimiri* and *A. ocellaris* still produce ONT eyeshine half as strong as that of the triplefins, suggests that head size is the dominant factor, followed by head structure. Pigmentation seems to be of secondary importance relative to these two. Ultimately, it is a constructive combination of all of these factors that produces the strongest ONT eyeshine, which seems to be the case in the two triplefins.

Extraocular optic nerve pathway and light guidance

The optic nerve (ON) is formed by all ganglion cell axons, which converge at the optic disc and form the intraocular ON that crosses the retina, RPE, choroid, and scleral eyecup. Outside the eye it continues as the extraocular ON, whose features and how they relate to the ONT eyeshine are discussed in this section. The intraocular nerve will be discussed in the next section.

The strength of ONT eyeshine will decrease with increasing distance between head surface and ON (referred to as ON depth from here on), because a longer light path increases losses through scattering and absorption. The measured ON depths of the specimens used for histology varied from 1.09 ± 0.05 mm ($n = 2$) in *T. melanurus* to 2.16 ± 0.31 mm ($n = 4$) in *A. ocellaris* (Table 1). Expressed relative to mean head diameter, however, ON depth was similar across all four species (Table 1). This suggests that *T. delaisi*'s small ON depth is not specifically adapted to enhance ONT eyeshine, but results from the correlation with its small head and body size, both of which can be small for many other reasons [11]. Comparing the ONT eyeshine intensity with ON depths shows no obvious correlation (Table 2). Hence, although a small ON depth certainly contributes to strong ONT eyeshine, its effect can be modulated by e.g., pigmentation as in the case of *T. melanurus*.

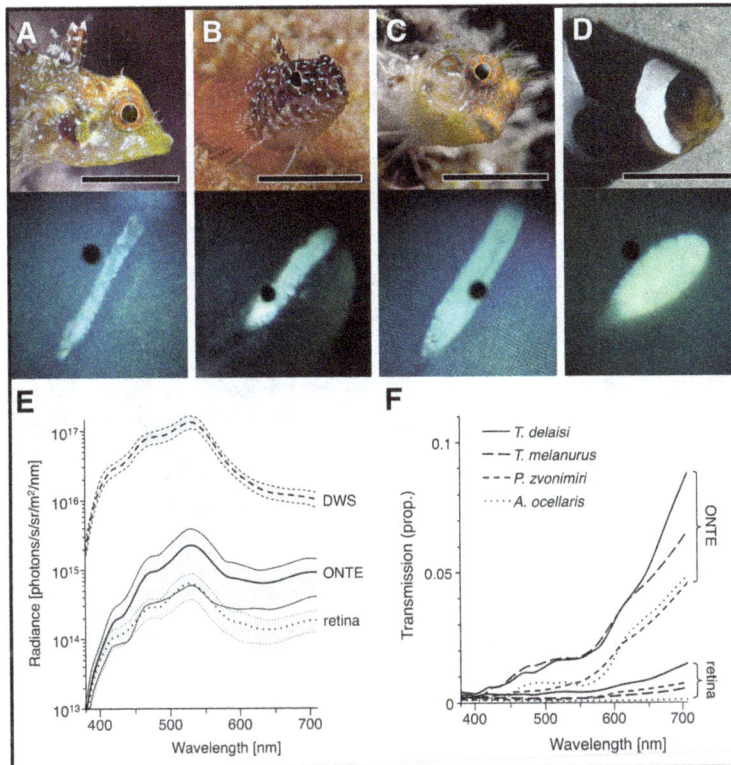

Fig. 3 Studied species and their ONT eyeshine. **a-d** Habitus and optic disc (OD) shape of *Tripterygion delaisi* (**a**), *Tripterygion melanurus* (**b**), *Parablennius zvonimiri* (**c**), and *Amphiprion ocellaris* (**d**). Fish images a, b, and c were taken in the field, d in the laboratory. *Scale bars* equal 1 cm. OD images were taken through a spectrophotometer with attached endoscope under dorsal illumination of the fish's head. The *black dot* is the detection area of the spectrophotometer and was positioned within the OD during measurements. The bluish tint is due to the cyan filter used for illumination. **e** Absolute radiance measurements ($n = 5$, mean $\pm$ SD) of a diffuse white reflectance standard (DWS), *T. delaisi*'s optic disc with ONT eyeshine (ONTE), and the adjacent retinal PET eyeshine, all under the same conditions (see Additional file 6). The DWS served as a proxy for illumination intensity. **f** Mean transmission efficiency of the study species' ONT eyeshine and retinal background relative to the DWS (*T. melanurus*: $n = 2$, all others: $n = 5$). Transmission is poor for short-wavelength light in all species, but consistently rises with increasing wavelength. *T. delaisi* features the highest proportional transmission, reaching about 8% at the red end of the visible spectrum

Many biological tissues exhibit light guiding properties under certain conditions and to varying extents [12–18]. We argue that this also applies to the ON. Light guidance is usually assumed to arise from total internal reflection of light within structures that exhibit a difference in refractive index between internal and external medium, as was the case for the light guidance properties of the dentine layer in vertebrate teeth [19]. More recent studies, however, suggest that the underlying mechanism is repeated anisotropic scattering along the low-refractive-index tubules embedded in the high-refractive-index dentine tissue, and that this phenomenon may also occur in neural tissue [15, 20]. A confirmatory finding showed that Müller cells appear to funnel light arriving at the inner retina towards individual photoreceptors [12, 14]. Whatever the mechanism, transitions in refractive indices are a key prerequisite for light guiding. The ON's structure creates such transitions on three different, yet interconnected, levels of anatomical detail.

1. ONs contain parallel-running myelinated axons. Due to their high lipid and protein content, myelin sheaths have a high refractive index ($n = 1.455$) compared to neural cell somata ($n = 1.358$) and the aqueous extracellular medium ($n \cong 1.335$) [14, 21]. Myelinated axons have been shown to trap and guide laser light through internal reflection [22], but they also constitute high-refractive-index tubular microstructures and thus fulfil the requirements for anisotropic scattering as described for dentine [15].

2. Nerves are interspersed with connective tissue, consisting of bundles and layers of collagen fibres. Their high refractive index, compared to the surrounding cytoplasm and extracellular matrix, and regular arrangement can make collagenous layers reflective enough to serve as *tapetum lucidum*, as they do in ungulates [4]. Thus, they might contribute to internal reflection in the ON as well.

3. Teleost ONs are usually ribbon-shaped and pleated [23], which creates alternating layers of nerve tissue and interstitial space, which constitutes yet another level of

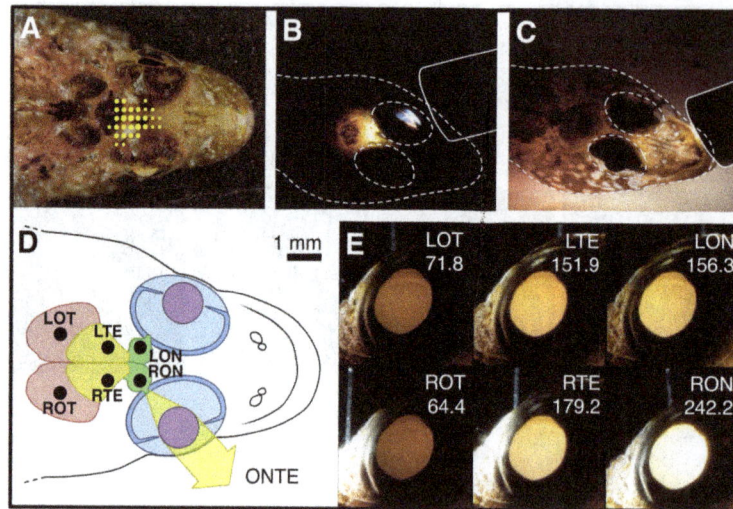

Fig. 4 ONT eyeshine excitation efficiency and light path in *T. delaisi*. **a** Excitation map showing area where ONT eyeshine can be induced by dorsal illumination with small light spots (ca. 1 mm²). *Dot size* corresponds with perceived intensity of ONT eyeshine in the right eye. **b** Reversing the light path, i.e., shining light through the (*left*) pupil of a recently sacrificed fish, produces glow corresponding to the maximum excitation field just behind the eye (compare a). Position of glass fibre is shown as *solid line*, head contour as *dashed line*. **c** Aiming the glass fibre light at the snout makes most of the head glow, except for the eyes. **d** Schematic of *T. delaisi*'s head in dorsal view. The *black dots* represent the positions of ONT eyeshine excitation, the result of which is shown in **e**). *LOT/ROT left/right* optic tectum, *LTE/RTE* left/right half of telencephalon, *LON/RON left/right* optic nerve, *ONTE* optic-nerve-transmitted eyeshine. **e** ONT eyeshine intensity in the right eye of a recently sacrificed *T. delaisi* with dorsal cranium removed. The tip of a single glass fibre was placed over different brain areas, as specified in **d**). The images show the correspondent ONT eyeshine emerging in the right eye. *Numbers* indicate pupil mean grey value as a proxy for ONT eyeshine brightness. All images were taken with a DSLR camera and identical settings

interfaces with changes in refractive index that might enhance internal reflection. We found this pleated-ribbon-like ON shape with 4–11 layers, each 30–60 μm thick, prominently present in all four study species (Fig. 6a1 to c1; Table 1). The ONs in *T. delaisi* and *T. melanurus* feature relatively widely and regularly spaced layers directly behind the eyes (Fig. 6a1). The ONs of *P. zvonimiri* feature few and well-separated layers, compared to *A. ocellaris*'s twice as many and more densely packed pleats (Table 1). The fact that their ONT eyeshine intensities are similar to each other, yet lower than those of the two triplefins (Table 2), could indicate that many layers with regular spacing are optimal for ONT eyeshine generation.

Additionally, the above features are non-exclusive in enabling and promoting light guidance. Combined they may act synergistically and generate a stronger effect than what the reflectivity of the individual structures would predict. An in-depth analysis and quantitative assessment of the extent of light guidance by the ON, however, is beyond the scope of this initial study, but should be addressed in future experiments.

Further factors that should impact the ON's capacity to capture and guide light are its cross-sectional shape and orientation, especially if the pleated ribbon structure plays a major role. In *T. delaisi*, *T. melanurus*, and *P. zvonimiri*, the ON widens immediately behind the eye, exposing a

larger area to incoming light (Fig. 5a2-3 and b2-3; Fig. 6b1). In contrast, the ONs of *A. ocellaris* have a round cross-section that does not change much in diameter upon leaving the eye (Fig. 5d2-3). The orientation of the ONs varied along their paths to the brain in all species examined (Table 3; Additional file 4). The rate of torsion (degrees per millimetre of ON) was similar when averaged at the genus level (78.2 to 93.4°/mm), but showed considerable variation between individuals, and even between the two nerves of the same individual. In the first third of the ON, immediately behind the eye, however, the pleats were close to horizontal in all species (Table 3). In a stack of horizontal layers, downwelling light would have to pass all reflective interfaces to make it through the nerve, thus maximising the chances and proportion of light to be reflected internally. With increasing tilt of the layers, however, downwelling light would pass fewer and fewer interfaces, thus reducing the chances of reflection. Additionally, the ON layers expose their maximal surface area to the downwelling light when they are horizontal. Given the torsion found in all study species, however, the orientation becomes suboptimal further down the ON's path to the brain. The combination of a widened ON and nearly horizontal nerve layers directly behind the eye locally maximizes the potential to capture incoming light and further supports why ONT eyeshine can be induced most strongly in that specific area.

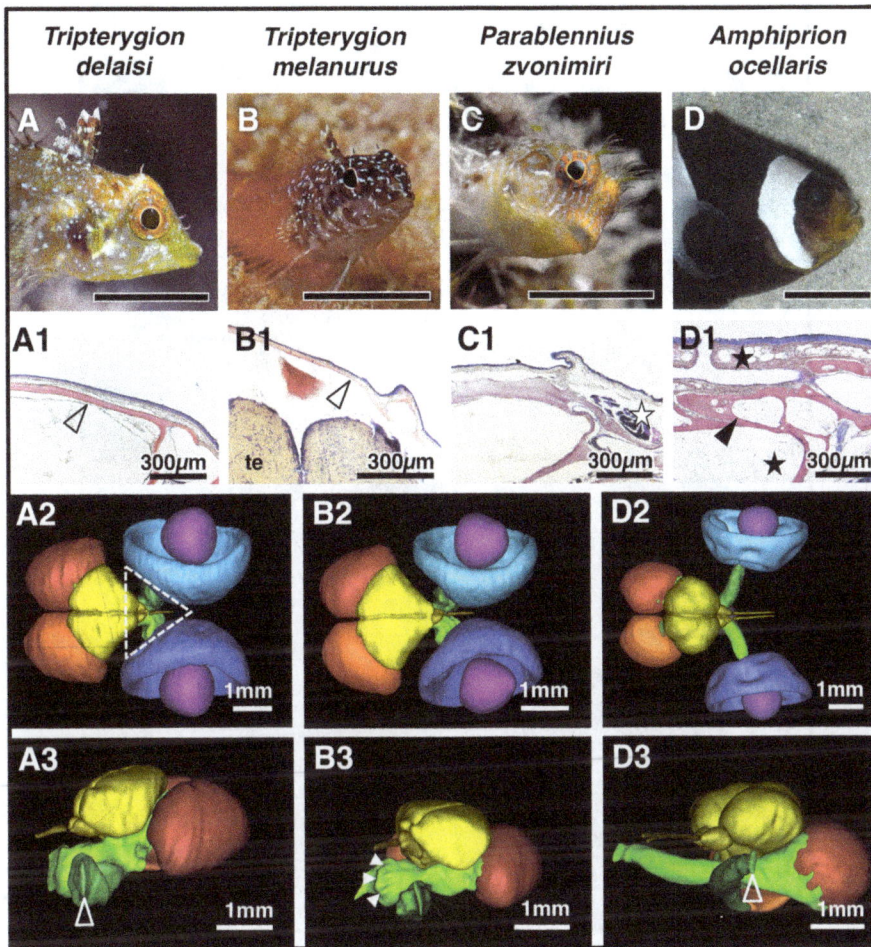

Fig. 5 Skull structure and visual system of *T. delaisi* (**a**), *T. melanurus* (**b**), *P. zvonimiri* (**c**), and *A. ocellaris* (**d**; *scale bars* 1 cm). **a1-d1** Close-up of the dorsal skull and dermis. Triplefins have delicate but solid bones (**a1-b1**, *open arrowheads*; *te* telencephalon) compared to the thicker, chambered bones (**d1**, *black arrowhead*) of the clownfish. The skull is only covered by a thin dermis in triplefins, while the blenny has muscles running between the dermis and the skull (**c1**, *white star*), and the clownfish has fatty tissue (**d1**, *black stars*) throughout its head. **a2-d3** MRI segmentations of inner head anatomy in *T. delaisi*, *T. melanurus*, and *A. ocellaris*: retinae (*blue*), lenses (*purple*), ONs and tracts (*green*), optic tecta (*red*), telencephalon, olfactory nerves and bulbs, and diencephalon (*yellow*). *T. delaisi* (**a2**, **a3**) and *T. melanurus* (**b2**, **b3**) are practically identical, except for differences in absolute size. **a2-d2** Dorsal view. The *dashed triangle* frames the exposed part of the ON, which coincides with the area of most efficient ONT eyeshine induction (see Fig. 4a). In *A. ocellaris* (**d2**, **d3**), the telencephalon covers a smaller part of the nerves, but these lie deeper in the head. **a3-d3** Frontolateral view onto the left optic disc. While narrow and elongated in triplefins, it is shorter and elliptical in the clownfish (*open arrowheads*). The ridges along the nerve surface in **b3** (*solid arrowheads*) result from the triplefins' loose ON pleats. The clownfish's ON is smooth in comparison, due to the denser pleating

Two observations support the hypothesised directional propagation of light in *T. delaisi*'s ONs. First, when we reversed the light path by shining light into the pupil, only a limited area of the head lit up (Fig. 4b) that corresponded closely with the area of ONT eyeshine excitation (Fig. 4a). By comparison, pointing the same light source at the same fish's snout made the whole head glow, demonstrating random scattering within non-neural tissue (Fig. 4c). Second, we illuminated specific areas of a dead specimen's brain to determine their contribution to the strength of the ONT eyeshine emanating from the right eye (Fig. 4d). If light propagation in neural tissue were completely isotropic, i.e., homogeneous, the distance between light

source and the right eye's optic disc should determine ONT eyeshine intensity. If light is scattered anisotropically, or even guided by the neural tissue as we assume, the neural pathway should correspond with the light's path to some extent and affect ONT eyeshine intensity accordingly. Since the ONs cross at the chiasm, the right ON is mostly covered by the tip of the right telencephalon, but projects to the left optic tectum (and vice versa for the left ON). Figure 4e shows the unedited images, all taken with the same, fixed camera settings, of the ONT eyeshine that results from illuminating the spots indicated in Fig. 4d. The numbers represent the average grey value of the pupil area as a rough but objective measure for

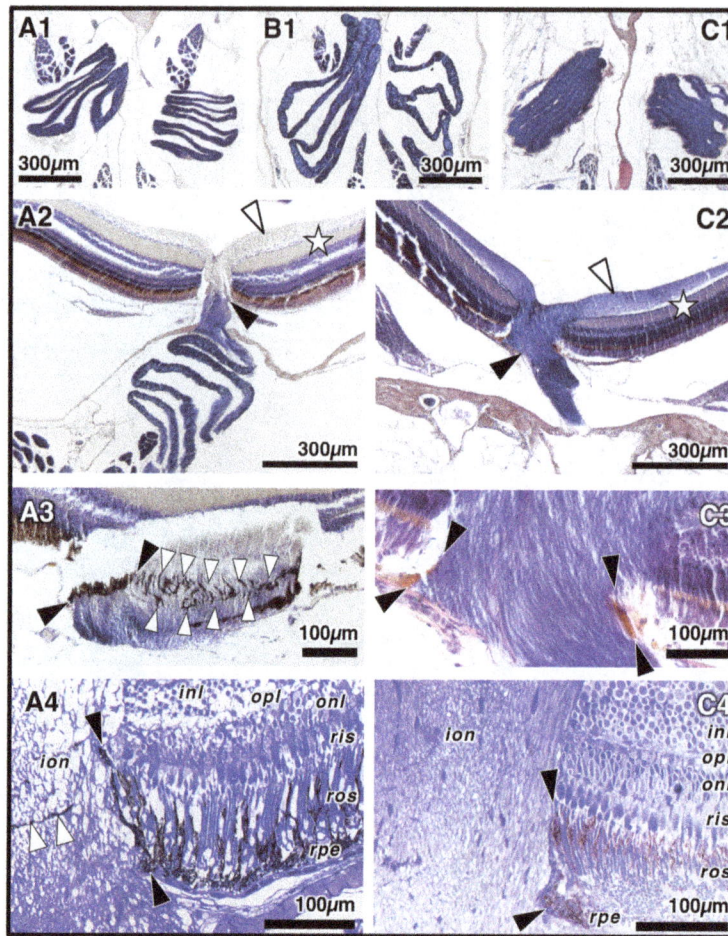

Fig. 6 Optic nerve characteristics of *T. delaisi* (**a1-a4**), *P. zvonimiri* (**b1**), and *A. ocellaris* (**c1-c4**). **a1-c1** Transverse sections showing elaborately pleated ONs that differ in pleat number, thickness, regularity, and spacing. In *T. delaisi* (**a1**), pleats are further apart than in *A. ocellaris* (**c1**) and more evenly spaced than in *P. zvonimiri* (**b1**). This regular pattern of alternating high-refractive-index neural tissue and low-refractive-index interstitial spaces may enable the ON to act as a biological light guide (*T. melanurus'* ON structure is identical to *T. delaisi*'s). **a2-c2** Coronal sections of the intra- and extraocular ON in *T. delaisi* (a2) and *A. ocellaris* (c2). The PTAH staining, which stains myelin dark blue, confirms the common pattern that axons are unmyelinated within the nerve fibre layer (NFL) of the retina (*white arrowhead*) and become myelinated outside the eyecup (*black arrowhead*) in *T. delaisi*. The axons of *A. ocellaris* stained dark blue throughout both the intraocular ON and the NFL of the retina (*black and white arrowhead*), indicating that at least some of them are myelinated within the NFL. **a3-c3** Detail of the intraocular ON (coronal section). The RPE forms a pigment sheath (*black arrowheads*) that surrounds the ON in both *T. delaisi* (**a1**) and *A. ocellaris* (**b1**). In *T. delaisi*, the sheath comprises a distinct band of pigmented processes that criss-crosses the entire intraocular ON (*white arrowheads*). **a4-c4** Semi-thin sections of the intraocular ON confirm that the pigment sheath (*black arrowheads*) is an extension of the RPE in both species. The sheath extends just far enough to shield the light-sensitive photoreceptor outer segments (ros) (between the *black arrowheads*). Within-nerve pigment processes (*white arrowheads*) seem to be only present in *T. delaisi*. Abbreviations: *inl* inner nuclear layer, *ion* intraocular optic nerve, *onl* outer nuclear layer, *opl* outer plexiform layer, *ris* receptor inner segments, *ros* receptor outer segments, *rpe* retinal pigment epithelium

brightness. The resulting ONT eyeshine was consistently brighter when the illuminated spot lay along the neural path of the right ON, compared to similarly distant other spots. Specifically, the left ON and right telencephalon sites had the same distance to the right optic disc, yet the ONT eyeshine from the right telencephalon was clearly brighter; the left telencephalon site was further away than the left ON site, yet both produced almost equally bright ONT eyeshine; the left optic tectum had a slightly greater distance but still produced brighter ONT eyeshine than the right optic tectum. All these observations support a non-random

component in the transmission of light through the neural tissue and the optic nerve in particular.

In summary, it seems plausible that ONs guide light. Whether this is achieved through internal reflection, anisotropic scattering, or a combination of both, is as yet unclear. The structures that allow for the above mechanisms are common to the optic nerve architecture in most teleosts [23]. Bearing in mind that all investigated species produced measurable ONT eyeshine, the slight differences in their optic nerve structure cannot be regarded the main determinant of ONT eyeshine occurrence and intensity.

Nevertheless, the relatively large light-exposed surface, as well as the regularly pleated, widely spaced, and horizontally oriented ON layers found in triplefins may increase the nerve's ability to capture and guide light, thus posing an additional contribution to the triplefins' strong ONT eyeshine.

Intraocular optic nerve and photoreceptor shielding

The intraocular ON of all four species is surrounded by a pigmented sheath formed by the RPE who's main functions are to prevent backscattering of light not absorbed by photoreceptors and to block light from reaching the photoreceptors through the fundus from the back of the eye. Apparently, that function extends to shielding the photosensitive outer segments of nearby photoreceptors from light leaking from the ON (Fig. 6a3 and c3). We found additional RPE processes within the intraocular ON in both *Tripterygion* species, but not in *A. ocellaris* (Fig. 6a4 and c4). This might be due to a greater need to shield the photoreceptors from the stronger ONT light in the two triplefins compared to *A. ocellaris*. For *P. zvonimiri* we lack the required semi-thin histological sections to assess this aspect.

The optic discs varied in shape from a tapered oval in *A. ocellaris* to a long and narrow strip in *T. delaisi* (Fig. 3a-d). Optic disk size increases with axon number and degree of myelination, being smaller with unmyelinated axons [24]. Being a blind spot, it seems reasonable to assume selection for a minimal size. This constraint and the retrograde myelination of fish ONs, during ontogeny [25] and regeneration [26], may explain why only the extraocular part of the axons is myelinated in most fishes. PTAH staining, which differentially dyes several tissue types and substances, including myelin in deep blue, confirmed this pattern in *T. delaisi*, *T. melanurus*, and *P. zvonimiri*, whose axons are unmyelinated in the intraocular ON and the axons were homogenously and densely packed (Fig. 6a2). From the perspective of light travelling through the ON towards the retina, both the axon's myelin sheaths and the regular structure with layers of alternating tissue types gradually disappear along the intraocular ON. This removes the differences in refractive index needed for light guiding and presumably allows ONT light to eradiate from the optic disc area and leave the eye as ONT eyeshine rather than being guided towards the photoreceptors. In contrast, the retina of *A. ocellaris* possesses at least some intraocularly myelinated axons (Fig. 6c2), which is similar to previous findings in other vertebrate species [27–29]. When the axons' myelin sheaths extend past the optic disc, light contamination may enter the retina. This could expose the photoreceptors to additional incident light and possibly affect the signal-to-noise ratio. Too much noise may impair perception (as discussed below). A lack of myelinated axons in the retina, as seen in *Tripterygion* and *Parablennius*, will reduce such noise interference.

Possible functions of ONT eyeshine

Thus far, we have merely described ONT eyeshine and the factors that determine its brightness. Our results suggest that the effect strongly depends on small body size and is modulated by head pigmentation, head anatomy, and ON structure. Given these simple requirements for the occurrence of ONT eyeshine, and that we did not find any unique adaptations that specifically allow or maximise ONT eyeshine, it follows that many sufficiently small and lightly pigmented species should exhibit and be affected by this phenomenon. Over the last few years of fieldwork and excursions, we indeed observed ONT eyeshine in more fishes than the four species that were investigated in detail here (Fig. 7), but such observations were sparse and capturing them on camera difficult. Several factors make it hard to spot ONT eyeshine in the field and may explain why it has been overlooked thus far: ONT light is emitted into the environment as a narrow beam in a forward direction. Benthic fishes usually avoid long eye contact, and most small fishes even turn away when approached. As a consequence, even if an observer spotted a fish at the right moment and from the perfect angle, the phenomenon is usually so ephemeral that it is likely dismissed as a passing reflection, if noticed at all. Therefore, it is understandable to question the significance of the effect and whether it is just a coincidental by-product that is inevitable in sufficiently small fishes. At this stage, we cannot exclude that option since none of *T. delaisi*'s anatomical features described in this study could be regarded as a specific adaptation solely evolved to enhance ONT eyeshine. Nevertheless, from the fish's perspective, the effect is constantly present, and even

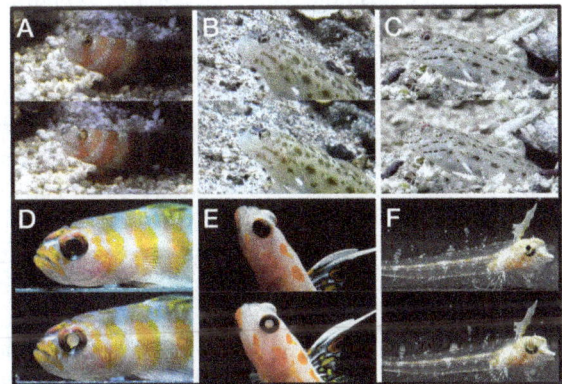

Fig. 7 ONT eyeshine occurrence in other fish species. Additional examples of small, benthic fish species observed to exhibit ONT eyeshine: **a** *Amblyeleotris periophthalma*, **b** *Ctenogobiops pomastictus*, **c** *Ctenogobiops maculosus*, **d** *Trimma cana*, **e** *Stonogobiops yasha*, **f** *Enneapterygius pusillus*. *Upper* and *lower* images show the same individuals only moments apart. Slight differences in eye orientation direct the beam of ONT light either away from the observer and let the eyes appear normal (*upper* images), or towards the observer and reveal the ONT eyeshine (*lower* images). Pictures a-c were taken in the field, d-f in field station facilities; all courtesy of NKM

a coincidental effect may have consequences and interact with other phenomena. Natural selection often exploits initially coincidental effects and recruits them for new functions, if they constitute a significant advantage to their bearers. Therefore, it is worthwhile not only to discuss the potential costs of ONT eyeshine's interference with the visual system, but also to speculate about possible benefits and functions it might serve. We discuss four options we think warrant attention and may be tested in future studies.

Interactions with the visual system

It is intuitively tempting to assume ONT eyeshine affects visual performance negatively, as ONT light enters the eye from the seemingly wrong direction. If such light reached the photoreceptors, it would arrive scrambled and unfocussed since it never went through an optical system. Hence, it could certainly not serve image formation, but would nevertheless interact with other light and affect vision. Depending on the circumstances, the interference would not necessarily have to be detrimental. As a direct benefit, ONT light could still convey general information such as changes in ambient light caused by a passing shadow from a predator swimming by overhead. ONT light could also stimulate intrinsically photosensitive retinal neuronal cells that were found in some teleost retinas and might regulate circadian rhythms, modulate receptive fields, or tune the retinal circuitry to the dynamically changing aquatic light environment [30–32]. The possible contribution of ONT light can only be minor in this context, however, since the light reaching the retina through the conventional pathway stimulates the photosensitive cells as well. Teleosts also possess several photosensitive brain areas, not only in the pineal but also throughout the deep brain, that are involved in many regulatory processes [33]. Both the translucency of the skull and the light guidance by the optic nerves could originally be adaptations that allow light to reach those brain areas. ONT eyeshine may have originated as a side-effect of this, as there is no a priori reason to assume the ON would guide light in only one direction.

Another possible effect relates to the extreme flicker on benthic surfaces that is generated by wavelets on the surface on sunny days and which is a challenge for human eyes and digital cameras alike. That flicker results in synchronous fluctuations in the intensity of ONT eyeshine (personal observations; see also Additional file 1). Hence, ONT light may provide feedback to prime visual perception for illumination variance in a visual scene [34], improving vision under these conditions. Allowing ONT light to reach the rods and cones to achieve the above also means adding an unspecific, uncorrelated component to the total visual stimulus, i.e., increasing receptor noise levels. Therefore, the amount of ONT light reaching the receptors should be kept low and well

regulated, which might be achieved by the unmyelinated axons inside the retina and by the RPE processes that extend into *T. delaisi*'s intraocular ON. Otherwise, the aforementioned benefits would come at the cost of reduced general image contrast and acuity. Investigating these effects was beyond the scope of this study, but would be of interest for future research.

Predator and/or prey detection

Could ONT eyeshine be used as a searchlight? Reflective targets in the environment may reflect the emitted ONT eyeshine back to *T. delaisi*, which then might use this information for target detection. Such targets could be eyes that exhibit reflection-based eyeshine [1], which is indeed featured by both prey (e.g., small crustaceans, [35]) and predators (e.g., scorpion fishes, [1]) of *T. delaisi*. Reflective eyeshine can indeed be induced by weak light sources, as long as the distance between sender and target is short, as in small benthic fish searching for prey, and the light source is coaxial to the viewing axis, which applies almost perfectly to ONT eyeshine. It has previously been suggested that reflective eyeshine can be conspicuous and might alert potential predators [36], or enable private communication with conspecifics [37]. The practical use of ONT eyeshine in such a context has one intrinsic limitation. Since the light passes through the triplefin's optical system both when emitted and returning, it is focused on the same retinal area where it originated, i.e., the optic disc, a blind spot. This restriction could be circumvented if the triplefin looked at a target with defocussed eyes, causing the target's image to blur on the retina and excite some of the receptors adjacent to the optic disc. Consequently, alternating accommodation would make the target's eye appear to blink.

Camouflage

Eyes are striking and attention-drawing structures, which is exemplified by the use of false eyes, i.e., eyespots, to deter or mislead predators and reduce predation, e.g., in butterflies [38] and fishes [39]. In cryptic animals, however, eyes can reduce camouflage [40] and some predators even specifically target eyes or eyespots [41]. Hence, the usual, round and black, shape of a pupil may be a conspicuous telltale sign for the presence of an otherwise cryptic species [8]. Similar to what has been suggested for corneal iridescence [3, 42], ONT eyeshine could reduce the contrast between iris and pupil by letting the latter appear bright and blend in with the luminance of the surrounding iris.

The narrow, beam-like shape of the ONT eyeshine, however, limits such a function. To achieve the assumed effect, *T. delaisi* would have to permanently point the ONT eyeshine, and thus its blind spot, at an observer and follow its movements, which could be done in principal, but with maximally two observers simultaneously, taking into account *T. delaisi*'s ability to move its eyes independently. Reflection-based and PET eyeshine may be more useful in a camouflage context, as they can be seen from much wider angles [1, 8, 43]. *T. delaisi* does also exhibit PET eyeshine, as demonstrated by the dimly green glowing retina when viewed through an endoscope (Figs. 2c and 3e-f). Both effects benefit from the translucent skull of *T. delaisi* and increase with ambient light levels and they are not mutually exclusive in their potential camouflage function.

Intraspecific communication

ONT eyeshine may also be a component of visual, intraspecific communication. To an observer, the ONT eyeshine does not differ from most other visual signals. Yet, two characteristics make it unique. First, sending a signal from the eye itself may generally pique an observer's attention [40]. Second, emitting a narrow beam from the eye allows the sender to target a specific receiver. Minute eye movements in the sender will turn the signal on and off from the receiver's perspective. To further assess this possible function, we modelled the perceived contrast between a pupil with and without ONT eyeshine as seen by another *T. delaisi*. While ONT eyeshine does not generate a perceivable chromatic contrast, there is a clearly visible achromatic contrast at both 5 and 20 m depth, with 4.85 ± 2.1 JNDs and 5.04 ± 2.2 JNDs, respectively (Additional file 5). ONT eyeshine could thus serve as an easily visible, yet non-interceptable, private signal between specific individuals. Furthermore, this scenario requires the light to travel the distance between sender and receiver only once, and hence the signal will be perceivable over greater distances compared to the two-way path needed for active photolocation.

Since the narrow spread of the beam works to the fish's advantage in this context, and the generated contrast is strong, this might be the most likely of the three discussed functions for ONT eyeshine.

Conclusions

Optic-nerve-transmitted eyeshine is a previously undescribed phenomenon that lets diurnal fishes' pupils glow. It results from collection of ambient light by the ON in the post-orbital head area. The light is then transmitted into the eye and emitted through the pupil into the environment. This one-way, transmission-based light path distinguishes ONT eyeshine from the conventional, reflection-based types of eyeshine. ONT eyeshine is strong in the triplefin *T. delaisi* due to the additive effects of several traits, the most important being the short distance between ONs and the head surface. The ONs of teleosts generally feature specific structural characteristics that suggest the potential to partially guide light. Leakage of light from ONT eyeshine to the photoreceptors is prevented by an extended pigment sheath around the

intraocular ON and by the typical, unmyelinated retinal axons. As a consequence, most of the light leaves the eye through the pupil. Given that many small fishes feature at least some of the required traits, ONT eyeshine may be common among a wide range of small-sized teleost species. The exact impact of ONT eyeshine on the affected species' visual ecology remains to be investigated.

Methods
Summary of experimental procedures
ONT eyeshine was discovered while observing small benthic fish in the Red Sea and Mediterranean in 2007–2012 and subsequently confirmed in the laboratory. The fishes used in this study were either wild-caught in cooperation with two Mediterranean marine research stations (*T. delaisi*, *T. melanurus*, and *P. zvonimiri*), or originated from commercial fish breeders (*A. ocellaris*). We carried out spectrophotometric measurements of the ONT eyeshine on five *T. delaisi*, two *T. melanurus*, five *P. zvonimiri*, and five *A. ocellaris*. An additional eight individuals of *T. delaisi* were used to demonstrate the ONT eyeshine's appearance and causes, test the contribution of components of the visual pathway, and to measure the angular extension of the emitted beam of light. To induce ONT eyeshine, we illuminated a fish's head either with a LCD projector, or a cold light source and fibre optics. Still and video footage was taken with a Nikon D4 (Nikon Corporation, Tokyo 100-8331, Japan) and quantitative measurements were made with a PR-740 spectroradiometer (Photo Research Inc., North Syracuse, NY 13212-3349, USA).

Another ten *T. delaisi*, four *T. melanurus*, one *P. zvonimiri*, and nine *A. ocellaris* were used for comparative anatomy. To obtain diverse and complementary data at different levels of detail, we split the available samples among three procedures: MRI (two *T. delaisi*, one *T. melanurus*, three *A. ocellaris*), paraffin-embedded thick histological sectioning (four *T. delaisi*, two *T. melanurus*, one *P. zvonimiri*, three *A. ocellaris*), and resin-embedded semithin sectioning (four *T. delaisi*, one *T. melanurus*, three *A. ocellaris*). The single *P. zvonimiri* was assigned to paraffin-embedded sectioning because this method was expected to produce more data than the other two approaches. Because data from the different approaches could usually not be combined, resulting in low independent sample sizes, analyses are restricted to descriptive statistics.

Ecology of the investigated species
Tripterygiidae occur worldwide and generally inhabit hard substrates in the shallow littoral zone from tropical to temperate regions [44]. The genus *Tripterygion* occurs in the East Atlantic and Mediterranean, and underwent a recent radiation leading to four recognised and possibly several cryptic species [45–47]. Our focal species, *Tripterygion delaisi* Cadenat and Blache 1970, can be found across the

genus' entire distribution range, while *T. melanurus* is endemic to the Mediterranean [47]. Both species co-locate, prefer rocky substrates and prey on benthic invertebrates [48]. What distinguishes them is that they occupy different depth and light niches, which may have led to their divergence [49]. *T. delaisi* occurs between about 3 and 50 m [50], and is most abundant at depths of 6 to 12 m [44]. It shuttles between shaded microhabitats under rocks and overhangs, and exposed, sunlit surfaces [49] (Pers. obs.). *T. delaisi*'s visual system features a fovea, a cone-dominated, regular, square-mosaic pattern, is trichromatic, and restricted to visible light, since its ocular media do not transmit ultra violet light (unpublished data). *T. melanurus* can be found at depths of about 1 to 12 m, mostly between 2 and 8 m, where it inhabits only dimly and indirectly lit microhabitats, such as caves, crevices and the underside of ledges, often in association with sponges [49, 51, 52]. Its visual system has not been investigated, but may be expected to be similar to *T. delaisi*'s.

Both species live in small groups of two to six individuals from different age classes and both sexes in a loosely defined "home area" of just a few m^2 (personal observation). Males establish and vigorously defend territories during the breeding season [44, 53, 54]. *T. delaisi* is highly cryptically coloured [49], except for breeding males, which feature a black head and a bright yellow body. Although *T. melanurus* is mostly bright red throughout the year, with a grey-to-black head, it also appears cryptic in shady crevices at depth. Most of the time, both species sit still and only move their eyes to scan the environment. They never swim continuously, but rather move in dashes and freeze again in their new position. Even when a potential predator approaches, they stay put at first, and wait for the threat to pass. Only if approached too closely or quickly, they dash away in one to several short darts, just to freeze again (Pers. obs.). *T. delaisi* is easy to catch and keep for laboratory observations and measurements.

Blenniidae and Tripterygiidae both belong to the Order Blenniiformes, although their specific relationship is not yet confirmed [55, 56]. Members of both families often form guilds in the littoral zone of the Mediterranean, which have been studied extensively, e.g., concerning depth distribution [52, 53, 57–59], trophic interactions [48, 60–63], and habitat characteristics [64]. Given these similarities, we wanted to see whether blennies, represented by *Parablennius zvonimiri*, have similar morphological features in relation to the ONT eyeshine. *P. zvonimiri* is endemic to the Mediterranean and the Black Sea, where it inhabits mostly the lower intertidal and upper subtidal zones down to about 6 m, but is most abundant between depths of 0.5 and 1.5 m [59, 65, 66]. It feeds predominantly on periphyton [65], but opportunistically also on small invertebrates (pers. obs.), and prefers vertical rock walls as a substrate, which is correlated to the availability of the piddock holes it commonly occupies [65, 67].

Amphiprion ocellaris, the false clown anemonefish, belongs to the family Pomacentridae and is only distantly related to the Tripterygiidae and Blenniidae families [55], all being within the order Perciformes. Anemonefishes live in hierarchically structured groups of 2–8 individuals in symbiosis with sea anemones [68]. The symbiosis with their hosts [69–71], as well as the social interactions among individuals in a group [72–74], have been studied extensively. Anemonefishes are protandrous hermaphrodites, benthic spawners with male egg guarding [75, 76], and feed mostly on zooplankton [76, 77] with a few exceptions [78]. We chose the black-and-white morph of *A. ocellaris* to assess the effect of pigmentation on the ONT eyeshine, mainly because it had the darkest pigmentation of all considered fishes in the size range of our study species.

Origin of specimens

All triplefins (23 *Tripterygion delaisi* and 6 *Tripterygion melanurus*) and blennies (6 *Parablennius zvonimiri*) were wild-caught at either the *Centro Marino Elba* research station (Loc. Fetovaia 72, I-57034 Campo nell'Elba, Italy) in June 2013, or at STARESO research station (Pointe Revellata, BP33 20260 Calvi, Corsica, France) in June 2010, 2011, and 2016. Fish were kept in the aquarium facilities of the *Animal Evolutionary Ecology* group at the University of Tübingen until euthanized.

The clownfish (*Amphiprion ocellaris*, black-and-white morphs) originated from fish breeders. The nine individuals used for comparative anatomy were acquired on 14 August 2013 from *Oceanreefs marine aquariums* (51/7 Buckingham Drive, Wangara WA 6056, Australia) and housed temporarily in the aquaria facility of the *Neuroecology* group at The University of Western Australia until euthanized on 16 August 2013. The five individuals used for spectrophotometric measurements of their ONT eyeshine were acquired on 6 September 2016 from *Riffwelt* (Thomas-Walch-Str. 45a, A-6460 Imst, Austria). They were sacrificed and had their ONT eyeshine measured upon arrival.

Spectrometry and emission angles of the ONT eyeshine

We conducted quantitative, spectrophotometric measurements of the ONT eyeshine for five *T. delaisi*, two *T. melanurus*, five *P. zvonimiri*, and five *A. ocellaris*. The low number of *T. melanurus* is the result from both difficulties to catch them in the wild and the death of several successfully caught individuals before we could measure them. All fish were euthanized in seawater containing a lethal dose of 500 mg/l tricaine methanesulfonate (MS-222), re-adjusted to its original pH with sodium hydroxide, immediately before taking measurements. After their opercular movement had ceased, the fish were transferred to an acrylic glass cylinder filled with more of the MS-222 solution to ensure euthanasia. Fish were fixed with pins on a piece of rubber foam. Piercing the fishes was avoided to prevent bleeding. A piece of foamed-PTFE diffuse white reflectance standard with 45°-tilted surface was pinned to the rubber foam bed next to the fish. See Additional file 6 for an overview of the set-up.

The tank was then placed on a platform that allowed vertical, rotational, and translational movements. A PR-740 SpectraScan® spectroradiometer (Photoresearch, Chatsworth CA 91311, USA), attached to a tripod, was positioned above the platform with the tank. Measurements were taken through an endoscope (No. 86190 CF) attached via a C-Mount adapter (No. 80591 C, Karl Storz GmbH & Co. KG, Industrial Group, 78532 Tuttlingen, Germany). All illumination was provided by a KL2500 LCD cold light source (Schott AG, 55127 Mainz, Germany), set to light intensity 5E and using the inbuilt cyan filter, via a 15 mm Ø optic cable (No. 250102, Schott AG). The spectroradiometer and light source remained stationary while the fishes' eyes and the white standard were brought into position by the sole movement of the tank and platform. Nine measurements, three of each type, were taken for each eye, always in the same order: the white standard under the same illumination conditions as the subsequent measurements, serving as a proxy for the incident light intensity, then the optic disc exhibiting ONT eyeshine, and finally the surrounding retina and its potential PET eyeshine. Photographs of each fish's optic discs were taken between measurements and directly through the spectroradiometer's eyepiece with a Nexus 5 mobile phone camera. Data were collected with SpectraWin® (version 2.3.7, Photo Research Inc., North Syracuse, NY 13212-3349, USA), and processed and analyzed in JMP® (version 11.1.1, SAS Institute Inc., Cary, NC 27513-2414, USA).

Angular measurements were carried out in 2011 (before the main study) using six specimens from Corsica (Stareso) that had been collected in 2010 and 2011. For the emission angle measurements, the fish were placed in a small water tank and illuminated from above. The fish's eye was then observed from a few meter distance through a monocular telescope attached to a mobile tripod. While the set-up was moved in an arch around the tank, the points where ONT eyeshine became visible and vanished again were noted. Measuring the lengths of all sides of the triangle between these two points and the fish's eye, correcting for refraction at the glass/air interface, and applying the law of cosines, yielded the horizontal angular width of the ONT eyeshine. This procedure was repeated three times for each eye of six fish and the data then averaged per fish. The vertical angle was measured analogously by moving the tripod up and down instead of sideways for both eyes in five out of the original six fish.

Video documentation and ONT eyeshine generation pathway

We used one *T. delaisi* to photo- and video-document its ONT eyeshine under controlled conditions. The fish was transferred to a cylindrical, acrylic glass observation tank that rested on a turntable. This allowed us to adjust the horizontal angle. The fish was illuminated by two KL2500 LCD cold light sources (Schott AG, 55127 Mainz, Germany). One provided diffuse background illumination of the whole animal via a 15 mm Ø optic cable (No. 250102, Schott AG), and the other induction illumination for ONT eyeshine via a 3 mm Ø optic cable (No. 155101, Schott AG) that ended approximately 5 mm above the animal's head and illuminated only a 4 mm wide area above and behind the eyes. This allowed us to change the colours of the background and induction independently, and thus highlight the origin of the light contributing to ONT eyeshine. All photographs and videos were taken with a Nikon D4 (Nikon Corporation, Tokyo 100-8331, Japan).

After all footage from the live fish had been taken, the animal was sacrificed in a MS-222 bath as described previously. The fish was kept in physiological saline for marine teleosts [79] and used to document light transmission under a reversed light path. To that end we shone a narrow beam of light from a 1 mm Ø optic cable (FC UV1000-2, Avantes BV, NL-7333 NS Apeldoorn, Netherlands) through the pupil of one eye onto the optic disc and recorded which areas of the skull lit up most. As a control, we also illuminated other parts of the fish's head to document light transmission through normal scattering. To test the different brain areas' contribution to ONT eyeshine, we removed the skin and skullcap (posterior frontals and parietals) between and directly behind the eyes. Then a single glass fibre was positioned directly above the area of interest, limiting direct illumination only to the tested structure. Brain parts tested in this way included cerebellum, optic tecta, left and right sides of the telencephalon, and left and right ONs. The resulting ONT eyeshine was recorded photographically with the Nikon D4, using the same, manually fixed settings for all images. The ONT eyeshine's brightness was later assessed by measuring the average grey values of the pupil area in the images using ImageJ (version 1.46q; https://imagej.nih.gov/ij/index.html).

In another *T. delaisi*, we mapped which area of the nape and interorbital region of the skull shows the highest efficiency in generating ONT eyeshine. For this purpose, an LCD projector was positioned 2 m above a small tank with a living fish in a dark room. Due to their cryptobenthic nature, *T. delaisi* individuals usually sit inactive in the dark with no signs of increased activity, breathing, or stress. A laptop was used to produce a single white pixel that could be moved around on the otherwise black screen using the touchpad of an IBM laptop computer. By focusing the pixel on the head of the fish, it was possible to move it across the fish head and scan the whole fish in approximately 1 mm^2 sized squares, covering the whole head from the tip to the first dorsal fin. At each position, the strength of the ONT eyeshine was visually assessed in three steps: 0 (no ONT eyeshine), 1 (weak ONT eyeshine) and 2 (bright ONT eyeshine). We assigned these values to the corresponding grid coordinates on a photograph of the fish. The data of the left eye was mirrored and combined with that of the right eye to produce the excitation map in Fig. 4a.

Histology I: semi-thin, resin-embedded sections

Semi-thin histological sections were produced from four *T. delaisi* (Td5-8), one *T. melanurus* (Tm3), and three *A. ocellaris* (Ao1-3). Euthanasia was carried out in a solution of seawater and 500 mg/l MS-222, as previously. Exposure lasted at least 2 min or until the following conditions were met: ceased opercular movement and complete loss of buoyancy control for 1 min. Thereafter, the spinal cord was severed to ensure euthanasia. The total length, head width, and head height were measured, then the fish were decapitated, and their heads trimmed to the respective regions of interest. These samples were then fixed and stored according to the protocols described below. Subsequent histological processing and MRI scans were carried out at UWA (Perth) and UQ (Brisbane). *Tripterygion* and *Parablennius* samples were therefore imported to Australia according to DAFF regulations (permit no. IP13006051). The lower jaw was removed and the remaining head cropped, leaving a region from the rostral edge of the eyes to the caudal edge of the cranium. The sample was subsequently immersion fixed in a modified Karnovsky's solution containing 2% paraformaldehyde (PFA), 1.5% glutaraldehyde (GA), and 1% dimethyl sulfoxide (DMSO) in 0.155 M, pH = 7.3, Sorensen's phosphate buffer (SPB). Then, samples were washed once with a glycine solution to block residual aldehyde groups (1% DMSO and 1% glycine in 0.155 M, pH = 7.3, SPB). Finally, samples were washed two more times and eventually stored in 0.155 M, pH = 7.3, SPB containing 1% DMSO and 0.1% sodium azide, which prevents bacterial and fungal growth. Each washing step lasted 24 h.

Sample preparation continued without decalcification in a Lynx EL automated tissue processor. To fit the baskets, all heads were split medially and further trimmed to the caudal region of the eye and the ON. The tissue processor dehydrated the samples through an ascending ethanol series and infiltrated them with epoxy resin, using propylene oxide as an intermediary solvent overnight. Infiltrated samples were embedded in resin blocks and polymerised at 60 °C for 72 h. The resin blocks were sectioned with a LKB Bromma Ultratome NOVA. Glass knives were used to skim through the blocks and position was controlled every 50 µm. Once within the target area, consisting of optic disk, surrounding retina and ON, three to four 1-µm-sections were cut, using a DiATOM ultra diamond knife (45° angle), before skimming

another 50 µm with a glass knife. Successful target sections were transferred to water droplets on a positively charged glass slide and put on a hotplate (90 °C) to expand and dry. Sections were stained with Toluidine blue after the resin was removed with alcoholic sodium hydroxide, and mounted using Entellan® (Merck Millipore, 64293 Darmstadt, Germany).

Histology II: thick, paraffin-embedded sections

Four *T. delaisi* (Td1-4), two *T. melanurus* (Tm1-2), one *P. zvonimiri* (Pz1), and three *A. ocellaris* (Ao4-6) were prepared for paraffin-embedded thick sections as follows:

Euthanasia and sample trimming were carried out as described above. The samples were then immersion-fixed in 0.155 M SPB at pH = 7.3, containing 4% PFA and 1% DMSO. The fixed samples were washed three to four times, each lasting 24 h, with 0.155 M, pH = 7.3, SPB containing 1% DMSO and 0.1% sodium azide. The same solution also served sample storage. To allow cutting with steel blades, the tissue was furthermore decalcified in citrate (7.5%) buffered formic acid (15%) for 48 h. Larger specimens were split medially to facilitate processing.

Samples were dehydrated in an ascending ethanol series, cleared in xylene and infiltrated with paraffin in a Shandon Citadel™ 2000 tissue-processing carousel. When embedding them in paraffin blocks, samples were oriented such that whole-head specimens were cut in the transverse plane, while one half of split-head specimens was cut coronally, the other cut sagittally. Blocks were sectioned on an AO® rotary microtome (model 820) at 10 µm section thickness using a steel blade. Sections (ribbons of 5–10) were floated in a water bath (43 °C, degassed and deionised water, 0.01% gelatine) to flatten and expand, and then transferred to standard glass slides and dried overnight at 60 °C in a heating cabinet.

Samples were then subjected to a modified phosphotungstic acid haematoxylin (PTAH) staining protocol [80], which stains several tissue types simultaneously and differently, e.g., muscles dark purple, connective tissue reddish-brown, and neuroglia (including myelin sheaths) deep-blue. We proceeded as follows: The sections were first dewaxed in xylene and rehydrated through a descending ethanol series. Then they were postfixed in 3% potassium dichromate solution for 20 min and rinsed with water afterwards. Next, the sections were immersed in 0.25% acidified potassium permanganate solution for 1 min, rinsed in water, and then bleached in 2% oxalic acid until clear and rinsed in water. Finally, sections were stained overnight in the PTAH solution. This solution needs to be prepared in advance as follows: For a batch of 500 ml, individually dissolve 0.5 g haematoxylin, 10 g phosphotungstic acid and 0.0625 g potassium permanganate in 100 ml, 375 ml and 25 ml distilled water, respectively; mix the component solutions and allow ripening at least overnight, best for a week; the solution's shelf life is several weeks. After staining, the sections were dehydrated through an ascending ethanol series and cleared once more in xylene, before mounting them with Entellan® (Merck Millipore, 64293 Darmstadt, Germany).

Magnetic resonance imaging

MRI was performed on two *T. delaisi* (Td9-10), one *T. melanurus* (Tm5), and three *A. ocellaris* (Ao7-9). After euthanizing the fishes as described previously, their heads were trimmed to upper jaw, eyes, and cranium. Samples were immersion-fixed for 24 h in phosphate buffered saline (PBS) containing 4% PFA and 0.25% of 1 M Gadovist® (Bayer AG, 51373 Leverkusen, Germany), and washed three to four times, 24 h each, in PBS with 0.1% sodium azide and 0.25% 1 M Gadovist® at pH = 7.3. Gadovist® is a gadolinium-based contrast agent for MRI scans. Limited availability of the MRI scanner led to several weeks delay between sample fixation and scanning. Specimens were stored in the final washing step solution at 4 °C for the intervening time.

All scans were acquired overnight on a 16.4 T Ultra-shield™ Plus 700 WB Avance NMR spectrometer running ParaVision® 5.1 software (both from Bruker BioSpin GmbH, 76287 Rheinstetten, Germany) at the Centre for Advanced Imaging, University of Queensland, Brisbane, Australia. For most fish (Ao7-9, Td9) individual scans were obtained and processed. Tm5 and Td10, however, were imaged and processed together, resulting in a few divergent parameters compared to the other specimens. Acquisition parameters of the T_1-weighted 3D FLASH scans were as follows: Repetition time 40 ms, echo time 4.8 ms (Tm5/Td10: 15.2 ms), number of excitations 9 (Tm5/Td10: 8), flip angle 45°, field of view $7 \times 7 \times 7.7$ mm (Tm5/Td10: $7 \times 7 \times 20.5$ mm), image matrix 464/464/512 (Tm5/Td10: 464/464/1216), isotropic resolution 15 μm (Tm5/Td10 slightly anisotropic $15 \times 15 \times 17$ μm).

We used the MRI data to create digital 3D segmentations of the structures we suspected to be involved in the ONT eyeshine by being part of the light path. To that end we used the segmentation programme ITK-SNAP, version 2.4.0 from 21 November 2012, developed by Yushkevich et al. [81]. ITK-SNAP features a user-guided automatic segmentation, which we used for a draft segmentation that we then corrected and refined manually, where necessary.

Digital image processing, measurements and statistical analysis

Both paraffin-embedded histological sections and MRI scans were used to obtain morphological data for between and within-individual comparisons as follows: All histological thick sections from the region of interest of each respective fish sample were digitalised using a Leica DM5000 B microscope equipped with a Leica DFC320 camera (Leica

Microsystems, 35578 Wetzlar, Germany). The resulting TIFF images were resized and centred in Adobe® Photoshop® CS4 11.0.2 for easier processing. Individual images were then aligned and stacked using the Elastic Alignment and Montage plugin [82] in ImageJ 1.48t (Fiji distribution package). MRI images for non-volumetric measurements were obtained by exporting individual images from ITK-SNAP via the built-in snapshot function.

All linear, angular and areal measurements from digital images were taken in ImageJ 1.48t (Fiji distribution package). Measurements along the ON were taken in 40–60 μm steps between optic disc and optic chiasm in both the transverse and sagittal planes. This resulted in 10 to 21 sampled sections per fish, depending on the length of the ON. In all such sections, the length and angle of each individual layer, as well as the overall cross-sectional area (CSA) of the ON were taken. These raw measurements were then averaged per sampled section, fish, and genus, depending on what was to be compared. Absolute measurements were corrected for size, i.e., divided by the estimated body volume or the mean head diameter, where possible and applicable. These relative values were corrected for allometric relationships with body size, where the appropriate allometric factor was available.

ON layer angles and torsion were derived from transverse sections and transverse projections of aligned stacks with a different section plane. With angle of an ON layer, we mean its tilt in relation to the horizontal. The change of average layer angle along the nerve in °/mm is referred to as ON torsion. The layers in the ON represent bidirectional axes rather than unidirectional vectors, i.e., both angles $\alpha = 180°$ and $\alpha' = 0°$ correspond to the same horizontal layer. Hence, we considered any angle $\alpha = x°$ equivalent to $\alpha' = (x \pm 180)°$ and transformed some of the raw angles accordingly such that the difference between smallest and largest individual angle in a sample, i.e., the circular range, was minimised. This procedure not only facilitated the calculation of mean angle and circular standard deviation, but also made the resulting means more reliable.

The following equations were used for all angular calculations:

Mean angle calculation:

$$\overline{a} = \arctan\left(\frac{\sin \overline{a}}{\cos \overline{a}}\right) \tag{1}$$

with

$$\sin \overline{a} = \sum_{i=1}^{n} \sin_{a_i} l_{r_i} \quad \text{and} \quad \cos \overline{a} = \sum_{i=1}^{n} \cos_{a_i} l_{r_i} \tag{2}$$

and

$$l_{r_i} = \frac{l_i}{\sum_{i=1}^{n} l_i} \qquad (3)$$

$\bar{a}$ Mean sample angle

l_r Relative layer length

Circular standard deviation calculation:

$$v = \sqrt[2]{-2\ln(\overline{R})} \quad (4)$$

with

$$\overline{R} = \sqrt[2]{(\sin \overline{a})^2 + (\cos \overline{a})^2} \quad (5)$$

V Sample circular standard deviation

$\overline{R}$ Sample mean resultant length (mean angle vector length).

Modified from [75] and [76].

The mean angles were calculated according to Eq. 1, weighting each individual angle by the relative length of the respective layer (Eqs. 2–3), and the circular standard deviation according to Eqs. 4-5. To quantify the torsion of the ON, we plotted mean angles of sampled transverse sections against distance from optic disc and fitted a linear function to the data. Assuming a continuous and constant torsion, we equivalently transformed mean angle values to reduce large steps between data points, where necessary. Where an estimated parameter of an individual linear fit was not significantly different from zero, we set its value to zero. We then averaged these parameters to obtain mean linear fits of the ON torsion per genus.

All previous calculations and statistical analyses of the resulting data were carried out in JMP® (version 11.1.1, SAS Institute Inc., Cary, NC 27513-2414, USA). Although we used a total number of 14 *A. ocellaris*, six *P. zvonimiri*, 21 *T. delaisi*, and six *T. melanurus* in this study, we could take some measurements only from a subset of these fish. Therefore, sample sizes may differ and are given separately for each type of measurements. Statistical tests were only applied when sample sizes allowed them. For most of the data, however, only descriptive statistics could be calculated and are given as mean ± standard deviation, unless stated otherwise.

Visual model of potential signalling function

To investigate the potential role of ONT eyeshine in intraspecific signalling, we assessed the effect of the ONT eyeshine on the achromatic and chromatic contrast between the triplefin pupil with and without the eyeshine, as it would be perceived by a conspecific. Our calculations are based on the receptor-noise colour discrimination model described by Vorobyev and Osorio [83] and implemented in the R package pavo [84]. In these neural

noise models, we included the photoreceptor sensitivity curves of *T. delaisi* (λmax: single cone – 468 nm, double cone – 516 and 530 nm, treated as trichromat, [85]), which were generated using the visual template equation developed by [86], the light transmission properties of the ocular media of *T. delaisi*, a photoreceptor density ratio of 1:4:4 (unpublished data) and a Weber fraction of 0.05 for the most abundant photoreceptor type. We further assumed that downwelling irradiance was the main contributor to the ONT eyeshine, and used the pigment-epithelium transmitted eyeshine as basis for comparison. We used the transmission values from the measurement with the overall brightest ONT eyeshine for each of the five triplefins, and the ambient irradiance values from the field. The calculation of the achromatic contrasts was based on the sensitivity curves of the double cone photopigment. Values of ΔS and ΔL greater than one just-noticeable-difference (JND) indicate that the contrast between two signals would be discriminable while values of less than 1 JND would indicate that the contrast between the two signals would not be discernible [83].

Data sources for visual model

Downwelling irradiance

Downwelling irradiance was measured in 0.5 m depth intervals on 15 June 2011 (around noon, sunny weather) between 0 and 10 m while scuba diving near Stareso field station, Calvi, Corsica. We used a calibrated PR 670 PhotoResearch radiospectrometer in a custom underwater housing (UK-Germany) and measured the radiance of (1) an exposed white reflectance standard and (2) a shaded white reflectance standard. Both standards had their surface positioned vertically and facing South, to simulate the combined direct and scattered light reaching and being radiated off the side of a fish. The shaded standard had a hood made of black anodised aluminium foil that blocked out direct light. Radiance spectra in Watts/sr/m^2/nm were transformed into photon irradiance (photons/s/m^2/nm) following [87], and used to calculate attenuation coefficients as explained in [88]. With these attenuation values we recalculated the expected irradiance at the required depth, using the irradiance value measured just below the water surface as the total amount of incoming light.

Photoreceptor sensitivities

The photoreceptor sensitivities of *T. delaisi* were supplied by Connor M. Champ and Shelby Temple (unpublished data), who used microspectrometry (MSP) on eyes from 7 individual fish in the laboratory of Christian Donner at the University of Helsinki, following the general method proposed by Govardovskii et al. [86].

Ocular media properties

NKM dissected the eyes of 8 individual *T. delaisi* in Stareso, Calvi, Corsica, in June 2012. The dermal (external) and scleral (internal) corneas, as well as the lens were individually placed in a petri dish with isotonic, marine Ringer buffer. The dish was placed under a stereomicroscope equipped with a PR 670 PhotoResearch Radiospectrometer, and illuminated from below with a KL 2500 (Schott) cold light source. The radiance of the light source was repeatedly and alternatingly measured through the Petri dish and buffer alone (L_0), and through dish, buffer, and ocular tissue sample (L_{OT}) in 1 nm steps. Wavelength-specific transmittance was calculated as $T(\lambda) = L_{OT}(\lambda)/L_0(\lambda)$. For the model, average transmittance was used.

Additional files

Additional file 1: Video of ONT eyeshine in *T. delaisi* in its natural environment. Footage of *T. delaisi* displaying ONT eyeshine under ambient light conditions in its natural habitat. The pupil is not completely dark because *T. delaisi*'s eyes also feature a certain degree of PET eyeshine. For the brief moment the ONT light shines directly at the camera, its much greater brightness and conspicuousness become obvious. Footage was taken in Corsica, 2012. (MP4 2899 kb)

Additional file 2: Video of ONT eyeshine in the laboratory under alternating illumination. A live *T. delaisi* was filmed while kept in a small observation tank, positioned so that the ONT eyeshine was oriented towards the camera. The fish was illuminated through two fibre optic cables connected to independent cold light sources. One provided blue, general illumination, while the other was specifically aimed at the top of the fish's head, and its illumination colour was changed using the inbuilt filters of the cold light source. The video shows that the colour of the ONT eyeshine corresponds to the light illuminating the head surface, proving that the ONT light must indeed be transmitted through the head and nerve tissue. Compare Fig. 3a. (MOV 8824 kb)

Additional file 3: Video of *T. delaisi*'s optic disc viewed through an endoscope. The optic disc and surrounding retinal region of a live *T. delaisi* were filmed through the fish's pupil using an endoscope attached to a Nikon D4 camera. The fish was only illuminated from above, and not through the endoscope. The greenish glow of the retina represents PET eyeshine, while the bright appearance of the optic disc is due to ONT light, which is externally perceived as ONT eyeshine. (MOV 1172 kb)

Additional file 4: Video of *T. delaisi*'s optic nerve torsion. Video produced from the aligned and stacked images of transverse serial sections of *T. delaisi*'s and *A. ocellaris*' heads (see Methods section for more details). The clip shows the ON's pleating, layer orientation, and trajectory from exiting the eyecup to past the optic chiasm, where the nerves become the optic tracts. The two species also differ in the surrounding tissue, especially thickness and structure of the skull. (MP4 7632 kb)

Additional file 5: Table with modelled effects of ONT eyeshine on perceived contrast of the eye. Comprehensive modelling data that show the contrast changes of pupil and iris caused by exhibiting ONT eyeshine, as perceived by another *T. delaisi*, to assess the potential role of ONT eyeshine in intraspecific communication and signalling. (PDF 59 kb)

Additional file 6: Figure showing overview of spectrophotometry set-up. Set-up used for the spectrophotometric measurements of the ONT eyeshine in the studied species. 1) spectroradiometer; 2) endoscope, attached via a C-Mount adapter; 3) tripod; 4) cold light source, using the inbuilt cyan filter; 5) optic cable; 6) platform that allowed for controlled vertical movements; 7) cylindrical acrylic glass tank that could be rotated and displaced horizontally on the platform; 8) rubber foam; 9) diffuse white reflectance standard made of foamed PTFE; 10) fish, euthanized and immobilised with pins; 11) laptop running SpectraWin®, version 2.3.7, for data collection. (PNG 282 kb)

Additional file 7: Table with comprehensive, absolute anatomical data, the summary of which appears in Table 1. (PDF 75 kb)

Additional file 8: Table with comprehensive, relative anatomical data, the summary of which appears in Table 1. (PDF 102 kb)

Additional file 9: Table with full optic nerve torsion data and parameters, on which the summary in Table 2 was based. (PDF 61 kb)

Abbreviations

CSA: Cross-sectioned area; CTR: Choroidal-tapetum-reflected; ICR: Iridescent-cornea-reflected; MRI: Magnetic resonance imaging; ON: Optic nerve; ONT: Optic-nerve-transmitted; PET: Pigment-epithelium-transmitted; PTAH: Phosphotungstic acid haematoxylin; RPE: Retinal pigment epithelium; RTR: Retinal-tapetum-reflected; SAR: Stratum-argenteum-reflected

Acknowledgements

For his early involvement with the project and preliminary spectrophotometric analysis of the ONT eyeshine we'd like to express our deepest gratitude to the late Dominic Amann, whose untimely passing robbed the world of a brilliant and enthusiastic student and future scientist. We further thank Hanna Braun for providing the data for the ONT eyeshine excitation map of *T. delaisi*'s head; Michael Archer, Tom Stewart, and Gary Cowen for their invaluable technical assistance; and Caroline Kerr, João Paulo Coimbra, and Eduardo Garza-Gisholt for their continuous, diverse support.

Funding

This project was funded by Koselleck grant Mi 482/13-1 from the German Science Foundation to NKM in addition to the Australian Research Council (ARC) and the West Australian Government to SPC. P-PB was funded by the Natural Sciences and Engineering Research Council of Canada, through a Postdoctoral Fellowship. We also thank the German Science Foundation and the University of Tübingen's Open Access Publishing Fund for their support in making this study openly available.

Authors' contributions

Conceptualization, RF and NKM; Funding Acquisition, NKM and SPC; Resources, JU, SPC, and NKM; Methodology, RF, JU, P-PB, SPC, and NKM; Investigation, RF and JU; Formal Analysis, RF, P-PB, and NKM; Writing – Original Draft, RF; Writing – Review & Editing, RF, JU, P-PB, SPC, and NKM; Supervision, SPC and NKM. All authors have read and approved the final manuscript.

Competing interests

The authors declare that they have no competing interests.

Author details

[1]Institute of Evolution and Ecology, University of Tübingen, 72076 Tübingen, Baden-Württemberg, Germany. [2]Centre for Advanced Imaging, University of Queensland, Brisbane 4072, Queensland, Australia. [3]Department of Neurology, Boston Children's Hospital & Harvard Medical School, Boston, MA 02115, USA. [4]School of Biological Sciences and the Oceans Institute, University of Western Australia, Crawley 6009, Western Australia, Australia.

References

1. Best A, Nicol J. Eyeshine in fishes. A review of ocular reflectors. Can J Zool. 1980;58:945–56.
2. Schwab IR, Yuen CK, Buyukmihci NC, Blankenship TN, Fitzgerald PG. Evolution of the tapetum. Trans Am Ophthalmol Soc. 2002;100:187–99. discussion 199–200.
3. Collin SP, Collin HB. The fish cornea: adaptations for different aquatic environments. In: Kapoor BG, Hara TJ, editors. Sensory biology of jawed fishes: new insights. Enfield: Plymouth, UK Science Publishers, Inc; 2001. p. 57–96.
4. Ollivier FJ, Samuelson DA, Brooks DE, Lewis PA, Kallberg ME, Komaromy AM. Comparative morphology of the tapetum lucidum (among selected species). Vet Ophthalmol. 2004;7:11–22. doi:10.1111/j.1463-5224.2004.00318.x.
5. von Helmholtz H. Handbuch der Physiologischen Optik. 1st edn. Leipzig: Leopold Voss; 1867

6. Denton E, Nicol J. The chorioidal tapeta of some cartilaginous fishes (Chondrichthyes). JMBA. 1964;44:219–58.

7. Fineran BA, Nicol JAC. Studies on photoreceptors of Anchoa-Mitchilli and Anchoa-Hepsetus (Engraulidae) with particular reference to cones. Phil Trans R Soc B. 1978;283:25–60. doi:10.1098/rstb.1978.0017.

8. Feller KD, Cronin TW. Hiding opaque eyes in transparent organisms: a potential role for larval eyeshine in stomatopod crustaceans. J Exp Biol. 2014;217:3263–73. doi:10.1242/jeb.108076.

9. Nicol JAC. Studies on the eyes of toadfishes opsanus - structure and reflectivity of the Stratum argenteum. Can J Zool. 1980;58:114–21.

10. Mourant JR, Fuselier T, Boyer J, Johnson TM, Bigio IJ. Predictions and measurements of scattering and absorption over broad wavelength ranges in tissue phantoms. Appl Optics. 1997;36:949–57. doi:10.1364/Ao.36.000949.

11. Munday PL, Jones GP. The ecological implications of small body size among coral-reef fishes. Oceanogr Mar Biol. 1998;36:373–411.

12. Agte S, Junek S, Matthias S, Ulbricht E, Erdmann I, Wurm A, Schild D, Kas JA, Reichenbach A. Muller glial cell-provided cellular light guidance through the vital guinea-pig retina. Biophys J. 2011;101:2611–9. doi:10.1016/J.Bpj.2011.09.062.

13. Claes JM, Dean MN, Nilsson DE, Hart NS, Mallefet J. A deepwater fish with 'lightsabers'–dorsal spine-associated luminescence in a counterilluminating lanternshark. Sci Rep. 2013;3:1308. doi:10.1038/srep01308.

14. Franze K, Grosche J, Skatchkov SN, Schinkinger S, Foja C, Schlid D, Uckermann O, Travis K, Reichenbach A, Guck J. Muller cells are living optical fibers in the vertebrate retina. PNAS. 2007;104:8287–92. doi:10.1073/Pnas.0611180104.

15. Kienle A, Hibst R. Light guiding in biological tissue due to scattering. Phys Rev Lett. 2006;97:018104.

16. Sundar VC, Yablon AD, Grazul JL, Ilan M, Aizenberg J. Fibre-optical features of a glass sponge. Nature. 2003;424:899–900.

17. Mandoli DF, Briggs WR. The photoperceptive sites and the function of tissue light-piping in photomorphogenesis of etiolated oat seedlings. Plant Cell Environ. 1982;5:137–45. doi:10.1111/1365-3040.ep11571543.

18. Yashunsky V, Marciano T, Lirtsman V, Golosovsky M, Davidov D, Aroeti B. Real-time sensing of cell morphology by infrared waveguide spectroscopy. PLoS One. 2012;7:e48454. doi:10.1371/journal.pone.0048454.

19. Walton RE, Outhwaite WC, Pashley DF. Magnification–an interesting optical property of dentin. J Dent Res. 1976;55:639–42. doi:10.1177/00220345760550041601.

20. Kienle A, Forster FK, Hibst R. Anisotropy of light propagation in biological tissue. Opt Lett. 2004;29:2617–9. doi:10.1364/ol.29.002617.

21. Antonov I, Goroshkov A, Kalyunov V, Markhvida I, Rubanov A, Tanin L. Measurement of the radial distribution of the refractive index of the Schwann's sheath and the axon of a myelinated nerve fiberin vivo. J Appl Spectrosc. 1983;39:822–4.

22. Kutuzov NP, Brazhe AR, Lyaskovskiy VL, Maksimov GV. Laser beam coupling into nerve fiber myelin allows one to assess its structural membrane properties. J Biomed Opt. 2015;20:50501. doi:10.1117/1.JBO.20.5.050501.

23. Scholes J. The design of the optic nerve in fish. Visual Neurosci. 1991;7:129–39. doi:10.1017/S0952523800011007.

24. Brooks DE, Komaromy AM, Kallberg ME. Comparative retinal ganglion cell and optic nerve morphology. Vet Ophthalmol. 1999;2:3–11.

25. Rahmann H, Jeserich G. Quantitative morphogenetic investigations on fine structural changes in the optic tectum of the rainbow trout (Salmo gairdneri) during ontogenesis. Wilhelm Roux's Archives. 1978;184:83–94.

26. Wolburg H. Growth and myelination of goldfish optic nerve fibers after retina regeneration and nerve crush. Z Naturforsch C. 1977;33:988–96.

27. Lillo C, Velasco A, Jimeno D, Lara J, Aijón J. Ultrastructural organization of the optic nerve of the tench (Cyprinidae, Teleostei). J Neurocytol. 1998;27:593–604. doi:10.1023/A:1006974311861.

28. Miyake E, Imagawa T, Uehara M. Fine structure of the retino-optic nerve junction in dogs. J Vet Med Sci. 2004;66:1549–54.

29. Wyse J, Spira A. Ultrastructural evidence of a peripheral nervous system pattern of myelination in the avascular retina of the guinea pig. Acta Neuropathol. 1981;54:203–10.

30. Cheng N, Tsunenari T, Yau K-W. Intrinsic light response of retinal horizontal cells of teleosts. Nature. 2009;460:899–903. doi:10.1038/nature08175.

31. Soni BG, Philp AR, Foster RG, Knox BE. Novel retinal photoreceptors. Nature. 1998;394:27–8. doi:10.1038/27794.

32. Davies WIL, Zheng L, Hughes S, Tamai TK, Turton M, Halford S, Foster RG, Whitmore D, Hankins MW. Functional diversity of melanopsins and their global expression in the teleost retina. Cell Mol Life Sci. 2011;68:4115–32.

33. Hang CY, Kitahashi T, Parhar IS. Neuronal organization of deep brain opsin photoreceptors in adult teleosts. Front Neuroanat. 2016;10:48. doi:10.3389/fnana.2016.00048.

34. Michael E, de Gardelle V, Summerfield C. Priming by the variability of visual information. PNAS. 2014;111:7873–8. doi:10.1073/pnas.1308674111.

35. Andersson A, Nilsson D-E. Fine-structure and optical-properties of an Ostracode (Crustacea) nauplius eye. Protoplasma. 1981;107:361–74. doi:10.1007/Bf01276836.

36. Denton E. Review lecture: on the organization of reflecting surfaces in some marine animals. Phil Trans R Soc B. 1970;258:285–313.

37. Howland HC, Murphy CJ, Mccosker JE. Detection of eyeshine by flashlight fishes of the family Anomalopidae. Vision Res. 1992;32:765–9.

38. Stevens M. The role of eyespots as anti-predator mechanisms, principally demonstrated in the Lepidoptera. Biol Rev. 2005;80:573–88.

39. . Kelley JL, Fitzpatrick JL, Merilaita S: Spots and stripes: ecology and colour pattern evolution in butterflyfishes. Proc R Soc B 2013;280. doi: 10.1098/rspb.2012.2730

40. Stevens M, Marshall KL, Troscianko J, Finlay S, Burnand D, Chadwick SL. Revealed by conspicuousness: distractive markings reduce camouflage. Behav Ecol. 2013;24:213–22.

41. Kjernsmo K, Merilaita S. Eyespots divert attacks by fish. Proc R Soc B. 2013;280:20131458.

42. Lythgoe J. The structure and function of iridescent corneas in teleost fishes. Proc R Soc B. 1975;188:437–57.

43. Johnson ML, Shelton PMJ, Gaten E, Herring PJ. Relationship of dorsoventral eyeshine distributions to habitat depth and animal size in mesopelagic decapods. Biol Bull. 2000;199:6–13. doi:10.2307/1542701.

44. Zander CD. Tripterygiidae. In: Whitehead PJP, Bauchot ML, Hureau JC, Nielsen J, Tortonese E, editors. Fishes of the north-eastern Atlantic and the Mediterranean. Volume 3. Paris: Unesco; 1986. p. 1118–21.

45. Carreras-Carbonell J, Macpherson E, Pascual M. Rapid radiation and cryptic speciation in Mediterranean triplefin blennies (Pisces : Tripterygiidae) combining multiple genes. Mol Phylogenet Evol. 2005;37:751–61. doi:10.1016/J.Ympev.2005.04.02.

46. Carreras-Carbonell J, Pascual M, Macpherson E. A review of the Tripterygion tripteronotus (Risso, 1810) complex, with a description of a new species from the Mediterranean Sea (Teleostei : Tripterygiidae). Sci Mar. 2007;71:75–86.

47. Domingues VS, Almada VC, Santos RS, Brito A, Bernardi G. Phylogeography and evolution of the triplefin Tripterygion delaisi (Pisces, Blennioidei). Mar Biol. 2007;150:509–19. doi:10.1007/s00227-006-0367-4.

48. Zander CD, Berg J. Feeding ecology of littoral Gobiid and Blennioid fishes of the Banyuls area (Mediterranean Sea) II. Prey selection and size preference. Vie Milieu. 1984;34:149–58.

49. Zander CD, Heymer A. Tripterygion-Tripteronotus (Risso, 1810) and Tripterygion-Xanthosoma N Sp, an Ecological Speciation (Pisces, Teleostei). Vie Milieu. 1970;21:363–94.

50. Louisy P. Guide d'identification des poissons marins: Europe de l'ouest et Méditerranée. 2nd ed. Paris: Les Editions Eugen Ulmer; 2002.

51. Fischer S, Patzner RA, Müller CH, Winkler HM. Studies on the ichthyofauna of the coastal waters of Ibiza (Balearic Islands, Spain). Rostock Meeresbiol Beitr. 2007;18:30–62.

52. Patzner RA. Habitat utilization and depth distribution of small cryptobenthic fishes (Blenniidae, Gobiesocidae, Gobiidae, Tripterygiidae) in Ibiza (western Mediterranean Sea). Environ Biol Fish. 1999;55:207–14.

53. Goncalves EJ, Almada VC. A comparative study of territoriality in intertidal and subtidal blennioids (Teleostei, Blennioidei). Environ Biol Fish. 1998;51:257–64.

54. Wirtz P. Zum Verhalten blennioider Fische, insbesondere der mediterranen Tripterygion Arten. Dissertation. Ludwig-Maximilian-University Munich, Department of Biology; 1977

55. Betancur RR, Broughton RE, Wiley EO, Carpenter K, Lopez JA, Li C, Holcroft NI, Arcila D, Sanciangco M, Cureton Ii JC, et al. The tree of life and a new classification of bony fishes. PLoS Curr 2013;5. doi: 10.1371/currents.tol.53ba26640df0ccaee75bb165c8c26288

56. Lin HC, Hastings PA. Phylogeny and biogeography of a shallow water fish clade (Teleostei: Blenniiformes). BMC Evol Biol 2013;13. doi: 10.1186/1471-2148-13-210

57. Faria C, Almada VC. Patterns of spatial distribution and behaviour of fish on a rocky intertidal platform at high tide. Mar Ecol Prog Ser. 2006;316:155–64. doi:10.3354/meps316155.

58. Illich IP, Kotrschal K. Depth distribution and abundance of northern adriatic littoral rocky reef blennioid fishes (Blenniidae and Tripterygion). PSZNI Mar Ecol. 1990;11:277–89. doi:10.1111/j.1439-0485.1990.tb00384.x.

59. Nieder J. Depth of sojourn and niche differentiation of benthic blennies (Pisces, Blenniidae) in the Mediterranean Sea (Catalonia, NE Spain and Italian coast of the Tyrrhenian Sea). Misc Zool. 2000;23:21–33.

60. Velasco E, Gómez-Cama M, Hernando J, Soriguer M. Trophic relationships in an intertidal rockpool fish assemblage in the gulf of Cádiz (NE Atlantic). J Mar Syst. 2010;80:248–52.

61. Zander CD. The distribution and feeding ecology of small-size epibenthic fish in the coastal Mediterranean Sea. In: Eleftheriou A, Ansell AD, Smith CJ, editors. Biology and ecology of shallow coastal waters. Fredensborg: Olsen & Olsen; 1995. p. 369–76.

62. Zander CD. Feeding ecology of littoral gobiid and blennioid fish of the Banyuls area (Mediterranean Sea). I. Main food and trophic dimension of niche and ecotope. Vie Milieu. 1982;32:1–10.

63. Zander CD, Hagemann T. Feeding ecology of littoral Gobiid and Blennioid fishes of the Banyuls area (France, Mediterranean Sea) III. Seasonal variations. Sientia Marina. 1989;53:441–50.

64. La Mesa G, Vacchi M. Analysis of the blennioid assemblages associated with different rocky shore habitats in the Ligurian Sea. J Fish Biol. 2005;66:1300–27. doi:10.1111/j.1095-8649.2005.00684.x.

65. Zander CD. Blenniidae. In: Whitehead PJP, Bauchot ML, Hureau JC, Nielsen J, Tortonese E, editors. Fishes of the north-eastern Atlantic and the Mediterranean. Volume 3. Paris: Unesco; 1986. p. 1096–112.

66. Duci A, Giacomello E, Chimento N, Mazzoldi C. Intertidal and subtidal blennies: assessment of their habitat through individual and nest distribution. Mar Ecol Prog Ser. 2009;383:273–83. doi:10.3354/Meps07986.

67. La Mesa G, Di Muccio S, Vacchi M. Structure of a Mediterranean cryptobenthic fish community and its relationships with habitat characteristics. Mar Biol. 2006;149:149–67. doi:10.1007/s00227-005-0194-z.

68. Allen GR. Damselfishes of the world. 1st ed. Melle: MERGUS Publishers; 1991.

69. Fautin DG. The anemonefish symbiosis - what is known and what is not. Symbiosis. 1991;10:23–46.

70. Godwin J, Fautin DG. Defense of host actinians by anemonefishes. Copeia. 1992;1992:902–8.

71. Porat D, Chadwick-Furman NE. Effects of anemonefish on giant sea anemones: ammonium uptake, zooxanthella content and tissue regeneration. Mar Freshw Behav Physiol. 2005;38:43–51. doi:10.1080/10236240500057929.

72. Buston P. Social hierarchies: size and growth modification in clownfish. Nature. 2003;424:145–6. doi:10.1038/424145a.

73. Kokko H, Johnstone RA. Social queuing in animal societies: a dynamic model of reproductive skew. Proc R Soc B. 1999;266:571–8.

74. Mitchell JS. Social correlates of reproductive success in false clown anemonefish: subordinate group members do not pay-to-stay. Evol Ecol Res. 2003;5:89–104.

75. Fricke H, Fricke S. Monogamy and sex change by aggressive dominance in coral-reef fish. Nature. 1977;266:830–2. doi:10.1038/266830a0.

76. Fautin DG, Allen GR, Fautin DG, Allen GR. Anemone fishes and their host sea anemones: a guide for aquarists and divers. Revised edition. Perth: Western Australian Museum; 1997.

77. Fricke HW. Öko-Ethologie des monogamen Anemonenfisches Amphiprion bicinctus (Freiwasseruntersuchung aus dem Roten Meer). Zeitschrift Tierpsychol. 1974;36:429–512.

78. Frédérich B, Fabri G, Lepoint G, Vandewalle P, Parmentier E. Trophic niches of thirteen damselfishes (Pomacentridae) at the Grand Récif of Toliara, Madagascar. Ichthyol Res. 2009;56:10–7. doi:10.1007/s10228-008-0053-2.

79. Wucherer MF, Michiels NK. A fluorescent chromatophore changes the level of fluorescence in a reef fish. PLoS One. 2012;7:e37913. doi:10.1371/journal.pone.0037913.

80. Mallory FB. A contribution to staining methods : I. A differential stain for connective-tissue fibrillae and reticulum. Ii. Chloride of iron haematoxylin for nuclei and fibrin. Iii. Phosphotungstic acid haematoxylin for neuroglia fibres. J Exp Med. 1900;5:15–20.

81. Yushkevich PA, Piven J, Hazlett HC, Smith RG, Ho S, Gee JC, Gerig G. User-guided 3D active contour segmentation of anatomical structures: significantly improved efficiency and reliability. Neuroimage. 2006;31:1116–28. doi:10.1016/J.Neuroimage.2006.01.015.

82. Saalfeld S, Fetter R, Cardona A, Tomancak P. Elastic volume reconstruction from series of ultra-thin microscopy sections. Nat Methods. 2012;9:717–20. doi:10.1038/nmeth.2072.

83. Vorobyev M, Osorio D. Receptor noise as a determinant of colour thresholds. Proc R Soc B. 1998;265:351–8.

84. Maia R, Eliason CM, Bitton PP, Doucet SM, Shawkey MD. pavo: an R package for the analysis, visualization and organization of spectral data. Methods Ecol Evol. 2013;4:906–13. doi:10.1111/2041-210x.12069.

85. Pignatelli V, Champ C, Marshall J, Vorobyev M. Double cones are used for colour discrimination in the reef fish, Rhinecanthus aculeatus. Biol Lett. 2010;6:537–9. doi:10.1098/Rsbl.2009.1010.

86. Govardovskii VI, Fyhrquist N, Reuter T, Kuzmin DG, Donner K. In search of the visual pigment template. Visual Neurosci. 2000;17:509–28. doi:10.1017/S0952523800174036.

87. Johnsen S. The optics of life: a biologist's guide to light in nature. 1st ed. Princeton: Princeton University Press; 2012.

88. Meadows MG, Anthes N, Dangelmayer S, Alwany MA, Gerlach T, Schulte G, Sprenger D, Theobald J, Michiels NK. Red fluorescence increases with depth in reef fishes, supporting a visual function, not UV protection. Proc R Soc B 2014;281. doi: 10.1098/rspb.2014.1211

Adaptive responses to salinity stress across multiple life stages in anuran amphibians

Molly A. Albecker*[iD] and Michael W. McCoy

Abstract

Background: In many regions, freshwater wetlands are increasing in salinity at rates exceeding historic levels. Some freshwater organisms, like amphibians, may be able to adapt and persist in salt-contaminated wetlands by developing salt tolerance. Yet adaptive responses may be more challenging for organisms with complex life histories, because the same environmental stressor can require responses across different ontogenetic stages. Here we investigated responses to salinity in anuran amphibians: a common, freshwater taxon with a complex life cycle. We conducted a meta-analysis to define how the lethality of saltwater exposure changes across multiple life stages, surveyed wetlands in a coastal region experiencing progressive salinization for the presence of anurans, and used common garden experiments to investigate whether chronic salt exposure alters responses in three sequential life stages (reproductive, egg, and tadpole life stages) in *Hyla cinerea*, a species repeatedly observed in saline wetlands.

Results: Meta-analysis revealed differential vulnerability to salt stress across life stages with the egg stage as the most salt-sensitive. Field surveys revealed that 25% of the species known to occur in the focal region were detected in salt-intruded habitats. Remarkably, *Hyla cinerea* was found in large abundances in multiple wetlands with salinity concentrations 450% higher than the tadpole-stage LC_{50}. Common garden experiments showed that coastal (chronically salt exposed) populations of *H. cinerea* lay more eggs, have higher hatching success, and greater tadpole survival in higher salinities compared to inland (salt naïve) populations.

Conclusions: Collectively, our data suggest that some species of anuran amphibians have divergent and adaptive responses to salt exposure across populations and across different life stages. We propose that anuran amphibians may be a novel and amenable natural model system for empirical explorations of adaptive responses to environmental change.

Keywords: Secondary salinization, Anuran amphibian, Sea level rise, Saltwater tolerance, Climate change, Complex life history

Background

Accumulating greenhouse gas concentrations are increasing the energy retained in the atmosphere, which is in turn causing global mean sea levels to rise through intensified ice sheet and glacier melting and thermal expansion of ocean water [1–4]. Sea levels have already risen 17-21 cm over the past 110 years, and current models forecast that sea levels could rise an additional 40–63 cm over the next century with additions expected if ice sheets on Greenland and West Antarctica collapse [2, 4–8]. Ancillary impacts of climate change on coastal wetlands include

alterations in the frequency and intensity of storm surges and coastal flooding, which may compound the effects of coastal erosion and saltwater inundation. The magnitude of sea level rise and impact on coastal ecosystems will vary depending on glacial isostatic adjustment, tectonic processes, oceanic circulation patterns, sediment compaction and accretion, wind patterns, and gravitational changes [4, 9–15], yet many areas are already being affected by sea level rise [16–20].

Rising salinities are broadly anticipated to negatively impact freshwater organisms inhabiting coastal regions by reducing both the quality and quantity of suitable habitat, lowering individual fitness (e.g., increased physiological stress, increased morphological deformities, reduced fecundity, and modifications to growth,

* Correspondence: albeckerm09@students.ecu.edu
Department of Biology, Howell Science Complex, East Carolina University, Greenville, NC, USA

development, and mortality), reducing population carrying capacity, and by altering biological interactions, disease risk, species movement, and community structure [21–24].

Osmoregulators require a wide variety of physiological, morphological, life historical, and behavioral traits to conserve water and expel enough excess ions to survive higher salinities. Although examples of adaptive responses across strong abiotic clines are multiplying quickly [25–30], adaptive responses might be slowed by an organism's life history strategy, amount of standing genetic variation, demographic constraints (e.g., competition), or decoupling of environmental cue from response [31–35]. For example, organisms with complex life cycles, such as amphibians, have different ontogenetic life stages that are typically marked by abrupt shifts in morphology, physiology, behavior, and often distinct changes in habitat use. Therefore, the same stressor may differently impact each life stage, and require multiple adaptive responses across life stages to successfully adapt to an emerging environmental stressor.

Amphibians are a classic model for exploring responses to environmental stressors such as salinity. Amphibians are widely regarded as important indicator species of wetland quality due to a life history tied to freshwater coupled with unique characteristics such as permeable skin, an inability to concentrate and excrete excess salts, and poor dispersal capabilities [36–39]. Additionally, amphibians comprise a significant proportion of the vertebrate biomass in wetland ecosystems [40, 41] and have been classified by the IUCN as "climate change susceptible" [23]. Most amphibians are obligatorily aquatic throughout the egg and larval period and become semi-terrestrial upon metamorphosis. Depending on the species, amphibians typically return to water as adults to breed or rehydrate.

A recent review identified ca. 140 anuran amphibian species that have been observed in saline habitats (ranging from tidal mangrove swamps to inland freshwater habitats contaminated with road deicing salts). Yet these species represent only 2% of all known species [38, 39], supporting the widely held belief that anurans are a generally salt-sensitive, freshwater order. A few notable species of amphibians such as *Fejervarya cancrivora* and *Bufo viridis* are known to tolerate brackish conditions [38, 39, 42–46], but these species still require freshwater habitats to complete their life cycles suggesting differential vulnerability to salt exposure across life stages even in specialist salt-tolerant species [47–50].

In addition to field observations, there are many published studies that experimentally explore embryonic, tadpole, or adult responses to salt stress. These studies typically evaluate how saltwater impacts anuran survivorship and behavior in a single life stage, and in doing

so, provide indispensable and informative data on expected responses across a range of salinities. Hopkins and Brodie published an extensive review of saltwater tolerance in amphibians [39], which provides a useful framework to better understand and predict how salinization affects anuran populations. Yet the data contained in these studies has not yet been coalesced to precisely quantify how salt tolerance changes across different life stages. Moreover, to best predict how anurans will respond to progressively increasing salinities, we not only need to define how salinity affects each life stage, but also how labile salt-tolerant responses are across populations.

In this study, we use multiple, complementary strategies to evaluate salt sensitivity in anurans generally, and substitute space for time to explore whether populations that inhabit coastal wetlands with a history of increasing salt exposure demonstrate adaptive responses across multiple life stages. First, we conducted a meta-analysis to establish an empirically derived quantitative framework of expected survivorship following exposure to saltwater in anuran amphibians for different life stages. Second, we performed a field survey of brackish and freshwater wetlands to describe and characterize amphibian distributions along a salt gradient in a coastal location predicted to be among the most impacted by sea level rise. Third, we substitute space for time in common garden experiments to investigate how exposure to saltwater across life stages differs among chronically salt-exposed (coastal) and salt-naïve (inland) anuran populations.

We focus on reproductive behaviors, egg hatching patterns, and post-hatching tadpole survival for our common garden experiments. During breeding events, male frogs amplex females and then she will transport the male to assess potential egg laying sites. Females of some species are highly discriminatory and choose among oviposition sites to avoid a variety of biotic and abiotic stressors [51, 52]. Oviposition site choice behaviors are under strong selection because her choice can considerably impact offspring survival and performance by affecting fertilization success, mortality risk to offspring, as well as resource availability to offspring [46, 51–55]. After eggs have been deposited, developing clutches are vulnerable to aquatic contaminants because frog eggs are enclosed by a permeable, jelly coat and lack a hard, protective shell [56, 57]. Upon hatching, the larvae of many frog species are obligatorily aquatic and cannot survive on land until the completion of metamorphosis. During this period, tadpoles respire and osmoregulate via gills that function similar to freshwater teleosts such that ions and salts are conserved and excess water is expelled [58–60]. We chose reproductive choices, embryo hatching success, and tadpole survival because these stages are key periods in the anuran life cycle that are highly vulnerable to external stressors, including saltwater, and strongly influence individual fitness and population persistence [46, 51, 61–65].

Methods
Study location
We conducted these studies in eastern North Carolina, USA. North Carolina's coastline, barrier islands, and coastal habitats are predicted to be among the most significantly impacted by sea level rise due to the geomorphology of the Northern coastal zone (Albemarle embayment), coastal subsidence rates (–1 mm ± 0.15 mm/yr.), and gently sloped coastal plains [15, 19, 66–68]. Indeed, the North Carolina coast has already seen intensified coastal flooding, and increased saltwater intrusion into coastal lowlands and freshwater aquifers making it an important location for investigating the impacts of sea level rise and increasing salinities on coastal organisms [11, 19, 69].

Meta-analysis
Literature search
We searched Google Scholar and Scopus databases for experimental studies evaluating the survivorship of anuran amphibians after experimental exposure to saltwater. We conducted the primary, exhaustive searches on December 16–20, 2014. Literature was checked again on July 14, 2015, September 23, 2015, February 25, 2016, and February 2, 2017 to ensure recently published work was included. We used the search terms (and all combinations of): "frog" OR "anuran" OR "amphibian" AND "saltwater" OR "salt" OR "salinity" OR "ocean" OR "NaCl" AND "mortality" OR "survivorship". Initial searches returned ~24,500 hits in total. These studies were further refined by scanning titles and abstracts. We excluded studies that did not mention survivorship or mortality of anurans and exposure to saltwater in the abstract. We also cross checked against the list of studies in Hopkins and Brodie's review of amphibian salt tolerance to ensure all appropriate studies were included [39].

Data extraction
After refining our database to 129 studies, each study was read in detail and data were extracted from the text or figures. We extracted data only on studies that experimentally and directly manipulated salt concentrations against known sample sizes (e.g., field observations and studies with incidental, non-targeted salt exposure were excluded). We used studies that exposed frogs to saltwater solutions comprised of sodium chloride (NaCl), (e.g. InstantOcean® or natural seawater) and excluded studies that exposed frogs to mixed salt solutions (e.g., mixed road salt solutions) [70]. In studies where multiple saltwater compositions (e.g., $MgCl_2$, KCl, $CaCl_2$)

were tested, we only used data from the trials that utilized NaCl. See Additional file 1 for detailed list of studies.

We used GraphClick® software version 3.0.3 (Arizona Software) to extract estimates from published figures and graphs. We report the mean survivorship (with error) for studies containing multiple replicates across salinities. For studies that compare survivorship across replicate populations, we present global averages across all populations tested. Although two studies report intra-specific differences in saltwater tolerance across different populations (e.g., [45, 71]), there were too few studies available to permit a meaningful formal analysis on population level differences in saltwater tolerance across studies or species. We recorded species identity, family, life stage (tadpole, egg, or adult), experimental salinity concentrations, sample size (N), survivorship (as proportion), the standard deviation of survivorship (converted from standard error when necessary), location of the study, and length of exposure (in hours) for each study. Because different studies reported salinity using different units, we used standard conversions to transform all salinity measurements to parts per thousand (ppt).

Field survey
Study sites
We monitored wetlands regularly to make sure species that breed at different times could be detected. We surveyed 55 salt and freshwater wetlands in eastern North Carolina between February and September of 2014 for the presence of anuran amphibians. We included bogs, retention areas, marshes, ponds, ditches, and swamps, but excluded estuaries, sea grass beds, and other large, open water habitats. The most southern and eastern location was Cape Hatteras National Seashore and the survey extended northward to the town of Nags Head. Along this transect, we surveyed wetlands along Rodanthe, New Inlet, Bodie Island, Oregon Inlet, and Pea Island National Wildlife Refuge. We also sampled wetlands along an east to west transect spanning from the outer banks of NC, across Roanoke Island, which lies between the inner and outer banks and throughout Alligator River National Wildlife Refuge located on the Albemarle peninsula. The geographic bounds of the study area are 35°55′7″N to 35°14′7″N, and between 75°48′43″W to 75°27′27″W, excluding the Atlantic Ocean and the Pamlico, Croatan, and Roanoke sounds.

Survey techniques
We used standard sampling methods to characterize anuran presence and relative abundance including auditory call surveys, standardized dip netting for larvae, and active searching for adults [72, 73]. Our primary

approach used auditory surveys to identify and locate frog populations, as well as to determine species identities and relative abundances of the anurans present. When frogs were detected via call, the site was geo-referenced using a Garmin® GPSMAP 60CSx GPS navigator (Garmin, Ltd., Olathe, KS) and salinity (in ppt) and the temperatures of the air and water were measured using YSI Professional Plus multiparameter meter (Xylem, Inc., Yellow Springs, OH). We returned the following day (auditory surveys occurred at night) to the geo-referenced sites to determine egg mass/larvae presence using fixed-effort dip netting, and visual transect surveys [72, 73]. To ensure that we thoroughly surveyed all wetlands for the presence of amphibians (and not just wetlands with detectable choruses), we used Google Maps® and visual surveys to identify additional wetlands that were not identified using call surveys, and sampled these wetlands using visual transect surveys and dip-netting for the presence of adult and/or larval anuran species. Tuberville et al. [74] conducted a thorough amphibian field survey along the North Carolina coast that included Cape Hatteras and Cape Lookout National Seashore and documented the current or historic presence of 17 anuran species, and we use the results of this study as a comparison for our own observations. Notably, the Tuberville study did not record salinity of locations in which anurans were observed.

Common garden experiments

We used *Hyla cinerea*, the American green tree frog (average size: 3.2–5.7 cm), for each of our common garden experiments, as this species is common across the Southeastern United States and has been repeatedly documented in saltwater intruded environments [38, 51, 75, 76]. These experiments were conducted between May and August 2015. To characterize and identify how responses to saltwater differ among populations, we compared individuals from chronically salt-exposed *Hyla cinerea* populations (hereafter referred to as "coastal" populations) against individuals from freshwater, salt-naive *Hyla cinerea* populations (hereafter referred to as "inland" populations). We located coastal and inland populations via the field survey. All coastal individuals were collected from sites in which salinities remained at or above 3 ppt over the course of the breeding season, and all inland individuals were collected from populations with salinities below 1 ppt. Coastal populations and inland populations were geographically separated from one another by at least 190 km, so we assume that pairs collected from populations within these locations are sufficiently distant both geographically and environmentally to provide an accurate assessment of population-level differences produced by the different salinity of their habitats.

Oviposition site choice and egg hatching

We tested oviposition site choice by collecting four amplexed pairs of *Hyla cinerea* from either coastal or inland populations. Each pair was placed into an 18-Liter clear bin, the bottom of which was lined with six pint cups. Three of the six cups contained 400 ml tap water (0 ppt) treated with API® Tap Water Conditioner (Chalfont, PA), and the remaining cups contained 400 ml saltwater prepared by mixing treated tap water with InstantOcean Sea Salt® (Blacksburg, VA). Each bin contained a single saltwater concentration that was either 4 ppt, 6 ppt, 8 ppt, or 12 ppt. In doing so, we presented each pair with a binary choice between laying eggs in freshwater or saltwater. The four salt concentration treatments collectively comprised a single replicate (i.e., four bins = one replicate). On nights when multiple replicates were conducted, each replicate was arranged in a spatial block at the site of collection.

Bins were left in situ overnight to allow pairs to complete breeding. The following morning, adult frogs were released, lids fastened to each cup, and bins were transported to the laboratory. Each cup was individually photographed, the salinity measured, and then monitored for hatching. Eggs hatched after 72–96 h, defined as the point in which individuals were no longer retained in egg matrix and have functional gills (Gosner stage 20 [77]). Hatchlings were counted and recorded.

Tadpole survivorship

To determine the effects of salinity on tadpole survival, we utilized the individuals hatched from eggs laid in freshwater during the previous oviposition experiments. Hatchlings were held in the laboratory that was maintained at 26.67 °C (~80 °F) and allowed to develop until reaching Gosner stage 25 (approximately 5 days) [77]. Several studies have indicated that acclimatizing anurans to elevated salinities reduces mortality [50, 52, 78], and natural salinity fluctuations typically do not exceed +/− 2 ppt per day, excluding an extreme event such as storm surge or flooding event. Therefore, to best mimic natural conditions and quantify survival, tadpoles were gradually acclimatized to a specified target salinity over 6 days. We chose five target salinities, 0.5 ppt, 4 ppt, 6 ppt, 8 ppt, and 12 ppt, which are representative of natural salinities observed in coastal wetlands. Freshwater treatments (0.5 ppt) were maintained at 0.5 ppt throughout the six-day acclimatization period. The 4 ppt treatments were raised by 0.67 ppt per day, 6 ppt treatments were raised by 1 ppt per day, 8 ppt treatments raised by 1.33 ppt per day, and 12 ppt treatments were raised by 2 ppt per day, with final target salinities reached on day 6.

We divided each clutch into five groups of fifty tadpoles, which were then randomly assigned to one of the five salinity treatments, replicated 8 times for each location. Each clutch divided into five groups comprised a single replicate block to account for potential parental effects. Groups of tadpoles were placed into 350 mL glass containers containing 300 mL of treated tap water (treated with API® Tap Water Conditioner (Chalfont, PA)) within a laboratory with 12-h light/dark cycle. After acclimatizing overnight, salinity was increased incrementally each day according to treatment. Prior to water changes each day, tadpole mortality in each cup was assessed and recorded, and deceased individuals were removed. Tadpoles were fed 0.01 g of Spirulina fish food flakes (Ocean Star International, Coral Springs, FL) each day following the water change. To perform water changes, tadpoles were carefully poured into a small holding container and returned after 300 mL of new, treated water with experimentally raised saltwater concentrations (InstantOcean Sea Salt® (Blacksburg, VA)) was poured into glass containers.

Statistical analyses

We use a Bayesian approach to analyze our data. For all statistical analyses we used JAGS interfaced with the R statistical programming environment, version 3.2.3 [79] via "R2jags" [80], "rjags" [81], and "coda" [82] packages. For each analysis, we ran 5000 iterations of three separate Markov Chain Monte Carlo (MCMC) chains with starting values that varied by an order of magnitude, each with a burn in of 2500 unless otherwise specified [83]. We used Gelman-Rubin diagnostics to assess model convergence in each analysis [83].

Meta-analysis

To estimate the probability of survival in saltwater for each life stage across anuran taxa and across salinities, we tested how increasing salinity affects anuran survivorship across clades for each life stage (e.g., egg, larvae, adult). We did not use phylogenetically corrected data because a recent review of all instances of amphibians in saline environments revealed no phylogenetic signal [39] and we detected no signal of phylogeny in the unexplained deviance from our analysis. We performed a Bayesian beta regression with an uninformative (relatively flat; mean = 0, std. dev. = 0.001) Gaussian prior. We chose the beta distribution because the data extracted for the meta-analysis were often only reported as "proportion survived" or "proportion killed" and lacked the necessary information (e.g., sample sizes and replicate numbers) required to back-calculate starting densities. In this analysis, survivorship and salinity were considered fixed effects, with individual studies treated as random effects.

Field survey

We utilized the posterior distribution from the meta-analysis of all anuran species to predict the probability of anuran survivorship across several salinities including the salinities where we observed coastal *Hyla cinerea* during field surveys. Specifically, we generated a survival curve (with uncertainty) across salinities ranging from 1 ppt (freshwater) up to 40 ppt, and estimated the expected probability and credible intervals for finding frogs in sites with salinity concentrations we found in our field observations. Although 40 ppt exceeds the salinity of natural seawater (35 ppt), Gordon and colleagues observed *Fejervarya cancrivora* tadpoles in 39 ppt water in 1961 [48]. While this particular observation was not included in our meta-analysis due to its non-experimental nature, we wanted to ensure that all possible salinities were considered in our meta-analysis.

Common garden experiments

We used ImageJ® software to quantify the number of eggs that were laid in each cup. Briefly, photograph files for each container were imported and changed to 8-bit images. The image background was subtracted, images were made binary, and files were converted to a mask. To separate groups of eggs that were clumped together, we used the watershed feature to demarcate individual egg boundaries. Outputs were visually inspected to ensure that all eggs were included and correctly counted.

We ran two-stage tests for both oviposition site-choice and hatching data. In the first step, we analyzed the data in binary form to ask if the probability of egg deposition or hatching changed as a function of the interaction between source population (e.g., coastal vs. inland) and salinity. In the second step, given that egg deposition or hatching occurred (i.e., excluding all cups in which zero eggs were laid or hatched), we analyzed the proportion of eggs deposited into freshwater and the proportion of offspring hatched as a function of the interaction between source population and salinity. These dual approaches answer distinct but complementary questions. Regarding oviposition, the first test asks if the probability of depositing eggs into saltwater or freshwater reflects a choice between salinities, while the second test reveals how parental investment differs according to salinity. Regarding hatching, the first test uncovers differences in the probability of complete loss due to salinity, while the second test reveals thresholds of sensitivity to salt.

To test the probability of oviposition, we ran Bernoulli regression to test for a relationship between egg presence or absence according to salinity and location (step one above). To test whether there were differences in investment (step two above), we ran a binomial regression to examine whether salinity and location affected the proportion of eggs deposited by a female into saltier water. For both of these analyses, we used uninformative Gaussian priors (mean = zero and precision as a decaying power function with exponent = −2). To test the probability of hatching and proportion that hatched, we use informed priors based on the posterior distribution produced by the egg stage meta-analysis. Similar to the oviposition analyses, we ran Bernoulli regression to determine the relationship between egg hatching and salinity and location. We then used a binomial regression to analyze differences in the proportion of eggs that hatched in each salinity and location. Each of these four models considers salinity and location (e.g., coastal or inland) as fixed effects with "bin" nested in location as a random effect to account for parental effects [84].

Tadpole survivorship

To quantify how salinity, location, and time (e.g., day) affect tadpole survivorship, we used a binomial regression with informed priors based on the posterior distribution produced by the tadpole stage meta-analysis. This model considers salinity and location (e.g., coastal or inland) as fixed effects with "clutch" included as a random effect to account for sibship [84]. For this analysis we ran four separate MCMC chains with 50,000 iterations, each with a burn in of 25,000 [83].

Results

Meta-analysis

Effects of salt on amphibian survivorship

We utilized data from 39 papers published between 1961 to early 2017 (see Additional file 1 for detailed information). Overall, the literature uniformly demonstrates that increasing saltwater concentrations lowers anuran survivorship across all three life-stages (Fig. 1). We found that across all studies included in this analysis, the lethal concentration of saltwater required to impose 50% mortality (LC_{50}) to anuran amphibian eggs is 4.15 ppt (95% Bayesian credible interval [BCI] = 2.25 to 6.25 ppt). The LC_{50} for larval anurans is 5.5 ppt (4.24–6.65 ppt BCI), while the LC_{50} for adults is 9.0 ppt (0–19.9 ppt BCI).

Field surveys

Species presence

In coastal freshwater habitats (<3 ppt) with no connection to saltwater influence (e.g., municipal retention ponds), we documented the regular presence of 16 of the 17 anuran species found in the Tuberville study including *Hyla cinerea, Hyla chrysoscelis, Hyla squirella, Hyla femoralis, Anaxyrus fowleri, Anaxyrus quercicus, Anaxyrus terrestris, Lithobates sphenocephalus, Lithobates clamitans, Lithobates virgatipes, Lithobates catesbeianus, Gastrophryne carolinensis, Pseudacris ocularis, Pseudacris crucifer,* and *Acris gryllus.* We did not detect

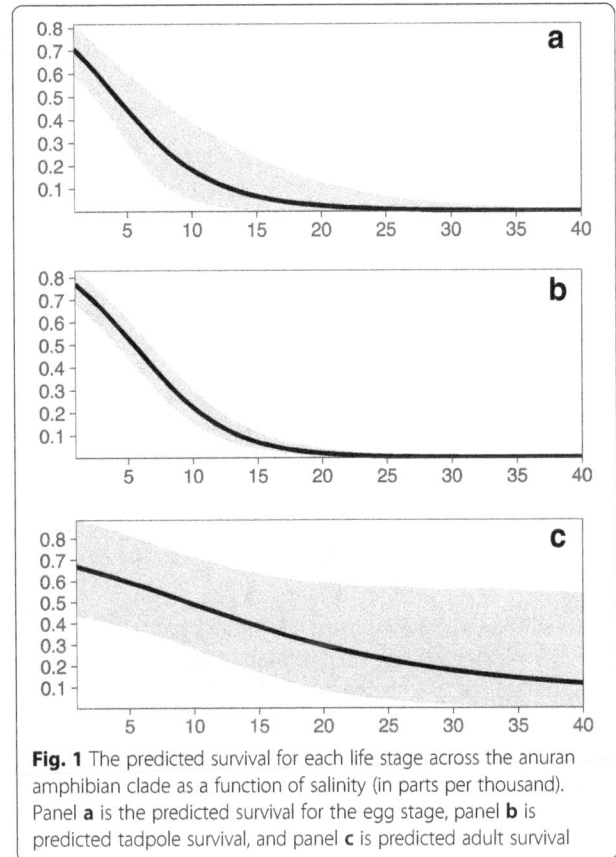

Fig. 1 The predicted survival for each life stage across the anuran amphibian clade as a function of salinity (in parts per thousand). Panel **a** is the predicted survival for the egg stage, panel **b** is predicted tadpole survival, and panel **c** is predicted adult survival

Scaphiopus holbrookii [74]. In salt-invaded wetlands (>3 ppt), we documented the presence of 4 of those 16 species (*Hyla cinerea, Gastrophryne carolinensis, Lithobates catesbeianus,* and *Lithobates sphenocephalus*) (Table 1).

Relative abundance

In general, we noted that relative abundances of all species (except *Hyla cinerea*) declined as wetlands grew more saline. *Hyla cinerea* demonstrated unique distribution patterns along North Carolina's coast as the most abundant species found within salt-invaded habitats along both the inner and outer banks. Notably, in some locations we observed that the relative abundance of *Hyla cinerea* actually increased with increasing salinity, a pattern not shared with any of the other species found in salt-invaded wetlands. We collected early and late stage *Hyla cinerea* tadpoles, metamorphs (between Gosner stages 31–39 [77]), and adults from multiple locations including from ponds and marshes with 3.9 ppt, 8.3 ppt, 11 ppt, 16.8 ppt, and 23.4 ppt water.

Probability of field findings

Using the posterior probability distributions from our meta-analysis we examined the relative probability of finding frogs in the observed salinities: 3.9 ppt, 8.3 ppt,

Table 1 The location and identity of the four anuran species observed in coastal, salt-invaded wetlands along with the highest salinity in which each species was observed

Species	Highest Salinity Observed	Occurrence	Location
Lithobates sphenocephalus	11 ppt	Abundant	Alligator River NWR
Hyla cinerea	23.4 ppt	Abundant	Cape Hatteras National Seashore
Gastrophryne carolinensis	3.9 ppt	Abundant	Alligator River NWR
Lithobates catesbeianus	6.2 ppt	Rare	Pea Island NWR

11 ppt, 16.8 ppt, and 23.4 ppt saltwater. The expected probability of survival for an individual anuran following exposure to a 3.9 ppt saltwater solution during the egg stage is 0.52 (0.39–0.66 95% BCI), 0.60 (0.51–0.68 BCI) for larval anurans, and 0.62 (0.39–0.84 BCI) for adults. The probability of survival in 8.3 ppt water for eggs is 0.25 (0.09–0.45 BCI), 0.32 (0.23–0.39 BCI) for larvae, and 0.53 (0.32–0.74 BCI) for adult frogs. At 11 ppt, the survivorship for eggs is 0.15 (0.03–0.35 BCI), larval survivorship is 0.18 (0.12–0.25 BCI), with adult survivorship predicted at 0.46 (0.25–0.70 BCI). Wetlands at 16.8 ppt have 0.04 (0.002–0.19 BCI) expected egg survivorship, 0.04 (0.02–0.07 BCI) expected larval survivorship, and 0.35 (0.13–0.61 BCI) expected adult survivorship. In 23.4 ppt wetlands, 0.01 (0.00–0.008 BCI) eggs are expected to survive, larval survivorship is 0.01 (0.002–0.02 BCI), and expected adult survivorship is 0.25 (0.05–0.57 BCI) (Table 2).

Common garden experiments

The oviposition site choice experiment utilized *Hyla cinerea* pairs collected from three geographically discrete populations from inland and coastal locations in eastern North Carolina. The subsequent egg hatching and tadpole survivorship experiments utilized the offspring of the collected pairs. For the coastal locations, we sampled three discrete populations along the inner and outer banks of North Carolina. We collected 1 replicate from a population near New Inlet bridge (35°41′11.5″ N, 75°29′03.92″W), 1 replicate from Coastal Studies

Institute on Roanoke Island (35°52′26.14″ N, 75°39′38.54″ W), and 2 replicates from Point Peter Road, Alligator River National Wildlife Refuge (35°46′13.1″ N, 75°44′30.1″ W). These populations are separated by the Croatan and/or Roanoke Sounds. For the inland locations, we sampled three discrete populations around Greenville, North Carolina. Specifically, we collected 1 replicate from a population near MacGregor Downs Road (35°37′15.8″ N, 77°26′45.29″ W), 1 replicate along Pactolus Highway (35°37′18.9″ N, 77°20′43.8″ W), and 2 replicates from a retention pond on 10th street (35°35′26.49″ N, 77°19′09.89″ W). Each inland population is at least 5 km apart from other populations with the Tar river and multiple highways between populations.

Oviposition site choice

We conducted four replicates in coastal and inland locations. Pairs successfully bred in every bin except one that contained a coastal pair. On average, females laid 1363 eggs (minimum = 713 eggs, maximum = 3039 eggs) per bin. We found that location (e.g., coastal vs. inland) and salinity both affected the probability that a female will lay her eggs in a particular pool (Fig. 2). As salinity increased, pairs from inland populations were less likely to deposit eggs in salinized water, while coastal females maintained a high probability of laying eggs in the higher salinity treatments (Fig. 2). For example, in the lower salinity treatments (4 ppt), females showed no

Table 2 Predicted survivorship (and Bayesian Credible Intervals) of anurans in various salinities based on the findings of the meta-analysis (Fig. 1). Each salinity concentration represents the salinity of a wetland in which frogs were observed along North Carolina's coast

Salinity (ppt) in which anurans were observed:	Predicted Egg Survivorship (+95% BCIs)	Predicted Larval Survivorship (+95% BCIs)	Predicted Adult Survivorship (+95% BCIs)
3.9	0.52 (0.39–0.66)	0.60 (0.51–0.68)	0.62 (0.39–0.84)
8.3	0.25 (0.09–0.45)	0.32 (0.23–0.39)	0.53 (0.32–0.74)
11	0.15 (0.03–0.35)	0.18 (0.12–0.25)	0.46 (0.25–0.70)
16.9	0.04 (0.002–0.19)	0.04 (0.02–0.07)	0.35 (0.13–0.61)
23.4	0.01 (0.00–0.008)	0.01 (0.002–0.02)	0.25 (0.05–0.57)

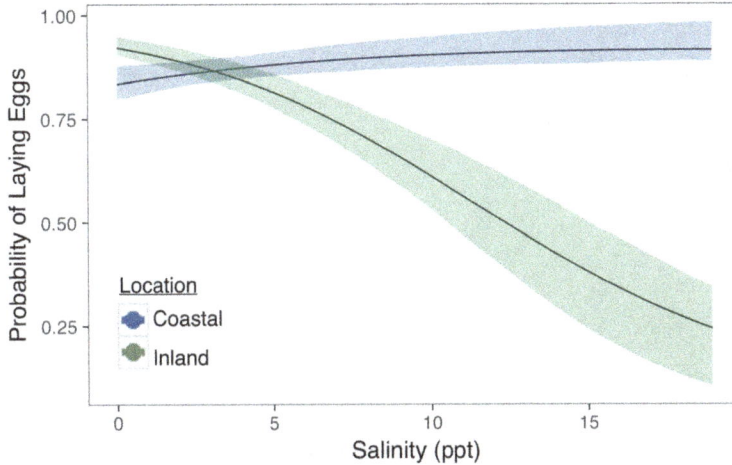

Fig. 2 Predicted probability of oviposition according to salinity and population location with 95% credibility envelopes. Green denotes the oviposition patterns from inland populations; *blue* indicates the oviposition patterns from coastal populations

divergence with inland females having 0.87 (0.85–0.91 BCI) probability of laying any eggs in the 4 ppt water, and coastal females having 0.84 (0.81–0.88 BCI) probability of laying eggs. Yet in the higher salinity treatments in which females chose between fresh or 12 ppt water, inland females had a 0.51 (0.41–0.61 BCI) probability of laying any eggs into 12 ppt water, while coastal females exhibited 0.91 (0.88–0.96 BCI) probability of laying eggs. Source population and salinity both affected the proportion of eggs laid in freshwater (Fig. 3). Pairs from both locations tended to lay the majority of their eggs into freshwater as salinity increased, but at 12 ppt, pairs from inland populations laid only 6% (0.04–0.07 BCI) into the saline water, while coastal pairs laid 16% (0.14–0.18 BCI) of their eggs in the saline water (Fig. 3).

Egg hatching

Salinity and source population affect the probability that any eggs would hatch out of a particular treatment (Fig. 4). At 4 ppt, the probability that an egg sourced from inland parents would hatch is 0.31 (0.24–0.38 BCI), while the probability that an egg laid by coastal parents would hatch is 0.54 (0.47–0.61 BCI). At higher salinities (10 ppt), eggs from both populations had an exceedingly low probability of hatching (inland probability: 0.02 (0.007–0.03 BCI); coastal probability: 0.04 (0.02–0.06 BCI)) (Fig. 4). We also observed that although the proportion of eggs that hatched in 3 ppt was similar across locations (inland proportion hatched: 0.33 (0.27–0.38 BCI); coastal proportion hatched: 0.36 (0.31–0.42 BCI)), 10% (0.07–0.11 BCI) of the coastal-sourced eggs hatched

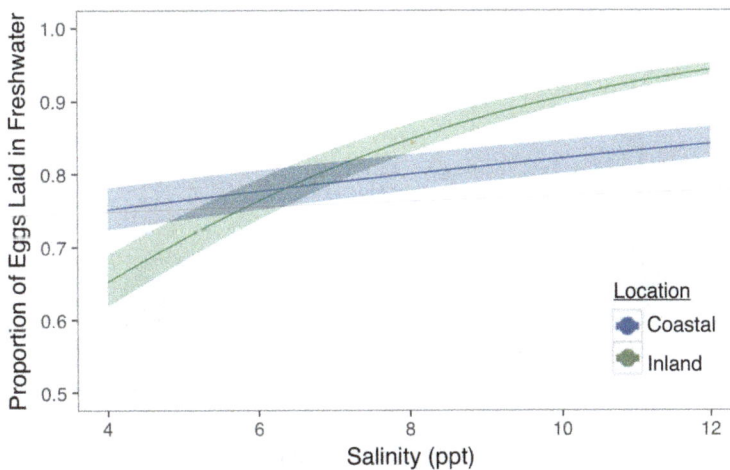

Fig. 3 The proportion of eggs laid in freshwater according to salinity and population location with 95% credible envelopes. Green denotes the proportion of eggs laid in freshwater by inland populations; *blue* indicates the proportion of eggs laid in freshwater from coastal populations

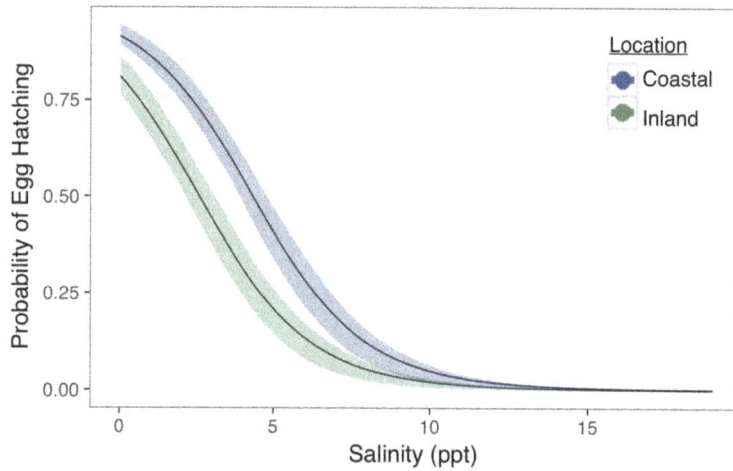

Fig. 4 Predicted probability of egg hatching according to salinity and population location with 95% credible envelopes. Green denotes the hatching patterns from inland populations; *blue* indicates hatching patterns from coastal populations

at 6 ppt compared to 3% (0.02–0.04 BCI) of the eggs sourced from inland populations (Fig. 5).

Tadpole survivorship

The predicted survival probability for coastal and inland *Hyla cinerea* tadpoles following a 6-day acclimation to freshwater (0.5 ppt) for coastal-sourced tadpoles is 0.98 (0.96–0.99 BCI) and 0.98 (0.96–0.99 BCI) for inland-sourced tadpoles (Fig. 6). At 4 ppt, predicted survivorship for coastal offspring is 0.96 (0.92–0.98 BCI) while inland offspring survivorship is 0.97 (0.95–0.99 BCI). Survivorship in 6 ppt treatments is 0.94 (0.90–0.97 BCI) for coastal tadpoles and 0.95 (0.89–0.98 BCI) from inland tadpoles. In the 8 ppt treatments, coastal tadpoles had higher survivorship at 0.97 (0.94–0.99 BCI) than inland tadpoles at 0.84 (0.73–0.92 BCI). At 12 ppt, we

again observed higher survivorship among coastal tadpoles with 0.24 (0.14–0.39 BCI) survivorship compared to inland tadpoles with 0.09 (0.04–0.16 BCI) survivorship. The random effect standard deviation representing parental influence is 0.17. Fixed effect slope and intercept estimates are listed in Additional file 2.

Discussion

We are at the precipice of dramatic environmental transformation as a result of global climate change, which provides the ideal canvas for exploring organismal responses to environmental change. Wetlands in coastal zones around the globe are among those anticipated to be most severely impacted from climate change due to increased frequency and intensity of coastal storms as

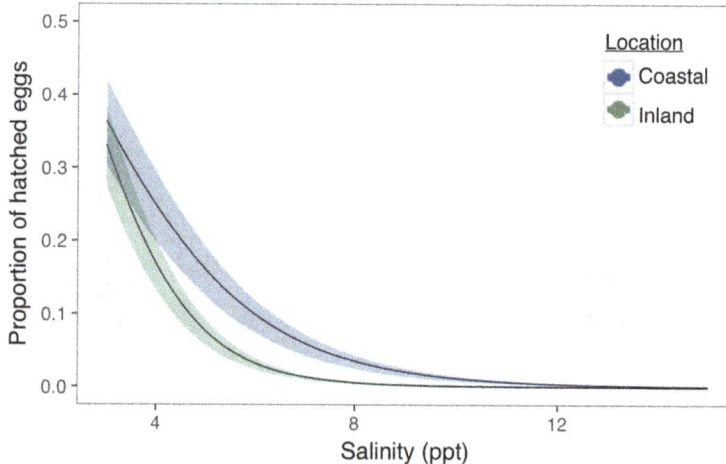

Fig. 5 The proportion of eggs that hatched according to salinity and population location with 95% confidence envelopes. Green denotes the proportion of eggs hatched from inland populations; *blue* indicates the proportion of eggs hatched from coastal populations

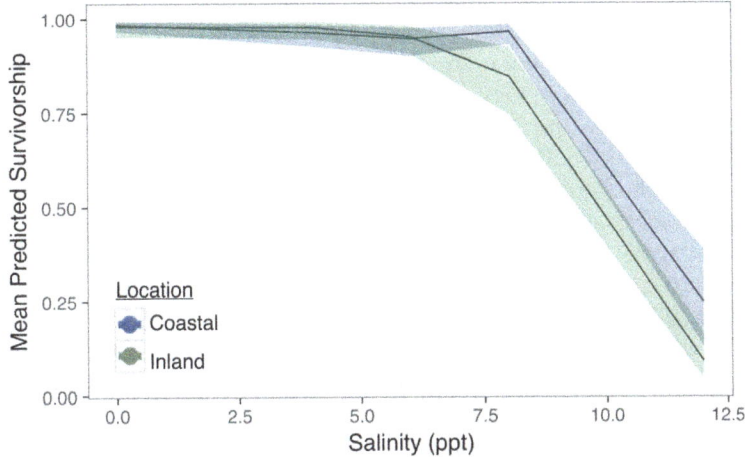

Fig. 6 Mean probability of tadpole survivorship according to salinity and population location with 95% credible envelopes. Green denotes the proportion of tadpoles sourced from inland populations; *blue* indicates tadpoles sourced from coastal populations

well as increased flooding and secondary salinization from sea level rise [1, 2, 8, 19, 66, 68, 85]. Yet despite the amount of cultural and research attention that climate change garners, a distressing deficiency exists in our empirical understanding of how rising salinities will impact coastal freshwater habitats and the animal communities sustained therein.

Ecological niche models aimed at understanding how environmental changes will impact affected populations typically predict that species that cannot emigrate to more suitable habitats are at risk of being locally extirpated as environmental quality degrades [3, 86–94]. This forecast is rational for freshwater organisms (like amphibians) that inhabit coastal wetlands given the lethal nature of osmotic stress [91, 94–100]. However, an important assumption inherent in most model predictions is that species either completely lack or have limited capacity to respond to environmental change – an assumption that can lead to overestimates of extinction rates or expected range contraction [24, 91, 94, 96, 98, 101–103]. Although adaptive evolution is increasingly well appreciated as a potential source of rescue for some, it is unclear whether organisms with complex life history strategies will be able to adapt to environmental change. In amphibians, we currently lack the ability to make more informed predictions that include adaptation for two main reasons. First, we do not know how sensitivity to salt stress varies across different life stages, and second, we know little about whether salt-tolerant responses are evolutionarily labile across life stages. In this paper, we address these gaps using a variety of tools (e.g., meta-analysis, field surveys, and common-garden experiments).

Meta-analysis and field surveys

Studies on amphibian responses to saltwater often begin with some variant of the statement, *it is well accepted that frogs do not belong in saline habitats*. These statements stem from long standing dogma that amphibians are not physiologically equipped to osmoregulate in non-freshwater environments. Nonetheless, we observed *Lithobates catesbeieanus*, *Lithobates sphenocephalus*, *Gastrophryne carolinensis*, and *Hyla cinerea* in brackish marshes in coastal North Carolina. These four species have been reported in brackish habitats previously [42, 75, 104–106] and the recurrence of these observations draws attention to the paucity of information explaining why some species are repeatedly observed inhabiting brackish wetlands while other closely related species are absent [39]. A particularly interesting contribution on this subject stems from our repeated field observations of abundant and thriving *Hyla cinerea* populations in salt marshes with salinities 450% higher than the expected larval LC_{50} concentration (as revealed by the meta-analysis). Indeed, these findings were inconceivable by the authors at the outset of the survey. While previous studies reported *Hyla cinerea* from saltmarshes along the Chesapeake Bay in Maryland in salinities up to 15 ppt [104], we found populations in salinities as high as 23 ppt, which is also the highest salinity that any North American frog species has been found to date (though Puerto Rican populations of *Rhinella marina*, *Eleutherodactylus coqui*, and *Lithobates grylio* come close at 20.5 ppt [107]).

Hopkins and Brodie (2015) recently updated Neill's 1958 review and provide a valuable and thorough review of all published observations of amphibians in saltwater [38, 39]. In their review, Hopkins and Brodie present a range of salinity tolerances revealed by experimental and field studies and suggest that the median maximum experimental salinity that can be tolerated by anuran amphibians falls between 9 ppt–12 ppt [39, 48, 108, 109]. Our meta-analysis refines and builds upon these estimates by providing an empirically derived range of survival probability estimates for each salinity and life stage. For example, at 9 ppt we may expect around 21% of eggs to survive, 27% of larvae to survive, and 50% of adults to survive – fundamental information for managing anuran populations across landscapes affected by salinization.

The meta-analysis underlines the fact that amphibians have different abilities to persist in saline environments according to life stage. Though most studies test the effects of salt on a single life stage, our meta-analysis integrates the findings of all of these studies to better understand how salt sensitivity changes through each life stage and provides a quantitative baseline and important context for our common garden experiments and field observations of anurans in salinities as high as 66% seawater. Broadly, all studies examined in our meta-analysis demonstrate declines in survivorship as salinity increased across each life stage, our analyses, which includes studies on 35 species representing 26 different genera across 10 families. We found that eggs are the most sensitive to osmotic stress across the anuran clade, followed by the larval stage, and adults are the least susceptible. The results of the meta-analysis indicate that the lethal experimental salt concentration in which 50% mortality (LC_{50}) is expected for eggs occurs at approximately 4.15 ppt for anuran eggs, 5.5 ppt for larvae, and 9.0 ppt for adults. Although the uncertainty in LC_{50} concentrations identified in the meta-analysis stems largely from differences in sample sizes (only three studies on adult frogs met our criteria for the meta-analysis), they might also reflect greater sensitivity during particular stages among species.

Embryos, for example, are expected to be more sensitive to external stressors than other stages because important developmental pathways are initiated during the early embryonic period and so perturbations at this stage may be teratogenic or fatal [110, 111]. It has been shown that pathogens (e.g., bacteria, endoparasites, or waterborne fungi), predators, ultraviolet radiation, and toxins can all have strong effects on embryonic survival, and induce effects that carry over to affect developmental outcomes in later life stages [57, 112–115]. Tadpoles are also expected to be more sensitive to water quality than are adults because they are obliged to the aquatic habitat. Larvae may be more tolerant of osmotic stress than embryos if they can increase the activity or concentration of ion pumps in the gills. However, tadpoles raised in saltwater tend to have stunted developmental rates and metamorphose at smaller sizes compared to freshwater-raised tadpoles [46, 52, 59, 64, 65, 116, 117], which can affect adult survival and reproductive success [118]. Adults, on the other hand, are less confined to aquatic environments and thus can reduce contact with stressful habitats via behavioral avoidance or dispersal. Additionally, adults can likely physiologically tolerate a greater degree of osmotic stress and/or desiccation by increasing urea in the blood [44, 119], altering cellular ion or water transport [58, 59, 120, 121], or adjusting the permeability of the skin [122, 123].

Common garden experiments

In the oviposition, hatching, and tadpole survivorship experiments, we find evidence for altered and adaptive responses to salinization across multiple life stages in *Hyla cinerea*. Specifically, we report differences in egg deposition patterns, hatching success, and tadpole survivorship between salt-exposed coastal and salt-naïve inland populations of North American green treefrogs (*Hyla cinerea*). We focus on reproductive behaviors, egg hatching, and tadpole viability because they are stages and traits that are highly vulnerable to environmental quality, and directly affect fitness and population viability [39, 48, 65, 124–128]. Female oviposition site selection directly affects the fitness of both the parents and the offspring, so decisions about oviposition sites should reflect an adaptive response. Therefore, we expected strong patterns of saltwater avoidance among both coastal and inland populations if salt were equally lethal to eggs and offspring from both inland and coastal populations [51, 61, 129, 130]. However, we found that coastal and inland frogs exhibited different patterns of oviposition site selection across the experimental salt gradient. Both inland and coastal pairs increasingly avoided saline water as salinity increased but inland frogs had greater response and did not deposit any eggs in salinities above ~12 ppt, whereas coastal pairs laid approximately 24% of their eggs in the highest salinities (Fig. 3). Additionally, eggs laid by coastal parents have higher probabilities of hatching in higher salinities and more coastal tadpoles survive in higher salinities when compared to inland-sourced conspecifics. Our inferences are based on experiments on three coastal and three inland populations and so should be extrapolated more broadly with caution. However, collectively our results provide evidence that some coastal populations of *Hyla cinerea* are responding adaptively to saltwater exposure across multiple life stages, which is contrary to expected outcomes given the general reputation of anuran amphibians as a highly salt sensitive order. Gomez-Mestre and Tejado report similar findings in *Bufo calamita*, the Natterjack Toad, in which embryos and tadpoles from brackish populations demonstrate higher survival compared to tadpoles from inland, freshwater populations

[45, 131]. Together these studies suggest that the ability to respond adaptively to saltwater exposure may be more possible than previously appreciated, and future studies may consider using comparative, common-garden approaches to not only determine how salt-exposure affects various endpoints, but also whether other species also exhibit population-level differences in salt tolerance across species and life stages.

The physiological mechanisms that explain why coastal pairs have relaxed salt avoidance behaviors, higher hatching success, and higher tadpole survivorship are likely to be numerous and spread across the different life stages. In adults, coastal male *Hyla cinerea* may have more viable and motile sperm in saline water. A recent study examined sperm survivorship and motility in *Hyla cinerea* located in Charleston, South Carolina (a coastal location) and found that 4 ppt saltwater reduced ability of sperm to survive and swim, but that study did not compare coastal and inland populations [51]. Alternatively, adult coastal females may increase the partitioning of yolk resources into eggs, or alter the egg coat matrix to provide additional protection against osmotic stressors compared to inland eggs. In tadpoles, coastal individuals may have an increased abundance of water channels (AQPs) and ion pumps (e.g., Na^+/K^+-ATPase) in the gills that enhance the ability to maintain internal water and ion balance, thus improving survival. Several studies have demonstrated that exposure to saltwater can increase the quantity and activity of sodium-potassium pumps in tadpole gills [58, 65, 132]. These hypotheses remain to be tested in coastal *Hyla cinerea*, leaving the exact mechanisms explaining the observed patterns undefined. Moreover, the adaptive processes that produce advantageous physiological responses also remain largely unknown.

There are three possible overlapping adaptive processes that may explain the divergence in responses that we observed between coastal and inland anuran populations; local adaptation, phenotypically plastic responses, and/or maternal effects. Local adaptation occurs when populations have higher fitness in their local environmental conditions compared to populations from other environments, and our results are consistent with expected outcomes if coastal populations are becoming locally adapted to tolerate elevated salt concentrations across different life stages [133]. Adaptive evolution is a well-appreciated process that can sustain or rescue populations facing strong selection gradients [134–138]. Yet several criteria must be met before local adaptation can be confirmed. "Adaptive" phenotypes must be shown to correlate positively with fitness, and the production of putatively adaptive phenotypes should be directly linked to specific environmental drivers, and studies on adaptive responses must demonstrate a genetic basis for differences observed among populations [139]. Our results are consistent with the expectations of the first two criteria, but we are not yet able to deduce whether there is a genetic basis for such changes.

Phenotypic plasticity (defined here as the ability to modulate phenotype in response to environmental cues) can also produce phenotypes that appear different and adaptive, yet may be genetically indistinguishable from other populations [139–141]. Because plasticity can promote adaptation, inhibit adaptation, or be the adaptive response itself, uncovering the role of phenotypic plasticity remains one of the most important challenges for understanding and predicting adaptive responses to climate change [31–34, 140, 142–145]. Indeed, some degree of phenotypic plasticity has been observed in nearly every trait that has been measured to date, which underlines the importance of examining the contribution of plasticity in studies of adaptive responses [32–34, 140, 144–146].

Maternal effects induced by environmental conditions experienced by the parents are also emerging as important factors that influence offspring fitness in different environments [147, 148]. Increased prevalence of maternally affected traits are expected when the environment experienced by the mother matches the environment experienced by the offspring [149], and in such situations, can explain up to 96% of the variation in improved offspring fitness in stressful environments [150].

The divergent responses that we present in this paper may be the production of either maternal effects, phenotypic plasticity, or local adaptation alone. However, some blend of these mechanisms is more likely. For example, exposure to saltwater during the ontogeny of coastal individuals may have initiated cascades of plastic responses that predisposed females from coastal populations toward salt tolerant responses. These responses may have transferred to offspring, which mixes plasticity with maternal effects. Alternatively, coastal individuals with increased ability to tolerate salt through enhanced plasticity may have been favored by selection. Presumably, selecting for more plastic individuals would gradually increase the overall amount of plasticity observed in coastal populations, which blends plasticity with genetic adaptation (*sensu* Baldwin effect) [151]. In reality, there are a multitude of possible mechanistic combinations as plasticity, local adaptation, and maternal effects can be reciprocal processes that serve as both the product and raw material for selection and adaptation. Future research should prioritize discerning how adaptive evolution, phenotypic plasticity, and maternal effects are interwoven to produce different responses to environmental stressors especially in organisms with complex life cycles. A more complete understanding of all contributing processes will help managers identify thresholds of tolerance, detect vulnerable populations, and determine which organisms are likely to successfully tolerate novel stressors and persist in their environments.

Despite the consistent differences in behavior, embryo, and larval survivorship we observed between inland and coastal populations, our results indicate that all populations and life stages of *Hyla cinerea* (coastal and inland populations) are salt-sensitive. Frog pairs laid the majority of eggs into freshwater in all populations; saltwater negatively affected hatching rates across all populations, and saltwater reduced survivorship for both coastal and inland tadpoles. While we have focused on the degree to which these responses differed among populations as indications of adaptive responses, we believe that it should be noted that anurans on the whole, remain an osmotically sensitive group of organisms even in chronically salt-exposed populations. The continued preference for, and higher performance in, freshwater, even among coastal populations, may indicate that thresholds of saltwater tolerance exist.

Conclusions

This study provides the following insights: First, our meta-analysis offers a quantitative baseline for salt tolerance in anurans and provides important context for future field observations and experimental studies exploring saltwater tolerance in anurans. The meta-analysis also shows that generally, anurans are salt-sensitive across species and across life stages and are therefore likely to be adversely affected by progressive salinization of freshwater systems. Second, we show different sensitivities and responses to salt stress across life stages and across populations, significant information for future studies and management. Third, we provide initial evidence that despite their sensitivity, some anuran species (*Hyla cinerea*) have populations that are able to respond adaptively to salt stress across different life stages. Though these findings are an encouraging indication that some frog populations may persist through salinization, our results also illuminate that much more remains to be known. Key unknowns include the physiological mechanisms and adaptive processes that underlie salt tolerance in anurans, determining whether we can expect adaptive responses to match the pace and intensity of environmental change (i.e., define the limits of tolerance and rates of adaptation), and exploring the factors that govern amphibian distributions across salt-invaded landscapes (i.e., why only 4 out of the 17 possible species occur in brackish wetlands).

Testing multiple mechanistic hypotheses about adaptive processes (e.g., maternal effects, genetic evolution, and phenotypic plasticity) in ecological time in wild macro-organisms has remained an empirical challenge. Yet identifying populations with complex life cycles that

demonstrate divergent responses to an environmental stressor across life stages (such as coastal frog populations adapting to saline environments) may provide unique and valuable opportunities to empirically address questions about the etiology of adaptive and non-adaptive responses, how novel adaptive phenotypes emerge, and how population and demographic dynamics interact with adaptive processes.

Additional files

> **Additional file 1:** Detailed list of studies included in the meta-analysis. (CSV 114 kb)
>
> **Additional file 2: Table S1.** Predicted *Hyla cinerea* tadpole survivorship after a six-day exposure to one of five salinity concentrations, along with slope and intercept estimates, each with 95% Bayesian credible intervals (L.C.I = Lower Credible Interval, U.C.I = Upper Credible Interval) (Fig. 6). (DOC 34 kb)

Acknowledgements
We thank members of McCoy lab group, K. McCoy, A. Stuckert, and J. Touchon for their thoughtful insights during the development of this work. We also extend thanks to A. Stuckert, T. McFarland and C. Thaxton for field and laboratory assistance. This manuscript was improved by helpful reviews by Gareth Hopkins and Ilaria Bernabò.

Funding
This work was funded by North Carolina Sea Grant (Project No. 2014-R/14-HCE-3), National Science Foundation Doctoral Dissertation Improvement Grant (#1701690) awarded to McCoy and Albecker, the Graduate Women in Science Nell Mondy Fellowship, as well as research grants from the North Carolina Herpetological Society, East Carolina University's Coastal Maritime Council, and Explorer's Club.

Authors' contributions
MAA and MWM conceived the study. MAA carried out experimentation and data extraction. MAA and MWM analyzed and interpreted results, and contributed to writing the manuscript. Both authors read and approved the final manuscript.

Competing interests
The authors declare that they have no competing interest.

References
1. Meehl GA, Washington WM, Collins WD, Arblaster JM, Hu A, Buja LE, Strand WG, Teng H. How much more climate change and sea level rise? Science. 2005;307:1769–72.
2. Domingues CM, Church JA, White NJ, Gleckler PJ, Wijffels SE, Barker PM, Dunn JR. Improved estimates of upper-ocean warming and multi-decadal sea-level rise. Nature. 2008;453:1090–3.
3. Nicholls RJ, Tol RS. Impacts and responses to sea-level rise: a global analysis of the SRES scenarios over the twenty-first century. Philos Trans A Math Phys Eng Sci. 2006;364:1073–95.
4. Church JA, Clark PU, Cazenave A, Gregory JM, Jevrejeva S, Levermann A, Merrifield MA, Milne GA, Nerem RS, Nunn PD, et al. IPCC fifth assessment report (AR5), climate change. Phys Sci Basis. 2013;2013:1–124.
5. Scavia D, Field JC, Boesch DF, Buddemeier RW, Burkett V, Cayan DR, Fogarty M, Harwell MA, Howarth RW, Mason C, et al. Climate change impacts on U. S. coastal and marine ecosystems. Estuaries. 2002;25:149–64.
6. Senior CA, Jones RG, Lowe JA, Durman CF, Hudson D. Predictions of extreme precipitation and sea-level rise under climate change. Philos Trans A Math Phys Eng Sci. 2002;360:1301 11.
7. Rahmstorf S, Perrette M, Vermeer M. Testing the robustness of semi-empirical sea level projections. Climate Dynam. 2012;39:861–75.

8. Kemp AC, Horton BP, Donnelly JP, Mann ME, Vermeer M, Rahmstorf S. Climate related sea level variations over the past two millennia. PNAS. 2011; 108:11017–22.

9. DaLaune RD, Pezeshki SR. The influence of subsidence and saltwater intrusion on coastal marsh stability: Louisiana gulf coast. J Coast Res. 1994;12:77–89.

10. Abrams PA. Implications of dynamically variable traits for identifying, classifying, and measuring direct and indirect effects in ecological communities. Am Nat. 1995:112–34.

11. Michener WK, Blood ER, Bildstein KL, Brinson MM, Gardner LR. Climate change, hurricanes and tropical storms, and rising sea level in coastal wetlands. Ecol Appl. 1997;7:770–801.

12. Loaiciga HA. Climate change and ground water. Ann Assoc Am Geogr. 2003;93:30–41.

13. Day JW, Christian RR, Boesch DM, Yáñez-Arancibia A, Morris J, Twilley RR, Naylor L, Schaffner L, Stevenson C. Consequences of climate change on the Ecogeomorphology of coastal wetlands. Estuar Coasts. 2008;31:477–91.

14. Meyssignac B, Cazenave A. Sea level- a review of present day and recent past changes and variability. J Geodyn. 2012;58:96–109.

15. Williams SJ. Sea level rise implications for coastal regions. J Coast Res. 2013; 63:184–96.

16. Baldwin AH, Mendelssohn IA. Effects of salinity and water level on coastal marshes: an experimental test of disturbance as a catalyst for vegetation change. Aquat Bot. 1998;61:255–68.

17. Williams K, Ewel KC, Stumpf RP, Putz FE, Workman TW. Sea-level rise and coastal forest retreat on the west coast of florida, USA. Ecology. 1999;80: 2045–63.

18. Geddes NA, Mopper S. Effects of environmental salinity on vertebrate florivory and wetland communities. Nat Areas J. 2006;26:31–7.

19. Kopp RE, Horton BP, Kemp AC, Tebaldi C. Past and future sea-level rise along the coast of North Carolina, USA. Clim Change. 2015;132:693–707.

20. Knighton AD, Mills K, Woodroffe CD. Tidal-creek extension and saltwater intrusion in northern Australia. Geology. 1991;19:831–4.

21. Morris JT, Sundarshwar PV, Nietch CT, Kjerfve B, Cahoon DR. Responses of coastal wetlands to rising sea level. Ecology. 2002;83:2869–77.

22. Hamer AJ, McDonnell MJ. Amphibian ecology and conservation in the urbanising world- a review. Biol Conserv. 2008;141:2432–49.

23. Foden WB, Mace GM, Vié J-C, Angulo A, Butchart SHM, DeVantier L, Dublin HT, Gutsche A, Stuart SN, Turak E. Species susceptibility to climate change impacts. In: Vie J-C, Hilton-Taylor C, Stuart SN, editors. Wildlife in a changing world: an analysis of the 2008 IUCN red list of threatened species. Volume 1. Barcelona, Spain; 2009. p. 77–88.

24. Reed TE, Schindler DE, Waples RS. Interacting effects of phenotypic plasticity and evolution on population persistence in a changing climate. Conserv Biol. 2011;25:56–63.

25. Anderson JT, Perera N, Chowdhury B, Mitchell-Olds T. Microgeographic patterns of genetic divergence and adaptation across environmental gradients in Boechera stricta (Brassicaceae). Am Nat. 2015;186:S60–73.

26. Brady SP. Road to evolution? Local adaptation to road adjacency in an amphibian (Ambystoma maculatum). Sci Rep. 2012;2

27. Fraser DJ, Weir LK, Bernatchez L, Hansen MM, Taylor EB. Extent and scale of local adaptation in salmonid fishes: review and meta-analysis. Heredity. 2011;106:404–20.

28. Lamichhaney S, Barrio AM, Rafati N, Sundström G, Rubin C-J, Gilbert ER, Berglund J, Wetterbom A, Laikre L, Webster MT. Population-scale sequencing reveals genetic differentiation due to local adaptation in Atlantic herring. Proc Natl Acad Sci. 2012;109:19345–50.

29. Mopper S, Strauss SY. Genetic structure and local adaptation in natural insect populations: effects of ecology, life history, and behavior: Springer eScience & Business Media; 2013.

30. Reznick DN, Ghalambor CK. The population ecology of contemporary adaptations: what empirical studies reveal about the conditions that promote adaptive evolution. Genetica. 2001;112:183–98.

31. Pfennig DW, Wund MA, Snell-Rood EC, Cruickshank T, Schlichting CD, Moczek AP. Phenotypic plasticity's impacts on diversification and speciation. Trends Ecol Evol. 2010;25:459–67.

32. Wund MA. Assessing the impacts of phenotypic plasticity on evolution. Integr Comp Biol. 2012;52:5–15.

33. Nonaka E, Svanbäck R, Thibert-Plante X, Englund G, Brännström Å. Mechanisms by which phenotypic plasticity affects adaptive divergence and ecological speciation. Am Nat. 2015;186:E126–43.

34. Hendry AP. Key questions on the role of phenotypic plasticity in eco-evolutionary dynamics. J Hered. 2015; esv060

35. Reed TE, Waples RS, Schindler DE, Hard JJ, Kinnison MT. Phenotypic plasticity and population viability: the importance of environmental predictability. Proc R Soc Lond B Biol Sci. 2010;277

36. Vitt LJ, Caldwell JP, Wilbur HM, Smith DC. Amphibians as harbingers of decay. Bioscience. 1990;40:418.

37. Carignan V, Villard M-A. Selecting indicator species to monitor ecological integrity: a review. Environ Monit Assess. 2002;78:45–61.

38. Neill WT. The occurrence of amphibians and reptiles in saltwater areas, and a bibliography. Bull Mar Sci. 1958;8:1–97.

39. Hopkins GR, Brodie JED. Occurrence of amphibians in saline habitats: a review and evolutionary perspective. Herpetol Monogr. 2015;29:1–27.

40. Gibbons JW, Winne CT, Scott DE, Willson JD, Glaudas X, Andrews KM, Todd BD, Fedewa LA, Wilkinson L, Tsaliagos RN, et al. Remarkable amphibian biomass and abundance in an isolated wetland: implications for wetland conservation. Conserv Biol. 2006;20:1457–65.

41. McCoy MW, Barfield M, Holt RD. Predator shadows: complex life histories as generators of spatially patterned indirect interactions across ecosystems. Oikos. 2009;118:87–100.

42. Christman SP. Geographic variation for salt water tolerance in the frog Rana sphenocephala. Copeia. 1974;1974:773–8.

43. Gibbons JW, Coker JW. Herpetofaunal colonization patterns of atlantic coast barrier islands. Am Midl Nat. 1978;99:219–33.

44. Balinsky JB. Adaptation of nitrogen metabolism to hyperosmotic environment in amphibia. J Exp Zool. 1981;215:335–50.

45. Gomez-Mestre I, Tejado M. Local adaptation of an anuran amphibian to osmotically stressful environments. Evolution. 2003;57:1889–99.

46. Wu C-S, Kam Y-C. Effects of salinity on the survival, growth, development, and metamorphosis of Fejervarya limnocharis tadpoles living in brackish water. Zoolog Sci. 2009;26:476–82.

47. Gordon MS, Tucker VA. Osmotic regulation in the tadpoles of the crab-eating frog (Rana cancrivora). J Exp Biol. 1965;42:437–45.

48. Gordon MS, Schmidt-Nielsen K, Kelly HM. Osmotic regulation in the crab-eating frog (Rana Cancrivora). J Exp Biol. 1961;38:659–78.

49. Gordon MS. Intracellular osmoregulation in skeletal muscle during salinity adaptation in two species of toads. Biol Bull. 1965;128:218–29.

50. Gordon MS. Osmotic regulation in the green toad (Bufo viridis). J Exp Biol. 1962;39:261–70.

51. Wilder AE, Welch AM. Effects of salinity and pesticide on sperm activity and Oviposition site selection in green Treefrogs, Hyla cinerea. Copeia. 2014; 2014:659–67.

52. Hsu W-T, Wu C-S, Lai J-C, Chiao Y-K, Hsu C-H, Kam Y-C. Salinity acclimation affects survival and metamorphosis of crab-eating frog tadpoles. Herpetologica. 2012;68:14–21.

53. Wu CS, Gomez-Mestre I, Kam YC. Irreversibility of a bad start: early exposure to osmotic stress limits growth and adaptive developmental plasticity. Oecologia. 2012;169:15–22.

54. Sanzo D, Hecnar SJ. Effects of road de-icing salt (NaCl) on larval wood frogs (Rana sylvatica). Environ Pollut. 2006;140:247–56.

55. Li N, Phummisutthigoon S, Charoenphandhu N. Low salinity increases survival, body weight and development in tadpoles of the Chinese edible frog Hoplobatrachus rugulosus. Aquacult Res. 2015;

56. Haramura T. Salinity tolerance of eggs of Buergeria japonica (Amphibia, Anura) inhabiting coastal areas. Zoolog Sci. 2007;24:820–3.

57. Touchon JC. Hatching plasticity in two temperate anurans: responses to pathogen and predation cues. Can J Zool. 2006;84:556–63.

58. Wu CS, Yang WK, Lee TH, Gomez-Mestre I, Kam YC. Salinity acclimation enhances salinity tolerance in tadpoles living in brackish water through increased Na(+), K(+) -ATPase expression. J Exp Zool A Ecol Genet Physiol. 2014;321:57–64.

59. Uchiyama M, Yoshizawa H. Salinity tolerance and structure of external and internal gills in tadpoles of the crab-eating frog, Rana Cancrivora. Cell Tissue Res. 1992;267:35–44.

60. Dietz TH, Alvarado RH. Na and Cl transport across gill chamber epithelium of Rana catesbeiana tadpoles. Am J Physiol. 1974;226:764–70.

61. Haramura T. Use of oviposition sites by a Rhacophorid frog inhabiting a coastal area in Japan. J Herpetol. 2011;45:432–7.

62. Smith MJ, Shreiber ESG, Scroggie MP, Kohout M, Ough K, Potts J, Lennie R, Turnbull D, Jin C, Clancy T. Associations between anuran tadpoles and salinity in a landscape mosaic of wetlands impacted by secondary salinisation. Freshw Biol. 2006;52:75–84.

63. Dougherty CK, Smith GR. Acute effects of road de-icers on the tadpoles of three anurans. Appl Herpetol. 2006;3:87–93.

64. Christy MT, Dickman CR. Effects of salinity on tadpoles of the green and golden bell frog (Litoria aurea). Amphibia-Reptilia. 2002;23:1–11.

65. Bernabò I, Bonacci A, Coscarelli F, Tripepi M, Brunelli E. Effects of salinity stress on Bufo Balearicus and Bufo bufo tadpoles: tolerance, morphological gill alterations and Na(+)/K(+)-ATPase localization. Aquat Toxicol. 2013;132-133:119–33.

66. Craft C, Clough J, Ehman J, Joye S, Park R, Pennings S, Guo H, Machmuller M. Forecasting the effects of accellerated sea-level rise on tidal marsh ecosystem services. Front Ecol Environ. 2009;7:73–8.

67. Kemp AC, Horton BP, Culver SJ, Corbett DR, Ovd P, Gehrels WR, Douglas BC, Parnell AC. Timing and magnitude of recent accelerated sea-level rise. Geology. 2009;37:1035–8.

68. Titus JG, Richman C. Maps of lands vulnerable to sea level rise: modeled elevations along the US Atlantic and gulf coasts. Climate Res. 2001;18:205–28.

69. Parkinson RW. Sea-level rise and the fate of tidal wetlands. J Coast Res. 1994;10:987–9.

70. Hintz WD, Relyea RA. Impacts of road deicing salts on the early-life growth and development of a stream salmonid: salt type matters. Environ Pollut. 2017;

71. Crother BI, Fontenot CL. Amphibian and reptile monitoring in the Ponchartrain-Maurepas region. In: Lake Pontchartrain Basin research program (PBRP); 2006. p. 35.

72. Heyer R, Donnelly MA, Foster M, Mcdiarmid R. Measuring and monitoring biological diversity: standard methods for amphibians: Smithsonian Institution; 2014.

73. Rader RB, Batzer DP, Wissinger SA. Bioassessment and management of north American freshwater wetlands: Wiley; 2001.

74. Tuberville TD, Willson JD, Dorcas ME, Gibbons JW. Herpetofaunal species richness of southeastern national parks. Southeast Nat. 2005;4:537–69.

75. Brown ME, Walls SC. Variation in salinity tolerance among larval anurans: implications for community composition and the spread of an invasive, non-native species. Copeia. 2013;2013:543–51.

76. Wells KD. The ecology and behavior of amphibians. Chicago: The University of Chicago Press; 2007.

77. Gosner KL. A simplified table for staging anuran embryos and larvae with notes on identification. Herpetologica. 1960:183–90.

78. Gordon MS, Tucker VA. Further observations on the physiology of salinity adaptation in the crab-eating frog (Rana Cancrivora). J Exp Biol. 1968;49:185–93.

79. R: a language and environment for statistical computing. In: R Core development team. 3.2.3 ed. Vienna: R Foundation for Statistical Computing; 2014.

80. Su Y-S, Yajima M. R2jags: using R to run 'JAGS': R package version 0.5–7; 2015.

81. Plummer M. Rjags: Bayesian graphical models using MCMC: R package version 3–15; 2015.

82. Plummer M, Best N, Cowles K, Vines K. CODA: convergence diagnosis and output analysis for MCMC. R News. 2006;6(1):7–11.

83. Gelman A, Carlin JB, Stern HS, Rubin DB. Bayesian data analysis. Texts in statistical science series. Boca Raton: Chapman & Hall/CRC; 2004.

84. Bennett JE, Racine-Poon A, Wakefield JC. MCMC for nonlinear hierarchical models. London, UK: Chapman and Hall; 1996.

85. Nicholls RJ, Cazenave A. Sea-level rise and its impact on coastal zones. Science. 2010;328:1517–20.

86. Bradshaw WE, Holzapfel CM. Climate change. Evolutionary response to rapid climate change. Science. 2006;312:1477–8.

87. Chen IC, Hill JK, Ohlemuller R, Roy DB, Thomas CD. Rapid range shifts of species associated with high levels of climate warming. Science. 2011;333:1024–6.

88. Davis MB, Shaw RG, Etterson JR. Evolutionary responses to changing climate. Ecology. 2005;86:1704–14.

89. Dawson TP, Jackson ST, House JI, Prentice IC, Mace GM. Beyond predictions: biodiversity conservation in a changing climate. Science. 2011;332:53–8.

90. Harley CD. Climate change, keystone predation, and biodiversity loss. Science. 2011;334:1124–7.

91. Moritz C, Agudo R. The future of species under climate change: resilience or decline? Science. 2013;341:504–8.

92. Parmesan C. Ecological and evolutionary responses to recent climate change. Annu Rev Ecol Evol Syst. 2006:637–69.

93. Walther G-R, Post E, Convey P, Menzel A, Parmesan C, Beebee TJC, Fromentin J-M, Hoegh-Guldberg O, Bairlein F. Ecological responses to recent climate change. Nature. 2002;416:389–95.

94. Thomas CD, Cameron A, Green RE, Bakkenes M, Beaumont LJ, Collingham YC, Erasmus BF, De Siqueira MF, Grainger A, Hannah L. Extinction risk from climate change. Nature. 2004;427:145–8.

95. Chown SL. Trait-based approaches to conservation physiology: forecasting environmental change risks from the bottom up. Philos Trans R Soc Lond B Biol Sci. 2012;367:1615–27.

96. Lewis OT. Climate change, species-area curves and the extinction crisis. Philos Trans R Soc Lond B Biol Sci. 2006;361:163–71.

97. Maclean IM, Wilson RJ. Recent ecological responses to climate change support predictions of high extinction risk. Proc Natl Acad Sci U S A. 2011; 108:12337–42.

98. Schwartz MW, Iverson LR, Prasad AM, Matthews SN, O'Connor RJ. Predicting extinctions as a result of climate change. Ecology. 2006;87:1611–5.

99. Stuart SN, Chanson JS, Cox NA, Young BE, Rodrigues AS, Fischman DL, Waller RW. Status and trends of amphibian declines and extinctions worldwide. Science. 2004;306:1783–6.

100. Traill LW, Lim ML, Sodhi NS, Bradshaw CJ. Mechanisms driving change: altered species interactions and ecosystem function through global warming. J Anim Ecol. 2010;79:937–47.

101. Davis MB, Shaw RG. Range shifts and adaptive responses to quaternary climate change. Science. 2001;292:673–9.

102. Holt R, Gomulkiewicz R. Conservation implications of niche conservatism and evolution in heterogeneous environments. In: Evolutionary conservation biology. Volume 2004: Cambridge University Press; 2004. p. 244–64.

103. Lawler JJ, Shafer SL, Bancroft BA, Blaustein AR. Projected climate impacts for the amphibians of the western hemisphere. Conserv Biol. 2010;24:38–50.

104. Hardy JD. Notes on the distribution of Mycrohyla carolinensis in southern Maryland. Herpetologica. 1953;8:162–6.

105. Hardy JDJ. Amphibians of the Chesapeake Bay region. Chesapeake Sci. 1972; 13:S123–8.

106. Gunzburger MS. Reproductive ecology of the green treefrog (Hyla cinerea) in northwestern Florida. Am Midl Nat. 2006;155:321–8.

107. Rios-López N. Effects of increased salinity on tadpoles of two anurans from a Caribbean coastal wetland in relation to their natural abundance. Amphibia-Reptilia. 2008;29:7–18.

108. Munsey LD. Salinity tolerance of the african Pipid frog, Xenopus laevis. Copeia. 1972;1972:584–6.

109. Ruibal R. The ecology of a brackish water population of Rana pipiens. Copeia. 1959;1959:315–22.

110. Wilbur HM. Complex life cycles. Annu Rev Ecol Syst. 1980;11:67–93.

111. Meteyer CU, Cole RA, Converse KA, Docherty DE, Wolcott M, Helgen JC, Levey R, Eaton-Poole L, Burkhart JG. Defining anuran malformations in the context of a developmental problem. J Iowa Acad Sci. 2000;107:72–8.

112. Burkhart JG, Helgen JC, Fort DJ, Gallagher K, Bowers D, Propst TL, Gernes M, Magner J, Shelby MD, Lucier G. Induction of mortality and malformation in Xenopus laevis embryos by water sources associated with field frog deformities. Environ Health Perspect. 1998;106:841.

113. Grant KP, Licht RL. Effects of ultraviolet radiation on life-history stages of anurans from Ontario, Canada. Can J Zool. 1995;73:2292–301.

114. Kiesecker JM, Blaustein AR. Influences of egg laying behavior on pathogenic infection of amphibian eggs. Conserv Biol. 1997;11:214–20.

115. Rohr JR, McCoy KA. A qualitative meta-analysis reveals consistent effects of atrazine on freshwater fish and amphibians. Environ Health Perspect. 2010; 2010:20–32.

116. Wood L, Welch AM. Assessment of interactive effects of elevated salinity and three pesticides on life history and behavior of southern toad (Anaxyrus terrestris) tadpoles. Environ Toxicol Chem. 2015;34:667–76.

117. Langhans M, Peterson B, Walker A, Smith GR, Rettig JE: Effects of salinity on survivorship of wood frog (Rana sylvatica) tadpoles. 2009.

118. Berven KA. Factors affecting population fluctuations in larval and adult stages of the wood frog (Rana sylvatica). Ecology. 1990;71:1599–608.

119. Shoemaker V, Hillman S, Hillyard S, Jackson D, McClanahan L, Withers P, Wygoda M. Exchange of water, ions, and respiratory gases in terrestrial amphibians: Environmental physiology of the amphibians; 1992. p. 125–50.

120. Uchiyama M, Konno N. Hormonal regulation of ion and water transport in anuran amphibians. Gen Comp Endocrinol. 2006;147:54–61.

121. Konno N, Hyodo S, Matsuda K, Uchiyama M. Effect of osmotic stress on expression of a putative facilitative urea transporter in the kidney and urinary bladder of the marine toad, Bufo marinus. J Exp Biol. 2006;209:1207–16.

122. McClanahan L Jr, Stinner JN, Shoemaker VH. Skin lipids, water loss, and energy metabolism in a south American tree frog (Phyllomedusa sauvagei). Physiol Zool. 1978;51:179–87.

123. Lillywhite HB. Water relations of tetrapod integument. J Exp Biol. 2006;209: 202–26.

124. Roberts J: Variations in salinity tolerance in the Pacific Treefrog, *Hyla regilla*. Oregon [dissertation] Corvallis, OR: Oregon State University 1970.

125. Chinathamby K, Reina RD, Bailey PC, Lees BK. Effects of salinity on the survival, growth and development of tadpoles of the brown tree frog, *Litoria ewingii*. Aust J Zool. 2006;54:97–105.

126. Brand AB, Snodgrass JW, Gallagher MT, Casey RE, Van Meter R. Lethal and sublethal effects of embryonic and larval exposure of *Hyla versicolor* to stormwater pond sediments. Arch Environ Contam Toxicol. 2010;58:325–31.

127. Petranka JW, Doyle EJ. Effects of road salts on the composition of seasonal pond communities: can the use of road salts enhance mosquito recruitment? Aquat Ecol. 2010;44:155–66.

128. Thirion J-M. Salinity of the reproduction habitats of the western spadefoot toad *Pelobates cultripes* (cuvier, 1829), along the atlantic coast of France. Herpetozoa. 2014;27:13–20.

129. Rieger JF, Binckley CA, Resetarits WJ Jr. Larval performance and oviposition site preference along a predation gradient. Ecology. 2004;85:2094–9.

130. Refsnider JM, Janzen FJ. Putting eggs in one basket: ecological and evolutionary hypotheses for variation in oviposition-site choice. Annu Rev Ecol Evol Syst. 2010;41:39–57.

131. Gomez-Mestre I, Tejado M. Adaptation or Exaptation? An experimental test of hypotheses on the origin of salinity tolerance in *Bufo calamita*. J Evol Biol. 2005;18:847–55.

132. Havird JC, Henry RP, Wilson AE. Altered expression of Na(+)/K(+)-ATPase and other osmoregulatory genes in the gills of euryhaline animals in response to salinity transfer: a meta-analysis of 59 quantitative PCR studies over 10 years. Comp Biochem Physiol Part D Genomics Proteomics. 2013;8:131–40.

133. Savolainen O, Lascoux M, Merila J. Ecological genomics of local adaptation. Nat Rev Genet. 2013;14:807–20.

134. Martin G, Aguile R, Ramsayer J, Kaltz O, Ronce O. The probability of evolutionary rescue: towards a quantitative comparison between theory and evolution experiments. Philos Trans R Soc B. 2013;368:20120088.

135. Bourne EC, Bocedi G, Travis JM, Pakeman RJ, Brooker RW, Schiffers K. Between migration load and evolutionary rescue: dispersal, adaptation and the response of spatially structured populations to environmental change. Proc R Soc Lond B Biol Sci. 2014;281:20132795.

136. Gonzalez A, Ronce O, Ferriere R, Hochberg ME. Evolutionary rescue: an emerging focus at the intersection between ecology and evolution. Philos Trans R Soc Lond B Biol Sci. 2013;368:20120404.

137. Carlson SM, Cunningham CJ, Westley PAH. Evolutionary rescue in a changing world. Trends Ecol Evol. 2014;29:521–30.

138. Bell G. Evolutionary rescue and the limits of adaptation. Philosophical Transactions of the Royal Society of London B: Biological Sciences. 2013;368: 20120080.

139. Merilä J, Hendry AP. Climate change, adaptation, and phenotypic plasticity: the problem and the evidence. Evol Appl. 2014;7:1–14.

140. Urban MC, Richardson JL, Reidenfelds NA. Plasticity and genetic adaptation mediate amphibian and reptile responses to climate change. Evol Appl. 2014;7:88–103.

141. Urban MC, Bocedi G, Hendry AP, Mihoub J-B, Pe'er G, Singer A, Bridle JR, Crozier LG, De Meester L, Godsoe W, et al. Improving the forecast for biodiversity under climate change. Science. 2016;353

142. Chevin LM, Lande R, Mace GM. Adaptation, plasticity, and extinction in a changing environment: towards a predictive theory. PLoS Biol. 2010;8: e1000357.

143. Lande R. Adaptation to an extraordinary environment by evolution of phenotypic plasticity and genetic assimilation. J Evol Biol. 2009;22:1435–46.

144. Whitman DW, Agrawal AA. What is phenotypic plasticity and why is it important? 2009. p. 1–63.

145. Murren CJ, Auld JR, Callahan H, Ghalambor CK, Handelsman CA, Heskel MA, Kingsolver J, Maclean HJ, Masel J, Maughan H. Constraints on the evolution of phenotypic plasticity: limits and costs of phenotype and plasticity. Heredity. 2015;115:293–301.

146. Forsman A. Rethinking phenotypic plasticity and its consequences for individuals, populations and species. Heredity. 2015;115:276–84.

147. Marshall DJ, Uller T. When is a maternal effect adaptive? Oikos. 2007;116: 1957–63.

148. Räsänen K, Kruuk L. Maternal effects and evolution at ecological timescales. Funct Ecol. 2007;21:408–21.

149. Kirkpatrick M, Lande R. The evolution of maternal characters. Evolution. 1989;1989:485–503.

150. Chirgwin E, Marshall DJ, Sgrò CM, Monro K. The other 96%: can neglected sources of fitness variation offer new insights into adaptation to global change? Evol Appl. 2016;10:267–75.

151. Crispo E. The Baldwin effect and genetic assimilation: revisiting two mechanisms of evolutionary change mediated by phenotypic plasticity. Evolution. 2007;61:2469–79.

PERMISSIONS

The contributors of this book come from diverse backgrounds, making this book a truly international effort. This book will bring forth new frontiers with its revolutionizing research information and detailed analysis of the nascent developments around the world.

We would like to thank all the contributing authors for lending their expertise to make the book truly unique. They have played a crucial role in the development of this book. Without their invaluable contributions this book wouldn't have been possible. They have made vital efforts to compile up to date information on the varied aspects of this subject to make this book a valuable addition to the collection of many professionals and students.

This book was conceptualized with the vision of imparting up-to-date information and advanced data in this field. To ensure the same, a matchless editorial board was set up. Every individual on the board went through rigorous rounds of assessment to prove their worth. After which they invested a large part of their time researching and compiling the most relevant data for our readers.

The editorial board has been involved in producing this book since its inception. They have spent rigorous hours researching and exploring the diverse topics which have resulted in the successful publishing of this book. They have passed on their knowledge of decades through this book. To expedite this challenging task, the publisher supported the team at every step. A small team of assistant editors was also appointed to further simplify the editing procedure and attain best results for the readers.

Apart from the editorial board, the designing team has also invested a significant amount of their time in understanding the subject and creating the most relevant covers. They scrutinized every image to scout for the most suitable representation of the subject and create an appropriate cover for the book.

The publishing team has been an ardent support to the editorial, designing and production team. Their endless efforts to recruit the best for this project, has resulted in the accomplishment of this book. They are a veteran in the field of academics and their pool of knowledge is as vast as their experience in printing. Their expertise and guidance has proved useful at every step. Their uncompromising quality standards have made this book an exceptional effort. Their encouragement from time to time has been an inspiration for everyone.

The publisher and the editorial board hope that this book will prove to be a valuable piece of knowledge for researchers, students, practitioners and scholars across the globe.

LIST OF CONTRIBUTORS

Martin Dohrmann
Department of Earth and Environmental Sciences, Palaeontology and Geobiology, Molecular Geo- and Palaeobiology Lab, Ludwig-Maximilians-University Munich, Richard-Wagner-Str. 10, 80333 Munich, Germany

Christopher Kelley
Hawaii Undersea Research Laboratory, University of Hawaii at Manoa, 1000 Pope Rd, MSB 229, Honolulu 96822, HI, USA

Michelle Kelly
Coasts and Oceans National Centre, National Institute of Water and Atmospheric Research (NIWA) Ltd, Private Bag 99940, Newmarket, Auckland 1149, New Zealand

Andrzej Pisera
Institute of Paleobiology, Polish Academy of Sciences, ul. Twarda 51/55, 00-818 Warszawa, Poland

John N. A. Hooper
Biodiversity and Geosciences Program,Queensland Museum, South Brisbane, QLD 4101, Australia.
Eskitis Institute for Drug Discovery, Griffith University, Nathan, QLD 4111, Australia

Henry M. Reiswig
Natural History Section, Royal British Columbia Museum, 675 Belleville Street, Victoria, BC V8W 9W2, Canada
Department of Biology, University of Victoria, 3800 Finnerty Road, Victoria, BC V8P 4H9, Canada
Matthias Teuscher, Nadi Ströhlein, Markus Birkenbach and Michael Schoppmeier
Department Biology, Developmental Biology Unit, Friedrich-Alexander-University Erlangen-Nuremberg, Staudtstr. 5, 91058 Erlangen, Germany

Dorothea Schultheis
Institute of Neuropathology, University Hospital Erlangen, Schwabachanlage 6, 91054 Erlangen, Germany

Janine M. Ziermann
Department of Anatomy, Howard University College of Medicine, 520 W St NW, Washington, DC 20059, USA

Renata Freitas
IBMC – Institute for Molecular and Cell Biology, Oporto, Portugal
I3S, Institute for Innovation and Health Research, University of Oporto, Oporto, Portugal

Rui Diogo
Department of Anatomy, Howard University College of Medicine, Washington, DC 20059, USA

Juan F. Masello, Julia Sommerfeld, Thomas Mattern and Petra Quillfeldt
Department of Animal Ecology and Systematics, Justus Liebig University Giessen, Heinrich-Buff-Ring 26, D-35392 Giessen, Germany

Akiko Kato
Centre d'Etudes Biologiques de Chizé, UMR7372 CNRS-Université La Rochelle, 79360 Villiers en Bois, France

Juan Antonio Martos-Sitcha, Josep Alvar Calduch-Giner and Jaume Pérez-Sánchez
Nutrigenomics and Fish Growth Endocrinology Group, Institute of Aquaculture Torre de la Sal, Consejo Superior de Investigaciones Científicas (IATS-CSIC), Ribera de Cabanes, E-12595 Castellón, Spain

Azucena Bermejo-Nogales
Nutrigenomics and Fish Growth Endocrinology Group, Institute of Aquaculture Torre de la Sal, Consejo Superior de Investigaciones Científicas (IATS-CSIC), Ribera de Cabanes, E-12595 Castellón, Spain
Endocrine Disruption and Toxicity of Contaminants, Department of Environment, INIA, Madrid, Spain

Xiaohui Sun, Zepeng Zhang, Yingying Sun, Jing Li, Shixia Xu and Guang Yang
Jiangsu Key Laboratory for Biodiversity and Biotechnology, College of Life Sciences, Nanjing Normal University, Nanjing 210023, China

Przemysław Gorzelak and Jarosław Stolarski
Institute of Paleobiology, Polish Academy of Sciences, Twarda 51/55, 00-818 Warsaw, Poland

Aurélie Dery and Philippe Dubois
Laboratoire de Biologie marine, Faculté des Sciences, Université Libre de Bruxelles, CP 160/15, av., F.D.Roosevelt, 50, B-1050 Bruxelles, Belgium

Krzysztof Kowalski and Leszek Rychlik
Department of Systematic Zoology, Institute of Environmental Biology, Adam Mickiewicz University, Umultowska 89, 61-614 Poznań, Poland

Paweł Marciniak and Grzegorz Rosiński
Department of Animal Physiology and Development, Institute of Experimental Biology, Faculty of Biology, Adam Mickiewicz University, Umultowska 89, 61-614 Poznań, Poland

Viktor V. Starunov
Department of Invertebrate Zoology, St-Petersburg State University, St-Petersburg 199034, Russia
Zoological Institute Rus, Acad. Sci, St-Petersburg 199034, Russia

Elena E. Voronezhskaya and Leonid P. Nezlin
Institute of Developmental Biology, Rus. Acad. Sci, Moscow 119991, Russia

Daniela Storch and Felix C. Mark
Alfred Wegener Institute, Helmholtz Centre for Polar and Marine Research, Integrative Ecophysiology, Am Handelshafen 12, D-27570 Bremerhaven, Germany

Elettra Leo, Matthias Schmidt and Hans-O. Pörtner
Alfred Wegener Institute, Helmholtz Centre for Polar and Marine Research,
Integrative Ecophysiology, Am Handelshafen 12, D-27570 Bremerhaven,Germany
University of Bremen, Fachbereich 2, NW 2/ Leobener Strasse,D-28359 Bremen, Germany

Kristina L. Kunz
Alfred Wegener Institute, Helmholtz Centre for Polar and Marine Research, Integrative Ecophysiology, Am Handelshafen 12, D-27570 Bremerhaven, Germany
University of Bremen, Fachbereich 2, NW 2/ Leobener Strasse,D-28359 Bremen, Germany
Alfred Wegener Institute, Helmholtz Centre for Polar and Marine Research, Bentho-Pelagic Processes, Am Alten Hafen 26, D-27568 Bremerhaven, Germany

Andrea C. Schuster, Uwe Zimmermann, Carina Hauer and Katharina Foerster
Department of Comparative Zoology, Institute for Evolution and Ecology, University of Tübingen, Auf der Morgenstelle 28, D-72076 Tübingen, Germany

Huanxi Zhu, Zhe Chen, Jianning Yu, Chuankun Wei, Zichun Dai and Zhendan Shi
Laboratory of Animal Improvement and Reproduction, Institute of Animal Science, Jiangsu Academy of Agricultural Sciences, Nanjing 210014, China

Xibin Shao
Sunlake Swan Farm, Changzhou 213101, China

Roland Fritsch, Pierre-Paul Bitton and Nico K. Michiels
Institute of Evolution and Ecology, University of Tübingen, 72076 Tübingen,Baden-Württemberg, Germany

Jeremy F. P. Ullmann
Centre for Advanced Imaging, University of Queensland, Brisbane 4072, Queensland, Australia
Department of Neurology, Boston Children's Hospital and Harvard Medical School, Boston, MA 02115, USA

Molly A. Albecker and Michael W. McCoy
Department of Biology, Howell Science Complex, East Carolina University, Greenville, NC, USA

Index

www.ingramcontent.com/pod-product-compliance
Lightning Source LLC
Chambersburg PA
CBHW082014190326
41458CB00010B/3187